Lecture Notes in Mathematics

Edited by A. Dold and B. Eckmann

703

# Equadiff IV

Proceedings, Prague, August 22–26, 1977

Edited by Jiří Fábera

Springer-Verlag
Berlin Heidelberg New York 1979

**Editor**
Jiří Fábera †
Mathematical Institute
Czechoslovak Academy of Sciences
Zitná 25
11567 Praha 1, Czechoslovakia

**Library of Congress Cataloging in Publication Data**

Czechoslovak Conference on Differential Equations and
Their Applications, 4th, Prague, 1977.
Equadiff IV.

(Lecture notes in mathematics ; 703)
Includes bibliographies and index.
1. Differential equations--Numerical solutions--Congresses. 2. Differential equations, Partial--Numerical solutions--Congresses. I. Fábera, Jiří. II. Title.
III. Series: Lecture notes in mathematics (Berlin) ; 703.
QA3.L28 no. 703 [QA372] 510'.8s [515'.35]
79-11103

AMS Subject Classifications (1970): 34A02, 35A02, 65A02, 93A02

ISBN 3-540-09116-5 Springer-Verlag Berlin Heidelberg New York
ISBN 0-387-09116-5 Springer-Verlag New York Heidelberg Berlin



Printing and binding: Beltz Offsetdruck, Hemsbach/Bergstr.
2141/3140-543210

Math
5 ep

# PREFACE

The Czechoslovak Conference on Differential Equations and Their Applications - EQUADIFF 4 - was held in Prague from August 22 to August 26, 1977. It continued the tradition of conferences held in Prague (1962), in Bratislava (1966) and in Brno (1972). The topic were differential equations in the broad sense including numerical methods and one of the goals of the conference was to stimulate cooperation between various branches in differential equations. The conference was organized by the Mathematical Institute of the Czechoslovak Academy of Sciences in cooperation with the Faculty of Mathematics and Physics of the Charles University in Prague, the Faculty of Sciences of the Comenius University in Bratislava, the Czech Technical University in Prague and the Technical University in Brno.

The Organizing Committee was presided by J.Fábera, the members being O.Borůvka, J.Brilla, O.Hajkr, J.Kurzweil, I.Marek, J.Moravčík, J.Nečas, B.Novák, M.Práger, M.Ráb, K.Rektorys, M.Švec, O.Vejvoda, M.Zlámal.

The scientific program of the conference comprised 5 invited addresses, 58 invited lectures and 157 scientific communications; the lectures and scientific communications were held simultaneously in three sections:

1. Ordinary Differential Equations,
2. Partial Differential Equations,
3. Numerical Methods and Applications.

The participants and their accompanying persons could enjoy a rich social program. The conference was attended by 167 participants from Czechoslovakia and 187 participants from abroad (Austria, Belgium, Berlin-West, Bulgaria, Canada, Egypt, F.R.G., France, G.D.R., Hungary, Italy, Japan, Netherlands, Poland, Roumania, Saudi Arabia, Sweden, Switzerland, U.S.A., U.S.S.R., United Kingdom, Yugoslavia). The number of accompanying persons was 49.

This volume contains the texts of plenary addresses and of invited lectures which were held in sections - with several exceptions.

We acknowledge the work of Vl.Doležal, J.Jarník and P.Přikryl in preparation of this volume.

Editors

## LIST OF INVITED ADDRESSES

### Plenary Session

| | |
|---|---|
| Borůvka, O. | Algebraic methods in the theory of global properties of the oscillatory equations $Y'' = Q(t)Y$ |
| Everitt, W.N. | Singular problems in the calculus of variations and ordinary differential equations |
| Marek, I. | Eigenvalues and bifurcations in reactor physics and chemistry |
| Nečas, J. | On the existence and regularity of weak solutions to variational equations and inequalities |
| Oleinik, O.A. | Energetičeskije ocenki, analogičnyje principu Saint-Venant i ich priloženija |

### Ordinary differential equations

| | |
|---|---|
| Antosiewicz, H.A. | Some remarks on the solution of boundary value problems |
| Bainov, D.D.: Bainov, D.D. and Milusheva, S.D. | Application of the averaging method for solving boundary problems for ordinary differential and integrodifferential equations (Presented by D.D.Bainov) |
| Bebernes, J.W. | Invariance and solution set properties for some nonlinear differential equations |
| Bihari, I. | Asymptotic invariant sets of some nonlinear autonomous systems of differential equations |
| Blagodatskich, V. | Some problems in the theory of differential inclusions |
| Coddington, E.A. | Differential subspaces associated with pairs of differential operators |
| Conti, R. | Control and the Van der Pol equation |
| Gamkrelidze, R.V. | Exponential representation of solutions of ordinary differential equations |
| Halanay, A. | Singular perturbations and linear feedback control |
| Kamenskii, G.A.: Kamenskii, G.A. and Myshkis, A.D. | Variational and boundary problems for differential equations with deviated argument (Presented by A.D.Myshkis) |
| Kiguradze, I.T. | O kolebljuščichsja i monotonnych rešenijach obyknověnnych differencial'nych uravněnij |

## Partial differential equations

## Numerical methods and applications

| | |
|---|---|
| Mika, J. | Asymptotic methods for the singularly perturbed differential equations in Banach spaces |
| Nashed, M.Z. | Iterative and projection methods for illposed boundary-value problems and operator equations |
| Nassif, N.: Descloux, J., Nassif, N. and Rappaz, J. | Numerical approximation of the spectrum of linear operators (Presented by J. Descloux) |
| Nedelec, J.C. | Finite element approximations of singular integral equations |
| Nedoma, J. | The solution of parabolic models by finite element space and A-stable time discretization |
| Nohel, J.A. | Volterra integrodifferential equations for materials with memory |
| Rappaz, J.: Descloux, J., Nassif, N. and Rappaz, J. | Numerical approximation of the spectrum of linear operators (Presented by J. Descloux) |
| Raviart, P.A. | Mixed finite element approximations of the Navier-Stokes equations |
| Rektorys, K. | Approximations of very weak solutions of the first biharmonic problem for multiply connected regions |
| Stetter, H.J. | The principle of defect correction and its application to discretization methods |
| Sultangazin, U.M. | Issledovanie rešenij simmetričeskich položitel'nych sistem metodom sferičeskich garmonik |
| Taufer, J.: Taufer, J. and Vitásek, E. | Numerical solution of evolution problems in Banach spaces (Presented by E.Vitásek) |
| Vitásek, E.: Taufer, J. and Vitásek, E. | Numerical solution of evolution problems in Banach spaces (Presented by E.Vitásek) |
| Zlámal, M. | Superconvergence in the finite element method |

LIST OF COMMUNICATIONS

Ordinary differential equations

| | |
|---|---|
| Angelov, V. | Suščestvovanije i edinstvennosť rešenija načalnoj zadači dla odnogo klassa differencial'nych uravněnij nejtralnogo tipa |
| Aulbach, B. | The method of Zubov in the case of an asymptotically stable limit cycle |
| Bobrowski, D. | On oscillation criteria |
| Brjuno, A.D. | Normal'naja forma i bifurkacii v dinamičeskich sistěmach |
| Brown, R.C. | Boundary value problems with general side conditions and their applications |
| Brunovský, P. | On the existence of regular synthesis of the optimal control |
| Čanturija, T. | O někotorych těoremach sravněnija dla obyknovennych differencial'nych uravněnij |
| Cerha, J. | Dirac function in Volterra equations |
| Chow, S. | Generalized Hopf bifurcation |
| Daneš, J. | On positive solutions of nonlinear operator equations |
| Dijksma, A. | On ordinary differential subspaces |
| Djaja, Č. | Někotoryje svojstva kvasi-počti periodičeskich dviženij dinamičeskich sistěm |
| Doležal, J. | On a certain type of discrete two-point boundary value problem arising in discrete optimal control |
| Elbert, Á. | A comparison theorem for first order nonlinear differential equations with delay |
| Erbe, L. | Some applications of change-of-variable techniques for linear differential equations |
| Farkas, M. | Estimates on the existence regions of perturbed periodic solutions |
| Gil'derman, J. | Kusočno-afinnyje dinamičeskije sistěmy |
| Golokvosčjus, P. | O strukture integralnoj matricy odnoj dvumernoj sistěmy differencial'nych uravněnij v okrestnosti regularnoj osoboj točki sistěmy |
| Habets, P. | On relaxation oscillations |
| Heil, E. | On an oscillation criterion of Hartman-Wintner-Potter |
| Hetzer, G. | Some remarks on periodic solutions of second order systems of ordinary differential equations |
| Hristova, S. | Asimptotičeskoje razloženije rešenija odnoj načalnoj zadači o singularno vozmuščennoj sistěme integro-differencial'nych uravněnij s zapazdyvanijem |

Partial differential equations

| | |
|---|---|
| Dragieva, N.A. | Une solution de l'équation parabolique dans un domaine non cylindrique |
| Gajewski, H.: Gajewski, H. and Zacharias, K. | On Vlasov's equation (Presented by K.Zacharias) |
| Göhde, D. | Eine Klasse singulär gestörter elliptischer Differentialgleichungen |
| Goncerzewicz, J. | On the weak solution of boundary value problem arising in theory of water percolation |
| Grabmüller, H. | Singular perturbations in linear integro-differential equations |
| Hernandez, J. | Positive solutions for a class of nonlinear eigenvalue problems |
| Herrmann, L. | Periodic solutions to abstract differential equations |
| Holaňová-Radochová, V. | Remark to the solution properties of some class of partial differential equations in the distributions space |
| Jäger, W. | A maximum principle and a uniqueness result for solutions of elliptic equations on Riemannian manifolds |
| Jarušek, J. | Ranges of certain type of nonlinear operators with applications in partial differential equations |
| Jasný, M. | Walter's method of lines applied to the equation $\partial w/\partial x = W'(x) + \nu\sqrt{w}\,\partial^2 w/\partial t^2$ |
| Kersner, R. | On nonlinear degenerate parabolic equations |
| Kolomý, J. | Structure of Banach spaces and solvability of nonlinear equations |
| König, M. | Zur Abschätzung der Lösung des Dirichletschen Aussenraumproblems für die Schwingungsgleichung |
| Kučera, M. | Eigenvalue problem for variational inequalities |
| Litewska, K. | Rothe methods for parabolic systems |
| Lovicar, V. | Periodic solutions of second order equations with dissipative terms |
| Marcinkowska, H. | On a class of mixed problems for linear hyperbolic equations |
| Maslennikova, V.N. | Asimptotičeskije svojstva rešenij sistěm gidrodinamiki vraščajuščejsja židkosti |
| Musiał, I. | Some problems of behaviour of solutions in Hilbert space |
| Muszyński, J. | About some integral equation |
| Naumann, J. | On a class of integro-differential equations |
| Netuka, I. | Potentials and boundary value problems for the heat equation |

Numerical methods and applications

| | |
|---|---|
| Adler, G. | Problems in the digital simulation of thermal secondary oil recovery by combustion |
| Agoškov, V. | Ispol'zovanie variacionnoj formy tožděstva G.I.Marčuka dla rešenija někotorych differencial'nych uravněnij |
| Bock, I. | Bending of viscoelastic plates with aging |
| Burda, P. | Finite elements and diffusion equations |
| Čermák, J. | Floquet's theory and neutron transport in a periodic slab lattice |
| Chocholatý, P.: Chocholatý, P. and Šlahor, L. | A remark on the solution of a boundary value problem with delay (Presented by L.Šlahor) |
| Dávid, A. | Orthonormal systems in quasiparabolic differential equations |
| Desperat, T. | Free boundary problems in the theory of fluid flow through porous media |
| Dryja, M. | Alternating-direction Galerkin method for Navier-Stokes equation |
| Farzan, R. | Někotoryje uslovija dla schodimosti i ustojčivosti čislennogo rešenija krajevoj zadači dla paraboličeskogo uravněnija so slaboj nělinějnost'ju |
| Feistauer, M. | On the study of subsonic rotational flow of an ideal gas |
| Galántai, A. | On the extensions of some stiff stability concept |
| Georg, K. | Unstable solutions of second order nonlinear boundary value problems |
| Godlewski, E.: Godlewski, E. and Raoult, A.P. | Multistep methods for $u'' + Bu' + Au = f$ (Presented by E.Godlewski) |
| Griepentrog, E. | Onestep methods for stiff differential equations |
| Gröger, K. | Evolution equations in the theory of plasticity |
| Haslinger, J. | Dual finite element analysis of variational inequalities |
| Kafka, J.M. | Seventh contribution to the numerical solution of parabolic partial differential equations |
| Kodnár, R. | Uravněnija uprugo-plastičeskich plastinok pri bol'šich progibach |
| Kuzněcov, Ju.: Kuzněcov, Ju. and Macokin, A. | Metod fiktivnych oblastěj i ego priměněnija (Presented by A.Macokin) |
| Lelek, V.: Lelek, V. and Šafář, J. | Numerical solution of diffusion equations on nonuniform net (Presented by J.Šafář) |

CONTENTS

# INVARIANT SETS FOR SEMILINEAR PARABOLIC AND ELLIPTIC SYSTEMS

H. Amann, Bochum

Let $\Omega$ be a bounded smooth domain in $\mathbb{R}^n$ and let $Q := \Omega \times (o,T)$ for some fixed $T > o$. Denote by $\partial/\partial t + A(x,t,D)$ a uniformly parabolic second order differential operator on $\overline{Q}$ with smooth coefficients, and let $B(x,D)$ be a (time independent) first order smooth boundary operator. We suppose that $B(x,D)$ is of the form $B(x,D)u = b(x)u + \delta(\partial u/\partial \beta)$, where either $\delta = o$ and $b(x) = 1$ (Dirichlet boundary operator) or $\delta = 1$ and $b(x) \geq o$ for all $x \in \partial\Omega$, and $\beta$ is a smooth outward pointing, nowhere tangent vector field on $\partial\Omega$ (Neumann or regular oblique derivative boundary operator).

We denote by $f : \overline{Q} \times \mathbb{R}^n \times \mathbb{R}^{nm} \to \mathbb{R}^m$ a Lipschitz continuous function, and consider parabolic initial boundary value problems of the form

$$(1) \qquad \begin{aligned} \frac{\partial u}{\partial t} + A(x,t,D)u &= f(x,t,u,Du) && \text{in } \Omega \times (o,T] , \\ B(x,D)u &= o && \text{on } \partial\Omega \times (o,T] , \\ u(.,o) &= u_o && \text{on } \overline{\Omega} , \end{aligned}$$

where $u = (u^1,\ldots,u^m)$. In other words, (1) is a "diagonal system" which is strongly coupled through the nonlinear function $f$. By a solution of (1) we mean a classical solution.

In order to obtain appropriate a priori estimates, we impose the following growth restriction for $f$, which we write in a self-explanatory symbolic form: namely we suppose that either

$$|f(x,t,u,Du)| \leq c(|u|)(1 + |Du|^{2-\varepsilon})$$

for some $\varepsilon > o$, or

$$|f^i(x,t,u,Du)| \leq c(|u|)(1 + |Du^i|^2)$$

for $i = 1,\ldots,m$, where $c \in C(\mathbb{R}_+,\mathbb{R}_+)$.

It is well known that (1) possesses a unique solution for every sufficiently smooth initial value $u_o$ satisfying appropriate compatibility conditions. However this solution may only exist for a small time interval and not in the whole cylinder $Q$. The existence of a *global* solution can be guaranteed provided an a priori bound for the maximum norm can be found. Unfortunately, establishing a priori bounds for the maximum norm is a rather difficult problem for systems since no good maximum principle is available.

Recently H. F. Weinberger [5] (and later Chueh, Conley and Smoller [3]) has given a weak substitute for a maximum principle which can be used for establishing a priori bounds. But these results presuppose a priori knowledge of the solution on the lateral boundary $\partial\Omega \times [o,T]$ of the cylinder $Q$ which is, in general, only available for the case of Dirichlet boundary conditions.

In this paper we present a global existence and uniqueness theorem for problem (1) *without assuming any a priori knowledge on the solution for* $t > o$. We emphasize the fact that our results apply to the case of boundary conditions of the third kind which are of particular importance in applications (to problems of chemical engineering, for example).

For an easy formulation of our results we introduce the following hypotheses and notations. Let $\mathbb{D}$ be a compact convex subset of $\mathbb{R}^n$ such that $o \in \mathbb{D}$. For every $\xi_o \in \partial\mathbb{D}$ let

$$N(\xi_o) := \{p \in \mathbb{R}^m \mid \langle p,\xi-\xi_o\rangle \leq o \;\; \forall \; \xi \in \mathbb{D}\} \quad ,$$

that is, $N(\xi_o)$ is the "set of outer normals" on $\partial\mathbb{D}$ at $\xi_o$. Finally, for $k = 1,2$, we let

$$C^k_B(\overline{\Omega},\mathbb{D}) := \{u \in C^k(\overline{\Omega},\mathbb{R}^m) \mid Bu = o \text{ on } \partial\Omega \text{ and } u(\overline{\Omega}) \subset \mathbb{D}\} \quad .$$

Then we impose the following *tangency condition:*

For every $v \in C^1_B(\overline{\Omega},\mathbb{D})$ and for every $x_o \in \overline{\Omega}$ with $v(x_o) \in \partial\mathbb{D}$, we suppose that

$$\text{(Tg)} \qquad \langle p,f(x_o,t,v(x_o),Dv(x_o))\rangle \leq o$$

for all $t \in [o,T]$ and all $p \in N(\xi_o)$, where $\langle .,.\rangle$ denotes the inner product in $\mathbb{R}^m$.

Condition (Tg) means geometrically that the vector $f(x_o,t,v(x_o),Dv(x_o))$, attached to $\partial\mathbb{D}$ at the point $v(x_o)$, lies in the cone which contains $\mathbb{D}$ and is described by the family of all supporting hyperplanes at $v(x_o)$. It is easily seen that (Tg) reduces to the condition introduced by Weinberger [5] (and also used by Chueh, Conley and Smoller [3]) in the case that $f$ is independent of $Du$ (the case studied in [3] and [5]). It is essentially the same condition as the one used by Bebernes [2]. We refer to [3] for a variety of examples satisfying (Tg). It is easy to give further examples in the case of nonlinear gradient dependence.

After these preparations we can give our basic existence and uniqueness theorem for problem (1) (cf. also [2] for the special case of Dirichlet and Neumann boundary conditions).

<u>*Theorem 1:*</u> *Let the growth condition and the tangency conditions be satisfied. Then*

*the initial boundary value problem* (1) *has a unique global solution* $u$ *for every initial value* $u_0 \in C^2_B(\overline{\Omega},\mathbb{D})$ , *and* $u(\overline{Q}) \subset \mathbb{D}$ .

*Proof:* By using the results of Kato, Tanabe, and Sobolevskii on abstract parabolic evolution equations as well as the results of Ladyzenskaja, Solonnikov, and Ural'ceva on the classical solvability of linear parabolic equations, it is shown that (1) is equivalent to the nonlinear evolution equation

$$\text{(2)} \qquad \begin{aligned} \dot{u} + A(t)u &= F(t,u) \quad , \quad o < t \leq T \ , \\ u(o) &= u_0 \end{aligned}$$

in $X := L_p(\Omega,\mathbb{R}^m)$ , where $p > 2$ is sufficiently large and $-A(t)$ is the infinitesimal generator of a holomorphic semigroup. We denote by $X_\alpha$ the domain of the fractional power $[A(o)]^\alpha$ , $o < \alpha < 1$ , and we let $\mathbb{M}_\alpha := L_p(\Omega,\mathbb{D}) \cap X_\alpha$ , endowed with the topology of $X_\alpha$ , where $\alpha$ is sufficiently close to $1$ . Then (2) is equivalent to the integral equation

$$\text{(3)} \qquad u(t) = U(t,o)u_0 + \int_0^t U(t,\tau)F(\tau,u(\tau))d\tau$$

in $C([o,T],X_\alpha)$ , where $U$ denotes the linear evolution operator associated with (2).

The maximum principle implies that $U(t,\tau)(\mathbb{M}_\alpha) \subset \mathbb{M}_\alpha$ for $o \leq \tau \leq t \leq T$ , and it is shown that the tangency condition implies that

$$\text{(4)} \qquad \mathrm{dist}_X(y + hF(t,y),\mathbb{M}) = o(h) \quad \text{as} \quad h \to o+$$

for each $y \in \mathbb{M}_\alpha$. Hence we are left with the problem of solving the integral equation on the *closed* bounded subset $\mathbb{M}_\alpha$ of the Banach space $X_\alpha$ . By employing a discontinuous Euler method as developed by R. H. Martin (e.g. [4] ), it can be shown that the Nagumo type condition (4) implies the existence of a unique local solution of (3) in $\mathbb{M}_\alpha$. Finally, by means of the growth condition, we obtain a priori estimates which guarantee that the local solution has a unique continuation to a global solution. □

Suppose now that $A$ and $f$ are independent of $t$ . Then, as a consequence of Theorem 1, it follows that (1) defines a nonlinear semigroup $\{S(t) \mid o \leq t < \infty\}$ on $\mathbb{M}_\alpha$ , where $S(t)u_0$ denotes the solution at time $t$ of the autonomous problem (1) with initial value $u_0 \in \mathbb{M}_\alpha$. On the basis of the integral equation (3) and by using appropriate a priori estimates, it can be shown that, for every $t > o$ , the nonlinear operator $S(t) : \mathbb{M}_\alpha \to \mathbb{M}_\alpha$ is continuous and has a relatively compact image.

For every $t \geq o$ , let

$$\mathfrak{F}_t := \{u \in \mathbb{M}_\alpha \mid S(t)u_0 = u_0\} \ ,$$

that is, $\mathfrak{F}_t$ is the fixed point set of $S(t)$ . Then, by Schauder's fixed point theorem, $\mathfrak{F}_t \neq \emptyset$ for every $t > o$ . Moreover, suppose that $t_1,\dots,t_m$ are positive numbers having $t > o$ as a common divisor. Then it is an easy consequence of the semigroup property (i.e., $S(t+\tau) = S(t)S(\tau)$) that

$$\mathcal{F}_t \subset \bigcap_{i=1}^{m} \mathcal{F}_{t_i} \quad .$$

This implies that the family $\{\mathcal{F}_t \mid t \in \mathbb{Q}_+\}$ has the finite intersection property. Hence, by compactness, $\cap\{\mathcal{F}_t \mid t \in \mathbb{Q}_+\} \neq \emptyset$. This shows that there exists an element $u_o \in \mathbb{M}_\alpha$ such that $S(t)u_o = u_o$ for all $t \in \mathbb{Q}_+$, that is, $u_o$ is a common fixed point of the family $\{S(t) \mid t \in \mathbb{Q}_+\}$. Finally, by using a continuity argument, it follows that $S(t)u_o = u_o$ for all $t \geq o$, that is, $u_o$ is a rest point of the flow $\{S(t) \mid t \geq o\}$, hence a solution of the stationary equation.

By this argument we obtain

*Theorem 2: Suppose that* $A(x,D)$ *is a strongly uniformly elliptic second order differential operator with smooth coefficients. Suppose that* $f$ *is independent of* $t$ *and satisfies the growth condition and the tangency condition. Then the semilinear elliptic system*

$$(5) \qquad \begin{aligned} A(x,D)u &= f(x,u,Du) && \text{in } \Omega \ , \\ B(x,D)u &= o && \text{on } \partial\Omega \end{aligned}$$

*has at least one solution* $u$ *such that* $u(\overline{\Omega}) \subset \mathbb{D}$.

It should be remarked that the assumption that in each single equation of the system (1) or (5) there occurs one and the same differential operator can be dropped if the conditions on $\mathbb{D}$ are strengthened. For further details, examples, and more detailed proofs we refer to [1].

## References


[1] H. Amann: Invariant sets and existence theorems for semi-linear parabolic and elliptic systems. J. Math. Anal. Appl., to appear.

[2] J. W. Bebernes: Solution set properties for some nonlinear parabolic differential equations. These Proceedings.

[3] K. Chueh, C. Conley, and J. A. Smoller: Positively invariant regions for systems of nonlinear diffusion equations. Indiana Math. J., 26 (1977), 373-392.

[4] R. H. Martin: Nonlinear Operators and Differential Equations in Banach Spaces. J. Wiley & Sons, New York 1976.

[5] H. F. Weinberger: Invariant sets for weakly coupled parabolic and elliptic systems. Rend. Math., 8 (1975), 295-310.


Author's address: Institut für Mathematik, Ruhr-Universität, D-4630 Bochum, Germany

ON THE NUMERICAL SOLUTION OF NONLINEAR PARTIAL DIFFERENTIAL EQUATIONS ON DIVERGENCE FORM

O. Axelsson, Göteborg

## 1. Introduction.

We will consider nonlinear partial differential equations on the form

$$(1.1) \qquad F(u,\nabla u) = -\text{div}\, A(u,\nabla u) + g(u,\nabla u) = 0, \qquad x \in \Omega \subset R^n$$

with given boundary conditions, Dirichlet conditions (for simplicity assumed to be homogeneous) on a set of measure > 0 on $\partial\Omega$. We have

$$A^T = (A_1,\dots,A_n), \qquad A_i = A_i(u,\nabla u) : R \times R^n \to R$$

$$\frac{\partial A}{\partial \xi} = \left(\frac{\partial A_i}{\partial \xi_j}\right), \qquad \xi = (\xi_1,\dots,\xi_n), \quad \xi_j = \frac{\partial u}{\partial x_j} .$$

The matrix $\frac{\partial A}{\partial \xi}$ is assumed to be uniformly positive definite (ellipticity)

$$\inf_{u,\nabla u} \xi^T \frac{\partial A}{\partial \xi} \xi \geq \rho|\xi|^2, \quad \rho > 0 \quad \forall\, \xi \in R^n .$$

In the first part of the talk we will consider a special class of problems with so called potential operators, for which optimization (minimization) algorithms may be applied to the corresponding energy functional. In the last part of the talk the more general problem (1.1) will be delt with by use of an embedding in a parabolic problem.

## 2. Potential operator problems.

Let us assume that the operator F is *potential* [1], i.e.

$$\exists f : V \to R \ni$$

$$(F,\eta) = (f'(u),\eta) = (\text{grad } f(u,\nabla u),\eta) \qquad \forall\, \eta \in V \subseteq H_0^1(\Omega) .$$

Then $F \in V^*$, the dual space of V and the variational (Galerkin) formulation of (1.1) is

$$(f'(u),\eta) = 0 \qquad \forall\, \eta \in V .$$

( , ) is the extended scalar product in $L^2$. A sufficiently regular operator is potential iff its Hessian f" has a symmetric bilinear form

$$(f''\eta,\zeta) = (f''\zeta,\eta) \qquad \forall\ \eta,\zeta \in V .$$

Here $f''\eta = F'(u,\nabla u)\eta$, the Gâteaux differential.

We will in particular consider potential operators on the form

$$F(u,\nabla u) = -\mathrm{div}\ A(\nabla u) + g(u)$$

where the matrix $\frac{\partial A}{\partial \xi}$ is symmetric. Then, apart from an integration constant,

(2.1) $$f(u) = \int_\Omega [\int_0^u F(v,\nabla v)dv]dx = \int_\Omega [\int_0^u A(\nabla v)d(\nabla v) + \int_0^u g(v)dv]dx$$

with

$$V = \overset{0}{H}{}^1(\Omega) = \{v \in H^1(\Omega);\ v \text{ satisfies ess.b.c's}\} .$$

In practice f often corresponds to the total energy in the system at hand. A particular example of practical importance is

(2.2) $$A(\nabla u) = a(|\nabla u|^2)\nabla u .$$

Then

$$F(u,\nabla u) = -\sum_{i=1}^{n} \frac{\partial}{\partial x_i}(a(|\nabla u|^2)\frac{\partial u}{\partial x_i}) + g(u)$$

We also assume that the Hessian is positive definite, i.e.

$$(f''(u)\eta,\eta) \geq \delta\|\nabla\eta\|^2, \qquad \delta > 0 \qquad \forall\ \eta \in V ,$$

where $\|\cdot\|$ is the norm in V. It is easily seen that this is satisfied if

$$\rho + \max\ (0, \frac{1}{\mu_1}\frac{\partial g}{\partial u}(u)) \geq \delta > 0 ,$$

where $\mu_1 = \mu_1(\Omega,-\Delta)$ is the smallest eigenvalue of the Laplacian operator $(-\Delta)$ on V and $\rho$ is the ellipticity constant. (In (2.2) we have

$$\rho = a(z) + 2\frac{\partial a}{\partial z}(z)\cdot z , \qquad z = |\nabla u|^2 .)$$

Thus f is strictly convex, proper and increasing. In practice its unique minimizer is approximated by the minimizer over a finite dimensional subspace, for instance a set of finite element functions, $V_N \subset H_0^1(\Omega)$, the Ritz method. This minimizer, $\hat{\underline{u}}$ may with practical efficiency be calculated by a preconditioned (scaled) conjugate gradient method, a Newton-Kantorovich method or probably most advantageously, by a combination of these.

## 3. Algorithms for potential operator problems.

To minimize the functional $f = f(u_1, u_2, \ldots, u_N)$ over $V_N$ we shall describe two algorithms, both of which use the Hessian matrix

$$H = \left[ \frac{\partial^2 f}{\partial u_i \partial u_j} \right] .$$

In general, it is too costly to update (recalculate) this matrix frequently, so we shall give means by which this can be avoided.

### 3.1 The Newton-Kantorovich method.

Let $\underline{u}^{\ell}$ be an approximation of $\hat{\underline{u}}$. Then we approximate f by the quadratic functional

$$f_{\ell}(\underline{u}) = f(\underline{u}^{\ell}) + (f'(\underline{u}^{\ell}), \underline{u} - \underline{u}^{\ell}) + \frac{1}{2}(f''(\underline{u}^{\ell})(\underline{u} - \underline{u}^{\ell}), \underline{u} - \underline{u}^{\ell}) ,$$

where the qradient and Hessian are evaluated at $\underline{u}^{\ell}$. Its minimizer, denoted by $\underline{u}^{\ell+1}$, satisfies

$$f'_{\ell}(\underline{u}) = f'(\underline{u}^{\ell}) + f''(\underline{u}^{\ell})(\underline{u}^{\ell+1} - \underline{u}^{\ell}) = 0 ,$$

and repeating the process for $\ell = 1,2,\ldots,\underline{u}^1$ given, we have arrived at the classical Newton-Kantorovich method for the solution of $f'(\underline{u}) = 0$. The quadratic convergence is assured if

$$\| \underline{u}^2 - \underline{u}^1 \| < 2\delta/K$$

(see e.g. [1] and [2]). Here K is an upper bound on the second Gateaux differential,

$$|F''\eta\zeta|, \quad \| \eta \| = \| \zeta \| = 1, \quad \eta,\zeta \in V .$$

At each Newton step we do not have to assemble the Hessian matrix, as would be the case in a direct LU-factorization method. This is of importance in particular in three-dimensional problems, $\Omega \subset R^3$ (see e.g. [3]). Instead we calculate the minimizer of $f_{\ell}$ by the preconditioned conjugate gradient (PCCG) method:

$$\underline{u} := \underline{u}^{\ell}; \quad \underline{g} := -f'(\underline{u}^{\ell});$$

$$C\underline{\gamma} := \underline{g}; \quad \underline{e} := -\underline{\gamma}; \quad \delta_0 := g^T\gamma; \quad \varepsilon := \varepsilon_0\delta_0;$$

$$\text{R:} \quad \lambda := -g^T e/e^T He;$$

$$\underline{u} := \underline{u} + \lambda\underline{e};$$

$$\underline{g} := \underline{g} + \lambda H\underline{e}; \quad C\underline{\gamma} := \underline{g};$$

$$\delta_1 := g^T\gamma; \quad \beta := \delta_1/\delta_0; \quad \delta_0 := \delta_1;$$

$$\underline{e} := -\underline{\gamma} + \beta\underline{e};$$

$$\text{IF} \quad \delta_1 > \varepsilon \quad \text{THEN GOTO R};$$

$$u := \underline{u}^{\ell} + u:,$$

If $V_N$ is spanned by N basis (or coordinate) functions $\phi_i(x) \in H_0^1(\Omega)$ with *local support* on a "small" element (the finite element method) the matrix-vector multiplication He is calculated as a sum of its contributions from each local element. In this way H does never have to be calculated, only the local finite element matrices are calculated.

The rate of convergence of the PCCG-algorithm is linear and the number of conjugate gradient steps, i.e. number of times He is calculated, is at most

$$p = \text{int}\,[\tfrac{1}{2}\sqrt{\mathcal{H}}\,\ln\frac{2}{\varepsilon_0} + 1]$$

where $\mathcal{H}$ is the spectral condition number, i.e. the quotient between the extreme eigenvalues, of $C^{-1}H$. C is usually a product of two sparse triangular matrices. In [3] it is shown that it is possible to choose C such that $\mathcal{H} = O(N^{1/n})$. Then $p \sim O(N^{1/2n})$, i.e. a small increase with the number of unknowns N.

## 3.2 Efficiency in handling the updating of the Hessian matrix.

Assume for simplicity that

$$A(\nabla u) = a(|\nabla u|^2)\nabla u\ .$$

The corresponding local element "stiffness" matrices are

$$k_{ij}^{(e)} = \int_{\Omega_e} a(|\nabla u|^2)\nabla\phi_i(x)\nabla\phi_j(x)\;dx\ ,$$

where $\Omega_e$ is the e'th element. Only basis functions with a common support over the element give non-zero elements (see figure).

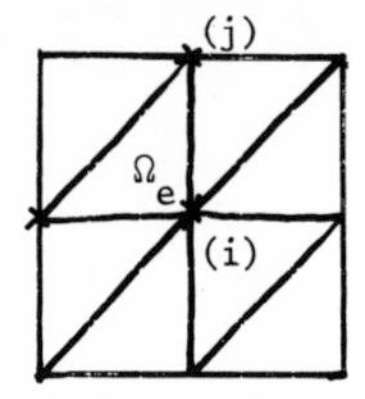


The corresponding part of the Hessian matrix has a similar form. The global matrices

$$K_{ij} = \sum_e k_{ij}^{(e)} \qquad (\text{and} \quad H_{ij})$$

do not have to be assembled (thereby avoiding possible cancellation of digits). We approximate

$$k_{ij}^{(e)} \simeq a(|\nabla u|_e^2) \int_{\Omega_e} \nabla\phi_i(x)\nabla\phi_j(x)\, dx \ .$$

For linear finite elements, this is exact, since then $|\nabla u|$ is locally constant. Only the first factor have to be reevaluated at each new Newton-step. The second factor is evaluated once and for all and stored, when the finite element mesh has been generated. This is so also for so called geometrically nonlinear problems (cf. Section 5). This is done as long as the relative change in the functional is large enough. When this change is small a true gradient should be calculated during the last Newton-steps (cf. [4]). This will give more accurate approximations of the solution.

## 3.3 PCCG with restart.

An alternative to the Newton-Kantorovich method is to use a preconditioned conjugate gradient method for the minimization of the generally nonquadratic functional (2.1). Then the only change in the PCCG-algorithm is that $\lambda$ has to be evaluated by some linesearch procedure (like Newtons modified method for one unknown variable) and the gradient g is evaluated as $g := f(\underline{u})$ at each c-g step. For problems with a strong nonlinearity, it may be advisable to restart the algorithm with a search along the negative gradient at every r'th step (i.e. $\beta := 0$ then). We observe that the number of iterations $p = O(\sqrt{\mathcal{H}})$ is still valid, if only $r \geq 2$. In the classical steepest descent method (where $r = 1$) we have however $p = O(\mathcal{H})$.

## 3.4 On preconditioning.

To explain the effect of preconditioning (or scaling) we consider a functional

$$\tilde{f}(\tilde{\underline{u}}) = f(E^T\tilde{\underline{u}})$$

where the variable is transformed by a simple (i.e. triangular matrix $E^T$). The minimizer of $\tilde{f}$ is $\tilde{\underline{u}} = E^{-T}\hat{\underline{u}}$. We get a new gradient

$$\tilde{g} = Eg(E^T\tilde{\underline{u}})$$

and Hessian

$$\tilde{H} = EH(E^T\tilde{\underline{u}})E^T \; .$$

If we work with untransformed quantities $\underline{u} := E^T\tilde{\underline{u}}$, then since $\tilde{\underline{u}} := \tilde{\underline{u}} + \lambda\tilde{g}$ etc, we get $g := E^T\tilde{g} = E^TEg$. Thus the only change in the classical c-g algorithm is $g \to E^TEg$, and we arrive at the PCCG-algorithm.

Apparently the best choice of E is such that $\mathcal{K}(\tilde{H}) \simeq 1$. If $E^{-1}E^{-T} \simeq H$, this is the case. Thus we may let $C = E^TE$ be an approximate factorization of H, (actually modified by a relaxation parameter). At each PCCG-step we have then to solve $C\gamma = g$, which is not costly since C is the product of two sparse triangular matrices.

Similar methods as described in 3.3 have been used by [5], [6], [7] among others.

## 4. Parabolic imbedding.

Consider now the problem (1.1) with

$$\rho_0 \leq \mu_1(\Omega,-\Delta)(\rho - \delta), \qquad \delta > 0 \; ,$$

where $\mu_1$ is the smallest eigenvalue of $-\Delta$ on $\Omega$ and

$$\rho_0 = \sup_{v,\nabla v} \{\frac{1}{2} \operatorname{div} [\frac{\partial A}{\partial u}(v,\nabla v) + \frac{\partial g}{\partial \xi}(v,\nabla v)] - \frac{\partial g}{\partial u}(v,\nabla v)\} \; .$$

Then it is easily seen that the operator

$$F = -\operatorname{div} A(u,\nabla u) + g(u,\nabla u)$$

is *strongly monotone*:

(4.1) $$a(u,w;v) = (A(u,\nabla u) - A(w,\nabla w),\nabla v) + (g(u,\nabla u) - g(w,\nabla w),v)$$

$$\geq \delta \| \nabla(u - w) \|^2 \qquad \forall\ u,w \in V\ .$$

Here $\|\cdot\|$ is the $L_2$-norm. The corresponding parabolic problem,

(4.2) $$u_t + F(u,\nabla u) = 0, \qquad t > 0, \qquad u(x,0) = u_0(x)$$

has then a unique solution for all $t > 0$ and is asymptotically stable, that is,

$$u(x,\infty) = \lim_{t\to\infty} u(x,t)$$

exists uniquely, independently of the initial function $u_0(x)$ (see [8]). Thus $u(x,t)$, $t$ large enough may be used as an approximation of $u(x,\infty)$. If one is only interested in this stationary solution, the initial function should be chosen as a smooth function satisfying the boundary conditions, so that higher order modes in the corresponding "Fourier series" have small components.

## 4.1 Discretization error estimates.

A semi-discrete approximation of (4.2) is achieved by Galerkins method. A variational formulation of (4.1) is

$$(u_t,v) + (A(u,\nabla u),\nabla v) + (g(u,\nabla u),v) = 0 \qquad \forall\ v \in H_0^1(\Omega)\ ,$$

and the corresponding Galerkin formulation is

$$(U_t,V) + (A(U,\nabla U),\nabla V) + (g(U,\nabla U),V) = 0 \qquad \forall\ V \in V_N \subset H_0^1(\Omega)\ .$$

Let $Z \in V_N$, for the moment be arbitrary and substract $(Z_t,V) + (A(Z,\nabla Z),\nabla V) + (g(Z,\nabla Z),V)$. Let $\mathcal{V} = U - Z$, $\eta = u - Z$. With $V = \mathcal{V}$ we get, by monotonicity (4.1)

(4.3) $$\frac{1}{2}\frac{d}{dt}(\mathcal{V},\mathcal{V}) + \delta\|\nabla\mathcal{V}\|^2 \leq |(\eta_t,\mathcal{V})| + |a(u,Z;\mathcal{V})|\ .$$

We have

(4.4) $$|(\eta_t,\mathcal{V})| \leq C\|\eta_t\|_{-1}^2 + \frac{\delta}{4}\|\nabla\mathcal{V}\|^2\ .$$

To estimate the second term we choose for each $t > 0$, $Z$ as an elliptic projection of u such that the elliptic projection errors $\|\eta\|$ and $\|\eta_t\|$ are simple to estimate and of optimal order (cf. [9], [10], [11] and [12]).

Two different approaches to this problem have been used. In [13], Z is defined as

$$\sum_{i=1}^{n} (A_i(u,\nabla u) - A_i(u,\nabla Z), \frac{\partial V}{\partial x_i}) + (g(u,\nabla u) - g(u,\nabla Z),V) = 0 \quad \forall\, V \in V_N,$$

which is still a nonlinear problem in Z. Here we use the linearized operator

$$\mathcal{A}(\omega,\nabla\omega) = \begin{bmatrix} \frac{\partial A}{\partial \xi}(\omega,\nabla\omega) & \frac{\partial A}{\partial u}(\omega,\nabla\omega) \\ \frac{\partial g}{\partial \xi}(\omega,\nabla\omega) & \frac{\partial g}{\partial u}(\omega,\nabla\omega) \end{bmatrix}$$

to define the bilinear form

$$b(u,\nabla u;\eta,V) = \int_\Omega [\nabla V^T,V]\mathcal{A}(u,\nabla u) \begin{bmatrix}\nabla\eta \\ \eta\end{bmatrix} dx = 0 \quad \forall\, V \in V_N$$

where $\eta = u - Z$ is the elliptic projection error. Since

$$b(u,\nabla u;V,V) \geq \delta \| \nabla V \|^2 \qquad \forall\, V \in V_N$$

b is a coercive form. Then it is possible to prove the quasioptimal error estimates (see [8])

$$\| \eta \|_j = Ch^{s-j} \| u \|_s, \quad j = 0,1, \quad 2 \leq s \leq r+1$$

$$\| \eta_t \|_1 \leq Ch^{s-1} [\| u \|_s + \| u_t \|_s]$$

and

$$\| \eta_t \|_{-1} \leq Ch^{r_0} [\| \eta_t \|_1 + \| \eta \|_1], \quad r_0 = \min(2,r)$$

where $C = C(u)$ and r is the degree of the continuous piecewise polynomials. These estimates are valid for all $t > 0$. By Cauchy-Schwarz inequality we now get from (4.3), (4.4)

$$\frac{d}{dt}(\boldsymbol{v},\boldsymbol{v}) + \delta \| \nabla \boldsymbol{v} \|^2 \leq C[\| \eta_t \|_{-1}^2 + \int_\Omega |\nabla\eta|^4 dx] .$$

By the inverse assumptions,

$$\| \nabla V \|_{L_\infty} \leq Ch^{-\nu_\infty} \| V \| , \quad V \in V_N$$

$$\inf_{\chi \in V_N} [\| \nabla(u-\chi) \|_{L_\infty} + h^{-\nu_\infty} \| u - \chi \|] \leq Ch^{s-\nu_\infty} \| u \|_{s+\varepsilon}, \quad 2 \leq s \leq r+1, \quad 0 < \varepsilon < 1,$$

where for quasiregular elements, $\nu_\infty = \frac{n}{2} + 1$, we get

$$\| \nabla \eta \|_{L_\infty} \leq C h^{s-\nu_\infty} \| u \|_{s+\varepsilon} .$$

Finally we then get

$$\frac{d}{dt}(\boldsymbol{v},\boldsymbol{v}) + \delta \| \nabla \boldsymbol{v} \|^2 \leq C h^{2r_0} \| \eta_t \|_1^2 + C[1 + \| u \|_{s+\varepsilon}] h^{2\min(r_0, s-\nu_\infty)} \| \eta \|_1^2 .$$

Thus by Gronwalls inequality

$$\| \boldsymbol{v}(\cdot,t) \| \leq \exp(-\frac{\delta}{2} t) \| \boldsymbol{v}(\cdot,0) \| + C \sup_{t>0} [h^{r_0} \| \eta_t \|_1 + h^{\min(r_0,s-\nu_\infty)} \| \eta \|_I]'$$

$2 \leq s \leq r + 1.$

The first term on the right hand side shows the independence of the error of the chosen initial function as $t \to \infty$. If $s = r + 1$, $r \geq \nu_\infty > 1$, $\nu_\infty = \frac{n}{2} + 1$, we have thus proven optimal order estimates

$$\| u - U \| \leq C h^{r+1} \sup_{t>0} [\| u \|_{r+1} + \| u_t \|_r], \quad t > 0 .$$

## 4.2 Time-integration.

To complete the discretization we choose a simple method, the so called $\theta$-method for the time-integration. Let W be the corresponding approximation, then

$$(4.5) \qquad (W(t + k) - W(t),V) + k[(A(\bar{W},\nabla\bar{W}),\nabla V) + (g(\bar{W},\nabla\bar{W}),V)] = 0 \quad \forall V \in V_N$$

where

$$\bar{W}(t) = \theta W(t) + (1 - \theta)W(t + k) \quad \text{and} \quad k > 0 \text{ is time-integration step.}$$

If $0 \leq \theta \leq \frac{1}{2} - |O(k)|$ one may prove the error estimate

$$\| u - W \| = O(h^{r+1}) + (\theta - \tfrac{1}{2})O(k) + O(k^2) = O(h^{r+1}) + O(k^2),$$

valid for all $t > 0$ (see [8]). Finally we have to linearize (4.5) and this is again done by help of the bilinear form b. Let Y be the solution of the linearized problem

$$(4.6) \qquad (Y(t + k) - Y(t),V) + kb(\tilde{Y},\nabla\tilde{Y}; \bar{Y} - \tilde{Y},V)$$
$$= -k[(A(\tilde{Y},\nabla\tilde{Y}),\nabla V) + (g(\tilde{Y},\nabla\tilde{Y}),V)] \qquad \forall V \in V_N.$$

Here $\tilde{Y} = Y$ or $\tilde{Y}(t) = \theta Y(t) + (1 - \theta)\tilde{\tilde{Y}}(t + k)$, $\tilde{\tilde{Y}}(t + k) = Y(t) + k \frac{d}{dt} Y(t)$. The error due to linearization is $O(k^2)$ if $k \leq ch^{\varepsilon+\nu_\infty/2}$, $\varepsilon > 0$ or $O(k^4)$, if $k \leq ch^{\varepsilon+\nu_\infty/4}$. We observe that (4.6) may be considered as a damped Newton-Kantorovich method for the numerical solution of

$$(F(U,\nabla U),V) = 0 \qquad \forall\, V \in V_N\,.$$

As $k \to \infty$ we get the (undamped) Newton-Kantorovich method. A numerical test of the above presented method is found in [4].

## 5. Applications.

Nonlinear monotone or even potential operators are found in many important practical applications. There are two classes of such examples,

(*i*) Problems with nonlinear material properties

(*ii*) Problems with nonlinear effects due to geometry.

Examples of the first kind are

$$(5.1)\quad a)\qquad -\nabla(a(|\nabla u|^2)\nabla u) + g(u) = 0 \quad \text{in } \Omega$$

with essential and/or natural boundary conditions on different parts of the boundary $\partial\Omega$ and are met in electromagnetic field equations and in torsion of a prismatic bar.

b) $\nabla(a(T)\nabla T) + g(u) = 0$ in $\Omega$

$$-\lambda\frac{\partial T}{\partial\nu} = \alpha(T - T_0) + \gamma(T^4 - T_0^4) \qquad \text{on } \partial\Omega,$$

$\alpha,\lambda,\gamma \geq 0$, a nonlinear heat convection equation.

Examples of the second kind are

a) (5.1) with $a(|\nabla u|^2) = 1/(1 + |\nabla u|^2)^{1/2}$ the minimal surface equation.

b) (5.1) with the function a as above, $g = Ku$ and boundary condition (see [14])

$$\frac{\partial u}{\partial\nu}/(1 + |\nabla u|^2)^{1/2} = c \qquad \text{on } \partial\Omega.$$

c) Large displacements theories like the von Karman model for a membrane

$$f(u) = \int_\Omega \{h[(2u_x + w_x^2)^2 + (2v_y + w_y^2) + 2(u_y + u_x + w_x w_y)^2$$

$$+ \frac{\nu}{1-\nu}(2u_x + 2v_y + w_x^2 + w_y^2)^2] + P\,\frac{w_x u + w_y u - w}{(1 + w_x^2 + w_y^2)^{1/2}}\}\partial\Omega$$

to be minimized over $[H_0^1(\Omega)]^3$, where u,v,w are the displacements in the x,y,z-directions, respectively.

References.

[1] M.M. Vainberg, Variational method and method of monotone operators in the theory of nonlinear equations, Wiley, 1973.

[2] O. Axelsson, U. Nävert, On a graphical package for nonlinear partial differential equation problems, Proc. IFIP Congress 77, B. Gilchrist (ed.), North-Holland Publ. Comp. (1977).

[3] O. Axelsson, A class of iterative methods for finite element equations, Comp. Meth. Appl. Mech. Engrg 9(1976), 123-137.

[4] O. Axelsson, T. Steihaug, Some computational aspects in the numerical solution of parabolic equations, in preparation.

[5] R. Bartels, J.W. Daniel, A conjugate gradient approach to nonlinear elliptic boundary value problems in irregular regions, CNA63, Center for Numerical Analysis, Austin, Texas, 1973.

[6] P. Concus, G.H. Golub, D. O'Leary, Numerical solution of nonlinear elliptic partial differential equations, STAN-CS-76-585, Computer Science Departmetn, Stanford University, 1976.

[7] D. O'Leary, Hybrid conjugate gradient algorithms, STAN-CS-76-548, Computer Science Department, Stanford University, 1976.

[8] O. Axelsson, Error estimates for Galerkin methods for quasilinear parabolic and elliptic differential equations in divergence form, Numer. Math. 28, 1-14,(1977).

[9] T. Dupont, Some $L^2$error estimates for parabölic Galerkin method *in* The mathematical foundations of the finite element method with applications to partial differential equations (A.K. Aziz, ed.) 491-504. Academic Press N.Y. 1972.

[10] I. Hlavàcek, On a semi-variational method for parabolic equations I, Aplikaee Matematiky 17(1972), 327-351.

[11] J. Douglas, T. Dupont, A Galerkin method for a nonlinear Dirichlet problem, Math. Comp. 29(1975), 689-696.

[12] M.F. Wheeler, A *priori* $L^2$error estimates for Galerkin approximations to parabolic partial differential equations, SIAM J. Numer. Anal. 10(1973), 713-759.

[13] J.E. Dendy, Galerkin's meghod for some highly nonlinear problems, SIAM J. Numer. Anal. 14(1977), 327-347.

[14] H.D. Mittelmann, Numerische Behandlung nichtlinearer Randwertproblemen mit finiten Elementen, Computing 18, 67-77 (1977).

Author's address: Chalmers University of Technology, Göteborg, Sweden.

# APPLICATION OF THE AVERAGING METHOD FOR THE SOLUTION OF BOUNDARY PROBLEMS FOR ORDINARY DIFFERENTIAL AND INTEGRO-DIFFERENTIAL EQUATIONS

D.D. Bainov, Sofia
S.D. Milusheva, Sofia

The averaging method first appeared in space mechanics. The basic technique of the averaging method is to replace the right hand side parts of complex systems of differential equations by averaged functions, the latter not containing explicitly time and fast - changing parameters of the system.

The averaging method found a strict mathematical justification in the fundamental works of N.M. Krilov, N.N. Bogolubov and J.A.Mitropolsky [13] , [15] , [20] . This method reached its further development and generalization in [1] , [14] , etc.

The period after 1960 was one of vigorous development of the averaging method. At that time 7 monographs on the averaging method were published where a number of schemes were displayed for its application to the solution of initial problems. In this way naturally arose the question of the justification of the averaging method for the solution of boundary problems for ordinary differential equations. The first results concerning the justification of the averaging method for the solution of boundary problems for ordinary differential equations were obtained by D.D. Bainov in 1964, and from 1970 on the authors of this survey achieved a number of new results. Some of these results are exposed in the present paper.

1. Solution of boundary problems by means of the averaging method on the basis of asymptotics constructed for the Cauchy problem. In [14]V.M. Volosov proposes the general averaging scheme for the solution of the Cauchy problem for the system, as follows:

$$\dot{x} = \varepsilon X(x,y,t,\varepsilon) = \varepsilon X_1(x,y,t) + \varepsilon^2 X_2(x,y,t) + \cdots$$
$$\dot{y} = Y(x,y,t,\varepsilon) = Y_0(x,y,t) + \varepsilon Y_1(x,y,t) + \varepsilon^2 Y_2(x,y,t) + \cdots \qquad (1.1)$$

with initial condition $x(t_0) = x_0$ , $y(t_0) = y_0$ , where $x, X \in R_n$ ; $y, Y \in R_m$ , while $\varepsilon > 0$ is a small parameter. In view of this the question of the possibility to apply the averaging method for solving boundary problems came to the fore. The paper [4] , namely, is devoted to the use of the averaging method to solve boundary problems for systems of the (1.1) type on the basis of asymptotics constructed for the Cauchy problem. An ordinary multipoint boundary problem and a multipoint boundary problem with boundary condition depending on several parameters are considered. Two theorems have been proved for each of these boundary problems. The first theorem points to conditions under which a formal asymptotics of the solution of the problem can be constructed. In the second theorem the existence and uniqueness of the solution of the boundary problem are proved. The paper [12] considers a boundary problem of the eigen-values for systems of ordinary differential equations with fast and slow variables. Theorems analogous to the ones in [4] have been proved.

2. Justification of the averaging method for the solution of two-point boundary problems for differential and integro-differential equations with fast and slow variables.

Consider the system of ordinary differential equations

(2.1) $\dot{x}(t)=\varepsilon X(t,x(t),y(t)),\ \dot{y}(t)=Y(t,x(t),y(t))$

with boundary condition

(2.2) $x(0)=x_0,\quad R[\lambda,y(0),y(T)]=0$

where $x,X\in R_n;\ y,Y,R\in R_m;\ \lambda\in\Lambda\subset R_m;\ T=L\varepsilon^{-1};\ L=\mathrm{const}>0$, while $\varepsilon>0$ is a small parameter. Together with the system (2.1), consider its degenerate system

(2.3) $\dot{y}(t)=Y(t,x,y(t)),\quad x=\mathrm{const}$

with boundary condition

(2.4) $R[\lambda,y(0),y(T)]=0$.

Assume that the solution of the problem (2.3), (2.4) is known and has the form $y=\psi(t,x,\lambda,T)$. Then, if along the integral curves $y=\psi(t,x,\lambda,T)$ of the boundary problem (2.3), (2.4), where $\lambda$ is considered as a vector parameter, there exists a non-dependent on $\lambda$ mean value

(2.5) $\lim\limits_{T\to\infty}\frac{1}{T}\int_0^T X(t,x,\psi(t,x,\lambda,T))\,dt=\overline{X}(x)$,

then the equation

(2.6) $\dot{\xi}(t)=\varepsilon\overline{X}(\xi(t))$

with initial condition

(2.7) $\xi(0)=x(0)$

will be called averaged equation of first approximation for the slow variables $x(t)$ of the system (2.1).

The following theorem for the proximity of the component $x(t)$ of the solution of boundary problem (2.1), (2.2) and the solution of the Cauchy problem (2.6), (2.7) holds.

THEOREM. Let us assume:

1. The functions $X(t,x,y)$ and $\frac{\partial}{\partial y}X(t,x,y)$ are continuous in the domain $\Omega(t,x,y)=\Omega(t)\times\Omega(x)\times\Omega(y)$, where $\Omega(t)=[0,\infty)$, $\Omega(x)$ and $\Omega(y)$ are certain open domains of the spaces $R_n$ and $R_m$, resp.

2. In the domain $\Omega(t,x,y)$ the following inequalities are satisfied $\|X(t,x,y)-X(t,x',y')\|\le\mu\|x-x'\|+\theta_1(t)\|y-y'\|$, $\|\frac{\partial}{\partial y}X(t,x,y)\|\le\theta_2(t)$, where $\mu$ is a positive constant, while $\theta_i(t)$ $(i=1,2)$ is a continuous non-negative function.

3. The unique integral curve of boundary problem (2.3), (2.4) corresponding to some value of the parameter $\lambda$, passes through every point of the domain $\Omega(t,x,y)$, and besides,

a. This curve is definite and lies inside the domain $\Omega(y)$ for any $t\ge 0$.

b. The vector functions $\Psi(t,x,\lambda,T)$ and $\frac{\partial}{\partial T}\Psi(t,x,\lambda,T)$ are continuous along the set of variables $t,x,\lambda,T$ and satisfy in the domain $\{\Omega(t,x)\times\Lambda=\Omega(t)\times\Omega(x)\times\Lambda,\ T\ge 0\}$ the inequalities

$$\|\Psi(t,x,\lambda,T)\|\le K,\quad \|\frac{\partial}{\partial T}\Psi(t,x,\lambda,T)\|\le g(t,T),$$

where $K$ is a positive constant, while $g(t,T)$ is a continuous non-negative function.

4. The boundary problem (2.1), (2.2) has a unique continuous solution $\{x(t),y(t)\}$, whose component $y(t)$ is bounded ($\|y(t)\|\le b=\text{const}$). (In (2.2) $\lambda$ means a certain fixed value of the parameter $\lambda$ from the domain $\Lambda$.)

5. For $t\ge 0$ and $T\ge 0$ the functions $\theta_i(t)$ $(i=1,2)$ and $g(t,T)$ satisfy the conditions

$$\lim_{t\to\infty}\frac{1}{t}\int_0^t\theta_1(\tau)\,d\tau=0,\quad \lim_{t\to\infty}\frac{1}{t}\int_0^t d\tau\int_0^\tau\theta_2(s)g(s,\tau)\,ds=0$$

6. For every $(x,\lambda)\in\Omega(x)\times\Lambda$ there exists a bound of (2.5) not depending on the parameter $\lambda$, and the boundary transition in (2.5) is accomplished uniformly with respect to the set $(x,\lambda)\in\Omega(x)\times\Lambda$. In the domain $\Omega(x)$ the function $\overline{X}(x)$ is continuous and satisfies the condition

$\|\overline{X}(x)\|\le M$, $\|\overline{X}(x)-\overline{X}(x')\|\le\nu\|x-x'\|$, where $M$ and $\nu$ are positive constants.

7. The solution $\xi=\xi(t)$ of the Cauchy problem (2.6), (2.7) for any $t\ge 0$ is bounded ($\|\xi(t)\|\le d=\text{const}$) and lies in the domain $\Omega(x)$ together with some $\rho$-neighbourhood ($\rho=\text{const}>0$).

Then, if $\{x(t),y(t)\}$ is a solution of the boundary problem (2.1), (2.2) and $\xi(t)$ is a solution of the Cauchy problem (2.6),

(2,7), then for any $\omega > 0$ and $L > 0$ such an $\varepsilon^0 > 0$ can be found that, for $0 \le \varepsilon \le \varepsilon^0$ on the cut $0 \le t \le L\varepsilon^{-1}$ the inequality $\| x(t) - \xi(t) \| < \omega$ will be satisfied.

PROOF. Introduce the function

$$(2.8)\quad v(t,x) = \int_{\Omega(x)} \Delta_\alpha(x-x') \Big\{ \int_0^t [X(\tau,x',\psi(\tau,x',\lambda,t)) - \overline{X}(x')] d\tau \Big\} dx',$$

where the smoothing kernel $\Delta_\alpha(x)$ has the form $\Delta_\alpha(x) = A_\alpha\left(1 - \frac{\|x\|^2}{\alpha^2}\right)^2$ for $\|x\| \le \alpha$ and $\Delta_\alpha(x) = 0$ for $\|x\| > \alpha$, $\alpha = \text{const} > 0$, while the positive constant $A_\alpha$ is determined by the condition $\int_{R_n} \Delta_\alpha(x)dx = 1$.

In view of the conditions of the theorem one can always construct such a monotonely decreasing function $\alpha(t)$ ($\alpha(t) \to 0$ for $t \to \infty$) that for every $x \in \Omega(x)$ the following inequality will hold

$$\Big\| \frac{1}{t}\int_0^t [X(\tau,x,\psi(\tau,x,\lambda,t)) - \overline{X}(x)] d\tau \Big\| \le \alpha(t).$$

Then, for $t \ge 0$, for any points $x$, $\alpha$ whose neighbourhood belongs to the domain $\Omega(x)$, the following inequalities will hold

$$(2.9)\quad \| v(t,x) \| \le t\alpha(t), \quad \Big\| \frac{\partial}{\partial x} v(t,x) \Big\| \le I_\alpha t \alpha(t),$$

where $I_\alpha = \int_{R_n} \| \frac{\partial}{\partial x} \Delta_\alpha(x) \| dx$.

Estimate the expression

$$P(t,x) = \frac{\partial}{\partial t} v(t,x) - X(t,x,\psi(t,x,\lambda,t)) + \overline{X}(x).$$

Since

$$\int_{\Omega(x)} \Delta_\alpha(x-x')dx' = \int_{\|x\| \le \alpha} \Delta_\alpha(x-x')dx' = 1,$$

then for $t \ge 0$, for any $x$, $\alpha$ whose neighbourhood belongs to the domain $\Omega(x)$, one obtains

$$(2.10)\quad \| P(t,x) \| \le (\mu+\nu)\alpha + 2K\theta_1(t) + \int_0^t \theta_2(\tau) g(\tau,t) d\tau.$$

Set $\tilde{x}(t) = \xi(t) + \varepsilon v(t,\xi(t))$. According to the conditions of the theorem $\xi(t)$ lies in the domain $\Omega(x)$ together with the $\rho$-neighbourhood, and hence for $\alpha < \rho$ the estimate (2.9) holds for the function $v(t,\xi(t))$.

Set $A(\varepsilon) = \sup_{0 \le \tau \le L} \tau\alpha(\tau\varepsilon^{-1})$. Obviously, $A(\varepsilon) \to 0$ for $\varepsilon \to 0$ and the following inequality will hold on the segment $0 \le t \le L\varepsilon^{-1}$ if $\varepsilon$ is sufficiently small:

$$(2.11)\quad \| \varepsilon v(t,\xi(t)) \| \le \varepsilon t \alpha(t) \le A(\varepsilon) < \frac{1}{2}\min(\rho,\omega).$$

Therefore, on the segment $0 \le t \le L\varepsilon^{-1}$ , $\tilde{x}(t)$ belongs to the domain $\Omega(x)$ together with the $\rho_1$ -neighbourhood ( $0<\rho_1=\mathrm{const}<\rho$ ) and $\|\tilde{x}(t)\| \le d_1$ , $d_1=\mathrm{const}>0$ .

Consider the difference

$$(2.12)\quad Q(t)=\frac{d\tilde{x}}{dt}-\varepsilon X(t,\tilde{x},y),$$

where $\tilde{x}=\tilde{x}(t)$ , and $y=y(t)$ is a component of the solution $\{x(t),y(t)\}$ of the boundary problem (2.1), (2.2).

Taking into consideration (2.9), (2.10), one obtains

$$(2.13)\quad \|Q(t)\| \le \varepsilon(\mu+\nu)a+\varepsilon(3K+b)\theta_1(t)+\varepsilon^2(I_a M+\mu)t\alpha(t)+\varepsilon\int_0^t \theta_2(\tau)g(\tau,t)\,d\tau .$$

It is easily verified that on the segment $0 \le t \le L\varepsilon^{-1}$ the component $x(t)$ of the solution $\{x(t),y(t)\}$ of the boundary problem (2.1), (2.2) does not leave the domain $\Omega(x)$ . Then on this segment one gets from (2.1) and (2.12)

$$\frac{d}{dt}\|x(t)-\tilde{x}(t)\| \le \varepsilon\mu\|x(t)-\tilde{x}(t)\|+\|Q(t)\|$$

whence, taking into account that $\tilde{x}(0)=x(0)$ , one finds

$$(2.14)\quad \|x(t)-\tilde{x}(t)\| \le \int_0^t \|Q(\tau)\|\exp\{\varepsilon\mu(t-\tau)\}\,d\tau .$$

Introducing the functions

$$\delta_1(t)=\frac{1}{t}\int_0^t \theta_1(\tau)\,d\tau \qquad , \delta_2(t)=\frac{1}{t^2}\int_0^t \tau\alpha(\tau)\,d\tau$$

$$\delta_3(t)=\frac{1}{t}\int_0^t d\tau\int_0^\tau \theta_2(s)g(s,\tau)\,ds\,,\ \delta_i(t)\to 0\,,\ t\to\infty\ (i=\overline{1,3})\ ,$$

for the right hand side of the inequality (2.14) on the segment $0 \le t \le L\varepsilon^{-1}$ one finds the estimate

$$(2.15)\quad \int_0^t \|Q(\tau)\|\exp\{\varepsilon\mu(t-\tau)\}\,d\tau \le \exp\{\mu L\}\{(\mu+\nu)La+$$
$$+(3K+b)L\delta_1(L\varepsilon^{-1})+(I_a M+\mu)L^2\delta_2(L\varepsilon^{-1})+L\delta_3(L\varepsilon^{-1})\}.$$

From (2.11) and (2.15) it follows that if $a$ and $\varepsilon$ are sufficiently small ( $0<a,\varepsilon \le \varepsilon^0$ ) then on the segment $0 \le t \le L\varepsilon^{-1}$ the inequality $\|x(t)-\xi(t)\| < \min(\rho,\omega)$ is satisfied. Thus, the theorem is proved.

In the papers [6] - [11] , [16] - [19] several variants of the averaging method have been justified for the solution of two-point boundary problems for integro-differential equations with fast and slow variables.

Consider the system of integro-differential equations

$$
(2.16)\qquad
\begin{aligned}
\dot{x}(t) &= \varepsilon X\Big(t, x(t), y(t), \int_0^t \varphi(t,s,x(s),y(s))\,ds\Big)\\
\dot{y}(t) &= Y\Big(t, x(t), y(t), \int_0^t \varphi_1(t,s,x(s),y(s))\,ds\Big)
\end{aligned}
$$

with boundary conditions $x(0)=x_0$ , $R[\lambda, y(0), y(T)]=0$ ,
where $x, X \in R_n$; $y, Y, R \in R_m$; $\varphi \in R_p$; $\varphi_1 \in R_q$; $\lambda \in \Lambda \subset R_m$ ,

$T = L\varepsilon^{-1}$, $L = \mathrm{const} > 0$ , while $\varepsilon > 0$ is a small parameter.

Assume that the degenerate system with respect to (2.16)

$$
(2.17)\quad \dot{y}(t) = Y\Big(t, x, y(t), \int_0^t \varphi_1(t,s,x,y(s))\,ds\Big)\ , \quad x = \mathrm{const}
$$

with boundary condition $R[\lambda, y(0), y(T)]=0$ , has a solution of the form

$$
(2.18)\quad y = \psi(t, x, \lambda, T)\ , \quad x = \mathrm{const}\ .
$$

Several schemes of the averaging method are possible. Here is one of them.

Let along the integral curves (2.18), where $\lambda$ is considered as a vector parameter, there exist mean values not depending on $\lambda$

$$
\lim_{T\to\infty} \frac{1}{T}\int_0^T X(t, x, \psi(t,x,\lambda,T), u)\,dt = \bar{X}_1(x,u)\ ,
$$

$$
\lim_{T\to\infty} \frac{1}{T}\int_0^T \varphi(t,s,x,\psi(s,x,\lambda,T))\,ds = \bar{\varphi}_1(t,x)\ .
$$

Then the equation

$$
\dot{\xi}(t) = \varepsilon \bar{X}_1\Big(\xi(t), \int_0^t \bar{\varphi}_1(t, \xi(s))\,ds\Big)
$$

with initial condition

$$
\xi(0) = x(0)
$$

will be called averaged equation of the first approximation for the slow variables $x(t)$ of the system (2.16).

This averaging scheme can be successfully applied when considering boundary problems for quasi-linear systems of the form

$$
(2.19)\qquad
\begin{aligned}
\dot{x}(t) &= \varepsilon\Big[\tilde{A}(x(t))y(t) + \tilde{B}(x(t)) + \int_0^t K(x(s))y(s)\,ds\Big]\\
\dot{y}(t) &= A(x(t))y(t) + B(x(t))\ ,\\
\alpha y(0) &+ \beta y(T) = y_0\ , \quad x(0) = x_0\ ,
\end{aligned}
$$

where $\alpha$ is a diagonal matrix whose first $p$ diagonal elements are units and the remaining $(m-p)$ are zeros; $\beta$ is an analogous matrix whose first $p$ elements are zeros and the rest are units;

$$A(x)=(a_{ij}(x))_1^m;\ \tilde{A}(x)=(\tilde{a}_{ij}(x))_{n,m};\ B(x)=(b_1(x),\dots,b_m(x));$$

$$\tilde{B}(x)=(\tilde{b}_1(x),\dots,\tilde{b}_m(x));\ K(x)=(k_{ij}(x))_{n,m};\ x\in R_n;\ y\in R_m,$$

while $\varepsilon>0$ is a small parameter.

Denote by $\nu_k=g_k(x)\ (k=\overline{1,m})$ the eigen-values of the matrix $A(x)$, and by $\Omega(x)=(\omega_i^{(k)}(x))_1^m$ denote the matrix whose columns are composed by the components $m$ of the linearly independent eigen-vectors of the matrix $A(x)$.

Under the assumption that in the considered domain $\Omega(x)$ the first $p$ eigen-values of the matrix $A(x)$ have negative real parts and the remaining $(m-p)$ eigen-values have positive real parts, and under the assumption that the elementary divisors of the matrix $A(x)$ are simple and that $\operatorname{Det} M(x).\operatorname{Det} \mathcal{N}(x)\neq 0$,

where $M(x)=(\omega_i^{(j)}(x))_1^p$, $\mathcal{N}(x)=(\omega_l^{(s)}(x))_{p+1}^m$

in the paper [19] it is shown that the averaged equation of the system (2.19 ) has the form

$$\dot{\xi}(t)=\varepsilon\Big[\tilde{B}(\xi(t))-\tilde{A}(\xi(t))A^{-1}(\xi(t))B(\xi(t))-$$
$$-\int_0^t K(\xi(s))A^{-1}(\xi(s))B(\xi(s))ds\Big].$$

REFERENCES

[1] Arnold V.I.:Uslovia primenimosti i otsenka pogreshnosti metoda usrednenia dla sistem kotorie v protsese evoljutsii prohodiat cherez resonansi. Dokl. Acad. Nauk USSR, 1965, 161, 9 - 12.

[2] Bainov D.D.:Metod usrednenia dla odnoi dvuhtochechnoi kraevoi zadachi. Matem. vestnik, 1968, 5 (20), No 2, 198 - 204.

[3] Bainov D.D.:Asymtoticheskie formuli dla odnoi kraevoi zadachi, Proc. of the International Conference on Nonlinear Oscillations, Kiev, 1970, I, 45 -53.

[4] Bainov D.D.:Reshenie nekotorih kraevih zadach metodom usrednenia na baze asymptotiki postroenoi dla zadaci Cauchy. Izv. Math. Inst. Bulg. Acad. Sci., 1974, 15, 5 - 20.

[5] Bainov D.D., Konstantinov M.M.:Metodat na usredniavaneto i negovoto prilozenie v tehnikata, "Nauka i Izkustvo", Sofia, 1973.

[6] Bainov D.D. Milusheva S.D.:Primenenie metoda usrednenia k odnoi dvuhtochechnoi kraevoi zadache dla system integro-diferentsialnih uravnenii tipa Fredholm. Bull.Math. de la Soc. Sci. Math. de la R.S. de Roumanie, 1973, 17 (65), No 1, 3 - 7.

[7] Bainov D.D., Milusheva S.D.:Metodi usrednenia dla odnoi dvuhtochechnoi kraevoi zadachi dla system integro-diferentsialnih uravnenii ne razreshenih otnositelno proizvodnoi. Publications de l'Institut Mathématique, Nouvelle serie, 1973, 16 (30), 13 - 23.

[8] Bainov D.D., Milusheva S.D.:O primenenii metoda usrednenia k odnoi dvuhtochechnoi kraevoi zadache dla system integro-diferentsialnih uravnenii tipa Fredholm. Glasnik Matematički, 1974, 9 (29), No 2, 251 - 265.

[9] Bainov D.D., Milusheva S.D.:Primenenie metoda usrednenia k odnoi dvuchtochechnoi kraevoi zadache dla system integro-diferentsialnih uravnenii tipa Fredholm. Diff. Eqs., 1974, 10, No II, 2042 - 2047.

[10] Bainov D.D., Milusheva S.D.:Metodi usrednenia dla odnoi dvuhtochechnoi kraevoi zadachi dla system integro-diferentsialnih uravnenii. Matem. vesnik, 1975, 12 (27), 3 - 17.

[11] Bainov D.D., Milusheva S.D.:Primenenie metoda usrednenia k odnoi dvuhtochechnoi kraevoi zadache dla nelineinih system integro-diferentsialnih uravnenii. Zagadnienia Drgan Nieliniowych, 1976, 17, 104 - 123.

[12] Bainov D.D., Sarafova G.H.:Reshenie dvuhtochechnoi kraevoi zadachi na sobsvenie znachenia metodom usrednenia na base asymptotiki, postroenoi dla zadachi Cauchy. Bull. de la Soc. des mathématiciens et des physiciens de la R.S. Macédonie, 1973, 24, 7 - 19.

[13] Bogolubov N.N., Mitropolsky J.A.:Asymptoticheskie metodi v teorii nelineinih kolebanii, "Fizmatgiz", Moscow, 1963.

[14] Volosov V.M.:Usrednenie v systemah obiknovenih diferentsialnih uravnenii. Uspehi mat. nauk, 1962, 17, No 6, 3 - 126.

[15] Krilov N.M., Bogolubov N.N.:Prilozenie metodov nelineinoi mehaniki k teorii statsionarnih kolebanii. Izd. Acad. Sci. Ukra. SSR, Kiev, 1934.

[16] Milusheva S.D.:Primenenie metoda usrednenia k odnoi dvuhtochechnoi kraevoi zadache dla system integro-diferentsialnih uravnenii tipa Volterra. Ukr. Math. Journal, 1974, 26, No 3, 338 - 344.

[17] Milusheva S.D.:O primenenii metoda usrednenia k odnoi dvuh-

tochechnoi kraevoi zadache dla system integro-diferentsialnih uravnenii tipa Volterra s bistrimi i medlenimi peremenimi. Nonlinear Vibration Problems, Zagadnienia Drgan Nieliniowych, 1975, 16, 155 - 163.

[18] Milusheva S.D., Bainov D.D.:Primenenie metoda usrednenia k odnoi dvuhtochechnoi kraevoi zadache dla system integro-diferentsialnih uravnenii tipa Volterra ne razreshenih otnositelno proizvodnoi. Mathematica Balkanica, 1973, 3, 347 - 357.

[19] Milusheva S.D., Bainov D.D.:Obosnovanie metoda usrednenia dla odnoi nelineinoi kraevoi zadachi. Izvestia VUZ, Mathematica, 1974, No 12, 19 - 28.

[20] Mitropolsky J.A.:Metod usrednenia v nelineinoi mehanike. "Naukova dumka", Kiev, 1971.

**Authors' addresses: Oborišče 23, 1504 Sofia 4, Bulgaria**
**ul. Dede-agač 11, 1408 Sofia, Bulgaria**

SOLUTION SET PROPERTIES FOR SOME NONLINEAR PARABOLIC DIFFERENTIAL EQUATIONS

J.W. Bebernes, Boulder

1. Introduction. This paper is concerned mainly with reporting some solution set properties for various classes or problems for nonlinear parabolic equations. Most of this work was done jointly with K. Schmitt. The example in section 6 is a special case of a class of problems being studied jointly with K.-N. Chueh and W. Fulks.

During the past few decades much work has been devoted to the problem of characterizing sets which are invariant with respect to a given ordinary differential equation. More recently several papers ([3], [5], [11], [13]) have considered the same question for nonlinear parabolic differential equations. In [5], the relationship between invariant sets and traveling wave solutions is noted. This relationship can be used to study the Fitzhugh-Nagumo and Hodgkin-Huxley equations, for example.

The assumptions which are sufficient for a given set to be invariant also yield existence of solutions for initial boundary value problems and yield the classical Kneser-Hukuhawa property, i.e., the set of solutions is a continuum in an appropriate function space. For scalar-valued problems, conditions sufficient for invariance give existence of maximal and minimal solutions [4].

2. Definitions and Notation. Let $\mathbb{R}^n$ denote n-dimensional real Euclidean space and let $\Omega$ be a bounded domain in $\mathbb{R}^n$ whose boundary $\partial\Omega$ is an (n-1)-dimensional manifold of class $C^{2+\alpha}$, $\alpha \in (0,1)$. Let $\pi = \Omega \times (0,T)$ and $\Gamma = (\partial\Omega \times [0,T)) \cup (\Omega \times \{0\})$. For $u: \bar{\pi} \to \mathbb{R}$, define the differential operators $L_k u$ by:

$$L_k u(x,t) = \sum_{i,j=1}^{n} a_{ij}^k(x,t) \frac{\partial^2 u}{\partial x_i \partial x_j} + \sum_{i=1} b_i^k(x,t) \frac{\partial u}{\partial x_i} + c^k(x,t)u - \frac{\partial u}{\partial t}$$

where $a_{ij}^k, b_i^k, c^k \in C^{\alpha,\alpha/2}(\bar{\Omega} \times [0,T])$, $0 < \alpha < 1$, $1 \le i, j \le n$, $1 \le k \le m$, and for all $k$, $c^k \le 0$. Here $C^{\alpha,\alpha/2}(\cdot)$ denotes the usual Hölder spaces of functions $u(x,t)$. For $u: \bar{\pi} \to \mathbb{R}^m$, let $L = (L_1, \ldots, L_m)$ be defined by $Lu = (L_1 u_1, \ldots, L_m u_m)$. Assume that $L$ is uniformly parabolic. Let $f: \bar{\pi} \times \mathbb{R}^m \times \mathbb{R}^{nm} \to \mathbb{R}^m$, defined by $(x,t,u,p) \to f(x,t,u,p)$ with $(x,t) \in \bar{\pi}$, $u = (u_1, \ldots, u_m) \in \mathbb{R}^m$, $p = (p_1, \ldots, p_m) \in \mathbb{R}^{nm}$, $p_i \in \mathbb{R}^n$ be a locally Hölder continuous function with Hölder exponents $\alpha, \alpha/2, \alpha, \alpha$ in the respective variables $x,t,u,p$.

Given $\psi: \Gamma \to \mathbb{R}^m$, consider the first initial boundary value problem $(IBVP)_1$:

$$(1) \qquad Lu = f(x,t,u,\nabla u), \quad (x,t) \in \pi$$

$$(2) \qquad u = \psi, \quad (x,t) \in \Gamma$$

where $\psi$ is continuous on $\Gamma$, may be extended to $\bar{\pi}$ so as to belong to $C^{2+\alpha,\,1+\alpha/2}(\bar{\pi})$, and satisfies compatibility conditions appropriate to (1) and (2).

For $\varphi\colon \bar{\Omega} \to \mathbb{R}^m$, the second initial boundary value problem $(\text{IBVP})_2$: (1) with

(3) $\quad u(x,0) = \varphi(x) \quad , \quad x \in \bar{D}$

(4) $\quad \dfrac{\partial u}{\partial \nu} = 0 \quad , \qquad (x,t) \in \partial D \times [0,T)$

where $\dfrac{\partial u(x,t)}{\partial \nu(x)} = (\nu \cdot \nabla u_1, \ldots, \nu \cdot \nabla u_m)$, $\nu(x)$ is an outer normal to $\Omega$ at $x$, and $\nu(x) \in C^{\alpha+1}(\partial\Omega)$ will also be considered.

3. <u>Positive Invariance</u>. A set $S \subset \mathbb{R}^m$ is <u>positively invariant</u> relative to $(\text{IBVP})_1$ ($(\text{IBVP})_2$) in case, given $\psi\colon \Gamma \to S$ ($\varphi\colon \Omega \to S$), every solution $u \in C^{(2,1)}(\bar{\pi})$ of $(\text{IBVP})_1$ ($(\text{IBVP})_2$) is such that $u\colon \bar{\pi} \to S$. A set $S \subset \mathbb{R}^m$ is weakly positively invariant relative to $(\text{IBVP})_1$ ($(\text{IBVP})_2$) in case, given any $\psi\colon \Gamma \to S$ ($\varphi\colon \Omega \to S$), there exists at least one solution $u \in C^{2,1}(\bar{\pi})$ of $(\text{IBVP})_1$ ($(\text{IBVP})_2$) such that $u\colon \bar{\pi} \to S$.

<u>Theorem 1</u>. Let $L_i = L_1$, $i = 1,\ldots,m$. Let $S \subset \mathbb{R}^m$ be a nonempty open bounded convex neighborhood of $0$ such that for each $u \in \partial S$, there exists an outer normal $n(u)$ to $S$ at $u$ with

(5) $\quad n(u) \cdot f(x,t,u,p) > 0$

for all $p = (p_1,\ldots,p_n)$, $p_i \in \mathbb{R}^m$, $1 \le i \le n$, such that $n(u) \cdot p_i = 0$, $i = 1,\ldots,n$, $(x,t) \in \bar{\pi}$ ($\nu(x) \cdot (p_1^j,\ldots,p_n^j) = 0$, $j = 1,\ldots,m$, $x \in \partial\Omega$, $t \in [0,T]$ or $p_i \cdot n(u) = 0$, $i = 1,\ldots,n$, $(x,t) \in \pi$). Then $S$ is positively invariant relative to $(\text{IBVP})_1$ ($(\text{IBVP})_2$).

This theorem is easily proven by standard maximum principal arguments.

Using the above theorem on positive invariance and assuming a Nagumo growth condition on $f$ with respect to $p$, a weak invariance result, i.e., existence of a solution which lies in the set, can be proven.

<u>Theorem 2</u>. Let $L_i = L_1$, $i = 1,\ldots,m$. Let $S \subset \mathbb{R}^m$ be a nonempty convex set such that for each $u \in \partial S$ and every out normal $n(u)$ to $S$ at $u$

(6) $\quad n(u) \cdot f(x,t,u,p) \ge 0$

for all $p = (p_1,\ldots,p_n)$, $p_i \in \mathbb{R}^m$, $1 \le i \le n$, such that $n(u) \cdot p_i = 0$, $i = 1,\ldots,n$, $(x,t) \in \bar{\pi}$ ($\nu(x) \cdot (p_1^j,\ldots,p_n^j) = 0$, $j = 1,\ldots,m$, $x \in \partial\Omega$, $t \in [0,T]$ or $p_i \cdot n(u) = 0$, $i = 1,\ldots,n$, $(x,t) \in \pi$). Furthermore, let there exist a positive, continuous, nondecreasing function $\varphi(s)$ satisfying $s^2/\varphi(s) \to \infty$ as $s \to \infty$ and $|f(x,t,u,p)| \le \varphi(|p|)$, $u \in S$, $(x,t) \in \pi$.

Then $S$ is weakly positively invariant relative to $(\text{IBVP})_1$ ($(\text{IBVP})_2$).

The growth condition imposed on $f$ is the Nagumo condition.

The details of the proof of this theorem can be found in [3]. To convey the idea of the proof for $(IBVP)_1$, first assume $S$ is an open convex neighborhood of $0$ and that the strict outer normal condition (5) is satisfied. Let $F: C^{1,0}(\bar{\pi}) \to C(\bar{\pi})$ be the continuous map taking bounded sets into bounded sets defined by

(7) $$(Fu)(x,t) = f(x,t,u(x,t), \nabla u(x,t)) .$$

Let $K: C(\bar{\pi}) \to C^{1,0}(\bar{\pi})$ be the compact bounded linear extension of the linear map $K: C^{\alpha,\alpha/2}(\bar{\pi}) \to C^{\alpha+2,1+\alpha/2}(\bar{\pi})$ defined as follows: for $v \in C^{\alpha,\alpha/2}(\bar{\pi})$, $Kv$ is the unique solution of

(8) $$LKv = v$$
$$Kv = 0 .$$

Let $g \in C^{\alpha+2,1+\alpha/2}(\bar{\pi})$ be the unique solution to

(9) $$Lg = 0$$
$$g = \psi .$$

For any $\lambda \in [0,1]$, $\lambda KF: C^{1,0}(\bar{\pi}) \to C^{1,0}(\bar{\pi})$ is a completely continuous map. For $\lambda \in [0,1]$, $g \in C^{2+\alpha,1+\alpha/2}(\bar{\pi})$ a solution of (9), $u \in C^{1,0}(\bar{\pi})$ is a solution of

(10) $$u = \lambda KFu + \lambda g$$

if and only if $u \in C^{2+\alpha,1+\alpha/2}(\bar{\pi})$ is a solution of

$(1_\lambda)$ $$Lu = \lambda f(x,t,u,\nabla u) \quad , \quad (x,t) \in \pi$$

$(2_\lambda)$ $$u(x,t) = \lambda\psi(x,t) \quad , \quad (x,t) \in \psi .$$

By the Nagumo growth condition imposed on $f$ in hypotheses of the theorem, if $u \in C^{\alpha+2,\alpha/2+1}(\bar{\pi})$ is a solution of $(1_\lambda)$ - $(2_\lambda)$ for any $\lambda \in [0,1]$ with $u: \bar{\pi} \to \bar{S}$, then there exists $M > 0$ such that $|\nabla u| \le M$.

The crux of the proof is to show that the continuous compact perturbation of the identity given by $I - \lambda(KF + g): \mathcal{O} \subset C^{1,0}(\bar{\pi}) \to C^{1,0}(\bar{\pi})$, $\lambda \in [0,1]$, where $\mathcal{O} = \{u \in C^{1,0}(\bar{\pi}) \mid u: \bar{\pi} \to S \ , \ |\nabla u(x,t)| < M+1\}$ is a nonempty bounded open subset in $C^{1,0}(\bar{\pi})$, has nonzero Leray-Schander degree at $0$ relative to $\mathcal{O}$. This can be accomplished by a homotopy argument. In this way, the existence of a solution for $(IBVP)_1$ is established for a strict outer normal condition on an open convex neighborhood of zero. By a perturbation argument, the weak outer normal condition (6) suffices. Finally, one shows that if $S$ is a compact convex set, then the result holds for $S_\varepsilon$, $\varepsilon$ - neighborhoods of $S$. By an approximating argument, the weak invariance of $S$ follows.

For certain sets $\Lambda \subset \mathbb{R}^n \times \mathbb{R} \times \mathbb{R}^m$ which have compact convex cross sections in $\mathbb{R}^m$ depending on $x$ and $t$, similar invariance results hold. For example, let $\alpha,\beta \in C^{2,1}(\bar{\pi})$ be given with $\alpha_i(x,t) < \beta_i(x,t)$ on $\bar{\pi}$ and define

$$(\alpha,\beta) = \{u \in \mathbb{R}^m : \alpha_i \le u_i \le \beta_i , \ i = 1,\dots,m\} .$$

Theorem 3. Assume that

$$(11) \quad \begin{cases} L_k\alpha_k - f_k(x,t,u_1,\dots,u_{k-1},\alpha_k,u_{k+1},\dots,u_m,\nabla u_1,\dots,\nabla u_{k-1},\nabla\alpha_k,\nabla u_{k+1},\dots,\nabla u_m) \\ \qquad\qquad \ge 0 \ge \\ L_k\beta_k - f_k(x,t,u_1,\dots,u_{k-1},\beta_k,\beta_{k+1},\dots,u_m,\nabla u_1,\dots,\nabla u_{k-1},\nabla\beta_k,\nabla u_{k+1},\dots,\nabla u_m) \end{cases}$$

for all $(x,t) \in \pi$ , $k = 1,\dots,m$ , and $\alpha_j \le u_j \le \beta_j$ , $k \neq j$ .

Furthermore, assume the Nagumo growth condition of theorem 2 relative to $(\alpha,\beta)$. Then $(\alpha,\beta)$ is weakly positively invariant relative to $(IBVP)_1((IBVP)_2)$ .

4. Funnel Properties. The classical Hukuhara-Kneser property for ordinary differential equations in $\mathbb{R}^n$ states that if all solutions of a given initial value problem exist on $[t_0,t_0+\delta]$ , then the set of solutions is a continuum (a compact connected set) in $C[t_0,t_0+\delta]$ . Krasnosel'skii and Sobolevskii [7] very elegently proved an abstracted version of this result for the set of fixed points of completely continuous operators defined in a normed linear space which also satisfy a certain approximation property. By using a modification of this result, the following theorem can be proven.

Theorem 4. Assume the hypotheses of Theorem 2, then the set $Q = \{u \in \mathcal{O} : u = KFu + g\}$ is a continuum in $C^{1,0}(\bar{\pi})$ .

Here $K$ , $f$ , $g$ , and $\mathcal{O}$ are as in Section 2.

5. Maximal and Minimal Solutions. When $m = 1$ , the invariance result given by theorem 3 can be used to establish the existence of maximal and minimal solutions for the scalar version of $(IBVP)_1$ and the Cauchy initial value problem for (1). In this section we report on the main result in [4].

In recent years a considerable amount of study has been devoted to establishing the existence of solutions for elliptic and parabolic problems provided upper and lower solutions of such problems exist. Much of this work has its basis in the fundamental paper of Nagumo [8] as carried further by Akô [1]. Keller [6] and Amann [2] constructed solutions between upper and lower solutions of elliptic problems using a monotone iteration scheme which was possible because of certain one sided Lipschitz continuity assumptions on the nonlinear terms and because the nonlinearities are assumed gradient independent. Sattinger [12], Pao [9], and Puel [10] extended Amann's results to parabolic initial boundary value problems using either monotone iteration techniques on the theory of monotone operators. While these procedures have certain computational advantages the permissible class of nonlinearities is quite restrictive.

Using a different approach patterned after methods employed by Akô, the existence of maximal and minimal solutions for the Cauchy initial value problem and the initial value problem for parabolic equations can be proven for a much larger class of nonlinearities.

A continuous function $v: \pi \to \mathbb{R}$ is called a <u>lower solution</u> of (1), (2) in case

$$(12) \qquad v(x,t) \le \psi(x,t) \quad , \quad (x,t) \in \Gamma$$

and if for every $(x_0,t_0) \in \pi$ there exists an open neighborhood $U$ of $(x_0,t_0)$ and a finite set of functions $\{v_r\}_{1 \le r \le s} \subset C^{2,1}(\bar{U} \cap \pi)$ such that

$$(13) \qquad Lv_r \ge f(x,t,v_r,\nabla v_r) \quad , \quad (x,t) \in \bar{U} \cap \pi \quad , \quad 1 \le r \le s ,$$

and

$$(14) \qquad v(x,t) = \max_{1 \le r \le s} v_r(x,t) \quad , \quad (x,t) \in \bar{U} \cap \pi .$$

If in the above definition the inequality signs in (12) and (13) are reversed and in (14) max is replaced by min , then $v$ is called an upper solution of (1) , (2) .

For such upper and lower solutions $\beta,\alpha$ of (1) - (2) respectively with $\alpha(x,t) \le \beta(x,t)$ , $(x,t) \in \pi$ , theorem 3 holds and hence $(IBVP)_1$ has a solution $u \in C^{2,1}(\bar{\pi})$ with $\alpha(x,t) \le u(x,t) \le \beta(x,t)$ for $(x,t) \in \bar{\pi}$ .

A solution $\bar{u}$ of the $(IBVP)_1$ with $f = f_1$ $(m = 1)$ is a <u>maximal solution</u> relative to a given pair of lower and upper solutions $\alpha$ and $\beta$ with $\alpha(x,t) \le \beta(x,t)$ , $(x,t) \in \bar{\pi}$ if $\alpha(x,t) \le \bar{u}(x,t) \le \beta(x,t)$ and if $u$ is any other such solution then $u(x,t) \le \bar{u}(x,t)$ for $(x,t) \in \bar{\pi}$ . Minimal solutions are defined analogously.

<u>Theorem 5</u>. Assume the hypotheses of theorem 3 for a given pair of upper and lower solutions $\beta$ and $\alpha$ with $\beta(x,t) \ge \alpha(x,t)$ , $(x,t) \in \bar{\pi}$ . Then $(IBVP)_1$ has a maximal and a minimal solution.

The proof of the existence of a maximal solution is obtained by considering the collection $\mathcal{L}$ of all lower solutions of $(IBVP)_1$ where $\mathcal{L} = \{v: \bar{\pi} \to \mathbb{R}: \alpha(x,t) \le v(x,t) \le \beta(x,t) \;,\; (x,t) \in \bar{\pi} \;,\; v \text{ is a lower solution of } (IBVP)_1\}$ and showing that

$$u_{max}(x,t) = \sup\{v(x,t): v \in \mathcal{L} \;,\; \alpha \le v \le \beta\}$$

so defined is the maximal solution using theorem 3.

This same result is true for $(IBVP)_2$ and for the Cauchy initial value problem.

<u>6. An Example</u>. To illustrate how invariance can be used to analyze a problem, consider

$$(15) \begin{cases} L_1 u \equiv a_1 u_{xx} - u_t = -uv^{\gamma} \\ L_2 v \equiv a_2 v_{xx} - v_t = +uv^{\gamma} \end{cases}$$

for $(x,t) \in \pi = (0,1) \times (0,\infty)$ where $\gamma > 0$ , together with the initial-boundary conditions

$$(16) \qquad \begin{matrix} u(x,0) = u_0(x) \ge 0 \;,\; v(x,0) = v_0(x) \ge 0 \quad \text{for} \quad x \in [0,1] \\ u(t,0) = 0 = u(t,1) \;,\; v(t,0) = 0 = v(t,1) \quad , \quad t \in (0,\infty) \;, \end{matrix}$$

where $u_0(x)$, $v_0(x) \in C[0,1]$.

Set $V = \max_{[0,1]} v_0(x)$ and $U = \max_{[0,1]} u_0(x)$, then $(\alpha,\beta) \subset \mathbb{R}^2$ as defined in section 2, where $\alpha(x,t) = (\alpha_1(x,t), \alpha_2(x,t)) = (0,0)$ and $\beta(x,t) = (\beta_1(x,t), \beta_2(x,t)) = (Ue^{V^\gamma t}, V)$, is a weakly positively invariant set by theorem 3. Hence, there exists at least one solution $(u(x,t), v(x,t)) \in (\alpha,\beta)$ for $(x,t) \in \bar{\pi}$. If $\gamma \geq 1$, then the solution to IBVP (15) - (16) is unique and one can obtain additional asymptotic properties.

Let $\varphi(x,t)$ be a solution of $L_2 v = 0$, the homogeneous heat equation, then $L_2\varphi(x,t) = 0 \leq u(x,t)\varphi^\gamma$ where $u(x,t)$ is the first component of the unique solution of (15) - (16). Hence, $\varphi$ is an upper solution of $L_2 v = u(x,t)v^\gamma$, and $\varphi(x,t) \geq v(x,t)$ for $(x,t) \in \bar{\pi}$. By standard estimates for the heat equation, $v(x,t) \leq \varphi(x,t) \leq 4/\pi\, Ve^{-a_2\pi^2 t}$. From this, $v(x,t) \to 0$ uniformly in $x$ as $t \to \infty$. For $v(x,t)$, there exists $T > 0$ such that, for all $t \geq T$, $|v^\gamma| < a_2\pi^2 - \varepsilon \equiv M$. Take $\psi(x,t)$ to be the solution of $L_1 u = -Mu$ with $\psi(x,T) = u(x,T)$, $\psi(1,t) = u(1,t)$ and $\psi(0,t) = u(0,t)$ for $t \geq T$. Then $L_1\psi = -M\psi < -(v(x,t))^\gamma\psi$ and $\psi$ is an upper solution. Hence $u(x,t) \leq \psi(x,t)$ for $t \geq T$. By again standard estimates, $u(x,t) \leq \psi(x,t) \leq 4/\pi\, Ke^{-\varepsilon t}$ for $t \geq T$. We conclude that $(u(x,t), v(x,t)) \to 0$ as $t \to \infty$.

References


[1] K. Akô, On the Dirichlet problem for quasi-linear elliptic differential equations of second order, J. Math. Soc. Japan 13 (1961), 45-62.

[2] H. Amann, On the existence of positive solutions of nonlinear elliptic boundary value problems, Ind. Univ. Math. J. 21 (1971), 125-146.

[3] J.W. Bebernes and K. Schmitt, Invariant sets and the Hukuhara-Kneser property for systems of parabolic partial differential equations, Rocky Mountain J. Math. 7 (1977), to appear.

[4] J.W. Bebernes and K. Schmitt, On the existence of maximal and minimal solutions for parabolic partial differential equations, submitted.

[5] K. Chueh, C. Conley, and J.A. Smoller, Positively invariant regions for systems of nonlinear diffusion equations, Indiana Math. J. 26 (1977), 373-392.

[6] H.B. Keller, Elliptic boundary value problems suggested by nonlinear diffusion processes, Arch. Rat. Mech. Anal. 5 (1969), 363-381.

[7] Krasnosel'skii, M., and Sobolevskii, Structure of the set of solutions of an equation of parabolic type, Ukranian. Math. J. 16 (1964), 319-333.

[8] M. Nagumo, On principally linear elliptic differential equations of second order, Osaka, Math. J. 6 (1954), 207-229.

[9] C.V. Pao, Positive solutions of a nonlinear boundary value problem of parabolic type, J. Diff. Eqs. 22 (1976), 145-163.

[10] J.P. Puel, Existence comportement a l'infini et stabilite dans certaines problemes quasilineares elliptiques et paraboliques d'ordre 2, Ann. Scuola Norm. Sup. Pisa, Sec. IV 3 (197 ), 89-119.

[11] R. Redheffer and W. Walter, Invariant sets for systems of partial differential equations, Arch. Rat. Mech. Anal., to appear.

[12] D.H. Sattinger, Monotone methods in nonlinear elliptic and parabolic boundary value problems, Ind. J. Math. 211 (1972), 979-1000.

[13] Weinberger, H., Invariant sets for weakly coupled parabolic and elliptic systems, Rend. Math. 8 (1975), 295-310.


Author's address: Department of Mathematics, University of Colorado, Boulder, Colorado 80309, U.S.A.

# ASYMPTOTIC INVARIANT SETS OF AUTONOMOUS DIFFERENTIAL EQUATIONS

I.Bihari, Budapest

Let us suppose that the solutions of the real autonomous system

$$\dot{x} = f(x), \quad x = (x_1,\ldots,x_n), \quad \dot{} = \frac{d}{dt} \tag{1}$$

are, in a domain D of $R_n$, uniquely determined by their initial values and exist for all t. Then the whole D is an invariant set of (1), but this is of no interest. We look for nontrivial invariant sets forming some interesting surfaces - perhaps certain curves - or investigate how the invariant surfaces of the linear equation

$$\dot{x} = Ax, \quad A = (a_{ik}) \tag{2}$$

will be deformed into the corresponding invariant surfaces of the nonlinear (perturbed) equation

$$\dot{x} = Ax + f(x), \quad F = (f_1,\ldots,f_n), \quad f_i = f_i(x) . \tag{3}$$

So we can seek asymptotically invariant surfaces, too, i.e. such invariant surfaces of (2) to which the corresponding invariant surface of (3) tends as $t \to \infty$. In a paper written jointly with A. Elbert [1] - restricted to $n=3$ and $A = \text{const}$ - a number of such problems were solved. We were faced there with the problem: The full set of paths of (3) depends on two parameters which need not be specified in detail - say u and v - both of which depend on three parameters $X_o$, $Y_o$, $Z_o$

$$u = u(X_o, Y_o, Z_o), \qquad v = v(X_o, Y_o, Z_o)$$

where

$$X_o = \lim_{t\to\infty} xe^{-\lambda t}, \quad Y_o = \lim_{t\to\infty} (ye^{-\lambda t} - X_o t),$$

$$Z_o = \lim_{t\to\infty} (ze^{-\lambda t} - Y_o t - \tfrac{1}{2} X_o t^2) .$$

These are the "end values" of the solutions which - conversely - determine them uniquely by means of the corresponding integral equations provided some appropriate supplementary conditions are introduced. - Now putting $X_o = 0$ it is plausible, however it must be proved, that it arises a one parameter family of paths, i.e. a surface. In the work referred to above this was done and the unique parameter $(Z_o/Y_o')$ upon which the family depended was determined as well as the corresponding invariant surface. Here $Y_o'$ means the value of $Y_o$ obtained by putting $X_o = 0$ .

In this lecture we give an example of an asymptotically invariant surface. Assume now in (2) - (3) $n=3$,

$$A = \begin{bmatrix} \lambda & 0 & 0 \\ 1 & \lambda & 0 \\ 0 & 1 & \lambda \end{bmatrix} \qquad \lambda < 0 ,$$

$$F = (f,g,h), \quad |f|,|g|,|h| \le r\omega(r), \quad r = \sqrt{x^2+y^2+z^2} ,$$

$$r < r_1 , \quad r_1 > 0$$

where $\omega(r)$ is nondecreasing continuous, $\omega(0) = 0$ and

$$\int_{+0} \frac{\omega(r)}{r}(\log r)^4 dr < \infty .$$

Let us determine all the quadratic invariant surfaces $\rho = 0$ of (2), where $\rho = x^*Bx$ is a quadratic form with $B = (b_{ik})$, $b_{ik}$ = const, $b_{ik} = b_{ki}$. The solutions of (2) are

$$x = x_0 e^{\lambda t}, \quad y = (y_0 + x_0 t)e^{\lambda t}, \quad z = (z_0 + y_0 t + \tfrac{1}{2} x_0 t^2)e^{\lambda t}$$

which have to satisfy $\rho = 0$ for every $t$. This condition gives necessarily

$$\rho \equiv ax^2 + b(y^2 - 2xz) \qquad (a^2+b^2 > 0)$$

where $a$ and $b$ are arbitrary parameters. Thus the invariant surfaces of (2) in question are

$$S(a,b) : \quad \rho = 0$$

and an easy consideration shows that these are conical surfaces (see Fig. 1) with the origin as vertex, symmetric with respect to

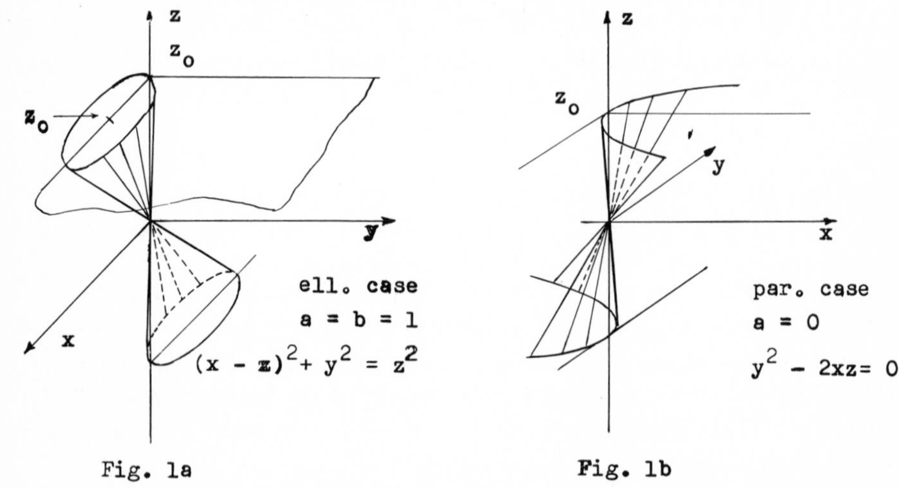


Fig. 1a

Fig. 1b

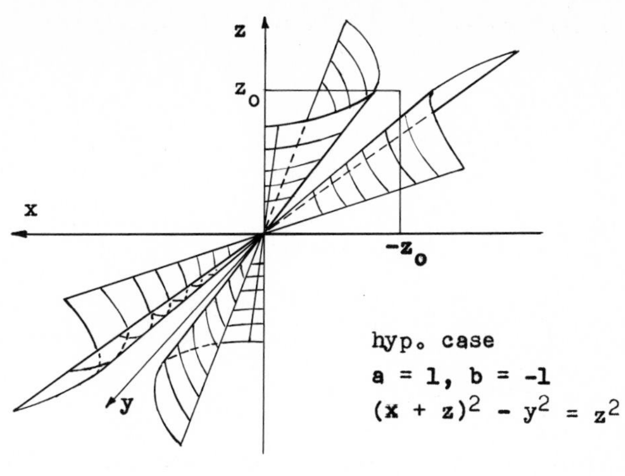


Fig. 1c

the plane (yz) and elliptic, hyperbolic or parabolic according to $ab \gtrless 0$ or $a=0$. Every element of S(a,b) contains the axis z which is an invariant line itself. Similarly, the plane (yz) is an invariant plane. Every point $P_o(X_o, Y_o, Z_o)$ of D - except the points of the axis z - is crossed by a member of the family S(a,b) the parameter a/b or b/a of which can be uniquely determined from

$$(3') \qquad \rho_o = ax_o^2 + b(y_o^2 - 2x_o z_o) = 0$$

or otherwise expressed. a and b can be uniquely determined from (3') up to a common factor. Also we can state that every path lies in a single member of S(a,b). By means of the integral equations - which we do not write here explicitly - it can be easily proved that the triple $(X_o, Y_o, Z_o)$ and the triple

$$(4) \qquad \begin{aligned} X &= X_o e^{\lambda t_1}, \\ Y &= (Y_o + X_o t_1) e^{\lambda t_1}, \qquad (t_1 \geq 0) \\ Z &= (Z_o + Y_o t_1 + \tfrac{1}{2} X_o t_1^2) e^{\lambda t_1} \end{aligned}$$

(taken as end values) determine the same path of (3). They have only a shift of parameter $t_1$ with respect to each other. However, in the space (X,Y,Z) (4) is the parametric equation of a curve $\sigma$. Thus the path p of (3) and the curves $\sigma$ are one to one. Since

$$aX^2 + b(Y^2-2XZ) = \left[aX_o^2 + b(Y_o^2-2X_oZ_o)\right] e^{2\lambda t_1},$$

the surface

(5) $$\sum : aX^2 + b(Y^2 - 2XZ) = 0$$

in this space is formed by the curves $\sigma$ provided $b/a$ is determined from

(6) $$aX_o^2 + b(Y_o^2 - 2X_oZ_o) = 0 .$$

Then the corresponding paths $p$ of (3) form an invariant surface $S'$ of (3). By (6) and the asymptotic form (not given here) of the solutions of (3) we have

$$ax^2 + b(y^2-2xz) = \left[aX_o^2 + b(Y_o^2-2X_oZ_o) + o(1)\right] e^{2\lambda t} = o(e^{2\lambda t}), \quad t \to \infty$$

$$ax^2 + b(y^2-2xz) = \left[aX_o^2 + b(Y_o^2-2X_oZ_o) + o(1)\right] e^{2\lambda t} = o\left[\frac{r^2}{(\log r)^4}\right], \quad r \to 0 .$$

The last expression is a consequence of the asymptotic formula $r \sim t^2 e^{\lambda t}$. The asymptotic invariant surface of (3) belonging to $p$ which has the end values $X_o, Y_o, Z_o$ is

(7) $$ax^2 + b(y^2 - 2xz) = 0$$

where $a$ and $b$ are given by (6) and $S'$ is situated between the surfaces

(8) $$ax^2 + b(y^2-2xz) = \pm F(x,y,z), \quad F = o\left[\frac{r^2}{(\log r)^4}\right], \quad r \to 0$$

(where $F$ is not determined in more detail) and approaches (7) as $t \to \infty$ which is an invariant surface of (2).

## Reference

[1] I.Bihari, A.Elbert: Perturbation theory of three-dimensional real autonomous systems, Periodica Math. Hung. 4 (4), (1973), pp. 233-302

Author's address: Math.Inst.Hung.Acad.Sci., 1053 Budapest
Reáltanoda-u. 13-15, Hungary

# ALGEBRAIC METHODS IN THE THEORY OF GLOBAL PROPERTIES OF THE OSCILLATORY EQUATIONS $Y'' = Q(t)Y$

O. Borůvka, Brno

CONTENTS. 

## I. INTRODUCTION

The theory of global properties of the ordinary 2nd order linear differential equations in the real domain has, in the last twenty years, made a remarkable progress. It originated from the problem of global equivalence of the nth order linear equations ($n \geqslant 2$) as the first step towards its solution. The question was, first, to solve the problem of global equivalence in the simplest case $n = 2$ and then, with acquired experience, get to the core of the problem in the general case. The result of this first step was quite satisfactory; in particular, the problem of global equivalence for 2nd order equations had been completely solved ([1], [2]).

The theory of global properties of the 2nd order equations in question consists of several partial theories concentrated about its most important notions. Among the latter is the theory of global properties of the oscillatory equations in the interval $\underline{R} = (-\infty, \infty)$ that deserves, for its algebraic character, a particular attention. It is exactly this theory, simply: the algebraic theory of the oscillatory equations, that forms the object of my lecture.

The basic notion of the theory in question is that of the group of phases, $\mathfrak{G}$. By the latter we understand the set composed of all phase functions together with the binary operation defined by composing functions. A phase function is any real function of class $C^3_{\underline{R}}$, unbounded on both sides, whose derivative of the first order is always different from zero. Various objects connected with oscillatory equations in $\underline{R}$ give rise to an algebraic structure of the group $\mathfrak{G}$. The latter is richly articulated and is, of course, in certain relations with the properties of the equations in question. The above algebraic objects are, in particular, the groups of dispersions of the single oscillatory equations, namely - let us note - groups formed by all phase functions transforming these equations into themselves and, furthermore, decompositions of the group $\mathfrak{G}$ into left and right cosets generated by groups of dispersions. The algebraic theory of the oscillatory equations is a study of the structure of the group $\mathfrak{G}$ in connection with the properties of the equations in question. We shall consider, in particular, equations with periodic coefficients. That opens a new way to Floquet's classical theory for 2nd order equations, leading to many new results in this field. This is a brief outline of the following lecture.

## II. GENERAL THEORY

1. Starting situation. Consider equations of the Jacobian form

$$(Q) \qquad Y'' = Q(t)Y, \quad t \in \underline{R}$$

and, from the beginning, suppose that we deal always with oscillatory equations. The coefficient Q, also called the carrier of (Q), is supposed to be of class $C^0_{\underline{R}}$ : $Q \in C^0_{\underline{R}}$. The equation (Q) is called oscillatory if each of its integrals has an infinite number of zeros towards $-\infty$ and $\infty$ . By an integral of (Q) we understand a solution of (Q) defined in the entire $\underline{R}$. The integral identically equal to zero is generally omitted.

As to the way of notation and terminology let us add the following remark: We often write Q instead of (Q); sometimes we speak, e.g., about integrals of the carrier Q instead of integrals of the equation (Q). The symbol $\underline{M}$ denotes the set of all carriers Q or all equations (Q).

2. Kummerian transformations. The general theorem about global transformations of 2nd order equations states that, in case of oscillatory equations in $\underline{R}$, every equation Q may be globally transformed into any equation R. That means: there exist phase functions X(t) such that, for every integral Y of Q, the function

$$Z(t) = \frac{c}{\sqrt{|X'(t)|}} \cdot Y[X(t)]$$

is an integral of R; c denotes an arbitrary constant ($\neq 0$). The function X is called transformator of Q into R, whereas $c:\sqrt{|X'(t)|}$ is the multiplicator of the transformation in question. The transformators X are exactly the integrals of the following equation which is, for historical reasons, called Kummer's equation

$$-\{X,t\} + Q(X)X'^2 = R(t). \tag{QR}$$

The set of these transformators is therefore the general integral $I_{QR}$ of (QR) ([1]).

The inverse of X, denoted $X^{-1}$, is a transformator of R into Q and therefore an integral of (RQ). Denoting by $I_{QR}^{-}$ the set formed by the inverses of the elements of $I_{QR}$, we have: $I_{RQ} = I_{QR}^{-}$ and $I_{QR} = I_{RQ}^{-}$.

The theory of Kummer's equation forms the kernel of our considerations concerning Kummerian transformations. It is based on a theorem on existence and unicity of integrals of the equation (QR), which is of Cauchyan type ([1]). The integrals of (QR) are called general dispersions of the equations Q,R so as to suggest their connection with the dislocation (dispersion) of zeros of the integrals of Q and R ([1]).

Finally, let us remark that, according to a classical result of P. Stäckel, the transformations of the form $Y(T) \to w(t)Y[X(t)]$ (w is, in our case, with regard to the special form of Jacobian equations, the previously mentioned multiplicator) are the most general transformations conserving the linearity and order ($n \geqslant 2$) of the transformed equations.

3. Groups of dispersions. The gateway leading to the realm of algebra is the notion of groups of dispersions.

By a dispersion (of the first kind) of the equation Q we understand any transformator of Q into Q or, in other words, any integral of the equation (QQ). We can show that the set of all dispersions of Q, i.e., the general integral $I_{QQ}$ of (QQ) is a group (the group operation being given by composing functions). It is called group of dispersions of Q. Notation: $\mathfrak{B}_Q$. The group $\mathfrak{B}_Q$ is evidently stationary as it does not change the equation Q.

Let us return to Kummer's equation (QR). A substantial progress in the theory of the latter is given by the fact that for any integral X of (QR) there holds ([3]):

$$I_{QR} = \mathfrak{B}_Q X = X\mathfrak{B}_R. \tag{1}$$

We see, first, that the group $\mathfrak{B}_R$ is the image of $\mathfrak{B}_Q$ by the inner automorphism of $\mathfrak{G}$, generated by any integral X of (QR); in other words, $\mathfrak{B}_R$ is conjugate with $\mathfrak{B}_Q$ by X:

$$\mathfrak{B}_R = X^{-1}\,\mathfrak{B}_Q X.$$

Hence, the system of subgroups of $\mathfrak{G}$, formed by the groups $\mathfrak{B}_R$ ($R \in \underline{M}$) is composed of groups **pairwise conjugate. We can also** show that the mapping $R \to \mathfrak{B}_R$ is bijective and the intersection of all the groups of dispersions is the group $\{t\}$.

Another important consequence of (1) is: $I_{QR}$ is the right coset of $\mathfrak{B}_Q$ containing X and, at the same time, the left coset of $\mathfrak{B}_R$ containing X. More precisely,

$$I_{QR} = \bar{Q}_d \cap \bar{R}_g; \tag{2}$$

$\bar{Q}_d$ denotes the right decomposition of $\mathfrak{G}$, formed by the right cosets of $\mathfrak{B}_Q$, and, similarly, $\bar{R}_g$ stands for the left decomposition of $\mathfrak{G}$, generated by $\mathfrak{B}_R$ ([3]).

Starting from (2) we are, by further considerations which we cannot deal with here in detail, led to results concerning common integrals of two Kummerian equations. Let us introduce at least some results concerning the notions of inverse equations and blocks of equations.

4. Inverse equations and blocks of equations. We consider the equation Q. The equation $R^-$ is called inverse to R ($\in \underline{M}$) with regard to Q or, simply: inverse to R, if there exist common integrals of the equations (QR), ($R^-Q$).

This notion is evidently symmetrical with regard to the equations R, $R^-$. We can show that the set of equations inverse to R is non-empty. The latter is called the block of equations inverse to R or, simply: a block. It can be proved that the set composed of all blocks is a decomposition of $\underline{M}$ (cf. the case $Q = -1$ in [7]). To each block $\bar{u}$ there corresponds precisely a block $\bar{u}^-$ called inverse to $\bar{u}$, formed by all equations inverse to any element of $\bar{u}$; there holds $(\bar{u}^-)^- = \bar{u}$. Two equations are inverse to each other iff they are contained in blocks inverse to each other. Every equation $R^*$ contained in the same block as R is called associated with R (with regard to Q).

Formulae expressing each equation inverse to R, $R^-$, or each equation associated with R, $R^*$, are

$$R^-(t) = Q(t) + [Q(X^{-1})-R(X^{-1})]\cdot X^{-1\prime 2}(t) \qquad \text{for } X \in I_{QR},$$

$$R^*(t) = Q(t) - [Q(X)-R(X)]\cdot X'^2(t) \qquad \text{for } X \in I_{QQ}.$$

5. Adjoint groups. With every equation Q we shall intrinsically associate a figure composed of four groups called adjoint to Q, briefly: adjoint groups:

$$\mathfrak{A}_Q \supset \mathfrak{B}_Q \supset \mathfrak{B}_Q^+ \supset \mathfrak{C}_Q ;$$

$\mathfrak{B}_Q$ is the group of dispersions of Q, $\mathfrak{B}_Q^+$ is its subgroup of all increasing dispersions, $\mathfrak{C}_Q$ the center of $\mathfrak{B}_Q^+$ and, finally, $\mathfrak{A}_Q$ is the normalizer of $\mathfrak{C}_Q$ in the group $\mathfrak{G}$ .

The structural properties of the adjoint groups and the stationary character of the latters are mainly these:

$\mathfrak{B}_Q$ is homomorphic with the group composed of all unimodular matrices 2x2 with real elements ([1]).

$\mathfrak{B}_Q^+$ is an invariant subgroup of $\mathfrak{B}_Q$ with the index 2. The coset of $\mathfrak{B}_Q^+$ in $\mathfrak{B}_Q$ is formed by all decreasing dispersions of Q. The normalizer of $\mathfrak{B}_Q^+$ in $\mathfrak{G}$ is $\mathfrak{B}_Q$ : $N\mathfrak{B}_Q^+ = \mathfrak{B}_Q$ (cf. the proof of $N\mathfrak{B}_{-1} = \mathfrak{B}_{-1}$ in [3]). The elements of $\mathfrak{B}_Q^+$ keep the orientation of the integral curves of Q.

$\mathfrak{C}_Q$ is an infinite cyclic group

$$\ldots < \Phi_{-2}(t) < \Phi_{-1}(t) < \Phi_0(t) = t < \Phi_1(t) < \Phi_2(t) < \ldots \quad (t \in \underline{R})$$

whose elements, called central dispersions of Q, may be constructively described as follows: the value $\Phi_\nu(t)$ is the $\nu$ th conjugate point with t lying to the right or to the left of t respectively for $\nu > 0$ or $\nu < 0$ ($\nu$ integer). The functions $\Phi_{-1}$ and $\Phi_1$ are the generators of $\mathfrak{C}_Q$; in particular $\Phi_1$ is called the fundamental dispersion of Q and there evidently holds: $\Phi_\nu(t) = \Phi_1^\nu(t)$. The group $\mathfrak{C}_Q$ is also called the center of Q; two equations with the same center are called concentric. The elements of $\mathfrak{C}_Q$ are exactly the dispersions of Q which transform every integral Y of Q into an integral linearly dependent on Y.

The elements of $\mathfrak{A}_Q$ are exactly the transformators of Q into some equation concentric with Q.

Proof. Let $\Phi$ be the fundamental dispersion of Q. Let R be the carrier of the transformed equation Q by some $X \in \mathfrak{G}$, and $\Psi$ its fundamental dispersion. Then we have ([1] p.176): $X\Psi = \Phi^{\operatorname{sgn} X'}X$. $X \in \mathfrak{A}_Q$ implies, by definition of $\mathfrak{A}_Q$ : $X\mathfrak{C}_Q = \mathfrak{C}_Q X$ and thus ([4]): $X\Phi = \Phi^{\operatorname{sgn} X'}X$. Hence it is $\Psi = \Phi$ and this yields $\mathfrak{C}_R = \mathfrak{C}_Q$. The rest of the assertion may be proved by similar arguments.

We call the elements of $\mathfrak{A}_Q$ <u>co-dispersions</u> of Q; $\mathfrak{A}_Q$ is of course the group of co-dispersions of Q.

We know that if R is, by the transformator X, the Kummerian image of Q, then $\mathfrak{B}_R$ is conjugate with $\mathfrak{B}_Q$ by the same X. In this case, any group adjoint to R is conjugate, by X, with the corresponding group adjoint to Q. We say that the Kummerian transformations of the equations $Q \in \underline{M}$ are accompanied by inner automorphisms acting on the corresponding adjoint groups.

6. <u>The inclusion theorems.</u> Groups adjoint to two equations Q,R are in certain mutual relations of partly dual character. These relations are described in the following three theorems that we call inclusion theorems. The symbols $\Psi$ and $\Phi$ denote the fundamental dispersions of R and Q, respectively.

<u>Theorem 1.</u> The relation $\mathfrak{A}_R \supset \mathfrak{C}_Q$ implies $\Psi\Phi = \Phi\Psi$ and vice versa.

Proof. If $\mathfrak{A}_R \supset \mathfrak{C}_Q$ then $\Phi\mathfrak{C}_R = \mathfrak{C}_R\Phi$ and this yields $\Phi\Psi = \Psi\Phi$. The second part of the proof is obvious.

<u>Corollary.</u> If $\mathfrak{A}_R \supset \mathfrak{C}_Q$ then $\mathfrak{A}_Q \supset \mathfrak{C}_R$.

<u>Theorem 2.</u> The relation $\mathfrak{A}_R \supset \mathfrak{B}_Q^+$ implies $\mathfrak{C}_R \subset \mathfrak{C}_Q$ and vice versa.

Proof. If $\mathfrak{A}_R \supset \mathfrak{B}_Q^+$ then $X\mathfrak{C}_R = \mathfrak{C}_R X$ for every $X \in \mathfrak{B}_Q^+$. Hence we conclude $X\Psi = \Psi X$ and this implies $\mathfrak{B}_Q^+\Psi = \Psi\mathfrak{B}_Q^+$. This yields $\Psi \in N\mathfrak{B}_Q^+$ $(= \mathfrak{B}_Q)$ and, since $\Psi$ increases, $\Psi \in \mathfrak{B}_Q^+$; because $\Psi$ commutes with every $X \in \mathfrak{B}_Q^+$ we find $\Psi \in \mathfrak{C}_Q$. The result: $\mathfrak{C}_R \subset \mathfrak{C}_Q$. The rest of the proof follows by similar arguments.

<u>Theorem 3.</u> The relation $\mathfrak{B}_R \supset \mathfrak{C}_Q$ implies $[R(\Phi) - Q(\Phi)]\Phi'^2(t) = R(t) - Q(t)$ $(t \in \underline{R})$ and vice versa.

Proof. $\Phi$ being the fundamental dispersion of Q it satisfies the equation (QQ). If $\mathfrak{B}_R \supset \mathfrak{C}_Q$ then $\Phi$ also satisfies the equation (RR) and the above relation follows. The rest of the proof is obvious.

The inclusion theorems are a source of interesting problems concentrated about the properties of equations R, Q satisfying the conditions given by the above theorems. Consider an arbitrary equation $Q \in \underline{M}$. Let $\mathscr{A}_Q$, $\mathscr{B}_Q$, $\mathscr{C}_Q$ be the classes of equations R characterized by the following properties:

Class $\mathscr{A}_Q$: The fundamental dispersion of any equation $R \in \mathscr{A}_Q$ is commuting with $\Phi$ .

Class $\mathscr{B}_Q$: The fundamental dispersion of any equation $R \in \mathscr{B}_Q$ is a central dispersion of Q.

Class $\mathscr{C}_Q$: Between the carriers $R \in \mathscr{C}_Q$, Q and the fundamental dispersion $\Phi$ we have the relation indicated in Theorem 3.

The above classes $\mathscr{A}_Q$, $\mathscr{B}_Q$ have been studied in case Q = -1 ([5], [6], [4]). As to the class $\mathscr{C}_{-1}$ let us remark that it is composed of all equations with $\pi$-periodic carriers.

## III. SPECIALIZED THEORY

7. The way of specialization. The general theory we have, so far, spoken about changes its aspect if we choose, arbitrarily, some equation $Q \in \underline{M}$, called canonical, as a representatión of the system $\underline{M}$. This equation Q and the right decomposition $Q_d$ of the group of phases, $\mathfrak{G}$ , generated by the group $\mathfrak{B}_Q$, enter the center of the theory: Any equation $R \in \underline{M}$ is a Kummerian image of Q. The transformators X of Q into R form an element of $\bar{Q}_d$; $\bar{Q}_d$ is composed of general integrals $I_{QR}$ of Kummer's equations (QR) associated with the single equations $R \in \underline{M}$. The groups adjoint to R arise from the corresponding groups adjoint to Q by inner automorphisms of $\mathfrak{G}$ , generated by the transformators $X \in I_{QR}$.

For Q, representing the system $\underline{M}$, it is convenient to choose the equation -1, namely $Y'' = -Y$ $(t \in \underline{R})$, whose simplicity yields an advantage in calculations.

8. The equation $Y'' = -Y$ $(t \in \underline{R})$. In the following formulae $\nu$ (integer), a,b,c denote constants; $t \in \underline{R}$.

Integrals: $\quad Y(t) = c \circ \sin(a+t)$; $0 \leq a < \pi$, $c \neq 0$.

Dispersions: $\epsilon(t) = {}_\nu\tan^{-1}(c\cdot\frac{\sin(a+t)}{\sin(b+t)})$;

$\epsilon(-a) = \nu\pi$ ; $0 \leqslant a,b < \pi$; $c(b-a) \neq 0$.

Increasing dispersions: The last formula with $c(b-a) > 0$.

Central dispersions: $\Phi_\nu(t) = t + \nu\pi$ .

Fundamental dispersion: $\Phi_1(t) = t + \pi$.

Co-dispersions: $h(t) = \delta t + d(t)$,

$\delta = \pm 1$; $d \in C^3_{\underline{R}}$; $d(t+\pi) = d(t)$; $-\delta\cdot d'(t) < 1$.

The adjoint groups of -1 are also denoted by $\mathfrak{H}$ , $\mathfrak{E}$ , $\mathfrak{E}^+$, $\mathfrak{Z}$ so that $\mathfrak{A}_{-1} = \mathfrak{H}$, $\mathfrak{B}_{-1} = \mathfrak{E}$, $\mathfrak{B}^+_{-1} = \mathfrak{E}^+$, $\mathfrak{C}_{-1} = \mathfrak{Z}$ .

$\mathfrak{E}$ is called the fundamental group; its elements $\epsilon$ are called special dispersions. $\mathfrak{H}$ is called the group of elementary phases; its elementa h are, of course, elementary phases.

9. Theory specialized by the choice $Q = -1$. In this case every equation $R \in \underline{M}$ is regarded as a Kummerian image of the equation -1. The transformators X, transforming -1 into R, form the element $I_{-1R} \in \bar{E}_d$, $\bar{E}_d$ naturally being the right decomposition of $\mathfrak{G}$, generated by the fundamental group $\mathfrak{E}$ . Since X are integrals of the equation (-1R): $-\{X,t\} - X'^2 = R(t)$, they coincide exactly with the phases of R. The decomposition $\bar{E}_d$ consists, therefore, of phases of single equations of the system $\underline{M}$. Every equation $R \in \underline{M}$ is a Kummerian image of -1 by the phases of R. That is the role of phases in the theory we are dealing with. Note that by a phase of R we understand any phase function A given by the formula $A(t) = {}_\nu\tan^{-1}(U(t):V(t))$; U, V denote linearly independent integrals of R. Note, in particular, that the phases of the equation R depend only on R; they originate, so to say, from the interior of R ([1]).

It is evident that the above objects associated with any equation $R \in \underline{M}$, e.g. integrals, adjoint groups, etc., may be expressed by the corresponding objects of the equation -1 and the phases of R.

Let A denote a phase of R. Then, for the integrals and the adjoint groups of R, we have:

$$Y(t) = \frac{c}{\sqrt{|A'(t)|}}\cdot\sin(a+A(t));\ 0 \leqslant a < \pi;\ c \neq 0\ (a,c = \text{const.})$$

$$\mathfrak{A}_R = A^{-1}\mathfrak{H}A,\quad \mathfrak{B}_R = A^{-1}\mathfrak{E}A,\quad \mathfrak{B}^+_R = A^{-1}\mathfrak{E}^+A,\quad \mathfrak{C}_R = A^{-1}\mathfrak{Z}A,$$

etc. Thus we find a powerful analytic instrument well adapted to research in the considered domain and functioning excellently.

Let us be satisfied with this information without a detailed consideration of the above theory.

## IV. EQUATIONS WITH $\pi$-PERIODIC CARRIERS

10. Introduction. In the above considerations we have met with equations with $\pi$-periodic carriers. There naturally arises a question concerning the relations between the classical theory of Floquet and the theory we have just exposed. As a matter of fact, these relations open a way to extend Floquet's theory in case of 2nd order equations. - We speak, simply, about $\pi$-periodic equations.

11. A brief outline of the algebraic theory of $\pi$-periodic equations. With regard to the above algebraic theory of oscillatory equations we may extend the classical theory of $\pi$-periodic equations in two directions: On one hand, by new notions, e.g., dispersions, inverse equations, etc., in case of $\pi$-periodic equations. On the other hand, by relations between Floquet's theory and the new notions we have just mentioned. In what follows we present a brief aspect of the region surrounding Floquet's theory in the case of 2nd order equations.

Let $\mathscr{C}$ (= $\mathscr{C}_{-1}$) be the class composed of all $\pi$-periodic carriers (equations). For $R \in \underline{M}$ let $R^-$ or $R^*$ be an inverse or an associated carrier (equation) of R with regard to -1, respectively.

Proposition 1. If $R \in \mathscr{C}$ then $\mathfrak{C}_R \subset \mathfrak{H}$ .

Proof. If $R \in \mathscr{C}$ then $\mathfrak{B}_R \supset \mathfrak{Z}$ , by the 3rd inclusion theorem; this and $\mathfrak{A}_R \supset \mathfrak{B}_R$ imply $\mathfrak{A}_R \supset \mathfrak{Z}$ . This implies $\mathfrak{C}_R \subset \mathfrak{H}$ , by the 1st inclusion theorem.

Proposition 2. If $R \in \mathscr{C}$ then $\mathfrak{C}_{R^-} \subset \mathfrak{C}$ and vice versa.

For the proof, see [5].

Proposition 3. If $R \in \mathscr{C}$ then $R^* \in \mathscr{C}$ .

For the proof, see [5].

Proposition 4 (the conservation law of periodicity factors). The periodicity factors of any two associated $\pi$-periodic equations are the same.

For the proof, see [7].

Proposition 5. The periodicity factors of any two equations inverse to concentric equations with center lying in $\mathfrak{C}$ , are the same.

For the proof, see [11] .

For more detail and results concerning the algebraic theory of $\pi$ -periodic equations, see [8], [9], [10].

## V. FINAL REMARK

Further development of the theory of differential linear equations will render it possible to judge the influence of the theory of global properties of 2nd order equations on the progress of the theory of linear equations. We mean, in particular, the progress of the theory of the 2nd order equations (complex domain, numerical methods, etc.) as well as the problem of global equivalence for $n > 2$. In any case the remarkable results in the field of the latter, presented in the recent papers of F. Neuman, are most encouraging ([2]).

## REFERENCES


O. Borůvka:

[1] Lineare Differentialtransformationen 2. Ordnung. Berlin 1967.
English Translation (F. M. Arscott):
Linear Differential Transformations of the Second Order. London 1971.

[2] Teorija global'nych svojstv obyknovennych linejnych differencial'nych uravnenij vtorogo porjadka. Diff. Ur.,12 (1976), 1347-1383.

[3] Über eine Charakterisierung der Allgemeinen Dispersionen linearer Differentialgleichungen 2. Ordnung. Math. Nachrichten, 38 (1968), 261-266.

[4] Sur quelques propriétés de structure du groupe des phases des équations différentielles linéaires du deuxième ordre. Rev. Roum. Math. p. et appl., XV (1970), 1345-1356.

[5] Sur la périodicité de la distance des zéros des intégrales de l'équation différentielle y" = q(t)y. Tensor, N. S., 26 (1972), 121-128.

[6] Über die Differentialgleichungen y" = q(t)y mit periodischen Abständen der Nullstellen ihrer Integrale. Wiss. Schriften-

reihe der Techn. Hochschule Karl-Marx-Stadt, 1975, 239-255.

[7] Sur les blocs des équations différentielles y" = q(t)y aux coefficients périodiques. Rend. Mat. (2), 8, S. VI (1975), 519-532.

[8] Sur quelques compléments à la théorie de Floquet pour les équations différentielles du deuxième ordre. Ann. Mat. p. ed appl., S. IV, CII (1975), 71-77.

[9] On central dispersions of the differential equation y" = q(t)y with periodic coefficients. Lecture Notes in Mathematics, 415, 1974, 47-60.

[10] Contribution à la théorie algébrique des équations Y" = Q(T)Y. Boll. U. M. I. (5) 13-B (1976), 896-915.

F. Neuman, S. Staněk:

[11] On the structure of second-order periodic differential equations with given characteristic multipliers. Arch. Math. (Brno), XIII (1977), 149-157.

Author's address: Mathematical Institute of the Czechoslovak
Academy of Sciences, Branch Brno
Janáčkovo nám. 2a, 662 95 Brno, Czechoslovakia

STABILITY PROBLEMS IN MATHEMATICAL THEORY OF VISCOELASTICITY

J.Brilla, Bratislava

## 1. Introduction

The analysis of non-linear stability problems in the mathematical theory of viscoelasticity has only recently begun to attract attention and it will take some time before our understanding of these problems reaches the maturity similar to that of analysis of elastic stability problems.

We shall start from governing equations of large deflection theory of viscoelastic plates

$$(1.1)\quad \frac{h^3}{12} K_{ijkl}(D)w_{,ijkl} = K(D)(q + h\,\epsilon_{ik}\,\epsilon_{jl}w_{,ij}F_{,kl}) ,$$

$$\epsilon_{im}\,\epsilon_{jn}\,\epsilon_{kr}\,\epsilon_{ls}L_{mnrs}(D)F_{,ijkl} = -\frac{1}{2}\epsilon_{ik}\,\epsilon_{jl}L(D)w_{,ij}w_{,kl} ,$$

where

$$(1.2)\quad K_{ijkl}(D) = \sum_{\nu=0}^{r} K_{ijkl}^{(\nu)} D^{\nu} ,$$

$$(1.3)\quad L(D) = \sum_{\mu=0}^{s} L_{\mu} D^{\mu} ,$$

are polynomials in $D = \frac{\partial}{\partial t}$, $s = r$ or $r = s + 1$ and

$$(1.4)\quad K(D)\,[K_{ijkl}(D)]^{-1} = L(D)^{-1} L_{ijkl}(D) .$$

$K_{ijkl}(D)$, $K(D)$, $L_{ijkl}(D)$ and $L(D)$ are differential operators of linear viscoelasticity. $w$ is the transverse displacement of the plate, $F$ - the stress function, $h$ - the thickness of the plate, $q$ - transverse loading and $\epsilon_{ij}$ - the alternating tensor.

We use the usual indicial notation. Latin subscripts have the range of integers 1,2 and summation over repeated Latin subscripts is implied. Subscripts preceded by a coma indicate differentiation with respect to the corresponding Cartesian coordinates.

In the case of real materials it holds

$$(1.5)\quad K_{ijkl}^{(\nu)}(D)\,\varepsilon_{ij}\,\varepsilon_{kl} \geq 0$$

for arbitrary values of $\varepsilon_{ij}$ and equality occurs iff $\varepsilon_{ij} = 0$ for all $i,j$. Further, the coefficients $K_{ijkl}^{(\nu)}$ are symmetric

$$(1.6)\quad K_{ijkl}^{(\nu)} = K_{jikl}^{(\nu)} = K_{ijlk}^{(\nu)} = K_{klij}^{(\nu)}$$

and polynomials (1.2)-(1.3) have real negative roots.

We assume that the domain of definition $\Omega$ is bounded with Lipschitzian boundary $\partial\Omega$. We shall consider the following boundary conditions

$$w = \frac{\partial w}{\partial n} = 0 \qquad \text{on} \quad \partial\Omega \, , \tag{1.7}$$

or

$$w = 0 \, , \; K_{ijkl}(D) w,_{ij} \, \nu_{kn} \, \nu_{ln} = 0 \qquad \text{on} \quad \partial\Omega \, , \tag{1.8}$$

where $\nu_{kn} = \cos(x_k, n)$ and $n$ is the outward normal to $\partial\Omega$ and

$$\frac{\partial F}{\partial n} = \frac{\partial^3 F}{\partial n^3} = 0 \qquad \text{on} \quad \partial\Omega \tag{1.9}$$

or

$$\frac{\partial^2 F}{\partial n^2} = \frac{\partial^2 F}{\partial s \partial n} = 0 \qquad \text{on} \quad \partial\Omega \, . \tag{1.10}$$

The initial condition may assume the form

$$\frac{\partial^\nu w}{\partial t^\nu} = w_\nu \qquad (\nu = 0,1,2,\ldots,r-1) \tag{1.11}$$

and

$$\frac{\partial^\nu F}{\partial t^\nu} = 0 \qquad (\nu = 0,1,2,\ldots,k-1) \tag{1.12}$$

where $k$ is the order of the operator $L_{ijkl}(D)$.

Simultaneously we shall consider the integrodifferential equations

$$\frac{h^3}{12} \int_0^t G_{ijkl}(t-\tau) \frac{\partial}{\partial\tau} w,_{ijkl}(\tau) d\tau = q + h \, \epsilon_{ik} \, \epsilon_{jl} w,_{ij} F,_{kl} \, , \tag{1.13}$$

$$\int_0^t \epsilon_{im} \, \epsilon_{jn} \, \epsilon_{kr} \, \epsilon_{ls} J_{mnrs}(t-\tau) \frac{\partial}{\partial\tau} F,_{ijkl}(\tau) d\tau = -\frac{1}{2} \epsilon_{ik} \, \epsilon_{jl} w,_{ij} w,_{kl}$$

with boundary conditions (1.7)- (1.10) respectively, and the first initial conditions (1.11)-(1.12). These equations are the governing equations of the large deflection theory of viscoelastic plates of a material of Boltzmann type. $J_{ijkl}(t-\tau)$ is given by the inversion of the constitutive equations corresponding to $G_{ijkl}(t-\tau)$.

## 2. Linearized stability problems

When dealing with stability problems of time dependent processes we have to consider perturbations from equilibrium state. When considering perturbations that are extremely small (infinitesimal), we may feel justified in neglecting non-linear terms in (1.1) as compared to the linear ones.

We assume that the plate is subject to a system of two-dimensional stresses $\lambda h \sigma^o_{ij} = -\lambda N^o_{ij}$, where $\lambda$ is a monotonically increasing factor of proportionality and the distribution of $N^o_{ij}$ is prescribed. We put $q = 0$. In the resulting linearized stability theory we have

$$\frac{h^3}{12} K_{ijkl}(D) w_{,ijkl} + \lambda N^o_{ij} w_{,ij} = 0 \tag{2.1}$$

with boundary conditions (1.6) or (1.7) and initial conditions

$$w = w_o \quad , \quad \frac{\partial^\nu w}{\partial t^\nu} = 0 \qquad (\nu = 1,2,\ldots,r-1) \quad . \tag{2.2}$$

Obviously the solution can be sought in the form

$$w(x,y,t) = e^{\mu t}\, u(x,y) \quad . \tag{2.3}$$

Then the function $u(x,y)$ has to satisfy the partial differential equation

$$\frac{h^3}{12} \sum_{\nu=0}^{r} K^{(\nu)}_{ijkl}\, \mu^\nu\, u_{,ijkl} + \lambda \sum_{\nu=0}^{s} K_\nu\, \mu^\nu\, N^o_{ij} u_{,ij} = 0 \quad . \tag{2.4}$$

Non-trivial solution for $u$ exists only if the parameter $\mu$ assumes special values $\mu = \mu_n$ which are generalized eigenvalues of the linearized problem.

We shall assume that $r=s$, then we can write

$$\sum_{\nu=0}^{r} \mu^\nu \left[\frac{h^3}{12} K^{(\nu)}_{ijkl}\, u_{,ijkl} + \lambda K_\nu N^o_{ij} u_{,ij}\right] = 0 \; . \tag{2.5}$$

If $\Phi_n$ are eigenfunctions of our problem, it holds

$$\sum_{\nu=0}^{r} \mu_n^\nu \left[\frac{h^3}{12} K^{(\nu)}_{ijkl} (\Phi_{n,ij}\, , \Phi_{n,kl}) - \lambda K_\nu N^o_{ij} (\Phi_{n,i}\, , \Phi_{n,j})\right] = 0 \; . \tag{2.6}$$

According to the assumptions the operator $\sum_{\nu=0}^{r} \mu^\nu K^{(\nu)}_{ijkl} (\cdot)_{,ijkl}$

is for positive values of $\mu$ positive definite. When dealing with stability problems we choose $N^o_{ij}$ in such a way that

$\sum_{\nu=0}^{r} \mu^\nu K_\nu N^o_{ij} (\Phi_{,i}\, , \Phi_{,j})$ is positive. Then for sufficiently small

it holds

$$(2.7)\qquad \sum_{\nu=0}^{r} \mu^{\nu}\left[\frac{h^3}{12} K^{(\nu)}_{ijkl}(\Phi,_{ij}\,,\Phi,_{kl}) - \lambda K_{\nu} N^{o}_{ij}(\Phi,_{i}\,,\Phi,_{k})\right] \geq K^2||\Phi||^2 \;.$$

The polynomial on the left hand side has for sufficiently small $\lambda$ positive coefficients and is a monotically increasing function for $\mu > 0$ . Roots of this polynomial are then negative or have negative real parts. For an operator corresponding to real materials it can be proved that its roots are negative.

If the roots of (2.6) are simple and $w_o$ in initial conditions (2.2) is considered as an initial perturbation, the solution of (2.1) can be written in the form

$$(2.8)\qquad w(x,y,t) = \sum_{n=1}^{\infty} \sum_{k=1}^{r} A_{nk} w_{on}\, \Phi_n(x,y) e^{-\mu_{nk} t} \;,$$

where we have denoted the roots of (2.6) (which are real for real materials) by $-\mu_{nk}$ ,

$$(2.9)\qquad w_{on} = (w_o, \Phi_n)$$

and

$$(2.10)\qquad A_{nk} = \frac{\mu_{n1}\,\mu_{n2}\cdots\mu_{n(k-1)}\,\mu_{n(k+1)}\cdots\,\mu_{nr}}{(\mu_{nk}-\mu_{n1})\cdots(\mu_{nk}-\mu_{n(k-1)})(\mu_{nk}-\mu_{n(k+1)})\cdots(\mu_{nk}-\mu_{nr})}$$

If all $\mu_{nk} > 0$ the solution is stable. If at least one $\mu_{nk} < 0$ the solution is unstable.

The left hand side of (2.6) is a continuous function of $\lambda$ and when $\lambda$ increases it reaches critical values for which the roots successively change their signs.

In order to determine critical values of $\lambda$ we have to analyse (2.6). We can write it in the form

$$(2.11)\qquad \sum_{\nu=0}^{r} \mu^{\nu} A_{n\nu}(\lambda) = 0 \;,$$

which can be rewritten as

$$(2.12)\qquad \prod_{i=1}^{r} A_{nr}(\lambda)(\mu + \mu_{ni}(\lambda)) = 0 \;.$$

As it holds

$$(2.13)\qquad \prod_{i=1}^{r} \mu_{ni}(\lambda) = A_{nr}(\lambda)^{-1} A_{no}(\lambda)$$

and $A_{nr}(\lambda)$, $A_{no}(\lambda)$ are continuous functions of $\lambda$, the change of the signs of roots $\mu_{ni}(\lambda)$, assuming that they are not multiple, occurs at such values of $\lambda$, which satisfy equations

$$(2.14)\qquad A_{no}(\lambda) = \frac{h^3}{12} K^{(o)}_{ijkl}(\Phi_{n,ij}, \Phi_{n,kl}) - \lambda K_o N^o_{ij}(\Phi_{n,i}, \Phi_{n,j}) = 0$$

and

$$(2.15)\qquad A_{nr}(\lambda) = \frac{h^3}{12} K^{(r)}_{ijkl}(\Phi_{n,ij}, \Phi_{n,kl}) - \lambda K_r N^o_{ij}(\Phi_{n,i}, \Phi_{n,j}) = 0 .$$

Applying Laplace transform to (2.1) and making use of Tauber's theorem on limit values of Laplace transform we find out that these values are eigenvalues of the equations

$$(2.16)\qquad \frac{h^3}{12} K^{(o)}_{ijkl}\, w_{,ijkl}(\infty) + \lambda N^o_{ij} K_o\, w_{,ij}(\infty) = 0$$

and

$$(2.17)\qquad \frac{h^3}{12} K^{(r)}_{ijkl} w_{,ijkl}(0) + \lambda N^o_{ij} K_r w_{,ij}(0) = 0 .$$

When $\lambda$ is an eigenvalue of (2.16) one of the roots (2.6) is equal to zero and when $\lambda$ increases above this value one $\mu_{nk}$ becomes negati-ve. When $\lambda$ is an eigenvalue of (2.17) one of the roots (2.6) has to be equal to infinity.

We call eigenvalues of (2.16) critical values for infinite critical time and denote them by $\lambda_{cr}$. For $\lambda < \min \lambda_{cr}$ each $\mu_{nk} > 0$ and and the basic solution of (2.1) is stable. For $\lambda = \min \lambda_{cr}$ we have neutral stability and for $\lambda > \min \lambda_{cr}$ at least one $\mu_{nk} < 0$ and the basic solution is unstable with infinite critical time.

Eigenvalues of (2.17) are critical values of instant instability or critical values for finite critical time and we denote them by $\lambda^o_{cr}$. When $\lambda$ reaches the value $\min \lambda^o_{cr}$ the plate becomes instantly unstable.

Now we have the following theorem:

Theorem 1. For $\lambda < \min \lambda_{cr}$, which is the minimum eigenvalue of (2.16), the basic solution of (2.1) is stable. For $\lambda = \min \lambda_{cr}$ this solution is neutral stable and for $\lambda > \min \lambda_{cr}$ the basic solution is unstable with infinite critical time. For $\lambda = \min \lambda^o_{cr}$, which is the minimum eigenvalue of (2.17) the basic solution of (2.1) becomes instantly unstable.

In the case of materials of Boltzmann type the corresponding critical values are eigenvalues of the equations

$$G_{ijkl}(\infty)w_{,ijkl}(\infty) + \lambda N^{o}_{ij}w_{,ij}(\infty) = 0 \tag{2.18}$$

and

$$G_{ijkl}(0)w_{,ijkl}(0) + \lambda N^{o}_{ij}w_{,ij}(0) = 0 \ . \tag{2.19}$$

## 3. Non-linear stability problems of viscoelastic plates

In the linear theory we assume that perturbations are arbitrarily small and in equations we neglect non-linear terms in the perturbation quantities as compared to the linear ones. In the case of instability with respect to infinitesimal perturbations we arrive at an apparent contradiction within the linearized theory, since we assume infinitesimal perturbations and find out that they grow without bounds. Therefore it is necessary to deal with non-linear analysis of stability problems. In studying non-linear problems we may find out that the magnitude of perturbation instead of growing without limit, tends to a finite value as time tends to infinity.

When dealing with nonlinear stability problems we restrict ourselves to an isotropic viscoelastic plate of a standard material. Thus we shall consider the generalized Karman equations

$$\begin{aligned} &K(1 + \alpha D)\Delta^2 w = h(1 + \beta D)(\lambda[F,w] + [f,w]) \ , \\ &(1 + \beta D)\Delta^2 f = -\tfrac{1}{2}E(1 + \alpha D)[w,w] \ , \end{aligned} \tag{3.1}$$

where

$$[f,w] = f_{,11}w_{,22} + f_{,22}w_{,11} - 2f_{,12}w_{,12} \ , \tag{3.2}$$

$F$ is the particular solution of the linear equation

$$\Delta^2 F = 0 \tag{3.3}$$

with boundary conditions corresponding to given boundary loading, $K$ is the bending stiffness of the plate, $E$ is Young modulus and $\alpha, \beta$ coefficients of viscoelastic properties.

We consider the boundary conditions (1.7 - 10) for $w$ and $f$ and initial conditions

$$w(x,0) = w_0(x) \ , \quad f(x,0) = 0 \ . \tag{3.4}$$

Then we can prove the following theorem

Theorem 2. Critical values with infinite critical time and the corresponding stationary solutions are given by the equations

$$K\,\Delta^2 w = h(\lambda[F,w] + [f,w]) \ , \tag{3.5}$$

$$\Delta^2 f = -\frac{1}{2} E[w,w] \ .$$

Thus critical values with the infinite critical time of the non-linear problem are equal to those critical values of the linearized problem.

This result we can get directly considering the stationary solution or in the following way.

Equations (3.1) can be rewritten in the form:

$$K\left[\frac{\alpha}{\beta}\Delta^2 w + \left(1-\frac{\alpha}{\beta}\right)\frac{1}{\beta}\int_0^t \Delta^2 w\, e^{-\frac{1}{\beta}(t-\tau)} d\tau - \frac{\alpha}{\beta}\Delta^2 w_o\, e^{-\frac{1}{\beta}t}\right]$$

(3.6)

$$= h\left(\lambda[F,w] + [f,w] - \lambda[F,w_o]\, e^{-\frac{1}{\beta}t}\right) \ ,$$

$$\frac{\beta}{\alpha}\Delta^2 f + \left(1-\frac{\beta}{\alpha}\right)\frac{1}{\alpha}\int_0^t \Delta^2 f\, e^{-\frac{1}{\alpha}(t-\tau)} d\tau = -\frac{1}{2}E\left([w,w]-[w_o,w_o]e^{-\frac{1}{\alpha}t}\right) \ .$$

Multiplyng $(3.6)_1$ scalarly by $w$ and expressing $[w,w]$ from $(3.6)_2$ after some transformation we arrive at

$$K\left\{\frac{\alpha}{\beta}(\Delta w,\Delta w) + \left(1-\frac{\alpha}{\beta}\right)\frac{1}{\beta}\left(\Delta w, \int_0^t \Delta w\, e^{-\frac{1}{\beta}(t-\tau)} d\tau\right)\right\} - h\lambda(F,[w,w]) + \frac{2}{E}h\left\{\frac{\beta}{\alpha}(\Delta f,\Delta f)\right.$$

$$\left. + \left(1-\frac{\beta}{\alpha}\right)\frac{-}{\alpha}\left(\Delta f, \int_0^t \Delta f\, e^{-\frac{1}{\alpha}(t-\tau)} d\tau\right)\right\}$$

(3.7)

$$= K\,\frac{\alpha}{\beta}(\Delta w_o,\Delta w)e^{-\frac{1}{\beta}t} - h\lambda(F,[w_o,w])\, e^{-\frac{1}{\beta}t} - h(f,[w_o,w_o])e^{-\frac{1}{\alpha}t} \ .$$

As the total energy is finite, for $t = \infty$ we get

$$K(\Delta w,\Delta w) - h\lambda(F,[w,w]) + \frac{2}{E}h(\Delta f,\Delta f) = 0 \ . \tag{3.8}$$

As $(\Delta f,\Delta f) \geq 0$ , the nonzero solution of (3.8) exists iff $\lambda$ is greater the critical values with the infinite critical time of the linearized problem, which are critical values of the non-linear problem (3.5) , too.

When there is no perturbation or $w_o = 0$ then the right hand side of (3.7) is equal to zero and for $t = 0$ one arrive at

$$K\alpha(\Delta w,\Delta w) - h\lambda\beta(F,[w,w] + \frac{2\beta^2}{E\alpha}h(\Delta f,\Delta f) = 0 \ . \tag{3.9}$$

The corresponding differential equations can be obtained from (3.6) . We get

$$K\alpha\Delta^2 w = \beta h(\lambda[F,w]+[f,w]) \,, \tag{3.10}$$
$$\beta\Delta^2 f = -\frac{1}{2}E\alpha[w,w] \,.$$

From (3.9) it is obvious that (3.9) and thus also (3.10) can have a non-zero solution iff

$$K\alpha(\Delta w,\Delta w) - h\beta\lambda(F,[w,w]) < 0 \tag{3.11}$$

and we can formulate the following theorem.

Theorem 3. The critical values for zero critical time, which are eigenvalues of (3.10) are critical values for zero critical time of the linearized problem.

The approximate solution of the problem which gives the exact values of critical values shows that critical values are bifurcation points. The bifurcation is continuous for $\lambda = \lambda_{cr}$ and discontinuous (by a jump) for $\lambda > \lambda_{cr}$ .

## References

[1] Brilla J.: Convolutional variational principles and stability of viscoelastic plates, Ing.Arch. 45 (1976), 273-280.

[2] Eckhaus W.: Studies in non-linear stability theory, Berlin--Heidelberg-New York, Springer 1965.

[3] Ržanicyn A.R.: Teorija polzučesti, Moskva 1968.

Author's address: Institute of Applied Mathematics and Computing Technique, Comenius University, Mlynská dolina, 816 31 Bratislava, Czechoslovakia

ON THE BRANCHING OF SOLUTIONS AND SIGNORINI'S PERTURBATION PROCEDURE IN ELASTICITY

G. Capriz, Pisa

## 1. Introduction

Let us formally express the traction boundary problem for large deformations of a hyperelastic body in the neighbourhood of a placement $\mathcal{B}_o$ under initial stress $S_o$ as follows

$$(1.1)\qquad \begin{aligned} -\operatorname{div} S(\nabla p) &= b & &\text{in } \mathcal{B}_o\,, \\ S(\nabla p)n_o &= s & &\text{in } \partial\mathcal{B}_o\,; \end{aligned}$$

where b is the force per unit volume and s the surface traction; **p** is the position vector in the equilibrium placement $\mathcal{B}$ and S is the Piola-Kirchhoff stress tensor, which is expressed by a constitutive relation in terms of the gradient $\nabla p$ of p; $n_o$ is the unit vector normal to the boundary $\partial\mathcal{B}_o$ of $\mathcal{B}_o$.

Assume that $\mathcal{B}_o$ (with position vector $p_o$) is an equilibrium placement under loads $(b_o, s_o)$:

$$(1.2)\qquad \begin{aligned} -\operatorname{div} S(\nabla p_o) &= b_o & &\text{in } \mathcal{B}_o, \\ S(\nabla p_o)n_o &= s_o & &\text{in } \partial\mathcal{B}_o, \text{ with } S(\nabla p_o) = S_o. \end{aligned}$$

Then the perturbation procedure of Signorini starts from the assumption that b and s are dead loads (i.e., they do not depend on the placement) and can be expressed as power series of a parameter

$$(1.3)\qquad b = b_o + \sum_{1}^{\infty}{}_h\, \varepsilon^h b_h\,, \qquad s = s_o + \sum_{1}^{\infty}{}_h\, \varepsilon^h {}_h\,,$$

proceeds with the hypothesis that the solution p of (1.1) is itself developable

(1.4) $$p = p_o + \sum_1^\infty {}_h\, \varepsilon^h u_h ,$$

and admits that the convergence properties of (1.4) are such that $u_n$ is the solution of the appropriate linear problem that can be deduced formally from (1.1), (1.4). Because of the non-linearity of the dependence of $S$ on $\nabla p$, the boundary problem to be satisfied by $u_h$, although expressed in terms of an operator which does not depend on h, involves "modified" extra loads $(b_h^*, s_h^*)$

(1.5) $$\begin{aligned} -\operatorname{div} \mathfrak{S}[H_h] &= b_h^*, && \text{in } \mathcal{B}_o; \\ \mathfrak{S}[H_h] n_o &= s_h^*, && \text{in } \partial\mathcal{B}_o,\ h = 1,2,\ldots \end{aligned}$$

Here $H_h = \nabla u_h$; $\mathfrak{S}$ is the fourth-order tensor of the elasticities

(1.6) $$\mathfrak{S}[H] = \nabla S\big|_{\nabla p=1} [H];$$

and the starred loads are defined as follows

(1.7) $$b_h^* = b_h + \operatorname{div} \mathcal{S}_h , \qquad s_h^* = s_h - \mathcal{S}_h n_o ,$$

where $\mathcal{S}_h$ can be specified in terms of $p_o$ and $H_k$ (k= 1,2,... h-1) (with $\mathcal{S}_1 \equiv 0$) because it is the quantity entering the expansion

(1.8) $$S(\nabla p(\varepsilon)) = S_o + \sum_1^\infty {}_h (\nabla S\big|_{\nabla p=1} [H_h] + \mathcal{S}_h (p_o; \{H_r\}_1^{h-1})) \varepsilon^h .$$

When $\mathcal{B}_o$ is a placement at ease (i.e., $S_o \equiv 0$), one can specify conditions on the shape of $\mathcal{B}_o$ and on the function $S(\nabla p)$ so that, when $\varepsilon$ is small enough, problem (1.1) has a solution of type (1.4), provided certain quantitative conditions on the loads are satisfied (Stoppelli's theorem). Signorini was mainly concerned with those side conditions, which have an interesting mechanical interpretation and some curious aspects [1]; in [3, 4] the conditions are also explored at length but without introducing assumptions on $S_o$. Here the hypothesis that the loads are dead is also abandoned; furthermore, in Sect.4, a dynamic analysis is pursued which clarifies the significance of certain failures of a purely static study.

## 2. Fredholm conditions.

We abandon here the hypothesis that the loads (b, s) are dead loads and do not exclude that they are of the "follower" type:

$$(2.1) \qquad b = b\,(p, \nabla p) \quad , \quad s = s\,(p, \nabla p),$$

so that the developments (1.3) must be substituted by

$$(2.2) \qquad \begin{aligned} b &= b_o + \sum_1^\infty {}_h\, \varepsilon^h \Big( \mathfrak{b}[H_h] + Bu_h + \mathfrak{b}_h (p_o, \nabla p_o, \{u_s\}_1^{h-1}, \{H_s\}_1^{h-1}) \Big), \\ s &= s_o + \sum_1^\infty {}_h\, \varepsilon^h \Big( \mathfrak{s}[H_h] + \Sigma\, u_h + \mathfrak{s}_h (p_o, \nabla p_o, \{u_s\}_1^{h-1}, \{H_s\}_1^{h-1}) \Big), \end{aligned}$$

where

$$(2.3) \qquad \begin{aligned} \mathfrak{b} &= \mathrm{grad}_{\nabla p}\, b \;, \quad B = \mathrm{grad}_p\, b \;, \\ \mathfrak{s} &= \mathrm{grad}_{\nabla p}\, s \;, \quad \Sigma = \mathrm{grad}_p\, s. \end{aligned}$$

At the same time the systems (1.5) become

$$(2.4) \qquad - \mathrm{div}\, \mathfrak{S}[H_h] - \mathfrak{b}[H_h] - Bu_h = \bar{b}_h^{*}, \qquad \text{in } \mathcal{B}_o,$$

$$\mathfrak{S}[H_h]n_o - \mathfrak{s}[H_h] - \Sigma\, u_h = \bar{s}_h^{*}, \qquad \text{in } \partial\mathcal{B}_o,$$

with

$$\bar{b}_h^{*} = \mathfrak{b}_h + \mathrm{div}\, \mathfrak{S}_h, \quad \bar{s}_h^{*} = \mathfrak{s}_h - \mathfrak{S}_h n_o, \qquad h = 1,2,\ldots$$

Apart from these qualifications which, at this stage, are of a formal character, a remark of substance is in order here. The loads (b, s) of problem (1.1) are balanced in the final placement

$$(2.5) \qquad \begin{aligned} &\int_{\mathcal{B}_o} b(p, \nabla p)\, d(\mathrm{vol}) + \int_{\partial\mathcal{B}_o} s(p, \nabla p)\, d(\mathrm{surf}) = 0, \\ &\int_{\mathcal{B}_o} p \times b(p, \nabla p)\, d(\mathrm{vol}) + \int_{\partial\mathcal{B}_o} p \times s(p, \nabla p)\, d(\mathrm{surf}) = 0, \end{aligned}$$

but these relations do not impose restrictions on (b,s); they are simply an expression of mutual consistency of the (data, solution) pair. Nor need they be verified when $p$ is reduced to $p_o$ in them; though the first reduced one must be true in Signorini's case. Further, if $s_o \equiv 0$, the choice of $\mathcal{B}_o$ is open among isometric placements; then Da Sylva's theorem assures us that among those placement there exists at least one where the moment is balanced. Such freedom in the choice of $\mathcal{B}_o$, however, is absent in the presence of follower loads or when $s_o \neq 0$. In conclusion the balance of loads (b, s) in the starting placement either can be trivially assured, and then it corresponds to a conventional choice of $\mathcal{B}_o$ among the many open choices, or must not be required.

Let us consider now the sequence of linear problems (2.4). The corresponding homogeneous problem

$$\text{(2.6)} \qquad \operatorname{div} \mathfrak{S}[H] + \mathfrak{b}[H] + Bu = 0, \quad \text{in } \mathcal{B}_o \qquad \mathfrak{S}[H]\, n_o - \sigma[H] - \Sigma u = 0, \quad \text{in } \partial\mathcal{B}_o,$$

admits a set $\mathcal{S}$ of non-trivial solutions: for instance it is well known that in the presence of dead loads and when $s_o \equiv 0$, $\mathcal{S}$ contains the set $\mathfrak{J}$ of all infinitesimal isometries. In any case the theorem of alternative states a prerequisite for the existence of solutions of (2.4): the loads $(\bar{b}^*_h, \bar{s}^*_h)$ must be orthogonal to all solutions of (2.6)

$$\text{(2.7)} \qquad \int_{\mathcal{B}_o} v.\bar{b}^*_h \, d(\text{vol}) + \int_{\partial\mathcal{B}_o} v.\bar{s}^*_h \, d(\text{surf}) = 0 , \qquad \forall\, v \in \mathcal{S} .$$

We will consider here only the case when $\mathcal{S}$ is finite-dimensional and non-trivial; after having chosen a basis for $\mathcal{S}$ ($v_i$ (i= 1,2,...r); $r = \dim \mathcal{S}$ ) one can write (2.7) as a set of r equations for $(\bar{b}^*_h, \bar{s}^*_h)$. Notice also that the solution of (2.4) is never unique: adding to any solution $\tilde{u}_h$ a linear combination of $v_i$ yields again a solution. This fact is the basis of the remark that the character of (2.7) for h=1 is completely different from that of (2.7) for $h \geq 2$.

In fact, let us examine first the case h=1; then $\bar{b}^*_1 = \mathfrak{b}_1(p_o, \nabla p_o)$, $\bar{s}^*_1 = \mathfrak{s}_1(p_o, \nabla p_o)$ and therefore

$$(2.8)\qquad \int_{\mathcal{B}_o} v.\,b_1 \, d(vol) + \int_{\partial\mathcal{B}_o} v.\,s_1 \, d(surf) = 0 \,, \quad \forall\, v \in \mathcal{S} \,,$$

represents a condition which is absent in the statement of the original non-linear problem and is of a technical character for the existence of solutions of (1.1) of type (1.4).

Signorini remarked that in his special case, when $\mathcal{S} \equiv \mathcal{T}$, (2.8) requires the balance of the first order loads in $\mathcal{B}_o$. This interesting remark is, however, misleading to some extent because it leads one to focus exclusive attention on classical balance conditions. An example of Ericksen and Toupin and an example of Bordoni (see [4], Sections 7 and 11) shows explicitly that the "balance conditions" (2.8) required of the first-order loads are much deeper than the classical ones.

## 3. An analysis of branching of solutions.

Suppose for the moment that (2.8) is satisfied, and a solution $\tilde{u}_1$ of (2.4) for h=1 has been found. Then system $(2.4)_1$ admits also a whole set of solutions

$$(3.1)\qquad u_1 = \tilde{u}_1 + \sum_1^r {}_k\, \gamma_1^k \, v_k ,$$

where $\gamma_1^k$ are r arbitrary real parameters. On the other hand, we have from (2.7) for h=2 the conditions

$$\int_{\mathcal{B}_o} v_k \cdot ( b_2 + \operatorname{div} S_2 ) \, d(vol) + \int_{\partial\mathcal{B}_o} v_k \cdot ( s_2 - S_2 n_o ) \, d(surf) = 0,$$

$$k = 1,2,\ldots.r;$$

here $b_2$, $s_2$, $S_2$ depend on ($p_o$, $\nabla p_o$ and) $u_1$, $\nabla u_1$, which can be expressed in turn through (3.1). The dependence on $u_1$, $\nabla u_1$ is algebraic of degree 2, so that finally we come up with an algebraic system of degree 2 for the coefficients $\gamma_1^i$ :

$$(3.2)\qquad c_{k,1} + \sum_1^r {}_i\, b_{ki}\, \gamma_1^i + \sum_1^r {}_{ij}\, a_{kij}\, \gamma_1^i \, \gamma_1^j = 0$$

here $\mathcal{C}_{k,1}$, $\mathcal{b}_{ki}$, $\mathcal{a}_{kij}$ have complicated expressions. For instance, in the case of dead loads one has

$$(3.3)\quad \begin{aligned} \mathcal{C}_{k,1} &= \int_{\mathcal{B}_o} (b_2\cdot v_k - (\nabla^2 s\,|_{\nabla p=1}[\nabla\tilde{u}_1])[\nabla u_1])\cdot \nabla v_k))\,d(vol) + \\ &\quad + \int_{\partial\mathcal{B}_o} s_2\cdot v_k\, d(surf), \\ \mathcal{b}_{k,i} &= -\int_{\mathcal{B}_o} ((\nabla^2 s\,|_{\nabla p=1}[\nabla\tilde{u}_1])[\nabla v_i])\cdot \nabla v_k\, d(vol), \\ \mathcal{a}_{kij} &= -\frac{1}{2}\int_{\mathcal{B}_o} ((\nabla^2 s\,|_{\nabla p=1}[\nabla v_i])[\nabla v_j])\cdot \nabla v_k\, d(vol). \end{aligned}$$

The set of r algebraic equations (3.2) may be used to determine the parameters $\gamma_1^k$. There are cases when such determination can be achieved and is unique; for instance, in Signorini's case all coefficients $\mathcal{a}_{kij}$ vanish because in that case the tensor $\nabla^2 s$ has certain properties of symmetry, whereas $\nabla v_i$ is necessarily skew as $v_i$ must represent an infinitesimal isometry; then the system becomes linear and the parameters $\gamma_1^k$ can be explicitly given, if det $\mathcal{b}_{ki} \neq 0$ (a condition which is equivalent to the requirement that the first order loads do not admit of axes of equilibrium).

In general the search for real solutions of (3.2) is more delicate and the variety of situations reflects the complexities of the cases of branching of solutions in the original non-linear problem. Suffice here to remark that the apparent indetermination observed for $u_1$ occurs also for all $u_h (h \geq 2)$ and that conditions (2.7) for $h \geq 2$ can be again invoked to overcome such indetermination. On the other hand there are cases where (3.2) cannot be satisfied at all; then the solution of $(2.4)_1$ which seems at stage 1 as a legitimate approximation of first order to a solution of the non-linear problem (1.1) must be rejected on the strength of the Fredholm conditions regarding stage 2; a similar discrepancy may occur at any stage. We devote therefore the next section to an interpretation and an analysis of these cases of "incompatibility".

## 4. Reformulation of the traction problem in elastodynamics.

Within the limits of a static study, the rejection of an alleged approximation of h-th order because of the failure of the Fredholm conditions at stage h+1 is absolute. Yet we can still give some significance to that approximation, relying on the fact that problem (1.1) can be considered as a special case of a more general dynamic problem ( $\rho$, density in $\mathcal{B}_o$)

$$(4.1)\qquad \begin{aligned} -\operatorname{div} S(\nabla p) &= b - \rho \ddot{p}, &&\text{in } \mathcal{B}_o,\\ S(\nabla p)n_o &= s, &&\text{in } \partial\mathcal{B}_o, \end{aligned}$$

which requires the assignement of appropriate initial conditions. As we shall see, we will be able to interpret the developments of the preceding Sections as the search for such initial conditions having special properties; our analysis will, at the same time, lead to the complete specification of an "acceptable" approximation (necessarily dynamic) of order h+1.

In fact, suppose that one of the initial conditions to be attached to (4.1) requires the vanishing of the velocity

$$(4.2)\qquad \dot{p}\big|_{t=0} = 0 \qquad\qquad \text{in } \overline{\mathcal{B}}_o;$$

and try to determine $p\big|_{t=0}$ as a function $\tilde{p}$ so that no motion ensues:

$$p \equiv \tilde{p}, \ \forall\, t.$$

Imagine that determination to proceed in successive stages of approximation, corresponding to the specification (2.2) of the loads.

If $b = b_o$, $s = s_o$, one solution is, by hypothesis, $\tilde{p} = p_o$. When $(b,s) \neq (b_o, s_o)$, to start the process at all, $(\mathfrak{b}_1, \mathfrak{s}_1)$ must satisfy (2.8) as we have already remarked.

But suppose for a moment that condition (2.8) is contravened; then, the following technique may be used to explore the main aspect of the ensuing dynamic phenomenon, and to provide for it an approximate description .

Multiply scalarly the members of

$$(4.3)\qquad \begin{aligned} &\operatorname{div}\mathfrak{S}[H] + \mathfrak{b}[H] + Bu - \mathfrak{b}_1 + \rho\ddot{u} = 0, \\ &\mathfrak{S}[H]n_o - \mathfrak{s}[H] - \Sigma u - \mathfrak{s}_1 = 0 , \end{aligned}$$

by $v_i$ (where,as before, $v_i$ form a basis for $\mathcal{S}$); integrate the result respectively over $\mathcal{B}_o$ and $\partial\mathcal{B}_o$ and sum. Finally try to solve the resulting global equation choosing for u the expression

$$(4.4)\qquad u^* = \sum_1^r {}_k\, \gamma_o^k(t)\, v_k .$$

It is sufficient for the functions $\gamma_o^k(t)$ to satisfy the ordinary differential system

$$(4.5)\qquad \sum_1^r {}_k\, J_{ik}\, \ddot{\gamma}_o^k = m_i ,$$

where the constants $J_{ik}$ are generalized coefficients of inertia

$$(4.6)\qquad J_{ik} = \int_{\mathcal{B}_o} \rho\, v_i \cdot v_k \, d(vol) ,$$

$m_i$ are generalized resultant forces

$$(4.7)\qquad m_i = \int_{\mathcal{B}_o} \mathfrak{b}_1 \cdot v_i \, d(vol) + \int_{\partial\mathcal{B}_o} \mathfrak{s}_1 \cdot v_i \, d(surf),$$

In Signorini's case, when $\mathcal{S} = \mathcal{T}$ , eqns (4.4) provide linear approximations to a rigid body motion of $\mathcal{B}$ ; in general, they describe a much more complex movement of our body:

$$(4.8)\qquad u^* = \frac{1}{2}\left(\sum_1^r {}_{ks}\, v_k ((\overset{-1}{J})^{sk}\, m_s)\right) t^2 .$$

At this point it is possible to return to (4.3) and determine $u_1 - u^*$ solving a compatible static boundary value problem where the loads $(\mathfrak{b}_1 - \rho \sum_1^r {}_k\, v_k ((\overset{-1}{J})^{sk}\, m_s), \mathfrak{s}_1)$ involve an apparent body force.

We have dealt in some detail with this almost trivial "case of incompatibility" because it is a simple model for slightly subtler cases of higher order. For instance, suppose that (2.8) applies, so that a solution of (2.4) for $h = 1$ exists, but (3.2) has no real solutions in $\gamma_1^i$; then one must return to the dynamic system

$$\text{(4.8)} \qquad \begin{aligned} &\operatorname{div} \mathfrak{S}[H] + \mathfrak{b}[H] + Bu - \bar{b}_2^* + \rho \ddot{u} = 0, \\ &\mathfrak{S}[H] h_o - \mathfrak{s}[H] - \Sigma u - \bar{s}_2^* = 0, \end{aligned}$$

and proceed as before to obtain a differential relation of the type (4.5), where now the generalized resultant forces are

$$\text{(4.9)} \qquad \begin{aligned} m_i = &\int_{\mathcal{B}_o} (\mathfrak{b}_2 \cdot v_i + \mathfrak{S}_2[H_1] \cdot \nabla v_i)\, d(\mathrm{vol}) + \\ &+ \int_{\partial \mathcal{B}_o} \mathfrak{s}_2 \cdot v_i \, d(\mathrm{surf}). \end{aligned}$$

An acceptable, necessarily dynamic, approximation to (4.8) is a new $u^*$ of type (4.4) with $\gamma_1^k$, solution of (4.5), but with the specification (4.9) for $m_i$, plus a time-independent solution of a static problem involving the appropriate apparent body force.

## 5. Conclusion.

Signorini's perturbation method is a special case of a general technique, particularly adapted to a study of stability and branching phenomena in hyperelasticity. The interpretation already advanced [2] of the phenomena of incompatibility discovered by Signorini can be extended to apply to more general cases where the ground state is stressed and follower loads are present. In particular the well-known arbitrariness in the choice of amplitude of the buckled shapes within the first approximation can be interpreted either as a temporary freedom soon to be mitigated by conditions of compatibility of higher order systems or as a real scope in the choice of initial placements within the class of placements whence a motion begins where the acceleration is of higher order.

## References

[1] A. Signorini , Trasformazioni termoelastiche finite. Ann. Mat. Pura Appl., IV 30 (1949), 1-72.

[2] G. Capriz, P. Podio Guidugli, On Signorini's perturbation method in finite elasticity. Arch. Rat. Mech. An. 57 (1974), 1-30.

[3] G. Capriz, P. Podio Guidugli, The rôle of Fredholm conditions in Signorini's perturbation method, to appear in Arch. Rat. Mech. An.

[4] S. Baratha, M. Levinson, Signorini's perturbation scheme for a general reference configuration in finite elastostatics. Arch. Rat. Mech. An. , 67 (1978), 365-394.

Author's address: Consiglio Nazionale delle Ricerche, Istituto di Elaborazione della Informazione, Via S.Maria 46, 56100 Pisa, Italy

DIFFERENTIAL SUBSPACES ASSOCIATED WITH PAIRS OF ORDINARY DIFFERENTIAL OPERATORS

E. A. Coddington, Los Angeles

1. Introduction. This is an account of some joint work in progress with H.S.V. de Snoo. It represents an attempt to place a study of boundary value and eigenvalue problems, associated with a pair of ordinary differential expressions $L$, $M$, in the general framework of two earlier papers by E.A. Coddington and A. Dijksma [7], [8]. In the first of these we showed how to describe very general eigenvalue problems, for the case when $M$ is the identity and $L$ is formally symmetric, and to obtain eigenfunction expansion results for these problems. In the second we described abstractly the adjoints of subspaces (multi-valued operators) in Banach spaces in terms of generalized boundary conditions, and applied these results to a study of boundary value problems with not necessarily formally symmetric differential expressions $L$.

There is a large literature devoted to problems for two expressions $L$, $M$. We mention the recent work by F. Brauer [2], [3], [4], F. Browder [5], [6], Å. Pleijel [9], C. Bennewitz [1]. We deal with systems, not necessarily formally symmetric $L$, and we do not assume that the order of $M$ is less than the order of $L$. From the point of view of subspaces, if a subspace $S$ is associated with a right definite $M$, then $S^{-1}$ is a problem associated with a left definite case. The set of Hilbert spaces which we allow differ from those considered by Bennewitz in [1].

We settle some notation matters. Let $\mathbb{R}$, $\mathbb{C}$ denote the real and complex numbers. We consider an open real interval $\iota = (a,b)$, and the set $F_m(\iota)$ of all vector valued functions $f : \iota \to \mathbb{C}^m$. By $C(\iota)$ we denote the set of all continuous $f \in F_m(\iota)$, and

$$C^k(\iota) = \{f \in F_m(\iota) \mid f^{(k)} \in C(\iota)\} ,$$

$$C_0^k(\iota) = \{f \in C^k(\iota) \mid \text{support of } f \text{ is compact}\} ,$$

$$C_0^\infty(\iota) = \bigcap_k C_0^k(\iota) .$$

By $L^2_{loc}(\iota)$ we mean the set of all $f \in F_m(\iota)$ such that

$$\int_J |f|^2 < \infty , \quad \text{each compact subinterval } J \subset \iota ,$$

where $|f|^2 = f^* f$, and we let

$$L^2(\iota) = \{f \in L^2_{loc}(\iota) \mid \int_\iota |f|^2 < \infty\} ,$$

$$L^2_0(\iota) = \{f \in L^2(\iota) \mid \text{support of } f \text{ is compact}\} .$$

If $f,g \in F_m(\iota)$, we use the notations

$$(f,g)_{2,J} = \int_J g^* f , \qquad (f,g)_2 = \int_\iota g^* f ,$$

if the components of $g^*f$ are integrable on the compact subinterval $J \subset \iota$, or on $\iota$, respectively. Note that we do not assume $f, g$ are in $L^2_{loc}(\iota)$ or $L^2(\iota)$.

2. <u>Hilbert spaces associated with positive differential expressions</u>. Let $M$ be the formal ordinary differential expression of order $\nu$

$$M = \sum_{k=0}^{\nu} Q_k D^k , \qquad D = d/dx ,$$

where the $Q_k$ are $m \times m$ complex matrix-valued functions whose columns are in $C^k(\iota)$, and $Q_\nu(x)$ is invertible for $x \in \iota$. We want to associate an inner product with this $M$ by first defining

$$(2.1) \qquad (\varphi, \psi) = (M\varphi, \psi)_2 , \qquad \varphi, \psi \in C_0^\infty(\iota) .$$

If this is to be an inner product on $C_0^\infty(\iota)$ we must have

$$(2.2) \qquad \begin{aligned} M = M^+ &= \sum_{k=0}^{\nu} (-1)^k D^k Q_k^* , \\ (M\varphi, \varphi)_2 &\geq 0 , \qquad \varphi \in C_0^\infty(\iota) , \end{aligned}$$

and we assume this. From this it follows that $\nu$ is even, $\nu = 2\mu$, and $(-1)^\mu Q_\nu(x) > 0$, $x \in \iota$, in the sense that

$$\xi^*(-1)^\mu Q_\nu(x)\xi \geq c(x)\xi^*\xi , \qquad \xi \in C^m ,$$

for some $c(x) > 0$. We can write such an $M$ in the form

$$M = \sum_{j=0}^{\mu} \sum_{k=j-1}^{j+1} (-1)^j D^j Q_{jk} D^k ,$$

where $Q_{jk}^* = Q_{kj}$, and $Q_{jj} \in C^j(\iota)$, $Q_{j+1\,j} \in C^{j+1}(\iota)$, $Q_{j\,j+1} \in C^{j+1}(\iota)$. Using this form for $M$ the formula (2.1) can be written as

$$(\varphi, \psi) = (M\varphi, \psi)_2 = \int_\iota \sum_{j=0}^{\mu} \sum_{k=j-1}^{j+1} (D^j\psi^*) Q_{jk} (D^k\varphi) , \qquad \varphi, \psi \in C_0^\infty(\iota) ,$$

and the right side is denoted by $(\varphi, \psi)_D$, the Dirichlet inner product.

The definition (2.1) gives an inner product $( \; , \; )$ on $C_0^\infty(\iota)$ under the assumption (2.2), and $\| \; \| = ( \; , \; )^{1/2}$ is a norm on $C_0^\infty(\iota)$. Let $\mathfrak{H}_M$ denote the completion of $C_0^\infty(\iota)$; it is a Hilbert space. In many cases $\mathfrak{H}_M$ can be imbedded into $L^2_{loc}(\iota)$, and this is assured if we assume:

($A_1$) for each compact subinterval $J \subset \iota$ there is a $c(J) > 0$ such that

$$\|\varphi\| \geq c(J)\|\varphi\|_{2,J} , \qquad \varphi \in C_0^\infty(\iota) .$$

Then the identity map on $C_0^\infty(\iota)$ has an extension which is an injection of $\mathfrak{H}_M$

into $L^2_{loc}(\iota)$, and we can identify $\mathfrak{H}_M$ as a subset of $L^2_{loc}(\iota)$. We have

$$(f,\varphi) = (f,M\varphi)_2 , \quad f \in \mathfrak{H}_M , \quad \varphi \in C_0^\infty(\iota) ,$$

$$\|f\| \geq c(J)\|f\|_{2,J} , \quad f \in \mathfrak{H}_M ,$$

and the injection $\mathfrak{H}_M \to L^2_{loc}(\iota)$ implies the existence of an injection $G_M$ : $L_0^2(\iota) \to \mathfrak{H}_M$ with the properties:

$$(2.3) \qquad \begin{aligned} (f,G_M h) &= (f,h)_2 , \quad f \in \mathfrak{H}_M , \quad h \in L_0^2(\iota) , \\ G_M M\varphi &= \varphi , \quad \varphi \in C_0^\infty(\iota) , \\ MG_M h &= h , \quad h \in L_0^2(\iota) , \\ (\mathfrak{R}(G_M))^c &= \mathfrak{H}_M , \end{aligned}$$

where $A^c$ denotes the closure of a set $A$, and $\mathfrak{R}(G_M)$ denotes the range of $G_M$.

An important special case is obtained if instead of $(A_1)$ we assume

$$(A_1') \qquad \|\varphi\| \geq c\|\varphi\|_2 , \quad \text{for some} \quad c > 0 .$$

Then $\mathfrak{H}_M \subset L^2(\iota)$ and $G_M$ has an extension, call it $G_M$ also, to an injection $G_M : L^2(\iota) \to \mathfrak{H}_M$ such that (2.3) is valid with $L_0^2(\iota)$ replaced by $L^2(\iota)$ everywhere. In fact, assuming $(A_1')$ we can identify $G_M$ more precisely. Let $M_0$ be the operator in $L^2(\iota)$ with domain $\mathfrak{D}(M_0) = C_0^\infty(\iota)$ given by $M_0\varphi = M\varphi$. It is a symmetric operator which is bounded below by $c > 0$ if $(A_1')$ holds, and thus has a Friedrichs extension which is a selfadjoint operator $M_F$ having the same lower bound $c$. Its inverse $M_F^{-1}$ exists on all of $L^2(\iota)$ and one can show that $G_M = M_F^{-1}$, and that $\mathfrak{H}_M$ is the domain $\mathfrak{D}(M_F^{1/2})$ of the positive square root $M_F^{1/2}$ of $M_F$.

There exist other Hilbert spaces $\mathfrak{H}$ having the essential properties of $\mathfrak{H}_M$. Let $\mathfrak{H}$ be any Hilbert space with inner product ( , ) and norm $\| \ \|$ satisfying:

$$(A_2) \qquad \begin{aligned} C_0^\infty(\iota) \subset \mathfrak{H} &\subset L^2_{loc}(\iota) , \\ (f,\varphi) = (f,M\varphi)_2 , \quad f &\in \mathfrak{H} , \quad \varphi \in C_0^\infty(\iota) , \\ \|f\| \geq c(J)\|f\|_{2,J} , \quad f &\in \mathfrak{H} , \quad c(J) > 0 , \end{aligned}$$

for each compact subinterval $J \subset \iota$. We have $(C_0^\infty(\iota))^c = \mathfrak{H}_M$, and in fact

$$\mathfrak{H} = \mathfrak{H}_M \oplus \mathfrak{N}_M ,$$

an orthogonal sum, where

$$\mathfrak{N}_M = \{f \in C^\nu(\iota) \cap \mathfrak{H} \mid Mf = 0\} .$$

Clearly $\dim \mathfrak{N}_M \leq \nu m$. As before there exists an injection $G : L_0^2(\iota) \to \mathfrak{H}$ such that:

$$
\begin{aligned}
(f,Gh) &= (f,h)_2\,, \quad f \in \mathfrak{H}\,, \quad h \in L_0^2(\iota)\,,\\
GM\varphi &= \varphi\,, \quad \varphi \in C_0^\infty(\iota)\,,\\
MGh &= h\,, \quad h \in L_0^2(\iota)\,,\\
(\mathfrak{R}(G))^c &= \mathfrak{H}\,,\\
G_M &= P_M G\,,
\end{aligned}
\tag{2.4}
$$

where $P_M$ is the orthogonal projection of $\mathfrak{H}$ onto $\mathfrak{H}_M$. If instead of $(A_2)$ we have

$$
\begin{aligned}
&C_0^\infty(\iota) \subset \mathfrak{H} \subset L^2(\iota)\,,\\
&(f,\varphi) = (f,M\varphi)_2\,, \quad f \in \mathfrak{H}\,, \quad \varphi \in C_0^\infty(\iota)\,,\\
&\|f\| \geq c\|f\|_2\,, \quad f \in \mathfrak{H}\,, \quad c > 0\,,
\end{aligned}
\tag{$A_2'$}
$$

then $G$ has an extension to all of $L^2(\iota)$ satisfying (2.4) with $L_0^2(\iota)$ replaced by $L^2(\iota)$.

3. Examples. Let $H$ be a positive selfadjoint extension of $M_0$ in $L^2(\iota)$ such that

$$
(Hf,f)_2 = (Mf,f)_2 \geq (c(J))^2(f,f)_{2,J}\,, \quad f \in \mathfrak{D}(H)\,, \quad c(J) > 0\,,
\tag{3.1}
$$

for each compact subinterval $J \subset \iota$, and let $\mathfrak{H}_H$ be the completion of $\mathfrak{D}(H)$ with

$$
(f,g) = (Mf,g)_2\,, \quad f,g \in \mathfrak{D}(H)\,.
$$

This is a Hilbert space, and it will be in $L_{loc}^2(\iota)$ if the following is assumed:

$(A_3)$ $f_n \in \mathfrak{D}(H)$, $\|f_n - f_m\| \to 0$, $\|f_n\|_{2,J} \to 0$ for each compact subinterval $J \subset \iota$, implies $\|f_n\| \to 0$.

Then $\mathfrak{H} = \mathfrak{H}_H$ satisfies $(A_2)$. As an example consider $M = -D^2$, $m = 1$, $\iota = (0,\infty)$. The maximal operator $M_{max}$ for $M$ in $L^2(\iota)$ has a domain $\mathfrak{D}_{max}$ consisting of all $f \in L^2(\iota)$ such that $f'$ is absolutely continuous on each compact subinterval $J \subset [0,\infty)$, and $Mf \in L^2(\iota)$. The selfadjoint extensions of $M_0$ are obtained from $M_{max}$ by imposing a homogeneous boundary condition at $0$. Let $H_h$ be the selfadjoint extension of $M_0$ given by

$$
\begin{aligned}
\mathfrak{D}(H_h) &= \{f \in \mathfrak{D}_{max} \mid f'(0) = hf(0)\}\,, \qquad h \in \mathbb{R}\,,\\
&= \{f \in \mathfrak{D}_{max} \mid f(0) = 0\}\,, \qquad h = \infty\,.
\end{aligned}
$$

We have for $f,g \in \mathfrak{D}(H_h)$

$$\begin{aligned}(H_h f,g)_2 &= hf(0)\overline{g}(0) + (f',g')_2^2 , \qquad & h \in \mathbb{R} ,\\ &= (f',g')_2^2 , & h = \infty .\end{aligned}$$

Only for $0 \leq h \leq \infty$ will $H_h$ satisfy $(H_h f,f)_2 \geq 0$ for $f \in \mathfrak{D}(H_h)$. In case $0 < h \leq \infty$ we can show that for each compact subinterval $J \subset [0,\infty)$ there is a $c(J) > 0$ such that

$$(H_h f,f)_2^{1/2} = \|f\| \geq c(J)\|f\|_{2,J} , \qquad f \in \mathfrak{D}(H_h) ,$$

and $(A_3)$ is valid. Then the Hilbert space completion $\mathfrak{H}_h$ of $\mathfrak{D}(H_h)$ is in $L^2_{loc}(\iota)$ and the form of the inner product persists, that is,

$$\begin{aligned}(f,g) &= hf(0)\overline{g}(0) + (f',g')_2 , \qquad & f,g \in \mathfrak{H}_h , \quad & 0 < h < \infty ,\\ (f,g) &= (f',g')_2 , & f,g \in \mathfrak{H}_h , \quad & h = \infty .\end{aligned}$$

Moreover it can be shown that $\mathfrak{N}_M = \text{span}\{1\}$ if $0 < h < \infty$ and $\mathfrak{N}_M = \{0\}$ if $h = \infty$. None of these $\mathfrak{H}_h$ are contained in $L^2(\iota)$, for there exists a sequence $\varphi_n \in C_0^2(\iota) \subset \mathfrak{D}(H_h)$ such that $\|\varphi_n\|^2 = (\varphi_n',\varphi_n') \to 0$ but $\|\varphi_n\|_2 \to +\infty$. In case $h = 0$ we get an inner product $(f,g) = (f',g')_2$ on $\mathfrak{D}(H_0)$, but the completion $\mathfrak{H}_0$ of $\mathfrak{D}(H_0)$ is not contained in $L^2_{loc}(\iota)$. There exists a sequence $\varphi_n \in \mathfrak{D}(H_0)$ such that $\|\varphi_n\| \to 0$ but $\|\varphi_n\|_{2,J} \to \infty$ on each proper compact subinterval $J \subset [0,\infty)$.

There may exist positive selfadjoint extensions $H$ of $M_0$ in $L^2(\iota)$ satisfying a global inequality:

$$(Hf,f)_2 = (Mf,f)_2 \geq c^2(f,f)_2 , \qquad f \in \mathfrak{D}(H) , \quad c > 0 .$$

If $\mathfrak{H}_H$ is the completion of $\mathfrak{D}(H)$ with $(f,g) = (Mf,g)_2$, $f,g \in \mathfrak{D}(H)$, then $\mathfrak{H}_H \subset L^2(\iota)$ and $\mathfrak{H} = \mathfrak{H}_H$ satisfies $(A_2')$. In fact $\mathfrak{H}_H = \mathfrak{D}(H^{1/2})$ and $G = H^{-1}$ in this case.

Another method of constructing an $\mathfrak{H}$ satisfying $(A_2)$ is as follows. Let $\mathfrak{N}_M$ be any linear subset of $N_M = \{f \in C^\nu(\iota) \mid Mf = 0\}$ with any inner product $(\ ,\ )_0$ such that

$$\|f_0\|_0 \geq c_0(J)\|f_0\|_{2,J} , \qquad f_0 \in \mathfrak{N}_M ,$$

for some $c_0(J) > 0$ and each compact subinterval $J \subset \iota$. Let $(\ ,\ )_1$, for the moment, denote the inner product on $\mathfrak{H}_M$. Define $\mathfrak{H} = \mathfrak{H}_M \oplus \mathfrak{N}_M$ with the inner product

$$(f,g) = (f_1,g_1)_1 + (f_0,g_0)_0 ,$$

$$f = f_1 + f_0 , \quad g = g_1 + g_0 , \quad f_1,g_1 \in \mathfrak{H}_M , \quad f_0,g_0 \in \mathfrak{N}_M .$$

Then $(A_2)$ is valid. As an example we could use $(f,g)_0 = (f,g)_2$, or $(f,g)_0 = (f,g)_D$.

4. Maximal and minimal subspaces. Let $M$ be as before, and let $L$ be another formal differential operator

$$L = \sum_{k=0}^{n} P_k D^k ,$$

where the $P_k$ are $m \times m$ complex matrix-valued functions on $\iota$ whose columns are in $C^k(\iota)$, and $P_n(x)$ is invertible for $x \in \iota$ if $n > \nu$. We consider any Hilbert space $\mathfrak{H}$ satisfying $(A_2)$. In $\mathfrak{H}^2 = \mathfrak{H} \oplus \mathfrak{H}$ we define the maximal linear manifolds

$$T = \{\{f,g\} \in \mathfrak{H}^2 \mid f \in C^r(\iota), \quad g \in C^\nu(\iota), \quad Lf = Mg\} ,$$

$$T^+ = \{\{f,g\} \in \mathfrak{H}^2 \mid f \in C^r(\iota), \quad g \in C^\nu(\iota), \quad L^+f = Mg\} ,$$

where $r = \max(n,\nu)$, and the minimal linear manifolds

$$S = \{\{\varphi, GL\varphi\} \mid \varphi \in C_0^\infty(\iota)\} ,$$

$$S^+ = \{\{\varphi, GL^+\varphi\} \mid \varphi \in C_0^\infty(\iota)\} .$$

Now $S$, $S^+$ are (the graphs of) operators, whereas $T$, $T^+$ need not be operators. In fact,

$$T(0) = \{g \in \mathfrak{H} \mid \{0,g\} \in T\} = T^+(0) = \mathfrak{N}_M ,$$

and this implies $S$, $S^+$ are densely defined if and only if $\mathfrak{N}_M = \{0\}$. It is clear that $S \subset T$, $S^+ \subset T^+$, and if we put $T_0 = S^c$, $T_1 = T^c$, $T_0^+ = (S^+)^c$, $T_1^+ = (T^+)^c$, we have $T_0 \subset T_1$, $T_0^+ \subset T_1^+$ and these are subspaces (closed linear manifolds) in $\mathfrak{H}^2$.

On $\mathfrak{H}^2 \times \mathfrak{H}^2$ we introduce the form $\langle\ ,\ \rangle$ given by

$$\langle u,v\rangle = (g,h) - (f,k) , \qquad u=\{f,g\},\ v=\{h,k\} \in \mathfrak{H}^2 .$$

If $Ju = \{g,-f\}$ then $\langle u,v\rangle = (Ju,v) = -(u,Jv)$. If $A$ is any linear manifold in $\mathfrak{H}^2$ its adjoint $A^*$ is the subspace defined by

$$A^* = \{v \in \mathfrak{H}^2 \mid \langle u,v\rangle = 0, \quad \text{all } u \in A\} .$$

The following result describes the adjoints of $S$, $T$, $S^+$, $T^+$ and their properties.

THEOREM. <u>We have</u>

(i) $S^* = \{\{f,g\} \in \mathfrak{H}^2 \mid (g,M\varphi)_2 = (f,L\varphi)_2,\ \underline{\text{all}}\ \varphi \in C_0^\infty(\iota)\} = T_1^+$ ,

(ii) $T_1^+ \ominus T_0^+ = T^+ \cap JT$ ,

(iii) $(S^+)^* = \{\{f,g\} \in \mathfrak{H}^2 \mid (g,M\varphi)_2 = (f,L^+\varphi)_2,\ \underline{\text{all}}\ \varphi \in C_0^\infty(\iota)\} = T_1$ ,

(iv) $T_1 \ominus T_0 = T \cap JT^+$ ,

(v) $T_1(0) = T_1^+(0) = T(0) = T^+(0) = \mathfrak{N}_M$ ,

(vi) $\nu(T_1^+ - \ell I) = \nu(T^+ - \ell I) = \{f \in \mathfrak{H} \cap C^r(\iota) \mid L^+f = \ell Mf\}$ ,

<u>where</u> $\ell \in \mathbb{C}$, $\quad n > 2\mu$ ,

$\ell \in \mathbb{C}\setminus\{0\}$, $\quad n < 2\mu$ ,

$$\ell \in \mathbf{C} \setminus \bigcup_{x \in \iota} \sigma(Q_{2\mu}^{-1}(x) P_{2\mu}^{*}(x)) , \quad n = 2\mu ,$$

(vii) $\nu(T_1 - \ell I) = \nu(T - \ell I) = \{f \in \mathfrak{H} \mid f \in C^r(\iota), \ Lf = \ell M f\}$ ,

<u>where</u> $\ell \in \mathbf{C}$ , $n > 2\mu$ ,

$\ell \in \mathbf{C} \setminus \{0\}$, $n < 2\mu$

$\ell \in \mathbf{C} \setminus \bigcup_{x \in \iota} \sigma(Q_{2\mu}^{-1}(x) P_{2\mu}(x))$ , $n = 2\mu$ .

In the above theorem, $I$ denotes the identity operator, $\nu(A)$ represents the null space of a linear manifold $A$,

$$\nu(A) = \{f \in \mathfrak{H} \mid \{f, 0\} \in A\} ,$$

and $\sigma(B)$ is the spectrum of a matrix $B$, that is, the set of its eigenvalues. This result shows that $T$, $T^+$ can be regarded as smooth versions of $(S^+)^*$, $S^*$, respectively, and that the only nonsmooth elements in the latter subspaces come from $T_0^+ \setminus S^+$ and $T_0 \setminus S$, respectively. Although $S$, $S^+$ are operators their closures $T_0$, $T_0^+$ need not be; they are operators if and only if $\mathfrak{D}(T^+)$, $\mathfrak{D}(T)$ are dense in $\mathfrak{H}$, respectively.

5. <u>Boundary value problems</u>. We are now in a position to apply the results in [8] to describe the subspaces $A$, $A^+$ satisfying

$$T_0 \subset A \subset T_1 , \quad T_0^+ \subset A^+ \subset T_1^+ .$$

Let $\dim(T_1 \ominus T_0) = \dim(T_1^+ \ominus T_0^+) = t \leq 2mr$. Then a sample result is the following.

THEOREM. <u>Let</u> $A$ <u>be a subspace satisfying</u>

(i) $$T_0 \subset A \subset T_1 , \quad \dim(A/T_0) = d .$$

<u>Then</u>

(ii) $$T_0^+ \subset A^* \subset T_1^+ , \quad \dim(A^*/T_0^+) = t - d ,$$

<u>and there exist subspaces</u> $M_1$, $M_1^+$ <u>such that</u>

(iii) $$M_1 \subset T_1 \ominus T_0 , \quad M_1^+ \subset T_1^+ \ominus T_0^+ ,$$
$$\dim M_1 = d , \quad \dim M_1^+ = t - d ,$$
$$M_1^+ \subset M_1^* ,$$

<u>and</u>

(iv) $$A = T_0 \oplus M_1 , \quad A^* = T_0^+ \oplus M_1^+ ,$$
$$A = T_1 \cap (M_1^+)^* , \quad A^* = T_1^+ \cap M_1^* .$$

<u>Conversely, if</u> $M_1$, $M_1^+$ <u>satisfy</u> (iii) <u>then</u> $A = T_0 \oplus M_1$ <u>satisfies</u> (i), <u>and</u> (ii), (iv) <u>are valid</u>.

The descriptions of $A$, $A^*$ given via $A = T_1 \cap (M_1^+)^*$, $A^* = (T_1^+ \cap M_1^*)$ show how $A$, $A^*$ are obtained from $T_1$, $T_1^+$ by the imposition of generalized boundary conditions. For example, we have

$$A = T_1 \cap (M_1^+)^* = \{w \in T_1 \mid \langle w, m_1^+ \rangle = 0\}$$

where $m_1^+ = (m_1^+, \cdots, m_{1\,t-d}^+)$ is a $1 \times (t - d)$ matrix whose elements form a basis for $M_1^+$.

It is important to note that $A$, $A^*$ will contain nonsmooth elements in general, for $T_0$, $T_0^+$ contain such elements. This even occurs in cases when $\mathfrak{H}$ satisfies the more stringent assumption $(A_2')$. However, there exist smooth versions of $A$, $A^*$, for we can show that if

$$\tilde{A} = T \cap (M_1^+)^*, \quad \tilde{A}^+ = T^+ \cap M_1^*,$$

then $(\tilde{A})^c = A$, $(\tilde{A}^+)^c = A^*$. Now $\tilde{A}$, $\tilde{A}^+ \subset C^r(\iota) \times C^\nu(\iota)$ and are obtained by restrictions defined by elements in $C^r(\iota) \times C^r(\iota)$. In case $(A_2')$ holds the boundary conditions, in some cases, can be reduced to conditions of the usual type for $L$ in $L^2(\iota) \times L^2(\iota)$.

More general problems can be treated. Let $B$, $B^+$ be subspaces in $\mathfrak{H}^2$ such that

$$\dim B = p < \infty, \quad \dim B^+ = p^+ < \infty,$$

and consider

$$A_0 = T_0 \cap (B^+)^*, \quad A_0^+ = T_0^+ \cap B^*,$$

where

$$A_0^* = T_1^+ \dot{+} B^+, \quad (A_0^+)^* = T_1 \dot{+} B$$

are algebraic direct sums. If $A_1^+ = A_0^*$, $A_1 = (A_0^+)^*$, then we have $A_0 \subset A_1$, $A_0^+ \subset A_1^+$, and we can characterize those $A$, $A^*$ satisfying

$$A_0 \subset A \subset A_1, \quad A_0^+ \subset A^* \subset A_1^+,$$

via generalized boundary conditions; see [8]. The major problem remaining is to see what these conditions reduce to in significant special cases.

6. <u>The symmetric case</u>. The minimal linear manifold $S$ is symmetric $(S \subset S^*)$ if and only if $L = L^+$, and we now assume this. Then $S$ has selfadjoint extensions $H = H^*$ in $\mathfrak{H}^2$ if and only if

$$\dim \nu(T - \ell I) = \dim \nu(T - \bar{\ell} I), \quad \text{some } \ell \in \mathbb{C} \setminus \mathbb{R}.$$

More generally, if $A_0 = T_0 \cap B^*$, $\dim B = p < \infty$, $B \subset \mathfrak{H}^2$, where $A_0^* = T_1 \dot{+} B$ is a direct sum, then $A_0$ is symmetric and has selfadjoint extensions in $\mathfrak{H}^2$ if and only if $S$ does. Now $A_0$ always has selfadjoint extensions $H$ in a larger space $\mathfrak{K}^2 \supset \mathfrak{H}^2$, $\mathfrak{K}$ a Hilbert space. If $P$ is the orthogonal projection of $\mathfrak{K}$ onto $\mathfrak{H}$, then $R(\ell)$ defined by

$$R(\ell)f = P(H - \ell I)^{-1} f, \quad f \in \mathfrak{H}, \quad \ell \in \mathbb{C} \setminus \mathbb{R},$$

is called a generalized resolvent of $A_0$ associated with the extension $H$.

We have

$$\{R(\ell)f,\ \ell R(\ell)f + f\} \in A_0^* = T_1 \dotplus B\ , \qquad f \in \mathfrak{H}\ ,$$

and we can show that $R(\ell)$ is an integral operator on $\mathfrak{R}(G)$:

$$R(\ell)Gh(x) = \int_\iota K(x,y,\ell)h(y)\ dy\ , \qquad h \in L_0^2(\iota)\ .$$

In the case $Mf = f$ this fact has been used to obtain an eigenfunction expansion result and Titchmarsh-Kodaira formula for the extension H. The carrying over of this method to the present case seems to require a special choice of basis for the solutions of $(L - \ell M)f = 0$. A second method for obtaining the eigenfunction expansion result in the case $Mf = f$ was presented in [7], and A. Dijksma and H.S.V. de Snoo have carried out this program in the present case, but a regularity result is required to complete the argument. We hope that both of these programs will be completed soon.

REFERENCES

[1] C. Bennewitz, Spectral theory for pairs of differential operators, Ark. Mat. 15 (1977), 33-61.
[2] F. Brauer, Singular self-adjoint boundary value problems for the differential equation $Lx = \lambda Mx$, Trans. Amer. Math. Soc. 88 (1958), 331-345.
[3] F. Brauer, Spectral theory for the differential equation $Lu = \lambda Mu$, Canad. J. Math. 10 (1958), 413-428.
[4] F. Brauer, Spectral theory for linear systems of differential equations, Pacific J. Math. 10 (1960), 17-34.
[5] F. Browder, Eigenfunction expansions for non-symmetric partial differential operators, I, Amer. J. Math. 80 (1958), 365-381.
[6] F. Browder, Eigenfunction expansions for non-symmetric partial differential operators, II, Amer. J. Math. 81 (1959), 1-22.
[7] E.A. Coddington and A. Dijksma, Self-adjoint subspaces and eigenfunction expansions for ordinary differential subspaces, J. Differential Equations 20 (1976), 473-526.
[8] E.A. Coddington and A. Dijksma, Adjoint subspaces in Banach spaces, with applications to ordinary differential subspaces, to appear in Ann. Mat. Pura Appl.
[9] Å. Pleijel, Spectral theory for pairs of formally selfadjoint ordinary differential operators, J. Indian Math. Soc. 34 (1971), 259-268.

Author's address: Mathematics Department, University of California
Los Angeles, CA 90024 U.S.A.

# CONTROL AND THE VAN DER POL EQUATION

R. Conti, Firenze

1.

After playing a central rôle in the theory of nonlinear ordinary differential equations for over 50 years, Van der Pol's equation

$$(E_0) \qquad \ddot{x} + \mu(x^2 - 1)\dot{x} + x = 0$$

recently attracted the attention also from people working in control. This is accounted for by the fact that a system (electronic oscillator, living organism, or whatever else) governed by $(E_0)$ cannot be brought to rest or to periodicity in a finite time by any change in the initial state $x(0)$, $\dot{x}(0)$. For that purpose one has to replace $(E_0)$ by

$$(E_u) \qquad \ddot{x} + \mu(x^2 - 1)\dot{x} + x = u(t) ,$$

where $u : t \mapsto u(t)$ denotes some appropriate external force acting as a control.

In 1969 E. Ya.Roitenberg [6] , as an application of a general theorem, proved that any solution of $(E_u)$ can be brought to rest in a prescribed time $T$ , provided that arbitrarily large controls $u$ are allowed.

A more realistic approach, suggested already in Lee - Markus' book ([5] , p. 391) of 1967, was adopted by Eleanor M. James in her 1972 PhD Thesis (published in a condensed version in 1974, [4] ), where $u$ is assumed to be bounded by some given $k > 0$ and $T$ is not fixed in advance, which gives rise to the minimum time problem.

The same point of view was adopted in the 1976-77 Thesis of my pupil Gabriele Villari , [7] , and, still more recently (1977), by N. K. Alekseev, [1] .

It should also be noted that controllability and minimum time problems for an equation

$$\ddot{x} + g(x,\dot{x}) = u(t)$$

are studied in the books of Lee - Markus, [5], and Boltyanskii, [2], under assumptions on $g$ which are not satisfied by $g(x,\dot{x}) = \mu(x^2-1)\dot{x}$.

To be more specific about $(E_u)$ let us denote, as usual, by $L^{\infty}_{loc}(\mathbb{R})$ the class of measurable, locally essentially bounded functions $u : t \mapsto u(t)$, $t \in \mathbb{R}$, $u(t) \in \mathbb{R}$, and by $U_k$ the set of "admissible" controls

$$U_k = \{u \in L^{\infty}_{loc}(\mathbb{R}) : |u(t)| \leq k , \quad \text{a.e.} \quad t \in \mathbb{R}\}$$

for a given $k > 0$.

Also, denote by $U'_k$ the subset of "relay" controls, i.e., the set of $u \in U_k$ taking only the two values $-k$, $k$, with a finite number of switches from one value to the other on every bounded interval.

If we write $(E_u)$ in the equivalent form

$$(S_u) \qquad \begin{cases} \dot{x} = y \\ \dot{y} = -x + \mu y - \mu x^2 y + u(t), \end{cases}$$

we know that, for $u = 0$, $(S_o)$ has a limit cycle $\Gamma_\mu$.

We shall then consider :

Problem $P_1$. Find some $u \in U_k$ such that the corresponding solution of $(S_u)$ joins $\Gamma_\mu$ with the rest point $0 = (0,0)$ in minimum time ;

Problem $P_2$. Find some $u \in U_k$ such that the corresponding solution of $(S_u)$ goes from $0$ to $\Gamma_\mu$ in minimum time.

A change of $\mu$ into $-\mu$ transforms one problem into the other. However we shall keep them distinct since we shall constantly assume $\mu > 0$.

## 2.

We shall deal first with problem $P_1$.

With fixed $\mu > 0$, $k > 0$, let us denote by $V(\mu,k)$ the set of points in the $(x,y)$-plane which can be transferred to $0$ along the solutions of $(S_u)$ by using $u \in U_k$. Let $V'(\mu,k)$ be the subset of $V(\mu,k)$ corresponding to $U'_k$.

According to [1], [4], [7], it can be shown that

$$(2.1) \qquad V'(\mu,k) = V(\mu,k)$$

is an open connected (not necessarily convex) set, symmetric with respect to 0 , containing the circle $x^2 + y^2 \leq k^2/\mu^2$ .

As a consequence of the presence of the nonlinear term $\mu x^2 y$ in $(S_u)$ there are pairs $(\mu,k)$ for which $V'(\mu,k) = \mathbb{R}^2$ , so that we can consider the two sets

$$\mathcal{E} = \{(\mu,k) : V'(\mu,k) = \mathbb{R}^2\},$$

$$\mathcal{N} = \{(\mu,k) : V'(\mu,k) \neq \mathbb{R}^2\}.$$

When $(\mu,k) \in \mathcal{N}$ , $V'(\mu,k)$ is bounded by an arc of an orbit of $(S_k)$ (i.e., $(S_u)$ with $u = k$) lying in the half-plane $y < 0$ and by the symmetric arc of an orbit of $(S_{-k})$ in the half-plane $y > 0$. A comparison of the vector fields defined by $(S_k)$, $(S_o)$ and $(S_{-k})$ shows that

$$(\mu,k) \in \mathcal{N} \iff V'(\mu,k) \text{ interior to } \Gamma_\mu$$

so that problem $P_1$ can have solutions only if $(\mu,k) \in \mathcal{E}$ and nothing changes if we replace $U_k'$ by the larger set $U_k$ , because of (2.1). Therefore it is important to recognize whether a given $(\mu,k)$ belongs to $\mathcal{E}$ or to $\mathcal{N}$ .

Now, for every $\mu > 0$ there exists

$$(2.2) \qquad k^*(\mu) = \max\{k : (\mu,k) \in \mathcal{N}\}$$

so that

$$\mathcal{E} = \{(\mu,k) : 0 < \mu,\ k^*(\mu) < k\},$$

$$\mathcal{N} = \{(\mu,k) : 0 < \mu,\ k \leq k^*(\mu)\},$$

but, unfortunately, no explicit formula giving the value of $k^*(\mu)$ for each $\mu > 0$ is known.

All that is known (again according to [1] , [4] , [7] , with some improvements) about $k^*$ can be summarized as follows :

$$k^*(\mu) \leq \min\{\sqrt{3}\,\mu, 1\} , \quad 0 < \mu ,$$

$$\sqrt{1 - 2/\mu} < k^*(\mu) , \quad 2 \leq \mu ,$$

$$4 \frac{\mu}{\sqrt{\mu+1}} \frac{\mu_2}{(3\mu_1)^{3/2}} \leq k^*(\mu) , \quad 0 < \mu ,$$

where

$$\mu_1 = (\mu^2 + 4 + \mu\sqrt{\mu^2 + 4})/2 \quad , \quad \mu_2 = (\mu^2 + 4 - \mu\sqrt{\mu^2 + 4})/2 .$$

Consequently,

$$\lim_{\mu \to 0} k^*(\mu) = 0 , \quad \lim_{\mu \to +\infty} k^*(\mu) = 1 ,$$

$$\frac{2\sqrt{2}}{9}\sqrt{3} \leq \liminf_{\mu \to 0} \frac{k^*(\mu)}{\mu} \leq \limsup_{\mu \to 0} \frac{k^*(\mu)}{\mu} \leq \sqrt{3} .$$

If we define $k^*(0) = 0$ then $k^*$ turns out to be lipschitzian on $[0, \bar{\mu}]$ for every $\bar{\mu} > 0$ ( [1] ).

In the absence of an explicit representation of $k^*(\mu)$ it might be interesting to ascertain whether $k^*$ is an increasing function and how much regular it is : for instance, $k^* \in C^{(1)}(\mathbb{R}_+)$ ?

More information about $k^*$ probably could be obtained by studying the behavior of the limit cycles of the systems $(S_k)$, $(S_{-k})$. It can be shown that the singular point $K = (k,0)$ is a global attractor for $(S_k)$ if $k \geq 1$ , whereas for $0 \leq k < 1$ there is at least one limit cycle $\Gamma_{\mu,k}$ of $(S_k)$ around $K$ . To decide whether $\Gamma_{\mu,k}$ is unique some ad hoc proof has to be found since the usual techniques fail for $k > 0$ because of the lack of symmetry of the orbits of $(S_k)$ with respect to $K$ . Taking the uniqueness for granted, we have

$$k \leq k^*(\mu) \iff V'(\mu, k) \subset G(\mu, k) ,$$

where $G(\mu,k)$ is the intersection of the two regions interior to $\Gamma_{\mu,k}$ and its symmetric $\Gamma_{\mu,-k}$ with respect to $0$ .

3.

Introducing $r^2 = x^2 + y^2$ we have

$$r \dot{r} = \mu (1 - x^2) y^2 + y\, u(t) \leq \mu r^2 + k r$$

along the solutions of $(S_u)$ with $u \in U_k$ , hence

$$r(t) \leq (k/\mu + r(0)) e^{\mu t} - k/\mu ,$$

i.e., there exists a uniform bound for all solutions initiating at $(x^o,y^o)$ for a finite time duration. Consequently (Cf. E. B. Lee - L. Markus, [5] , Th.4 , p. 259) we can go from any point $(x^o,y^o) \in V(\mu,k)$ to 0 by means of $u \in U_k$ in a minimum time $T(x^o,y^o)$.

Further, the function $T : (x^o,y^o) \mapsto T(x^o,y^o)$ is lower semicontinuous on $V(\mu,k)$ so that, if $(\mu,k) \in \mathcal{C}$ ,it takes its minimum value $T_{\mu,k}$ on the compact set $\Gamma_\mu$ .

Therefore, <u>problem $P_1$ has solutions for every</u> $(\mu,k) \in \mathcal{C}$ .

To determine such solutions one can use the techniques derived from Pontryagin's maximum principle (Cf. Lee- Markus' book, Chapter 7). In fact, if $(x,y) : t \mapsto (x(t),y(t))$ is the solution of $(S_u)$ , $x(0)=x^o$, $y(0) = y^o$, corresponding to a minimizing control $u$ ,then there exists a solution $(\eta_1,\eta_2) : t \mapsto (\eta_1(t),\eta_2(t))$ of the linear system

$$\begin{cases} \dot{\eta}_1 = (1 + 2\mu\, x(t)\, y(t))\, \eta_2 \\ \dot{\eta}_2 = -\eta_1 + \mu\,(x^2(t) - 1)\eta_2 \end{cases}$$

such that $\eta_2(t)\, u(t) = \max_{|v| \leq k} \eta_2(t)\, v$ , so that

$$u(t) = k \text{ sign } \eta_2(t) .$$

Therefore, minimum time controls are of relay type.

The maximum number $\nu$ of switches depends on $(\mu,k)$ according to the map shown in Fig. 1.

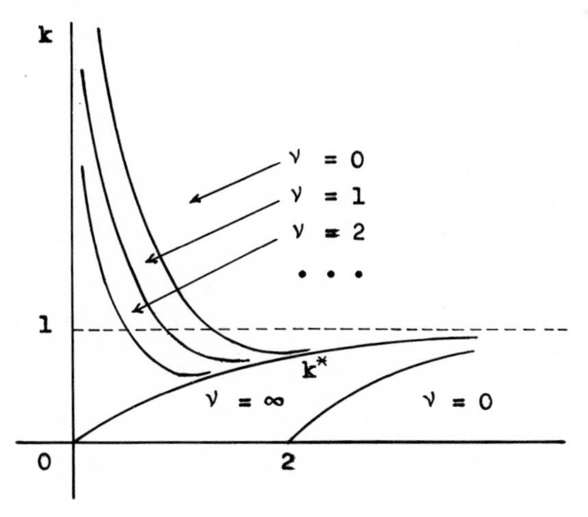

Such map is obtained by the construction of the switching locus by a combination of geometrical, comparison and computational methods (See [4] , [7] ).

Unfortunately, no explicit formulas are known to represent the "hyperbolas" in the $\mathcal{C}$ region.

4.

The next question is that of locating the points of $\Gamma_\mu$ at which the minimum $T_{\mu,k}$ is attained. This is a difficult question, because no analytical representation of $\Gamma_\mu$ is presently (1977) known, so we cannot expect to have exact solutions. On the other hand $\Gamma_\mu$ can be enclosed within an annulus whose inner and outer boundaries have simple enough analytical representations and may be made satisfactorily close to $\Gamma_\mu$ (Cf. R. Gomory - D. E. Richmond, [3] ). This, and the fact that also a good approximation of the switching locus can be obtained, suggest that substantial aid to the location of minimizing points can be expected from numerical methods.

The transversality condition is also of some help. In our case, such condition means that the vector of components $\eta_1(0)$, $\eta_2(0)$ is orthogonal to the tangent vector to $\Gamma_\mu$ at a minimizing point $M = (x,y)$, so that

$$\eta_1(0)\, y + \eta_2(0)\left[-x + \mu y - \mu x^2 y\right] = 0 . \tag{4.1}$$

Therefore the points $M$ are among the intersections of $\Gamma_\mu$ with the cubic (4.1). Since $\Gamma_\mu$ can be locally represented by an analytic function $x \mapsto y(x)$ or $y \mapsto x(y)$, like every other orbit of $(S_0)$, the number of intersections is finite. It is an open question whether there can be more than one pair of (symmetric) intersections.

5.

To deal with problem $P_2$ one has to replace $V(\mu,k)$ by the set $W(\mu,k)$ of $(x,y)$ points which can be attained from $0$ along the solutions of $(S_u)$ by using $u \in U_k$. Correspondingly, $V'(\mu,k)$ is replaced by $W'(\mu,k)$ and it can be shown that

$$W'(\mu,k) = W(\mu,k)$$

is an open connected set symmetric with respect to 0 . The effect of the term $\mu x^2 y$ in $(S_u)$ is that $W(\mu,k)$, unlike $V(\mu,k)$, is bounded for all pairs $\mu > 0$, $k > 0$ , whereas, in the absence of such term, the corresponding set $W(\mu,k)$ would be $= \mathbb{R}^2$. However,

$$\Gamma_\mu \subset W'(\mu,k) \ , \mu > 0, \ k > 0 \ ,$$

and by the Weierstrass - Baire theorem we see that problem $P_2$ has solutions for all pairs $\mu > 0$, $k > 0$ .

To determine the solutions offers the same difficulties as in the case of problem $P_1$.

The construction of the switching locus shows that, depending on $\mu$, k , either one can go from 0 to any point in $W'(\mu,k)$ in minimum time with one switch at most, or, for every positive integer N there are points in $W'(\mu,k)$ such that the corresponding number of switches is > N.

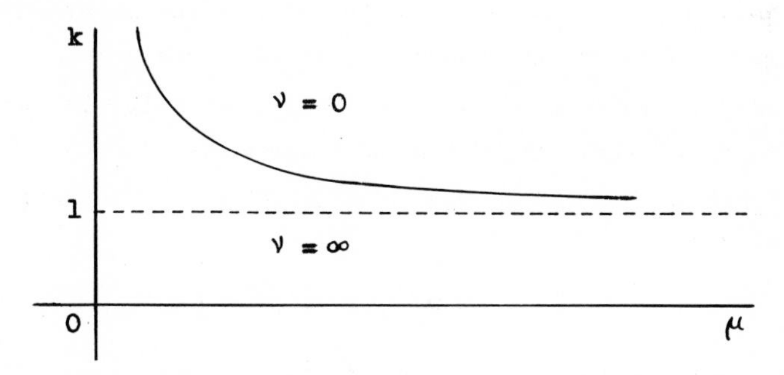


The map in Fig. 2 shows the dependence of the maximum number $\nu$ of switches on $\mu$, k .

Again, no explicit representation of the "hyperbola" separating the two zones is known.

The number of minimizing points, in pairs, is still finite, but uniqueness and their location on $\Gamma_\mu$ are open questions.

## References.


[1] N. K. Alekseev, Some questions of controllability for two-dimensional systems (Russian), Diff. Uravnenyia, 13 (1977), 387-397 ;

[2] V. G. Boltyanskii, Mathematical methods of optimal control, Holt-Rinehart-Winston, 1971 ;

[3] R. Gomory - D. E. Richmond, Boundaries for the limit cycle of Van der Pol's equation, Quart. Appl. Math., 9 (1951), 205 - 209 ;

[4] E. M. James, Time optimal control and the Van der Pol oscillator, J. Inst. Maths. Appls., 13 (1974), 67-81 ;

[5] E. B. Lee - L. Markus, Foundations of optimal control theory, J. Wiley and Sons, 1967 ;

[6] E. Ya. Roitenberg, On a sufficient condition of controllability for nonlinear systems (Russian), Vestnik Moskov. Univ., 1 (1969), 28-33 ;

[7] Gabriele Villari, Controllo del ciclo di Van der Pol, Tesi di Laurea in Matematica, Univ. di Firenze, 1976-77.



Author's address : Istituto Matematico U. Dini, Viale Morgagni 67/A
50134 Firenze, Italy.

# ON PROPERTIES OF SPECTRAL APPROXIMATIONS

J. Descloux, N. Nassif, J. Rappaz, Lausanne

In this paper, we want to discuss connections between some conditions used in the theory of spectral approximation. For the sake of simplicity we shall restrict ourselves to the following framework: $X$ is a complex Banach space with norm $\|\cdot\|$; $X_n$, $n \in \mathbb{N}$, is a sequence of finite dimensional subspaces of $X$; $\Pi_n: X \to X$ are linear projectors with range $X_n$ which converge strongly to the identity; $A: X \to X$ is a linear bounded operator; the linear operators $B_n: X \to X$, uniformely bounded, with range in $X_n$, are supposed to approximate $A$; $A_n: X_n \to X_n$ is then defined as the restriction of $B_n$ to $X_n$ (or, given the $A_n$'s, one can, for example, define $B_n = A_n\Pi_n$); $B_n$ will be called the "Galerkin approximation of A" if $B_n = \Pi_n A$. Remark that $B_n$ is compact and has the same eigenvalues and eigensubspaces as $A_n$ (with the exception of o).

We shall use the following notations. If $Y$ and $Z$ are closed subspaces of $X$, then, for $x \in X$, $\delta(x,Y) = \inf_{y\in Y}\|x-y\|$, $\delta(Y,Z) = \sup_{y\in Y,\|y\|=1} \delta(y,Z)$, $\hat{\delta}(Y,Z) = \max(\delta(Y,Z),\delta(Z,Y))$.
For a linear operator $C$ defined on $X$ or $X_n$, with range in $X$, we set $\|C\|_n = \sup_{x\in X_n,\|x\|=1} \|Cx\|$.

Let us introduce some properties of approximations of $A$ by $A_n$ or $B_n$:
U) $\lim_{n\to\infty}\|A-B_n\| = o$; A1) $\lim_{n\to\infty} B_n = A$ strongly; A2) $\{B_n x \mid \|x\| \leq 1,\ n \in \mathbb{N}\}$ is relatively compact; Z) $\lim_{n\to\infty}\|A-A_n\|_n = o$; R) $\lim_{n\to\infty} \sup_{x\in X_n,\|x\|=1} \delta(Ax,X_n) = o$; V1) $x_n \in X_n$, $\lim_{n\to\infty} x_n = x$ $\Rightarrow \lim_{n\to\infty} A_n x_n = Ax$; V2) for any bounded sequence $x_n \in X_n$, $\{(A-A_n)x_n\}$ is relatively compact ; G) for any $\lambda \in \rho(A)$, for any subsequence $\{x_\alpha\}$ of any bounded sequence $x_n \in X_n$ such that $(A_\alpha-\lambda)x_\alpha$ converges, there exists a converging subsequence $\{x_\beta\}$ of $\{x_\alpha\}$ such that $A(\lim_\beta x_\beta) = \lim_\beta A_\beta x_\beta$.

A2 means that $\{B_n\}$ is collectively compact in the sense of Anselone [1]; Z and R has been studied by the authors in [2]; R means that $X_n$ is "almost" an invariant subspace of $A$; V1 and V2 imply that $A_n$ is a compact approximation in the sense of Vainikko [8]; G is used, in a more general context, by Grigorieff and others in particular in [4],[5]. Since $B_n$ is compact, note that U or {A1,A2} implies that $A$ is compact.

In the following $\sigma(A)$, $\rho(A)$, $\sigma(A_n)$, $\rho(A_n)$, $\sigma(B_n)$, $\rho(B_n)$ will denote the spectrum and the resolvant sets of $A$, $A_n$ and $B_n$. $R_z(A) = (A-Z)^{-1}: X \to X$ and $R_z(A_n) = (A_n-z)^{-1}: X_n \to X_n$ are the resolvent operators of $A$ and $A_n$ defined respectively for $z \in \rho(A)$ and $z \in \rho(A_n)$.

Let $\Gamma \subset \rho(A)$ be a Jordan curve; we set $P = -(2\pi i)^{-1} \int_\Gamma R_z(A)dz$ and, if $\Gamma\ \rho(A_n)$, $P_n = -(2\pi i)^{-1} \int_\Gamma R_z(A_n)dz: X_n \to X_n$. $P$ and $P_n$ are the spectral projectors and $E = P(X)$, $E_n = P_n(X_n)$ are the invariant subspaces of $A$ and $A_n$ relative to $\Gamma$.

Consider now some spectral properties: S1) for any $z \in \rho(A), \exists N_z \in \mathbb{N}$ and $M_z$ such that $\| R_z(A_n)\|_n \leq M_z$, $n > N_z$; S2) $\forall x \in E$, $\lim_{n\to\infty} \delta(x,E_n) = 0$; S3) $\lim_{n\to\infty} \delta(E_n,E) = 0$; S4) if $E$ is finite dimensional, then $\lim_{n\to\infty} \hat{\delta}(E_n,E) = 0$. If $X$ is a Hilbert space and if $A$ and $A_n$ are selfadjoint, for an interval $I \subset \mathbb{R}$, define $E_I$ as the invariant subspace of $A$ relative to $I$ and $E_{In} \subset X_n$ as the invariant subspace of $A_n$ relative to $I$; we then introduce the property SH): for all intervals $I$ and $J$, the closure of $I$ being a subset of the interior of $J$, one has $\lim_{n\to\infty} \delta(E_{In},E_J) = 0$.

S1, which is a property of stability, implies the upper semi-continuity of the spectrum and garantees the meaningfullness of the approximated spectrum $\sigma(A_n)$. S2 has little importance for application; however S3 garantees the meaningfullness of all the elements of the approximate invariant subspace $E_n$. If $\Gamma$ contains only an eigenvalue $\lambda \in \sigma(A)$ of algebraic finite multiplicity, S1 and S4 imply that $\lambda$ is stable in the sense of Kato ([6],p.437). For the selfadjoint case, SH is a refinement of S3.

*Proposition 1:* a) $U \Rightarrow \{A1, A2, Z, R, V1, V2, G, S1, S2, S3, S4\}$; b) $\{A1, A2\} \Rightarrow \{R, V1, V2, G, S1, S2, S4\}$; $\{A1, A2\} \not\Rightarrow S3$; if $A$ and $B_n$ are selfadjoint $\{A1, A2\} \Rightarrow U$; c) $Z \Rightarrow \{R, V1, V2, G, S1, S2, S3, S4\}$; for the selfadjoint case, $Z \Leftrightarrow SH \Leftrightarrow \{V1, V2\}$; d) if $A_n$ is the Galerkin approximation of $A$, $R \Leftrightarrow Z \Leftrightarrow V2$; e) $\{V1, V2\} \Rightarrow \{G, S1, S2, S4\}$, $V2 \Rightarrow R$; $\{V1, V2\} \not\Rightarrow S3$; f) $G \Leftrightarrow \{V1, S1\}$; $G \not\Rightarrow S2$; $G \not\Rightarrow R$, $G \not\Rightarrow S3$; $G \not\Rightarrow S4$.

Most statements of Proposition 1 can be obtained directly or with little work from known results in the litterature; for b), see Anselone [1]; for c), d), see Descloux, Nassif, Rappaz [2],[3]; for e), see Vainikko [8]; for f), see, for example, Grigorieff[4], Jeggle [5]. However let us verify in e) that $V2 \Rightarrow R$: suppose $R$ false; $\exists\, \varepsilon > 0$, the sequence $x_n \in X_n$, $n \in \mathbb{N}$, $\|x_n\| = 1$ and a subsequence $\{x_\alpha\}$ of $\{x_n\}$ such that $\delta(Ax_\alpha,X_\alpha) \geq \varepsilon$; V2 implies the existence of $y \in X$ and of a subsequence $\{x_\beta\}$ of $\{x_\alpha\}$ such that $\lim_\beta (A-A_\beta)x_\beta = y$; setting $Z_\beta = A_\beta x_\beta + \Pi_\beta y \in X_\beta$, one has $\lim_{\beta\to\infty}(Ax_\beta - Z_\beta) = 0$, which is a contradiction. We verify in c) that $\{V1, V2\} \Rightarrow Z$ in the selfadjoint case: suppose $Z$ false; there exist $\varepsilon > 0$, the sequence $x_n \in X_n$, $n \in \mathbb{N}$, $\|x_n\| = 1$ and a subsequence $\{x_\alpha\}$ of $\{x_n\}$ such that $\|(A-A_\alpha)x_\alpha\| \geq \varepsilon$; V2 implies the existence of $y \in X$ and of a subsequence $\{x_\beta\}$ of $\{x_\alpha\}$ such that $\lim_\beta (A-A_\beta)x_\beta = y$; denoting by $(,)$ the scalar product in $X$, one has by V1: $\varepsilon^2 \leq \|y\|^2 = \lim_\beta((A-A_\beta)x_\beta,\Pi_\beta y) = \lim_\beta(x_\beta,(A-A_\beta)\Pi_\beta y) = 0$;

contradiction. Note that the last property we have verified is in fact a particular case of the following result: let $X^*$, $X_n^*$, $A^*$, $A_n^*$, $\Pi_b^*$ be the adjoint spaces of X, $X_n$ and the adjoint operators of A, $A_n$, $\Pi_n$; $X_n^*$ is identified as a subspace of $X^*$ by the map $\varphi_n \in X_n^* \to \varphi \in X^*$ with $\varphi(x) = \varphi_n(\Pi_n x)$ $\forall x \in X$; then the three properties V2, $\Pi_n^*$ converges strongly to the idendity in $X^*$, for all converging sequences $x_n \in X_n^*$ one has $\lim_{n\to\infty} A_n^* x_n^* = A^*(\lim_{n\to\infty} x_n^*)$, imply Z.

We also prove the negative statements of Proposition 1 by examples. Let $X = \ell^2$ with scalar product (,) and canonical basis $e_1, e_2, ..$; note $Y_n = \mathrm{span}(e_1, e_2, .., e_n)$; $\Pi_n$ will be the orthogonal projector on $Y_n$. We show that {A1, A2} $\not\Rightarrow$ S3 (and consequently {V1, V2} $\not\Rightarrow$ S3, G $\not\Rightarrow$ S3); set $X_n = Y_n$; the operators $Ax = (x,e_1)e_1$ and $B_n x = (x, e_1 + e_n)e_1$ verify {A1, A2}; but $e_1 - e_n$ is an eigenvector of $A_n \equiv B_n$ (restricted to $X_n$) for the eigenvalue o. The following example will show that even in the Galerkin selfadjoint case, G $\not\Rightarrow$ R and G $\not\Rightarrow$ S4; set $X_n = Y_{2n}$, $Ax = \sum_{n=1}^{\infty} (x, e_{2n})e_{2n+1} + (x, e_{2n+1})e_{2n}$, $A_n = \Pi_n A$ (restricted to $X_n$); clearly property R is not verified; furthermore $\sigma(A) = \{-1, o, 1\}$ where o is an eigenvalue of multiplicity 1 of A, $\sigma(A_n) = \sigma(A_n)$ $(n \geq 2)$ where o is an eigenvalue of multiplicity 2 of $A_n$ so that S4 is not verified; since $A_n$ is selfadjoint $\| R_z(A_n) \| = 1/(\text{distance } (z, \sigma(A_n))$, S1 is verified; since $A_n$ is a Galerkin approximation, V1 is satisfied and by proposition 1f, one has also G. (An example of a differential operator illustrating the same situation is contained in Rappaz [7] p. 71).

*Remarks*: Condition Z appears as a generalization of U, whereas {V1, V2} is generalization of {A1, A2}. G is essentially equivalent to the stability conditions S1. For practical applications, {A1, A2} has been used in connection with integral operators (see Anselone [1]), {V1, V2} and G have been used in connection with finite difference methods for compact operators (see Vainikko [9], Grigorieff [4]; condition Z has been verified in connection with Galerkin finite element methods for non compact operators of plasma physics (see Descloux, Nassif, Rappaz [2]).

Proposition 1 does not exhaust the list of relations between the different properties we have introduced. We mention another one.

*Proposition 2*: Let X be a Hilbert space, $\Pi_n$ be the orthogonal projector from X onto $X_n$, A be compact. $A_n$ is given and we set $B_n = A_n \Pi_n$; then Z $\Rightarrow$ U.

*Proof*: From the realation $A - B_n = (A - A_n)\Pi_n + A(I - \Pi_n)$, one has $\|A - B_n\| \leq \| A - A_n\|_n + \|A(I - \Pi_n)\|$;

by Z, $\lim_{n\to\infty} \| A - A_n \|_n = o$, since A and consequently its adjoint $A^*$ are compact, since $\lim_{n\to\infty} \Pi_n = I$ strongly, one has $\lim_{n\to\infty} \| A(I - \Pi_n) \| = \lim_{n\to\infty} \| (I - \Pi_n)A^* \| = o$.

Finally, we show for the typical situation of integral operators with continuous kernel that the properties {A1, A2} can be "transformed" in uniform convergence. To be specific, let $K: [o,1] \times [o,1] \to \mathbb{C}$ be a continuous kernel, X be either $C^o[o,1]$ or $L^2(o,1)$, $A: X \to X$ be the integral operator defined by $(Ax)(t) = \int_o^1 K(t,\tau)x(\tau)d\tau$. Let for $n \in \mathbb{N}$, $h = 1/n$, $t_i = ih$; for $X = C^o[o,1]$, we approximate A by the trapezoïdal rule and define $B_n: X \to X$ by $(B_n x)(t) = \sum_{j=1}^{n} \frac{h}{2}\{K(t,t_{j-1})x(t_{j-1}) + K(t,t_j)x(t_j)\}$. A and $B_n$ then satisfy properties $\{A_1, A_2\}$ (see Anselone [1]).

*Proposition 3:* For the above situation, there exists the operator $C_n: X \to X$, where $X = L^2(o,1)$ such that $\sigma(C_n) = \sigma(B_n)$ and $\lim_{n\to\infty} \| A - C_n \| = o$.

*Proof:* By proposition 2, it suffices to construct a subspace $X_n \subset L^2(o,1)$ and an operator $A_n: X_n \to X_n$ such that $\sigma(A_n) \cup \{o\} = \sigma(B_n)$ and $\lim_{n\to\infty} \| A - A_n \|_n = o$. Choose $X_n$ as the set of continuous piecewise linear function relative to the mesh $\{t_i\}$; for $x \in X_n$, $A_n x$ is then defined as the interpolant of $B_n x$ in $X_n$; using the uniform continuity of K, one obtains easily that $\lim_{n\to\infty} \| A - A_n \|_n = o$. (For more details see Descloux, Nassif, Rappaz [3]).

*Remark:* Proposition 3 is still valid when $B_n$ is obtained by other classical integration formulae, for example Newton cotes or Gauss-Legendre; one has only to define convenient subspaces $X_n$.

REFERENCES:

[1] P.M. Anselone. Collectively compact approximation theory. Prentice Hall, Englewood Cliffs, N.J. (1971).

[2] J. Descloux, N. Nassif, J. Rappaz.On spectral approximation; part 1: the problem of convergence. To appear in RAIRO.

[3] J. Descloux, N. Nassif, J. Rappaz. Various results on spectral approximation. Rapport, Dept. Math. EPFL 1977.

[4] R.D. Grigorieff. Diskrete Approximation von Eigenwertproblemen. Numerische Mathematik;part I: 24, 355-374 (1975); part II: 24, 415-433 (1975); part III: 25, 79-97 (1975).

[5] H. Jeggle. Über die Approximation von linearen Gleichungen zweiter Art und Eigenwertprobleme in Banach-Räumen. Math.Z. 124, 319-342 (1972).

[6] T. Kato. Perturbation theory of linear operators. Springer-Verlag 1966.

[7] J. Rappaz. Approximation par la méthode des éléments finis du spectre d'un opérateur non compact donné par la stabilité magnétohydrodynamique d'un plasma. Thèse EPF-Lausanne, 1976.

[8] G.M. Vainikko. The compact approximation principle in the theory of approximation methods. U.S.S.R. Computational Mathematics and Mathematical Physics, Vol. 9, Number 4, 1-32 (1969).

[9] G.M. Vainikko. A difference method for ordinary differential equations. U.S.S.R. Computational Mathematics and Mathematical Physics, Vol. 9, Number 5, 1969.

Authors' address:

Dept. Math. EPFL, Av. de Cour 61, CH 1007 Lausanne, Switzerland.

# SINGULAR PERTURBATIONS AND LINEAR FEEDBACK CONTROL

V.Dragan and A.Halanay, Bucharest

## 1. Introduction

The results reported here center around the classical regulator problem: a linear control system with a quadratic cost function. We shall consider two situations; the first one corresponds to the case of fast variables in the control system, the second one to "cheap control". A crucial point in solving these problems is the behaviour of the optimal cost and to study it one has to consider singularly perturbed matrix Riccati differential equations. References for these problems are [1] , [2] , [3] .

## 2. The control problems and the associated Riccati differential equations

A. Let the control system be

$$\varepsilon\dot{x} = A(t)x + B(t)u, \quad x(t_o) = x_o$$

$$J(u) = x^*(T)Gx(T) + \int_{t_o}^{T}\left[x^*(t)F(t)x(t)+u^*(t)H(t)u(t)\right]dt$$

$$G \geq 0, \; F(t) \geq 0, \; H(t) > 0 \; .$$

The matrix Riccati equation giving the optimal cost is

$$\dot{P} = -\frac{1}{\varepsilon}A^*(t)P - \frac{1}{\varepsilon}PA(t) + \frac{1}{\varepsilon^2}PB(t)H^{-1}(t)B^*(t)P - F(t)$$

$$P(T, \varepsilon) = G$$

and for $P(t, \varepsilon) = \varepsilon R(t, \varepsilon)$ we get

(1) $$\varepsilon\dot{R} = -A^*(t)R - RA(t) + RM(t)R - F(t), \quad R(T, \varepsilon) = \frac{1}{\varepsilon}G \; .$$

The problem is to study the behaviour of

$$R(t, \varepsilon) \quad \text{as} \quad \varepsilon \to 0 \; .$$

B. The "cheap control" problem is defined by

$$\dot{x} = A(t)x + B_o(t)u_o + B_1(t)u_1 \; , \quad x(t_o) = x_o$$

$$J(u) = x^*(T)Gx(T) +$$

$$+ \int_{t_o}^{T}\left[x^*(t)F(t)x(t) + u_o^*(t)H_o(t)u_o(t) + \right.$$

$$\left. + \; \varepsilon^2 u_1^*(t)H_1(t)u_1(t)\right]dt \; .$$

The associated Riccati equation is

$$\varepsilon^2\dot{P} = - \varepsilon^2\left[A^*(t)P + PA(t) - PM_o(t)P + F(t)\right] +$$

$$+ PB_1(t)H_1^{-1}(t)B_1(t)P \; ,$$

$$P(T, \varepsilon) = G \; .$$

The problem is again to study the behaviour of $P(t,\varepsilon)$ as $\varepsilon \to 0$.

3. The singularly perturbed Riccati equations

A. Let $M(t) = D^*(t)D(t)$, $F(t) = C^*(t)C(t)$, $(A^*(t),C(t))$ completely controllable, $(A(t),D^*(t))$ stabilizable, $A,C,D$, Lipschitz. Let $\hat{R}(t)$ be the unique positive definite solution of the equation

$$A^*(t)R + RA(t) - RM(t)R + F(t) = 0 .$$

Let $R(t,\varepsilon)$ be the solution of the Cauchy problem (1). Then $\lim_{\varepsilon \to 0} R(t,\varepsilon) = \hat{R}(t)$ for $t < T$.

To prove this result we first consider the solution $\tilde{R}(t,\varepsilon)$ of the equation in (1) with $\tilde{R}(T,\varepsilon) = \hat{R}(T)$ and the solution $R_0(t,\varepsilon)$ of the same equation with $R_0(T,\varepsilon) = 0$.

Denote by $\hat{C}(s,t,\varepsilon)$ the fundamental matrix solution of the system $\varepsilon x' = \hat{A}(s)x$ where $\hat{A}(t) = A(t) - M(t)R(t)$ is Hurwitz for every $t$.

We use the representation formulae

$$\begin{aligned}\tilde{R}(t,\varepsilon) &= \hat{C}^*(T,t,\varepsilon)\hat{R}(T)\hat{C}(T,t,\varepsilon) + \\ &+ \frac{1}{\varepsilon}\int_t^T \hat{C}^*(s,t,\varepsilon)\left[F(s) + \hat{R}(s)M(s)\hat{R}(s)\right]\hat{C}(s,t,\varepsilon)ds - \\ &- \frac{1}{\varepsilon}\int_t^T \hat{C}^*(s,t,\varepsilon)\left[\tilde{R}(s,\varepsilon) - \hat{R}(s)\right]M(s) . \\ &\cdot\left[\tilde{R}(s,\varepsilon) - \hat{R}(s)\right]\hat{C}(s,t,\varepsilon)ds ,\end{aligned}$$

$$\begin{aligned}\hat{R}(t) &= \exp\left(\hat{A}^*(t)\frac{T-t}{\varepsilon}\right)\hat{R}(t)\exp\left(\hat{A}(t)\frac{T-t}{\varepsilon}\right) + \\ &+ \frac{1}{\varepsilon}\int_t^T \exp\left(\hat{A}^*(t)\frac{s-t}{\varepsilon}\right)\left[F(t) + \hat{R}(t)M(t)\hat{R}(t)\right] . \\ &\cdot\exp\left(\hat{A}(t)\frac{s-t}{\varepsilon}\right)ds\end{aligned}$$

and

$$\hat{C}(s,t,\varepsilon) = \exp\left(\hat{A}(t)\frac{s-t}{\varepsilon}\right) + \omega(s,t,\varepsilon) ,$$

$$|\omega(s,t,\varepsilon)| \le \frac{\varepsilon \ell k^2}{2}\left(\frac{s-t}{\varepsilon}\right)^2 \exp\left(-\alpha\left(\frac{s-t}{\varepsilon}\right)\right)$$

to obtain $|\tilde{R}(t,\varepsilon) - \hat{R}(t)| \le k\varepsilon$.

Denote next $S(t,\varepsilon) = R_0(t,\varepsilon) - \tilde{R}(t,\varepsilon)$, and let $\tilde{C}(t,t_0,\varepsilon)$ be the fundamental matrix solution of the system $\varepsilon\dot{x} = \tilde{A}(t,\varepsilon)x$, where $\tilde{A}(t,\varepsilon) = A(t) - M(t)\tilde{R}(t,\varepsilon)$.

We obtain the representation formula

$$S(t,\varepsilon) = -\widetilde{C}^*(T,t,\varepsilon)\Big[\hat{R}^{-1}(T) - \frac{1}{\varepsilon}\int_t^T \widetilde{C}(T,s,\varepsilon)M(s)\cdot$$
$$\cdot\widetilde{C}^*(T,s,\varepsilon)ds\Big]^{-1}\widetilde{C}(T,t,\varepsilon)$$

and a lemma in singular perturbations [1] gives

$$\lim_{\varepsilon\to 0} \frac{1}{\varepsilon}\int_t^T \widetilde{C}(T,s,\varepsilon)M(s)\widetilde{C}^*(T,s,\varepsilon)ds = \hat{L}$$

where $\hat{A}(T)\hat{L} + \hat{L}\hat{A}^*(T) = -M(T)$ .

The representation formula gives then the estimate

$$|S(t,\varepsilon)| \le k' \exp\left(-\alpha'\left(\frac{T-t}{\varepsilon}\right)\right) .$$

To perform the last step denote $S_0(t,\varepsilon) = R(t,\varepsilon) - R_0(t,\varepsilon)$ and let $C_0(s,t,\varepsilon)$ be the fundamental matrix solution of the system $\varepsilon x'(s) = A_0(s,\varepsilon)x(s)$ with $A_0(t,\varepsilon) = A(t) - M(t)R_0(t,\varepsilon)$ .

Then a representation formula for $S_0$ gives

$$S_0(t,\varepsilon) \le C_0^*(T,t,\varepsilon)\,\frac{1}{\varepsilon}GC_0(T,t,\varepsilon)$$

and since $|C_0(s,t,\varepsilon)| \le k\exp\left(-\alpha\left(\frac{s-t}{\varepsilon}\right)\right)$ the result is proved.

B. For the "cheap control" problem assume

$$\widetilde{F}(t) = B_1^*(t)F(t)B_1(t) > 0, \quad \widetilde{G} = B_1^*(T)GB_1(T) > 0 .$$

Define

$$Q(t,\varepsilon) = \frac{1}{\varepsilon}B_1^*(t)P(t,\varepsilon), \quad R(t,\varepsilon) = \frac{1}{\varepsilon}B_1^*(t)P(t,\varepsilon)B_1(t),$$
$$\hat{F}(t) = B_1^*(t)F(t), \quad \hat{G} = B_1^*(T)G .$$

Then $P(t,\varepsilon)$, $Q(t,\varepsilon)$, $R(t,\varepsilon)$ is a solution of the problem

$$\dot{P} = -A^*(t)P - PA(t) + PM_0(t)P + QH_1^{-1}(t)Q - F(t),$$
$$\varepsilon\dot{Q} = RH_1^{-1}(t)Q - \varepsilon QA(t) + \varepsilon QM_0(t)P - B_2^*(t)P - \hat{F}(t),$$
$$\varepsilon\dot{R} = RH_1^{-1}(t)R - \widetilde{F}(t) - \varepsilon B_2^*(t)Q - \varepsilon QB_2(t) +$$
$$+ \varepsilon^2 QM_0(t)Q,$$
$$P(T,\varepsilon) = G, \quad Q(T,\varepsilon) = \frac{1}{\varepsilon}\hat{G}, \quad R(T,\varepsilon) = \frac{1}{\varepsilon}\widetilde{G} .$$

We start with the equation

$$\varepsilon\dot{R} = RH_1^{-1}(t)R - \widetilde{F}(t) .$$

Denote $\widetilde{R}(t)$ the unique positive definite stabilizing solution of the algebraic equation $RH_1^{-1}(t)R = \widetilde{F}(t)$ .

Denote $R_0(t,\varepsilon)$ the solution of the differential equation with $R_0(T,\varepsilon) = 0$ and $\widetilde{R}_0(t,\varepsilon)$ the solution of the same equation with $\widetilde{R}_0(T,\varepsilon) = \frac{1}{\varepsilon}\widetilde{G}$ .

We have the representation

$$\widetilde{R}_0(t,\varepsilon) - R_0(t,\varepsilon) = C_0^*(T,t,\varepsilon)U^{-1}(t,\varepsilon)C_0(T,t,\varepsilon) ,$$
$$U(t,\varepsilon) = \varepsilon\widetilde{G}^{-1} + \frac{1}{\varepsilon}\int_t^T C_0(T,\tau,\varepsilon)H_1^{-1}(\tau)C_0^*(T,\tau,\varepsilon)d\tau$$

where $C_o$ is defined by

$$\varepsilon \frac{dC}{ds} = -H_1^{-1}(s)R_o(s,\varepsilon)C, \quad C_o(t,t,\varepsilon) = E .$$

A crucial point in the proof is the estimate

$$|U^{-1}(t,\varepsilon)| \le \frac{2\beta}{2\beta\rho\varepsilon + 1 - \exp\left(-2\alpha\left(\frac{T-t}{\varepsilon}\right)\right)} .$$

Define $\widetilde{Q}_o$ as the solution of the Cauchy problem

$$\varepsilon\dot{Q} = \widetilde{R}_o(t,\varepsilon)H_1^{-1}(t)Q - \widehat{F}_1(t) , \quad \widetilde{Q}_o(T,\varepsilon) = \frac{1}{\varepsilon}\widetilde{G} .$$

We prove that

$$\lim_{\varepsilon\to 0} \widetilde{Q}_o(t,\varepsilon) = \widehat{Q}_o(t) \quad \text{for} \quad t < T$$

where $\widehat{Q}_o(t) = H_1(t)\widetilde{R}^{-1}(t)\widehat{F}_1(t)$ .

Consider now $\widetilde{P}$ defined by

$$\varepsilon\frac{d}{dt}\widetilde{P}(t,\varepsilon) = \widetilde{Q}_o^*(t,\varepsilon)H_1^{-1}(t)\widetilde{Q}_o(t,\varepsilon) - \widehat{Q}_o^*(t)H_1^{-1}(t)\widehat{Q}_o(t).$$

A long series of estimates give finally

$$|\widetilde{P}(t,\varepsilon)| \le \mu + \frac{\nu}{2\beta\rho\varepsilon + 1 - \exp\left(-2\alpha\left(\frac{T-t}{\varepsilon}\right)\right)} .$$

Let us state now the final result.

Define $P_o$ from the Cauchy problem

$$\begin{aligned}\dot{P}_o = &-[A^*(t) - \widehat{F}^*(t)\widetilde{F}^{-1}(t)B_2^*(t)]P_o - P_o[A(t) - \\ &- B_2(t)\widetilde{F}^{-1}(t)\widehat{F}_1(t) + P_o[M_o(t) + B_2(t)\widetilde{F}^{-1}(t)B_2^*(t)]P_o - \\ &- [F(t) - \widehat{F}^*(t)\widetilde{F}^{-1}(t)\widehat{F}(t)] ,\end{aligned}$$

$$P_o(T) = G - \widehat{G}^*\widetilde{G}^{-1}\widehat{G} .$$

Let $\widetilde{Q}_o$ and $\widetilde{P}$ be defined as above with $\widehat{F}_1$ replaced by $B_2^*(t)P_o(t) + \widehat{F}(t)$ .

Then

$$\begin{aligned}P(t,\varepsilon) &= P_o(t) + \varepsilon\widetilde{P}(t,\varepsilon) + \varepsilon P_1(t,\varepsilon)\\ Q(t,\varepsilon) &= \widetilde{Q}_o(t,\varepsilon) + \sqrt{\varepsilon}Q_1(t,\varepsilon)\\ R(t,\varepsilon) &= \widetilde{R}_o(t,\varepsilon) + \varepsilon R_1(t,\varepsilon)\end{aligned}$$

where $P_1$, $Q_1$, $R_1$ are bounded for $t \le T$ .

To prove this result we denote

$$\mathcal{P} = \begin{pmatrix} P_1 & Q_1^* \\ Q_1 & R_1 \end{pmatrix}$$

and show that $\mathcal{P}$ is the solution of the problem

$$\varepsilon\dot{\mathcal{P}} = -\mathcal{A}^*(t,\varepsilon)\mathcal{P} - \mathcal{P}\mathcal{A}(t,\varepsilon) + \mathcal{P}\mathcal{M}(t,\varepsilon)\mathcal{P} - \mathcal{F}(t,\varepsilon),$$

$$\mathcal{P}(T,\varepsilon) = 0$$

where

$$\mathcal{A} = \begin{pmatrix} \mathcal{A}_{11} & \mathcal{A}_{12} \\ \mathcal{A}_{21} & \mathcal{A}_{22} \end{pmatrix}, \ \mathcal{M} = \begin{pmatrix} \mathcal{M}_{11} & 0 \\ 0 & \mathcal{M}_{22} \end{pmatrix}, \ \mathcal{F} = \begin{pmatrix} \mathcal{F}_{11} & \mathcal{F}_{12} \\ \mathcal{F}_{12}^* & \mathcal{F}_{22} \end{pmatrix},$$

$$\mathcal{A}_{11}(t,\varepsilon) = \varepsilon[A(t) - M_0(t)(P_0(t) + \varepsilon\widetilde{P}(t,\varepsilon))],$$
$$\mathcal{A}_{12}(t,\varepsilon) = \sqrt{\varepsilon}[B_2^*(t) - M_0(t)\widetilde{Q}_0^*(t,\varepsilon)],$$
$$\mathcal{A}_{21}(t,\varepsilon) = \sqrt{\varepsilon}H_1^{-1}(t)\widetilde{Q}_0(t,\varepsilon),$$
$$\mathcal{A}_{22}(t,\varepsilon) = -H_1^{-1}(t)\widetilde{R}_0(t,\varepsilon),$$
$$\mathcal{M}_{11}(t,\varepsilon) = \varepsilon^2 M_0(t), \qquad \mathcal{M}_{22}(t,\varepsilon) = \varepsilon H_1^{-1}(t),$$
$$\begin{aligned}\mathcal{F}_{11}(t,\varepsilon) = {} & \varepsilon A^*(t)\widetilde{P}(t,\varepsilon) + \varepsilon\widetilde{P}(t,\varepsilon)A(t) - \\ & - \varepsilon\widetilde{P}(t,\varepsilon)M_0(t)(P_0(t) + \varepsilon\widetilde{P}(t,\varepsilon)) - \\ & - \varepsilon P_0(t)M_0(t)\widetilde{P}(t,\varepsilon),\end{aligned}$$
$$\begin{aligned}\mathcal{F}_{12}(t,\varepsilon) = {} & \sqrt{\varepsilon}B_2^*(t)\widetilde{P}(t,\varepsilon) - \sqrt{\varepsilon}\widetilde{Q}_0(t,\varepsilon)[M_0(t)P_0(t) + \\ & + \varepsilon M_0(t)\widetilde{P}(t,\varepsilon) - A(t)],\end{aligned}$$
$$\begin{aligned}\mathcal{F}_{22}(t,\varepsilon) = {} & B_2^*(t)\widetilde{Q}_0^*(t,\varepsilon) + \widetilde{Q}_0(t,\varepsilon)B_2(t) - \\ & - \varepsilon\widetilde{Q}_0(t,\varepsilon)M_0(t)\widetilde{Q}_0^*(t,\varepsilon).\end{aligned}$$

We write for $\mathcal{P}$ a nonlinear integral equation, which shows $\mathcal{P}$ is a fixed point of a certain nonlinear integral operator and we get the estimates for $\mathcal{P}$ by proving that this operator maps a ball into itself.

Let us mention also the esimate

$$|\widetilde{R}_0(t,\varepsilon) - \widetilde{R}(t)| \le \varepsilon k_0 + k_1 \exp\left(-\alpha\left(\frac{T-t}{\varepsilon}\right)\right) + \frac{k_2 \exp\left(-2\alpha\left(\frac{T-t}{\varepsilon}\right)\right)}{2\beta\rho\varepsilon + 1 - \exp\left(-2\alpha\left(\frac{T-t}{\varepsilon}\right)\right)},$$

$$|\widetilde{Q}_0(t,\varepsilon) - \hat{Q}_0(t)| \le \frac{k_3 \exp\left(-\alpha\left(\frac{T-t}{\varepsilon}\right)\right)}{2\beta\rho\varepsilon + 1 - \exp\left(-2\alpha\left(\frac{T-t}{\varepsilon}\right)\right)},$$

$$\hat{Q}_0(t) = \widetilde{R}^{-1}(t)[B_2^*(t)P_0(t) + \hat{F}(t)].$$

We obtained in this way the required information concerning the asymptotic behaviour of $P(t,\varepsilon)$. Remark that O'Malley [5] considered the case $\widetilde{F}(t) > 0$, $G = 0$; this case is much simpler since it implies $Q(T,\varepsilon) = 0$, $R(T,\varepsilon) = 0$.

## 4. Complementary remarks

In the "cheap control" problem if we want to use a suboptimal control that might be simpler to compute we have to consider the behaviour of the solutions of a system of the form

$$\varepsilon\dot{x} = [\varepsilon A(t) + B(t)K^*(t)]x$$

where $K^*(t)B(t)$ is Hurwitz for fixed $t$.

Let $C_A(t,s)$ be the fundamental matrix solution of the system

$\dot{x} = A(t)x$ and let $y(t, \varepsilon) = C_A(s,t)x(t, \varepsilon)$ . Then

$$\varepsilon \dot{y}(t, \varepsilon) = C_A(s,t)\left[\varepsilon A(t) + B(t)K^*(t)\right]x(t, \varepsilon) - \\ - \varepsilon C_A(s,t)A(t)x(t, \varepsilon) = \\ = C_A(s,t)B(t)K^*(t)C_A(t,s)y(t, \varepsilon)$$

hence

$$\varepsilon \dot{y}(t, \varepsilon) = \widetilde{B}(t)\widetilde{K}^*(t)y(t, \varepsilon)$$

where

$$\widetilde{B}(t) = C_A(s,t)B(t), \ K^*(t) = K^*(t)C_A(t,s)$$

hence

$$\widetilde{K}^*(t)\widetilde{B}(t) = K^*(t)B(t) \quad \text{is Hurwitz.}$$

Denote by $C(t,s, \varepsilon)$ the fundamental matrix of the given system and by $\widetilde{C}(t,s, \varepsilon)$ the fundamental matrix of the transformed one.

Then $C(t,s, \varepsilon) = C_A(t,s)\widetilde{C}(t,s, \varepsilon)$ .

We next prove that

$$|\widetilde{K}^*(t)\widetilde{C}(t,s, \varepsilon)| \leq \varepsilon k_2 + (k_1 - \varepsilon k_2) \exp\left(-\alpha\left(\frac{t-s}{\varepsilon}\right)\right)$$

and then

$$|\widetilde{C}(t,s, \varepsilon) - \widetilde{L}(t,s)| \leq \varepsilon k + M \exp\left(-\alpha\left(\frac{t-s}{\varepsilon}\right)\right)$$

where $\widetilde{L}(t,s)$ is the solution of the Cauchy problem

$$\dot{y} = -\widetilde{B}(t)\left[K^*(t)B(t)\right]^{-1}\widetilde{K}^*(t)y$$

$$\widetilde{L}(s,s) = E - B(s)\left[K^*(s)B(s)\right]^{-1}K^*(s) .$$

If we denote $L(t,s) = C_A(t,s)L(t,s)$

we have

$$|C(t,s, \varepsilon) - L(t,s)| \leq \varepsilon k' + M \exp\left(-\alpha\left(\frac{t-s}{\varepsilon}\right)\right)$$

where $L(t,s)$ is the solution of the Cauchy problem

$$\dot{x} = \left[A(t) - B(t)(K^*(t)B(t))^{-1}K^*(t)\right]x , \quad L(s,s) = \widetilde{L}(s,s).$$

Let us mention finally the formula

$$\exp\left(\frac{1}{\varepsilon} BK^*(t-s)\right) = E - B(K^*B)^{-1}K^* + \\ + B(K^*B)^{-1} \exp\left(\frac{1}{\varepsilon} K^*B(t-s)\right)K^* .$$

## References

[1] V.Dragan, A.Halanay: Suboptimal linear controller by singular perturbations techniques, Rev.Roum.Sci.Techn. Electrotechn. et Energ. 21 (1976), 4, 585-591

[2] R.E.O'Malley Jr., C.F.Kung: On the matrix Riccati approach to a singularly perturbed regulator problem, Journal of diff. equations vol. 16 (1974), No.3, 413-427

[3] V.I.Glizer, M.G.Dmitriev: Singular perturbations in the linear control problem with a quadratic functional, Diff.Uravnenia XI, (1975), 11, 1915-1921 (Russian)

[4] P.V.Kokotovic, R.E.O'Malley Jr., P.Sannuti: Singular perturbations and order reduction in control theory - An overview, Automatica, vol. 12 (1976), 123-132

[5] R.E.O'Malley Jr.: A more direct solution of the nearly singular linear regulator problem, SIAM Journal on Control and Optimization 14 (1976), No.6, 1063-1077

Author's address: University of Bucharest, Faculty of Mathematics,
Str. Academiei 14, 70109 Bucharest I, Romania

# ON SOME INVERSE PROBLEMS FOR PARTIAL DIFFERENTIAL EQUATIONS

S. Dümmel, Karl-Marx-Stadt

## 1. Introduction

By an inverse problem for a differential equation we understand any problem in which the coefficients or the right-hand side of the differential equation are to be determined from some information on the solutions of this equation. We confine ourselves to two special cases of second order linear parabolic equations. For hyperbolic equations we refer to the book of V. T. Romanov [15].

Let $u$ be a solution of a Cauchy problem or of an initial-boundary value problem. We shall investigate the question what further information on $u$ is sufficient for the uniqueness of the unknown coefficient. For the case that the right hand side of the parabolic equation is unknown such investigations can be found e. g. in the following papers: W. T. Ivanov, G. P. Smirnov, F. W. Lubyšev [7], A. Fasano [4], W. M. Isakov [5] and in the book Lavrentiev, Romanov, Vasiliev [10].

The case that the unknown function is the coefficient at $u$ in the parabolic equation is considered in several papers of M. M. Lavrentiev and K. G. Resnizkaja ([8], [9], [12], [13], [14]). These authors assumed that the unknown coefficient is only a function of one space variable. Coefficients of several variables are considered e. g. by A. D. Iskenderov [6] and I. Ja. Besnočenko [1], where the unknown coefficients are functions of n-1 space variables and of the time and $u$ is a function of n space variables and of the time.

In our lecture we shall consider the question of uniqueness for the parabolic equations

$$u_t(x,t) - q(x)\Delta u(x,t) = 0 \tag{1.1}$$

and

$$u_t(x,t) - \sum_{i=1}^{n} \frac{\partial}{\partial x_i}(p(x)u_{x_i}(x,t)) = 0 \tag{1.2}$$

with $x = (x_1, x_2, \ldots, x_n)$ $(n = 1,2,3,\ldots)$, where $q(x)$ and $p(x)$ are unknown. For the equation (1.2) there exist some papers by G. Chavent ([2], [3]) who determines $p$ by a gradient method.

We use the following notations: $R^n$ is the n-dimensional Euclidean space $(n = 1,2,3,\ldots)$, $x = (x_1, x_2, \ldots, x_n) \in R^n$, $D$ a bounded region of $R^n$ with a sufficiently smooth boundary $\partial D$. For $T > 0$ we define $Z_T = D \times (0,T)$, $\Gamma_T = D \times [0,T]$.

## 2. Uniqueness theorems with additional conditions on u for a fixed time

We consider the following initial-boundary value problem

$$(2.1) \qquad u_t(x,t) - q(x)\Delta u(x,t) = 0 \quad \text{for } (x,t) \in Z_T,$$

$$(2.2) \qquad u(x,0) = \varphi(x) \quad \text{for } x \in D,$$

$$(2.3) \qquad u(x,t) = \psi(x,t) \quad \text{for } (x,t) \in \Gamma_T,$$

where $q \in C(\overline{D})$, $q(x) > 0$ for all $x \in \overline{D}$, $u \in C^2(\overline{Z}_T)$,
$u \in C^3(Z_T)$, $|\Delta u_t(x,t)| \leqq K$ for all $(x,t) \in Z_T$,
$\varphi \in C^2(\overline{D})$, $\Delta\varphi(x) = 0$ for all $x \in \overline{D}$, $\psi \in C(\Gamma_T)$,
$\psi_t \in C(\Gamma_T)$, $\varphi(x) = \psi(x,0)$ for all $x \in \partial D$.

If q is known and u is a solution of the problem (2.1) (2.2) (2.3), then u is unique. Now let q be unknown. Then for the uniqueness of q in addition to (2.1) (2.2) (2.3) we need a further information on u. We demand that for a fixed $t_1$ with $0 < t_1 < T$ there is a function h with $h \in C^2(\overline{D})$, $|\Delta h(x)| \geqq t_1^{\alpha}$ $(0 < \alpha < \frac{1}{2})$ for all $x \in \overline{D}$ such that

$$(2.4) \qquad u(x,t_1) = h(x) \quad \text{for } x \in \overline{D}.$$

Then we obtain

Theorem 1. If $\varphi$, $\psi$, h are given functions with the above properties, if (q,u) and $(\overline{q}, \overline{u})$ are two pairs of functions satisfying (2.1) - (2.4) and if

$$(2.5) \qquad 0 < t_1 < \left(\frac{2\lambda_1}{K}\right)^{\frac{1}{1-2\alpha}},$$

where K and $\alpha$ have been introduced above and $\lambda_1$ is the smallest eigenvalue of the Dirichlet problem in D for the elliptic operator $q\Delta u$, then $q = \overline{q}$ and $u = \overline{u}$.

Proof. Let (q,u) and $(\overline{q}, \overline{u})$ be two pairs of functions satisfying (2.1) - (2.4). We introduce the notation $w = u_t$, $\overline{w} = \overline{u}_t$, $\widetilde{w} = w - \overline{w}$, $\widetilde{q} = q - \overline{q}$. Then it can be shown that

$$\widetilde{w}_t - q\Delta\widetilde{w} = \widetilde{q}\Delta\overline{w} \quad \text{in } Z_T,$$
$$\widetilde{w}(x,0) = 0 \quad \text{on } \overline{D},$$
$$\widetilde{w}(x,t) = 0 \quad \text{on } \Gamma_T .$$

By Fourier's separation method $\widetilde{w}$ can be represented in the form

$$\widetilde{w}(x,t) = \sum_{k=1}^{\infty} \int_0^t \int_D \widetilde{q}(y)\Delta w(y,\tau) g_k(y) dy e^{-\lambda_k(t-\tau)} d\tau \, g_k(y)$$

where $\{g_k\}$ is a complete orthogonal system (in $L^2(D)$) of corresponding eigenfunctions and $\{\lambda_k\}$ the system of corresponding eigenvalues. If we denote the norm in $L^2(D)$ by $\|\cdot\|$, then we can show that

$$\|\widetilde{w}(x,t)\|^2 \leq \frac{1}{2\lambda_1} \int_0^t \int_D (\widetilde{q}(y)\Delta\overline{w}(y,\tau))^2 dy d\tau \leqq \frac{K}{2\lambda_1} t\|q\|^2 .$$

For $t = t_1$ we obtain

$$\tilde{w}(x,t_1) = \tilde{q}(x)\,\Delta h(x).$$

Hence
$$\|\tilde{q}\|^2 \leqq \frac{1}{t_1^{2\alpha}} \|\tilde{q}\Delta h\|^2 \leqq \frac{K}{2\lambda_1} t_1^{1-2\alpha}\|q\|^2 .$$

By (2.5) we have $\frac{K}{2\lambda_1}t_1^{1-2\alpha}<1$ and thus $\|\tilde{q}\| = 0$. This completes the proof of the theorem.

The method of the proof of Theorem 1 can also be used in the case of the equation (1.2) if $n = 1$. Thus now let $n = 1$. We consider

$$\text{(2.6)} \qquad u_t(x,t) - \frac{\partial}{\partial x}(p(x)\, u_x(x,t)) = 0 \qquad \text{for } (x,t) \in Z_T$$

with the initial condition (2.2) and the boundary condition (2.3). $u$ and $\psi$ shall satisfy the same hypotheses as before. Furthermore we suppose that $D = (a,b)$, $\varphi$ is constant, $p \in C^1([a,b])$, $p(x) > 0$ for all $x \in [a,b]$ and (2.4) holds with $h \in C^2([a,b])$, $|h'(x)| \geqq t_1^{\alpha}$ $(0 < \alpha < \frac{1}{4})$ for all $x \in [a,b]$. Finally we demand that $p(a)$ is known:

$$\text{(2.7)} \qquad p(a) = c.$$

Now we obtain a theorem analogous to Theorem 1, where we omit the exact bound for $t_1$.

<u>Theorem 2.</u> If $\varphi$, $\psi$, $h$ are given functions with the above properties, if $(p,u)$ and $(\bar{p},\bar{u})$ are two pairs of functions satisfying (2.6), (2.2) - (2.4), (2.7) and if $t_1$ is sufficiently small, then $p = \bar{p}$ and $u = \bar{u}$.

<u>Proof</u>. Using the analogous notation and the same method as in the proof of Theorem 1 we obtain

$$\|\tilde{w}(x,t)\|^2 = M\,t\,(\|\tilde{p}\| + \|\tilde{p}'\|)^2,$$

where $M$ is a constant and $\tilde{p}' = \frac{d\tilde{p}}{dx}$.

For $t = t_1$ we have

$$\tilde{w}(x,t_1) - \frac{d}{dx}(\tilde{p}(x)\, h'(x)) = 0.$$

Hence

$$\text{(2.8)} \qquad \tilde{p}(x) = \frac{1}{h'(x)} \int_a^x \tilde{w}(\xi,t_1)$$

and

$$\text{(2.9)} \qquad \tilde{p}'(x) = \frac{h''(x)}{h'(x)}\,\tilde{p}(x) + \frac{1}{h'(x)}\,\tilde{w}(x,t_1) .$$

Using (2.8) and (2.9) one can estimate $\|\tilde{p}\|$ and $\|\tilde{p}'\|$ by $\|\tilde{w}(x,t_1)\|$. Then by an analogous conclusion as in the proof of Theorem 1 we obtain the assertion of Theorem 2.

## <u>3. Uniqueness theorems for the case that u(x,t) is analytic in t</u>

Now we shall use another method for proving the uniqueness of $p(x)$ in (2.1). This proof is due to H. P. Linke. For simplicity we formulate the following considerations for the case $n = 1$. But it is also possible to treat the case of several variables in a similar manner.

Thus let $n = 1$ and $Z_\infty = R^1 \times (0, \infty)$. We consider the following Cauchy problem:

$$u_t(x,t) - \frac{\partial}{\partial x}(p(x)\, u_x(x,t)) = 0 \quad \text{for } (x,t) \in Z_\infty \tag{3.1}$$

$$u(x,0) = \varphi(x) \quad \text{for } x \in R^1, \tag{3.2}$$

where $p(x) = \sum_{k=0}^{\infty} p_k x^k$, $p(x) > 0$ for all $x \in R^1$ and $\varphi \in C^\infty(R^1)$. Let $c > 0$ and

$$p(0) = p_o = c. \tag{3.3}$$

Furthermore we assume that $u$, $u_x$, $u_{xx}$ are representable in the form

$$u(x,t) = \sum_{k=0}^{\infty} u_k(x)\, t^k, \quad u_x(x,t) = \sum_{k=0}^{\infty} u_k'(x)\, t^k,$$
$$u_{xx}(x,t) = \sum_{k=0}^{\infty} u_k''(x)\, t^k$$

for $(x,t) \in \overline{Z}_\infty$, where $u_k \in C^2(R^1)$ and $u_k'$, $u_k''$ are the derivatives of $u_k$. Finally we suppose that there is a function $g \in C^\infty([0, \infty))$ such that

$$u(0,t) = g(t) \quad \text{for } 0 \leqq t. \tag{3.4}$$

Then the following theorem holds.

<u>Theorem 3.</u> Let $\varphi$ and $g$ be given functions and $p$ and $u$ unknown functions with the above properties and $c$ a given positive number. If in addition $p_{2\nu} = 0$ for $\nu = 1,2,3,\ldots$ and $\varphi'(0) = 0$, then there exists at most one $p$ such that (3.1) - (3.4) are fulfilled.

<u>Proof</u>. From (3.1) we obtain

$$\sum_{k=0}^{\infty} (k+1)\, u_{k+1}(x)\, t^k = \sum_{k=0}^{\infty} \frac{d}{dx}(p(x)\, u_k'(x))\, t^k$$

and

$$k\, u_k(x) = \frac{d}{dx}(p(x)\, u_{k-1}'(x)) \qquad (k = 1,2,3,\ldots)$$

with

$$u_o(x) = \varphi(x).$$

Since by (3.4) $u(0,t)$ is known, also all $u_k(0)$ are known. Using the equality

$$p_k = \frac{p^k(0)}{k!}$$

and the additional assumptions of the theorem, one can show that

$$p_{2\nu-1} = F_\nu(u\,(0),\, p_o,\, p_1, \ldots, p_{2\nu-2},\, \varphi'(0), \ldots, \varphi^{(2\nu)}(0)),$$

where the $F_\nu$ are known functions and $\nu = 1,2,3,\ldots$ . Thus all $p_k$ are uniquely determined by the given conditions and the theorem is proved.

It is also possible to prove an analogous theorem for the case that $p_{2\nu-1} = 0$ for all $\nu = 1,2,3,\ldots$ .

## 4. Reduction of a one-dimensional problem to an inverse Sturm-Liouville problem

At last we consider the following one-dimensional problem:

$$u_t(x,t) - \frac{\partial}{\partial x}(p(x)\, u_x(x,t)) = 0 \quad \text{for } a < x < b, \quad 0 < t, \tag{4.1}$$

$$u(x,0) = \delta(x - a) \quad \text{for } a \leqq x \leqq b, \tag{4.2}$$

$$\left.\begin{array}{l} u_x(a,t) - h\, u(a,t) = 0 \\ u_x(b,t) - H\, u(b,t) = 0 \end{array}\right\} \quad \text{for } 0 < t, \tag{4.3}$$

where $\delta$ is the Dirac delta function, h and H are real numbers, $p(x) > 0$ for all $x \in [a,b]$, $p \in C^2(a,b)$ and u is a (generalized) solution of (4.1) - (4.3). Furthermore we suppose that

$$p(a) = c_1, \quad p'(a) = c_2, \tag{4.4}$$

where $c_1$, $c_2$ are real numbers. Then we obtain the following theorem.

Theorem 4. Let h, H, $c_1$, $c_2$ be given real numbers. If in addition $u(a,t)$ is known for all t with $0 < t$, then there exists at most one function p with the above properties such that (4.1) - (4.4) are fulfilled.

The proof of this theorem can be given by reduction of the stated problem to an inverse Sturm-Liouville problem (comp. H. P. Linke [11]).

### References


[1] I. Ja. Besnočenko, On the determination of the coefficients in parabolic equations (Russ.), Diff. Uravnenija 10, 24-35 (1974).

[2] G. Chavent, Sur une méthode de resolution du problème inverse dans les equations aux dérivées partielles paraboliques, C. R. Acad. Sc. Paris 269, 1135-1138 (1969).

[3] ---, Une méthode de resolution de problème inverse dans les équations aux dérivées partielles, Bull. Acad. Polon. Sc., Série sci. techn. 8, 99-105 (1970).

[4] A. Fasano , Sulla inversione di un classico problema di diffusione del calore, Le Matematiche 16, 50-61 (1971).

[5] W. M. Isakov, Uniqueness theorems for inverse problems of heat potentials (Russ.), Sibir. Mat. Žurn. 17, 259-272 (1976).

[6] A. D. Iskenderov, Multidimensional inverse problems for linear and quasi-linear parabolic equations (Russ.), Dokl. Akad. Nauk USSR 225, 1005-1008 (1975).

[7] W. T. Ivanov, G. P. Smirnov, F. W. Lubyšev, On local direct and inverse problems for the heat equation (Russ.) Diff. Uravnenija 11, 19-26 (1975).

[8] M. M. Lavrentiev, K. G. Resnizkaja, Uniqueness theorems for some non-linear inverse problems at equations of parabolic type (Russ.) Dokl. Akad. Nauk USSR 203, 531-532 (1973).

[9] ---, ---, On the uniqueness of an inverse problem in the theory

of heat conduction (Russ.), Math Problems of Geophysics 6 , 117-121 (1975).

[10] Lavrentiev, Romanov, Vasiliev, Multidimensional inverse problems for differential equations, Springer-Verlag Berlin 1970.

[11] H. P. Linke, Über ein inverses Problem für die Wärmeleitungsgleichung, Wiss. Z. Techn. Hochsch. Karl-Marx-Stadt 18, 445-447 (1976).

[12] K. G. Resnizkaja, An existence and uniqueness theorem for a non-linear one-dimensional inverse problem of the theory of heat conduction (Russ.), Math. Problems of Geophysics 4, 131-134 (1973).

[13] ---, Uniqueness of the solution of a non-linear inverse problem of the theory of heat conduction in the analytic case (Russ.), Math. Problems of Geophysics 4, 135-139 (1973).

[14] ---, An existence and uniqueness theorem for a non-linear inverse problem of the theory of heat conduction - the method of Newton-Kantorovic (Russ.), Math. Problems of Geophysics 3, 152-163 (1972).

[15] V. T. Romanov, Certain inverse problems for equations of hyperbolic type (Russ.), Izdat. "Nauka" Sibirsk. Otdel. Novosibirsk 1972.

Author's address: Sektion Mathematik Technische Hochschule DDR 90

Karl-Marx-Stadt, PSF 964, G.D.R.

# NONLINEAR NONCOERCIVE BOUNDARY VALUE PROBLEMS

S. Fučík, Praha

## 1. Introduction

The main purpose of this lecture is to formulate some results and open problems concerning the solvability of nonlinear equations. The results are not presented in full generality and only the ideas are presented. For more general results see the references. But, of course, the list of references is not complete.

Let $\Omega$ be a bounded domain in $\mathbb{R}^N$ with the boundary $\partial\Omega$ sufficiently smooth if $N > 1$. Let $g: \mathbb{R}^1 \longrightarrow \mathbb{R}^1$ be a continuous function. We are concerned with the solvability of the Dirichlet problem

$$(1) \qquad \begin{cases} -\Delta u(x) - g(u(x)) = f(x) & \text{in } \Omega \\ \qquad\qquad u(x) = 0 & \text{on } \partial\Omega . \end{cases}$$

We shall discuss the existence of a weak solution of (1) under the assumption that there exist limits

$$\lim_{\xi\to\infty} \frac{g(\xi)}{\xi} = \mu , \qquad \lim_{\xi\to-\infty} \frac{g(\xi)}{\xi} = \nu ,$$

and we shall consider various configurations of $\mu, \nu \in [-\infty, \infty]$.

## 2. The case $\mu = \nu = \lambda \in \mathbb{R}^1$

For simplicity suppose that

$$(2) \qquad g(\xi) = \lambda\xi - \psi(\xi) , \quad \xi \in \mathbb{R}^1 ,$$

where $\psi : \mathbb{R}^1 \longrightarrow \mathbb{R}^1$ is a bounded and continuous function.

2.1. Regular case

Suppose that the problem

$$(3) \qquad \begin{cases} -\Delta u(x) - \lambda u(x) = 0 & \text{in } \Omega \\ u(x) = 0 & \text{on } \partial\Omega \end{cases}$$

has only trivial weak solutions.

Then (according to the Schauder fixed point theorem) the problem (1) with g given by (2) is weakly solvable for arbitrary $f \in L_2(\Omega)$. This is the so-called regular (or coercive) case since the operator

$$(4) \qquad L : u \longmapsto -\Delta u - \lambda u$$

is invertible in the Sobolev space $W_0^{1,2}(\Omega)$. For references on results of such type see e.g. [3].

2.2. Resonance case

2.2.1. Consider the problem (1) (with g given by (2)) under the assumption that the operator L defined by (4) is noninvertible, i.e. the problem (3) has nontrivial weak solutions. Such a case of (1) was considered first in [15]. It is included in the theory of the solvability of nonlinear operator equations of the type

$$(5) \qquad Lu = Su ,$$

where L is linear noninvertible and Fredholm operator between Banach spaces X and Z, and $S: X \longrightarrow Z$ is a nonlinear completely continuous mapping. It is possible to obtain the solvability of the nonlinear equation (5) using the Schauder fixed point theorem (see e.g. [4]); the method of fixed point theorems was used in the special case of differential operators in many papers e.g.

by the authors: J.Cronin, D.G. De Figueiredo, P.Hess, J.Nečas, L. Nirenberg, V.P.Portnov, M.Schechter, S.A.Williams, ... . It is also possible to use Mawhin's coincidence degree theory (for the exposition, numerous applications and further references see [13]) or a certain variational approach (see e.g. [11], [12]) .

The results mentioned above, applied to (1) , yield the following theorem.

2.2.2. Suppose that we have, for an arbitrary nontrivial solution w of (3),

$$(6)\qquad \limsup_{\xi\to\infty} \psi(\xi)\int_\Omega w^+(x)dx - \liminf_{\xi\to-\infty} \psi(\xi)\int_\Omega w^-(x)dx$$
$$< \int_\Omega f(x)\, w(x)\, dx <$$
$$\liminf_{\xi\to-\infty} \psi(\xi)\int_\Omega w^+(x)dx - \limsup_{\xi\to\infty} \psi(\xi)\int_\Omega w^-(x)dx \ ,$$

where $w^+$ and $w^-$ are the positive and negative parts of the function w , respectively.

Then the problem (1) is weakly solvable.

The existence of weak solutions of (1) follows also in the case when in (6) the reverse inequalities hold.

2.2.3. If

$$\psi(\infty) < \psi(\xi) < \psi(-\infty) \ , \ \xi\in \mathbb{R}^1$$

( $\psi(\infty) = \lim_{\xi\to\infty}\psi(\xi)$ , $\psi(-\infty) = \lim_{\xi\to-\infty}\psi(\xi)$) , then (6) is a necessary and sufficient condition for the weak solvability of (1).

## 2.3. Vanishing nonlinearities

2.3.1. The set of the right hand sides $f$ satisfying (6) may be empty e.g. in the case

$$\psi(-\infty) = \psi(\infty) = 0 . \tag{7}$$

Then, 2.2.2 gives no existence result.

The idea how to prove the existence of weak solutions of (1) with $g$ given by (2) where $\psi$ satisfies (7) is based on the so--called method of truncated equations: we change the function outside a sufficiently large interval $(-A, A)$ such that for the function $\tilde{\psi}$ obtained, condition (6) gives sense. The main part of the proof of existence is to show that an arbitrary weak solution $u$ of (1) with $g$ given by

$$g(\xi) = \lambda\xi - \tilde{\psi}(\xi) , \ \xi \in \mathbb{R}^1$$

satisfies

$$\|u\|_C < A . \tag{8}$$

Thus we obtain the following results under the assumption that $\psi$ is odd.

2.3.2 (see [9]). Consider (1) with $N = 1$ . Let $a > 0$ and let

$$\lim_{\xi\to\infty} \xi^2 \min_{\tau\in[a,\xi]} \psi(\tau) = \infty \tag{9}$$

and

$$\int_\Omega f(x)\, w(x)\, dx = 0 \tag{10}$$

for an arbitrary weak solution of (3) .

Then (1) has at least one weak solution.

2.3.3 ( see [10]). Consider (1) with $N > 1$. Let $a > 0$ and let

(11) $$\lim_{\xi\to\infty} \xi \min_{\tau\in[a,\xi]} \psi(\tau) = \infty \quad .$$

Suppose $f \in L_\infty(\Omega) \cap C(\Omega)$ and (10) .

Then (1) has at least one weak solution.

2.3.4. In [14] the result from 2.3.3 is extended under the assumption that

(12) $$\liminf_{\xi\to\infty} \xi \min_{\tau\in[a,\xi]} \psi(\tau) > 0$$

(instead of (11)) and that an arbitrary weak solution of (3) has the so-called "unique continuation property". For further generalization see [11] .

2.3.5. OPEN PROBLEM. To prove the apriori estimate (8) in the case of (1) with $N > 1$ it is necessary to estimate

(13) $$\sup \int_{\Omega_\varepsilon(w)} |w(x)| \, dx \leqq \text{const.} \quad \varepsilon^{1+\rho} \quad ,$$

where the supremum is taken over all the solutions of (3) with $\|w\|_C = 1$ and where

$$\Omega_\varepsilon(w) = \{ x \in \Omega \; ; \; 0 < |w(x)| < \varepsilon \} \quad .$$

Obviously, (13) holds with $\rho = 0$ ; thus we obtain the sufficient condition (11) in 2.3.3 . If $N = 1$ then (13) holds with $\rho = 1$ and hence we obtain the sufficient condition (9) in 2.3.2 . If (13) were true with some $\rho > 0$ , it would be possible to replace (11) by

$$\lim_{\xi\to\infty} \xi^{1+\rho} \min_{\tau\in[a,\xi]} \psi(\tau) = \infty \quad .$$

Probably, a proof of (13) with $\rho > 0$ may be based on the investigation of the nodal lines of the solutions of (3) with using a version of the maximum principle. Unfortunately, we are not aware of any correct result from this field.

2.3.6. OPEN PROBLEM. If the condition (12) is not satisfied (e.g. if $\psi$ has a compact support) nothing is known about the existence of solutions of (1) in the resonance case.

## 2.4. Expansive nonlinearities

Using the same method as in 2.3 we can investigate also the weak solvability of (1) with $g$ given by (2) in the case that there exists none of the limits $\psi(\infty)$, $\psi(-\infty)$ .

2.4.1. A bounded odd continuous and nontrivial function $\psi$ : $\mathbb{R}^1 \longrightarrow \mathbb{R}^1$ is said to be expansive if for each p with

$$0 \leqq p < \sup_{\xi \in \mathbb{R}^1} \psi(\xi)$$

there exist sequences $0 < a_k < b_k$ with

$$\lim_{k \to \infty} b_k a_k^{-1} = \infty$$

such that

$$\lim_{k \to \infty} \min_{\xi \in [a_k, b_k]} \psi(\xi) > p .$$

A typical example of an expansive function which has none of the limits $\psi(\infty)$ , $\psi(-\infty)$ is $\psi(\xi) = \sin \xi^{1-\varepsilon}$ with $1 > \varepsilon > 0$.

2.4.2 (see [9], [10]). Considering (1) with $g$ given by (2) with $\psi$ expansive and $f \in L_\infty(\Omega) \cap C(\Omega)$ if $N > 1$ we see that

$$(14) \qquad \left| \int_{\Omega} f(x)\, w(x)\, dx \right| < \sup_{\xi \in \mathbb{R}^1} \psi(\xi) \int_{\Omega} |w(x)|\, dx$$

for an arbitrary nontrivial solution $w$ of (3) is a necessary and sufficient condition for the weak solvability of (1).

2.4.3. OPEN PROBLEM. It seems that nontrivial conditions for the weak solvability of (1) with $g$ given by (2) where $\psi(\xi) = \sin \xi$ (or an analogous periodic function) in the resonance case are so far unknown.

## 3. Jumping nonlinearities

3.1. $\mu, \nu \in \mathbb{R}^1$, $\mu \neq \nu$

3.1.1. Consider (1) with $N = 1$ and $\Omega = (0, \pi)$. Suppose that

$$(16) \qquad g(\xi) = \mu \xi^+ - \nu \xi^- - \psi(\xi)$$

where $\psi : \mathbb{R}^1 \to \mathbb{R}^1$ is bounded and continuous.

3.1.2 (see [8], [2], [6]). Let the assumptions from 3.1.1 be satisfied. Put

$$\mathcal{M} = \{(\mu, \nu) \in \mathbb{R}^2 ;\ \mu < 1,\ \nu < 1\} \cup \bigcup_{k=0}^{\infty} \{(\mu, \nu) \in \mathbb{R}^2 ;$$

$$\mu^{1/2} > k + 1,\quad \omega_k(\mu^{1/2}) < \nu^{1/2} < \gamma_{k+1}(\mu^{1/2})\}$$

$$\cup \bigcup_{k=1}^{\infty} \{(\mu, \nu) \in \mathbb{R}^2 ;\quad \mu^{1/2} > k,\ \gamma_k(\mu^{1/2}) < \nu^{1/2}$$

$$< \vartheta_k(\mu^{1/2})\},$$

where

$$\vartheta_k(\tau) = \begin{cases} \dfrac{(k+1)\tau}{\tau - k} , & \tau \in (k, 2k+1] \\ \\ \dfrac{k\tau}{\tau-(k+1)} , & \tau \in (2k+1, \infty) \end{cases} ,$$

$$\omega_k(\tau) = \begin{cases} \dfrac{k\tau}{\tau-(k+1)} , & \tau \in (k+1, 2k+1] \\ \\ \dfrac{(k+1)\tau}{\tau - k} , & \tau \in (2k+1, \infty) \end{cases} ,$$

$$\gamma_k(\tau) = \frac{k\tau}{\tau - k} , \quad \tau \in (k, \infty) .$$

(i) If $(\mu,\nu) \in \mathfrak{M}$ , (1) has at least one weak solution for each f .

(ii) If $(\mu,\nu) \notin \overline{\mathfrak{M}}$ , there exists an $f \in C^{\infty}$ for which (1) has no weak solution.

3.1.3. OPEN PROBLEMS. If $N>1$ then for the weak solvability of (1) with g satisfying (16) we can characterize the parameters $(\mu, \nu)$ only in the following cases:

a) $(\mu, \nu)$ are close to the diagonal $\{(\lambda,\lambda) \; ; \lambda \in \mathbb{R}^1\}$ (see [2],[6]) ;

b) $\mu<\lambda_1, \nu<\lambda_1$ ( $\lambda_1$ is the first eigenvalue of $-\Delta$) - this case is included in the theory of pseudomonotone operators;

c) $\mu, \nu$ are close to $\lambda_1$ (see [1], [5] and the papers of A.Ambrosetti - G.Mancini , M.S.Berger, E.N.Dancer, E.Podolak,...);

d) $\mu, \nu$ are close to a simple eigenvalue of $-\Delta$ ( see [2], [6], and the papers of A.Ambrosetti - G.Mancini , E.Podolak,...) .

For a better description of the solvability of (1) with respect to the values $\mu, \nu$ it would be important to solve the following problems:

(i) Find all $(\mu, \nu) \in \mathbb{R}^2$ for which

$$\begin{cases} -\Delta u - \mu u^+ + \nu u^- = 0 & \text{in } \Omega \\ u = 0 & \text{on } \partial\Omega \end{cases}$$

has nontrivial solutions (even in the case of special domains $\Omega$).

(ii) If $S$ is a linear completely continuous operator in the Hilbert space with a certain "good" cone (semiordering), consider the mapping

$$T_{(\mu,\nu)} : u \longmapsto u - \mu S u^+ + \nu S u^- , \quad u \in H.$$

Is it true that if $T_{(\mu,\nu)} u = 0$ has only trivial solution then the Leray-Schauder degree of the mapping $T_{(\mu,\nu)}$ is nonzero if and only if $T_{(\mu,\nu)}$ is onto H ?

The answer is affirmative if (1) with $N = 1$ is considered, unknown in the case $N > 1$ .

3.2. $\nu = \lambda_1$ , $\mu = -\infty$

3.2.1. A typical example of this case is the Dirichlet problem

$$(17) \qquad \begin{cases} -u'' - \lambda u + e^u = f & \text{in } (0, \pi) \\ u(0) = u(\pi) = 0 \end{cases}$$

with $\lambda = 1$. It is possible to prove that (17) has a weak solution for $f \in L_1(0,\pi)$ provided

$$\int_{\Omega} f(x) \sin x \, dx > 0 .$$

3.2.2. OPEN PROBLEM. The solvability of (17) with $\lambda > 1$ is an open problem. It is easy to see that there exists a right hand side $f$ for which (17) is not solvable.

## 4. Rapid nonlinearities

4.1. Consider (1) with $N = 1$. If

$$\lim_{|\xi| \to \infty} \frac{g(\xi)}{\xi} = \infty \tag{18}$$

then (1) has infinitely many solutions (see [7]).

4.2. OPEN PROBLEM. The solvability of (1) with $N > 1$ under the assumption (18), e.g. the weak solvability of

$$\begin{cases} -\Delta u - |u|^{\varepsilon} u = f & \text{in } \Omega \\ u = 0 & \text{on } \partial\Omega \end{cases}$$

if $\varepsilon > 0$ is sufficiently small, seems to be terra incognita.

## References

[1] A.Ambrosetti-G.Prodi: On the inversion of some differentiable mappings with singularities between Banach spaces. Annali Mat. Pura Appl. 93, 1973, 231-247.

[2] E.N.Dancer: On the Dirichlet problem for weakly nonlinear elliptic partial differential equations. Proc. Royal Soc.Edinburgh (to appear).

[3] S.Fučík-J.Nečas-J.Souček-V.Souček: Spectral Analysis of Nonlinear operators. Lecture Notes in Mathematics No 346. Springer

Verlag 1973.

[4] S.Fučík: Nonlinear equations with noninvertible linear part. Czechosl. Math.J. 24, 1974, 467-495.

[5] S.Fučík: Remarks on a result by A.Ambrosetti and G.Prodi. Boll. Unione Mat.Ital. 11, 1975, 259-267.

[6] S.Fučík: Solvability and nonsolvability of weakly nonlinear equations. Proceedings of the Summer School "Theory of Nonlinear Operators" held in September 1975 at Berlin (to appear).

[7] S.Fučík-V.Lovicar: Periodic solutions of the equation $x''(t)+g(x(t))=p(t)$. Časopis Pěst. Mat. 100, 1975, 160- 175.

[8] S.Fučík: Boundary value problems with jumping nonlinearities. Časopis Pěst.Mat. 101, 1976, 69-87.

[9] S.Fučík: Remarks on some nonlinear boundary value problems. Comment.Math.Univ.Carolinae 17, 1976, 721 - 730.

[10] S.Fučík-M.Krbec: Boundary value problems with bounded nonlinearity and general null-space of the linear part. Math. Z. 155, 1977, 129-138.

[11] S.Fučík:Nonlinear potential equations with linear part at resonance. Časopis Pěst.Mat. (to appear).

[12] S.Fučík: Nonlinear equations with linear part at resonance: Variational approach. Comment.Math.Univ.Carolinae (to appear).

[13] R.E.Gaines-J.L.Mawhin: Coincidence Degree,and Nonlinear Differential Equations. Lecture Notes in Mathematics No 568. Springer Verlag 1977.

[14] P.Hess: A remark on the preceding paper of Fučík and Krbec. Math.Z. 155,1977,139-141.

[15] E.M.Landesman-A.C.Lazer:Nonlinear perturbations of linear boundary value problem at resonance. J.Math.Mech. 19,1970,609-623.

Author's address: Department of Mathematical Analysis, Faculty of Mathematics and Physics, Charles University, Sokolovská 83, 18600 Prague, Czechoslovakia

# ON THE ITERATIVE SOLUTION OF SOME NONLINEAR EVOLUTION EQUATIONS

H. Gajewski, Berlin

The purpose of this paper is to show by three examples of nonlinear evolution equations arising from mathematical physics how a priori estimates can be used to establish globally convergent iteration processes. An important feature of these iteration processes is that one proceeds by solving <u>linear</u> evolution equations with <u>constant</u> coefficients.

We shall start our discussion with Burgers' equation. As further examples the spatially two-dimensional Navier-Stokes equations and the nonlinear Schrödinger equation will be considered. We shall conclude with some remarks concerning the numerical realisation of the iteration processes.

At first we introduce some notations. Let $X$ be a Banach space and $S=[0,T]$ a bounded time interval. Then $C(S;X)$ is the Banach space of continuous mappings from $S$ into $X$ provided with the maximum norm. $L^p(S;X)$ , $1\leq p\leq\infty$ , denotes the Banach space of Bochner integrable functions $u:(0,T)\rightarrow X$ with the norms

$$\left(\int_S \|u(t)\|_X^p dt\right)^{1/p} , \quad 1\leq p<\infty , \qquad \operatorname*{ess\,sup}_{t\in S}\|u(t)\|_X , \quad p=\infty .$$

## 1. Burgers' equation

Let $H=L^2(0,1)$ , $V=H_0^1(0,1)$ and $V^*=H^{-1}(0,1)$ be the usual spaces with the norms $|\cdot|$ , $\|\cdot\|$ and $\|\cdot\|_*$ , respectively. We consider the initial-boundary value problem

$$\begin{aligned} &u_t - \nu u_{xx} + u u_x = f \quad \text{in} \quad (0,T)\times(0,1) , \\ (1.1)\qquad &u(0,x)=a(x) , \ x\in(0,1) , \quad u(t,0)=u(t,1)=0 , \ t\in(0,T] . \end{aligned}$$

Here the subscripts $t$ and $x$ indicate partial differentiation, $\nu$ is a positive constant. We suppose up to the end of this section that

$$(1.2)\qquad f\in L^2(S;V^*) , \quad a\in H .$$

Then, as is well known, the problem (1.1) has a unique solution $u\in L^2(S;V)\cap C(S;H)$ with $u_t\in L^2(S;V^*)$ satisfying the a priori estimate (cf. /3/)

(1.3) $$\|u\|_{C(S;H)} \leq r \quad , \quad r^2 = |a|^2 + \frac{1}{2\nu}\|f\|^2_{L^2(S;V^*)} \ .$$

For constructing the solution of (1.1) Carasso /1/ proposed the following iteration procedure

(1.4) $$u^j_t - \nu u^j_{xx} = f - u^{j-1}u^{j-1}_x \ , \quad j=1,2,\ldots, \ u^o=0 \ ,$$
$$u^j(0,x) = a(x) \ , \ u^j(t,0) = u(t,1) = 0 \ .$$

A corresponding method has been used by Fujita and Kato /2/ as a means of proving existence and uniqueness theorems for the Navier-Stokes equations. Carasso /1/ stated the following sufficient convergence condition for (1.4)

(1.5) $$\left(\frac{64T}{\nu}\right)^{1/2} \left(\|a\| + \int_S \|f(t)\| \, dt\right) < 1 \ .$$

Possibly this condition could be weakened but it cannot be replaced by a global condition because counter-examples show (cf. /1/) that the convergence of the procedure (1.4) is in fact only local in time, even if the global solution of (1.1) is smooth.

We want now to show that the iteration method (1.4) can be easily modified in such a way that we get a globally convergent process. For that we define the projector of $H$ onto the r-ball in $H$ by

(1.6) $$P\,v = \begin{cases} v & \text{if} \quad |v| \leq r \\ r\frac{v}{|v|} & \text{if} \quad |v| > r \ , \end{cases}$$

where $r$ is the constant from (1.3). We suggest replacing (1.2) by

(1.7) $$u^j_t - \nu u^j_{xx} = f - (Pu^{j-1})u^{j-1}_x \ , \quad j = 1,2,\ldots,$$
$$u^j(0,x) = a(x) \ , \quad u^j(t,0) = u^j(t,1) = 0 \ .$$

The following global convergence theorem holds.

<u>Theorem 1.</u> Let $u$ be the solution of (1.1) and $u^o \in L^2(S;V) \cap C(S;H)$ an arbitrary starting function. Then the sequence $(u^j)$ defined by (1.7) converges to $u$ in $C(S;H)$ and $L^2(S;V)$ .

<u>Proof.</u> First we note the simple inequalities

$$|P\,v - P\,w| \leq |v - w| \ , \ v, \ w \in H \quad \text{and} \quad |v|_\infty \leq 2\,|v|\,\|v\| \ , \ v \in V \ ,$$

where $|\cdot|_\infty$ is the norm in $L^\infty(0,1)$ . Next we define by

$$\|v\|^2_{C,k} = \sup_{t\in S} \left(e^{-k(t)}|v(t)|^2\right) \quad , \quad \|v\|^2_{X,k} = \frac{1}{2}\|v\|^2_{C,k} + \sup_{t\in S}\left(e^{-k(t)}\int_0^t \|v\|^2 ds\right)$$

norms being equivalent to the basic norms in $C(S;H)$ and $X = C(S;H) \cap L^2(S;V)$ , respectively. Here the function $k$ is defined by

$$k(t) = 2\int_0^t (\|u(s)\|^2 + \frac{16}{\nu}(\frac{r^2}{\nu}+1))\,ds\,.$$

Now we see from (1.3) and (1.6) that $Pu(t) = u(t)$ for $t \in S$. Consequently, (1.2) may be written in the form

$$u_t - \nu u_{xx} = f - (Pu)\,u_x\,. \tag{1.8}$$

Denoting the scalar product in $H$ by $(.,.)$ and setting $v^j = u^j - u$, we obtain from (1.7) and (1.8)

$$\begin{aligned}
(\tfrac{1}{2}|v^j|^2)_t + \nu\|v^j\|^2 &\leq |(Pu^{j-1}u_x^{j-1} - Pu\cdot u_x\,,\, v^j)| \\
&= |(Pu^{j-1}v_x^{j-1} + (Pu^{j-1} - Pu)u_x\,,\, v^j)| \\
&\leq (|Pu^{j-1}|\,\|v^{j-1}\| + |Pu^{j-1} - Pu|\,\|u\|)\,|v^j|_\infty \\
&\leq (r\|v^{j-1}\| + |v^{j-1}|\,\|u\|)\,|v^j|_\infty \\
&\leq \tfrac{\nu}{8}\|v^{j-1}\|^2 + \tfrac{2r^2}{\nu}|v^j|_\infty^2 + \tfrac{1}{8}\|u\|^2|v^{j-1}|^2 + 2|v^j|_\infty^2 \\
&\leq \tfrac{\nu}{8}\|v^{j-1}\|^2 + \tfrac{1}{8}\|u\|^2|v^{j-1}|^2 + 4(\tfrac{r^2}{\nu}+1)|v^j|\,\|v^j\| \\
&\leq \tfrac{\nu}{8}\|v^{j-1}\|^2 + \tfrac{1}{8}\|u\|^2|v^{j-1}|^2 + \tfrac{8}{\nu}(\tfrac{r^2}{\nu}+1)^2|v^j|^2 + \tfrac{\nu}{2}\|v^j\|^2
\end{aligned}$$

or

$$\begin{aligned}
(|v^j|^2)_t + \nu\|v^j\|^2 &\leq \tfrac{\nu}{4}\|v^{j-1}\|^2 + \tfrac{1}{4}\|u\|^2|v^{j-1}|^2 + \tfrac{16}{\nu}(\tfrac{r^2}{\nu}+1)^2|v^j|^2 \\
&\leq \tfrac{\nu}{4}\|v^{j-1}\|^2 + \tfrac{k'}{2}(\tfrac{1}{4}\|v^{j-1}\|^2 + |v^j|^2)\,.
\end{aligned}$$

Integration with respect to $t$ yields

$$\begin{aligned}
|v^j(t)|^2 + \nu\int_0^t\|v^j\|^2 ds &\leq \int_0^t (\tfrac{\nu}{4}\|v^{j-1}\|^2 + (\tfrac{1}{8}|v^{j-1}|^2 + \tfrac{1}{2}|v^j|^2)e^{-k}k'e^k)\,ds \\
&\leq \tfrac{\nu}{4}\int_0^t \|v^{j-1}\|^2 ds + (\tfrac{1}{8}\|v^{j-1}\|_{C,k}^2 + \tfrac{1}{2}\|v^j\|_{C,k}^2)(e^{k(t)} - 1)\,.
\end{aligned}$$

We divide by $e^{k(t)}$ and obtain

$$e^{-k(t)}(\|v^j(t)\|^2 + \nu\int_0^t \|v^j\|^2 ds) \leq \tfrac{1}{4}\|v^{j-1}\|_{X,k}^2 + \tfrac{1}{2}\|v^j\|_{C,k}^2$$

and hence

$$\|v^j\|_{X,k}^2 \leq \tfrac{3}{4}\|v^{j-1}\|_{X,k}^2 \leq \ldots \leq (\tfrac{3}{4})^j\|v^0\|_{X,k}^2\,.$$

From this our theorem follows.

Remark 1.1. Of course, the constant $r$ in (1.6) can be replaced by any other $C(S;H)$ a priori estimate for $u$. So in the special case $f=0$, $a \in L_\infty(0,1)$ one can set $r = \|a\|_\infty$ because of the maximum principle. If $f \in L^2(S;H)$, it is easy to see that $r = \sqrt{2}(|a| + \sqrt{2T}\|f\|_{L^2(S;H)})$

is a suitable bound. It is worth noticing that both these estimates are independent of the viscosity $\nu$ .

Remark 1.2. Evidently, there are other possibilities to introduce a projector like $P$ in order to obtain a globally convergent version of the Kato-Fujita method. However, the operator $P$ defined by (1.6) turns out to be favourable with respect to the numerical realisation of the iteration process.

## 2. The Navier-Stokes equations in two space dimensions

Let $G$ be a bounded domain in $R^2$ with smooth boundary $\Gamma$ and let $L^2(G)$ , $H_0^1(G)$ , $H^2(G)$ , $H^{-1}(G)$ be the usual Hilbert spaces. We set

$$H = (H_0^1(G))^2 = H_0^1(G)\times H_0^1(G) \ , \ V = (H^2(G)\cap H_0^1(G))^2 \ , \ V^* = (L^2(G))^2$$

and use again the symbols $|\cdot|$ , $\|\cdot\|$ , $\|\cdot\|_*$ to denote the norms in $H$ , $V$ and $V^*$ , respectively.

Let us consider the spatially two-dimensional Navier-Stokes equations

$$(2.1) \qquad \begin{aligned} & u_t - \nu\Delta u + u\cdot\nabla u + \nabla p = f \ , \ \nabla\cdot u = 0 \quad \text{in } G \ , \\ & u(0,x) = a(x) \ , \quad u|_\Gamma = 0 \ . \end{aligned}$$

Throughout this section we assume that

$$f\in L^2(S;V^*) \ , \quad a\in H \ , \quad \nabla\cdot a = 0 \ .$$

Then (cf. /5/), (2.1) has a unique solution $(u,p)$ with

$$u\in L^2(S;V)\cap C(S;H) \ , \ u_t\in L^2(S;V^*) \ , \ p\in L^2(S;V^*)$$

and the following a priori estimate holds

$$(2.2) \qquad \|u\|_{C(S;H)} \leq r \ , \quad r^2 = c\,(\,|a|^2 + \|f\|^2_{L^2(S;V^*)}) \ ,$$

where the constant $c$ depends only on $\nu$ and $G$ .

We now turn to the formulation of a globally convergent iteration procedure for solving (2.1). To this purpose we introduce the projector of $V^*$ onto the r-ball in $V^*$ which is defined by

$$P\,v = \begin{cases} v & \text{if } \|v\|_* \leq r \\ \dfrac{r\,v}{\|v\|_*} & \text{if } \|v\|_* > r \ , \end{cases}$$

where $r$ is the constant from (2.2). Now we are able to present the announced iteration procedure

$$(2.3)\quad \begin{aligned} & u^j_t - \nu\Delta u^j + \nabla p^j = f - u^{j-1}\cdot P\nabla u^{j-1}\ , \ \nabla\cdot u^j = 0\ , \ j=1,2,\ldots, \\ & u^j(0,x) = a(x)\ , \quad u^j\big|_\Gamma = 0\ . \end{aligned}$$

<u>Theorem 2.</u> Let $(u,p)$ be the solution of (2.1), $u^0 \in L^2(S;V)\cap C(S;H)$ an arbitrary starting function and $((u^j,p^j))$ the iteration sequence defined by (2.3). Then the following assertions hold

$$u^j \to u \ \text{ in } \ C(S;V^*) \ \text{ and } \ L^2(S;H)\ , \qquad \nabla p^j \to \nabla p \ \text{ in } \ (L^2(S;H^{-1}))^2.$$

<u>Proof.</u> We need the following well known inequalities

$$\|Pv - Pw\|_* \le \|v - w\|_*\ , \ v,w\in V\ , \qquad \|v\|_4^2 \le c\|v\|_*\,|v|\ , \quad v\in H\ .$$

Here the constant $c$ depends on $G$ and $\nu$, $|\cdot|_4$ denotes the $L^4(G)$-norm. Let $(.,.)$ be the scalar product in $V^*$. Then, using $P\nabla u = \nabla u$ and setting $v^j = u^j - u$, we find from (2.1) and (2.3)

$$\begin{aligned} (\tfrac{1}{2}\|v^j\|_*^2)_t + \nu|v^j|^2 &= |(v^{j-1}P\nabla u^{j-1} + u\cdot(P\nabla u^{j-1} - P\nabla u)\ ,\ v^j)| \\ &\le (\|v^{j-1}\|_4\|P\nabla u^{j-1}\|_* + \|u\|_4\|P\nabla v^{j-1} - P\nabla u\|_*)\|v^j\|_4 \\ &\le c_1(\|v^{j-1}\|_4 + \|\nabla v^{j-1}\|_*)\,\|v^j\|_4 \\ &\le c_1(\|v^{j-1}\|_*^{1/2}|v^{j-1}|^{1/2} + |v^{j-1}|)\|v^j\|_*^{1/2}|v^j|^{1/2} \\ &\le \tfrac{\nu}{2}(\tfrac{1}{4}|v^{j-1}|^2 + |v^j|^2) + \tfrac{c_2}{4}(\tfrac{1}{4}\|v^{j-1}\|_*^2 + \|v^j\|_*^2\ ) \end{aligned}$$

or

$$(\|v^j\|_*^2)_t + \nu|v^j|^2 \le \tfrac{\nu}{4}|v^{j-1}|^2 + \tfrac{c_2}{2}(\tfrac{1}{4}\|v^{j-1}\|_*^2 + \|v^j\|_*^2)\ .$$

Now we introduce the norms

$$\|v\|_{C,k} = \sup_{t\in S}(e^{-k(t)}\|v(t)\|_*)\ , \qquad \|v\|_{X,k}^2 = \tfrac{1}{2}\|v\|_{C,k}^2 + \nu\sup_{t\in S}(e^{-k(t)}\int_0^t|v|^2\,ds\ ,$$

being equivalent to the usual norms in $C(S;V^*)$ and $X=C(S;V^*)\cap L^2(S;H)$, respectively. Here $k(t) = c_2t$. As in the proof of Theorem 1 we then obtain

$$\|v^j\|_{X,k} \le (\tfrac{3}{4})^j\|v^0\|_{X,k}$$

and hence $u^j \to u$ in $C(S;V^*)$ and $L^2(S;H)$. Using (2.1) and (2.3), we conclude from the last convergence statement firstly $u^j_t \to u_t$ in $L^2(S;H^{-1}(G))$ and after that $\nabla p_j \to \nabla p$ in $L^2(S;H^{-1}(G))$.

## <u>3. The nonlinear Schrödinger equation</u>

In this section $L^2(0,1)$ denotes the space of complex-valued quadratically integrable functions on $(0,1)$. We set

$$H = L^2(0,1)\ , \quad V = \{\ v\in H \ |\ v_x \in H\ , \ \ v(0) = v(1)\}$$

and use now the symbol $\|.\|$ to denote the norm in $H$, whereas $|z|$

is the modulus of the complex number $z$ .

We consider the nonlinear Schrödinger equation with spatially periodic boundary conditions

$$
(3.1) \qquad \begin{aligned} & i\,u_t + u_{xx} + k|u|^2 u = 0 \;, \quad i^2 = -1 \;, \\ & u(0,x) = a(x) \;,\; u(t,0) = u(t,1) \;,\; u_x(t,0) = u_x(t,1) \;. \end{aligned}
$$

Here $k$ is a real constant.

We suppose $a \in V$ . Then (3.1) has a unique solution $u \in C(S;H) \cap L^\infty(S;V)$ with $u_t \in L^\infty(S;V^*)$ . Moreover, $u$ satisfies the a priori estimate (cf. /4/)

$$
\|u\|_{L^\infty((0,T)\times(0,1))} \leqq r \,,
$$

where

$$
r^2 = \|a\|(\|a\| + 2(|k|\,\|a\|^4(1 + |k|\,\|a\|^2) + \left|2\|a_x\|^2 - k\|a\|^4_{L^4(0,1)}\right|)^{1/2}) \;.
$$

This time we choose as the operator $P$ the projector of the complex plane onto the r-circle, i.e.

$$
P\,z = \begin{cases} z & \text{if } \;|z| \leqq r \\ r\frac{z}{|z|} & \text{if } \;|z| > r \;. \end{cases}
$$

Now we can formulate a globally convergent iteration method for solving (3.1).

$$
(3.2) \qquad \begin{aligned} & i\,u^j_t + u^j_{xx} = -k\,|Pu^{j-1}|^2 u^{j-1} \;, \qquad j=1,2,\ldots, \\ & u^j(0,x) = a(x) \;,\; u^j(t,0) = u^j(t,1) \;,\; u^j_x(t,0) = u^j_x(t,1) \;. \end{aligned}
$$

<u>Theorem 3.</u> Let $u$ be the solution of (3.1), $u^0 \in L^2(S;V) \cap C(S;H)$ an arbitrary starting function. Then the sequence $(u^j)$ defined by (3.2) converges to $u$ in $C(S;H)$ .

The proof of this theorem as well as proofs of further convergence statements concerning the iteration process (3.2) may be found in /4/.

## 4. Numerical realisation

The iteration processes under consideration reduce the problem of solving nonlinear evolution equations to the successive solution of sequences of <u>linear</u> evolution equations with <u>constant</u> coefficients. Nevertheless for numerical purposes it is necessary to combine them with other approximation methods. We have made some good numerical experience by combining iteration processes with a time-discrete Galerkin method. Let us briefly discuss this point. We confine ourself to Burgers' equation and use the notation introduced in section 1.

As basis functions we choose

$$h_l = h_l(x) = \sqrt{2}\sin l\pi x\ , \quad l=1,2,\dots\ .$$

The initial value $a$ has then the representation

$$a = \sum_{l=1}^{n} a_l h_l\ , \quad a_l = \int_0^1 a h_l\, dx\ .$$

We set $u_n = \sum_{l=1}^{n} c_l h_l$ and determine the coefficients $c_l = c_l(t)$ according to Galerkin's method by the following system of nonlinear ordinary differential equations

$$(4.1)\qquad \begin{aligned} & c_l' + p_l c_l + \int_0^1 (u_n(u_n)_x - f)\, h_l\, dx = 0\ , \quad p_l = \nu(l\pi)^2\ , \\ & c_l(0) = a_l\ , \quad l = 1,\dots,n\ . \end{aligned}$$

Taking into account (1.2), it is easy to show that the sequence $(u_n)$ of Galerkin approximations converges to the solution $u$ of Burgers' equation in $L^2(S;H_0^1(0,1))$ and $C(S;L^2(0,1))$.

In order to calculate $u_n$ we use an iteration process like (1.7). We set $u_n^j = \sum_{l=1}^{n} c_l^j h_l$ and determine the coefficients $c_l^j = c_l^j(t)$ by the system of linear ordinary differential equations

$$\begin{aligned} & (c_l^j)' + p_l c_l^j = \int_0^1 (f - Pu_n^{j-1}(u_n^{j-1})_x) h_l\, dx\ , \quad j = 1,2,\dots\ , \\ & c_l^j(0) = a_l\ , \quad l = 1,\dots,n\ . \end{aligned}$$

The solution of this system is

$$(4.2)\qquad c_l^j(t) = \exp(-p_l t)\Big(a_l + \int_0^t \exp(p_l s) F_l^{j-1}(s)\, ds\Big)\ , \quad l = 1,\dots,n\ ,$$

where

$$F_l^{j-1}(s) = F_l(s, u_n^{j-1}(s))$$

and the function $F_l(s,.)$ is defined by

$$F_l(s,v) = \int_0^1 (f(s) - vv_x)\, h_l\, dx \cdot \begin{cases} 1 & \text{if } |v| \le r \\ \frac{r}{|v|} & \text{if } |v| > r\ . \end{cases}$$

Here $|.|$ denotes the norm in $H = L^2(0,1)$ and $r$ is the a priori bound given in (1.3). We see that in order to get $u_n^j$ from $u_n^{j-1}$ we have only to calculate definite integrals. This can be done by using suitable rules for numerical integration. In our calculations it turned out to be adventageous to divide the time interval in smaller intervals $S_k$ , $S = \bigcup_{k=1}^{m} S_k$, and to carry out the iteration successively in $S_k$ , $k=1,\ldots,m$ .

## References

[1] Carasso, A., Computing Small Solutions of Burgers' Equation backw. in Time, Jnl. Math. Anal. Appl. 59, 169-209 (1977).
[2] Fujita, H., T. Kato, On the Navier-Stokes initial value problem, Arch. Rat. Mech. Anal. 16, 269-315(1965).
[3] Gajewski, H., Zur globalen Konvergenz eines modifizierten Kato-Fujita Verfahrens, ZAMM (to appear).
[4] Gajewski, H., Über Näherungsverfahren zur Lösung der nichtlinearen Schrödinger Gleichung, Math. Nachr. (to appear).
[5] Von Wahl, W., Instationary Navier-Stokes Equations and Parabolic Systems, Universität Bonn, Sonderforschungsbereich 72, preprint no. 112 (1976).

Author's address: Akademie der Wissenschaften der DDR,
Zentralinstitut für Mathematik und Mechanik,
DDR - 108 Berlin, Mohrenstr. 39

# EXPONENTIAL REPRESENTATION OF SOLUTIONS OF ORDINARY DIFFERENTIAL EQUATIONS

R. Gamkrelidze, Moscow

I shall describe here a kind of calculus for solutions of ordinary differential equations developed jointly with my collaborator A.Agrachev. This calculus is based on the exponential representation of the solutions and reflects their most general group-theoretic properties. In deriving the calculus we were strongly influenced by problems of control and optimization and it is shaped according to the needs of these theories. Nevertheless it might be considered, as I believe, not merely as a technical tool for dealing with control problems only but could also be of more general interest. This may justify my choice of the topic for the Equadiff conference.

## 1. Differential equations considered

Let us consider a differential equation in $\mathbb{R}^n$

$$\dot{z} = X_t(z) \tag{1}$$

where $X_t(z)$ is a $C^\infty$-function in $z \in \mathbb{R}^n$ for $\forall t \in \mathbb{R}$, measurable in $t$ for $\forall z \in \mathbb{R}^n$ and satisfying the condition

$$\|X_t\|_k \leq \mu_k(t), \quad \int_{\mathbb{R}} \mu_k(t)dt < \infty, \quad k=0,1,\dots \tag{2}$$

where $\|\cdot\|_k$ denotes the norm of the uniform convergence in $\mathbb{R}^n$ up to the k-th derivative.

Our first goal is to find a suitable representation of the flow defined by (1), that is, of a family of $C^\infty$-diffeomorphisms $F_t$, $t \in \mathbb{R}$, of $\mathbb{R}^n$ satisfying the equation

$$\frac{d}{dt} F_t x = X_t(F_t x), \quad F_0 = Id \quad \forall x \in \mathbb{R}^n. \tag{3}$$

The existence of $F_t$ is guaranteed by (2).

## 2. Transforming (3) into a linear "operator equation"

There is a procedure transforming the nonlinear equation (3) into a certain linear "operator equation" for $F_t$. To describe it let me introduce some standard notions.

$\Phi$ will denote the algebra of all $C^\infty$-scalar functions $f, g, \dots$ on $\mathbb{R}^n$ with the topology of the uniform convergence on compact sets for every derivative. $\mathcal{A}$ stands for the associative algebra of all continuous linear transformations of $\Phi$. The composition of two elements $A_1, A_2$ in $\mathcal{A}$ will be denoted by $A_1 \circ A_2$. The operators from $\mathcal{A}$ can be applied also to vector-valued functions on $\mathbb{R}^n$. Denote by $\Theta$ the identity mapping of $\mathbb{R}^n$ : $\Theta(x) \equiv x$. We shall say that a sequence of operators $A_1, A_2, \ldots$ from $\mathcal{A}$ is convergent to $A$ iff the sequence of functions $A_1\Theta, A_2\Theta, \ldots$ converges in $\Phi$ to $A\Theta$. Every diffeomorphism $F$ of $\mathbb{R}^n$ will be considered as an element of $\mathcal{A}$: $Ff(x) = f(Fx)$, $\forall x \in \mathbb{R}^n$, and the set of all $C^\infty$-diffeomorphisms of $\mathbb{R}^n$ will be denoted by $\mathcal{D}$. By $\mathcal{L}$ we shall denote the Lie algebra of all $C^\infty$-vector fields on $\mathbb{R}^n$, which is a subspace of $\mathcal{A}$ characterized by the differentiation rule $X(fg) = (Xf)g + f(Xg)$ $\forall X \in \mathcal{L}$, $\forall f,g \in \Phi$. The Lie bracket of two fields will be denoted as usual by $[X,Y] = X \circ Y - Y \circ X = (\operatorname{ad} X)Y$. The following important relation holds:

$$(4) \qquad F \circ X \circ F^{-1} \overset{\text{def}}{=} (\operatorname{Ad} F)X \in \mathcal{L} \ \forall X \in \mathcal{L}, \ \forall F \in \mathcal{D}.$$

Consider $X_t$, $t \in \mathbb{R}$, in (1) as a nonstationary (time-dependent) vector field on $\mathbb{R}^n$. It is not difficult to show that (3) is equivalent to the linear "operator equation" for the flow $F_t$

$$(5) \qquad \frac{d}{dt} F_t = F_t \circ X_t, \quad F_0 = \mathrm{Id} \Leftrightarrow F_t = \mathrm{Id} + \int_0^t F_\tau \circ X_\tau \, d\tau,$$

where the operations of differentiation and integration in $t$ should be understood in the "weak" sense: first apply the operator to an arbitrary function from $\Phi$ and then differentiate or integrate. The equivalence between (3) and (5) should be understood literally - the existence of a unique solution of (3) implies the existence of a unique solution $F_t$, $t \in \mathbb{R}$, for (5) which at the same time necessarily turns out to be a flow and vice versa. Certainly we can always consider the flow $F_t$ only for values of $t$ sufficiently close to zero since the equation $F_t = F_{t_0} + \int_{t_0}^t F_\tau \circ X_\tau \, d\tau$, $t_0$- arbitrary fixed, has exactly the same properties as (5), which permits to restore the whole flow $F_t$, $t \in \mathbb{R}$.

Call the formal series

$$(6) \qquad \mathrm{Id} + \sum_{i=1}^{\infty} \int_0^t d\tau_1 \int_0^{\tau_1} d\tau_2 \cdots \int_0^{\tau_{i-1}} d\tau_i X_{\tau_i} \circ X_{\tau_{i-1}} \circ \cdots \circ X_{\tau_1},$$

arising when solving the linear equation (5) formally, the Volterra series corresponding to (5).

3. Exponential representation of the flow

Suppose the field $X_t$ analytic on $\mathbb{C}^n$ and subject to the condition (2), where the norms $\|\cdot\|_k$ should be understood (in this case) as norms of the uniform convergence in a certain complex $\sigma_k$--neighbourhood ($\sigma_k > 0$) of $\mathbb{R}^n \subset \mathbb{C}^n$. Then the Volterra series (6) converges (in the above defined sense) for every $t$ rendering the integral $\int_0^t \mu_0(\tau)d\tau$ a sufficiently small value to an analytic diffeomorphism $x \mapsto F_t x$, and the obtained flow is the unique solution of the equation (5) (proof by the method of majorants).

The fields $X_t$ generally do not commute for different values of $t$, hence the order of the factors in the term $\int_0^t d\tau_1 \int_0^{\tau_1} d\tau_2 \dots \int_0^{\tau_{i-1}} d\tau_i X_{\tau_i} \dots X_{\tau_1}$ could not be changed and the corresponding times $\tau_j$ increase from left to right: $0 \leq \tau_i \leq \dots \leq \tau_1 \leq t$. Adopting the terminology used by physicists we call the flow $F_t$ to which the series (6) converges the right chronological exponent of $X_t$ and denote it by

$$F_t = \overrightarrow{\exp} \int_0^t X_\tau d\tau \quad , \tag{7}$$

the arrow indicating the direction of growth of the $\tau_j$-s in the successive terms of the "right" Volterra series (6).

In the general $C^\infty$-case the series (6) is not convergent, however, we can call the unique solution of (5) (which exists and is a flow according to the standard existence theorem for (3)) the right chronological exponent of $X_t$ and denote it with the same symbol (7). The following basic asymptotics may justify this convention:

$$\Big\| \Big\{ \overrightarrow{\exp} \int_0^t X_\tau d\tau - \Big(Id + \sum_{i=1}^k \int_0^t d\tau_1 \int_0^{\tau_1} d\tau_2 \dots \int_0^{\tau_{i-1}} d\tau_i X_{\tau_i} \circ \dots \circ X_{\tau_1}\Big)\Big\} f \Big\|_{Q,0} \leq C_{Q,k} \Big( \int_0^t \|X_\tau\|_{\hat{Q},k+1} d\tau \Big)^{k+1} \|f\|_{\hat{Q},k+1} \quad \forall f \in \Phi ,$$

$$k = 1, 2, \dots ,$$

where $\|\cdot\|_{Q,j}$ , $j=0,1,2,\ldots$ denotes here and in the sequel the norm of the uniform convergence of all derivatives up to the order $j$ on an arbitrary compact set $Q \in \mathbb{R}^n$, $\hat{Q}$ - a compact neighbourhood of $Q$ of radius $\int_0^t \mu_0(\tau)d\tau$ .

However, we can go even further in interpreting the symbol (7) and describe a sort of "summation procedure" which enables us to "sum up" the Volterra series (6) in the general $C^\infty$-case to a family of operators $F_t$ which turns out to be a flow satisfying the equation (5), and thus give an existence proof for (5). Uniqueness is an easy consequence of the fact that $F_t$ is a flow.

To describe the "summation procedure" take the "$\delta$-type" analytic mollifier

$$\omega_\varepsilon = \frac{1}{(\sqrt{\pi}\,\varepsilon)^n} e^{-(\frac{z}{\varepsilon})^2} \qquad (\varepsilon \to 0)$$

and consider the convolution

$$X_t^{(\varepsilon)}(z) = \omega_\varepsilon * X_t = \frac{1}{(\sqrt{\pi}\,\varepsilon)^n} \int_{\mathbb{R}^n} e^{-(\frac{z-w}{\varepsilon})^2} X_t(w)dw \quad .$$

The obtained field $X_t^{(\varepsilon)}$ is an entire-analytic field on $\mathbb{C}^n$ for every $\varepsilon > 0$ subject to (2) (up to a constant factor for $\mu_k(t)$), thus the corresponding Volterra series is convergent to a flow $F_t^{(\varepsilon)}$ which satisfies the equation (5). It turns out that $F_t^{(\varepsilon)}$ $(\varepsilon \to 0)$ is a Cauchy family of flows (in the topology of the uniform convergence on compact sets of $\mathbb{R}^n$ for every derivative) and converges to a flow $F_t$ which is the unique solution of (5). We consider the flow $F_t$ , $t \in \mathbb{R}$ as the "generalized sum" of the right Volterra series (6) and call it the right chronological exponent of $X_t$ :

$$\frac{d}{dt} \overrightarrow{\exp} \int_0^t X_\tau \, d\tau = \overrightarrow{\exp} \int_0^t X_\tau \, d\tau \circ X_t \, .$$

The left Volterra series and the corresponding left chronological exponent could be considered in a completely symmetrical way

$$G_t = \overleftarrow{\exp} \int_0^t X_\tau \, d\tau = \mathrm{Id} + \sum_{i=1}^{\infty} \int_0^t d\tau_1 \int_0^{\tau_1} d\tau_2 \ldots \int_0^{\tau_{i-1}} d\tau_i X_{\tau_1} \circ \ldots \circ X_{\tau_i} \, .$$

The flow $\overleftarrow{\exp}\int_0^t -X_\tau\, d\tau$ satisfies the "adjoint operator equation" for (5)

$$\frac{d}{dt} G_t = -X_t \circ G_t \,, \quad G_0 = Id \,,$$

which is equivalent to the linear partial differential equation of the first order in $\mathbb{R}^n$

$$\frac{\partial w(t,x)}{\partial t} + \sum_{i=1}^{n} X_t^i(x) \frac{\partial w(t,x)}{\partial x^i} = \frac{\partial w}{\partial t} + X_t w = 0 \,,$$

$w(t,x) = G_t f(x)$, $f(x) = G_0 f(x) = w(0,x)$ - the initial function. Evidently

$$\overrightarrow{\exp}\int_0^t X_\tau\, d\tau \circ \overleftarrow{\exp}\int_0^t -X_\tau\, d\tau = \overleftarrow{\exp}\int_0^t -X_\tau\, d\tau \circ \overrightarrow{\exp}\int_0^t X_\tau d\tau = Id \,.$$

In the "commutative case" that is if $\left[X_t \,, \int_0^t X_\tau\, d\tau\right] = 0$ $\forall\, t \in \mathbb{R}$ we have

$$\overrightarrow{\exp}\int_0^t X_\tau\, d\tau = \overleftarrow{\exp}\int_0^t X_\tau\, d\tau = Id + \sum_{i=1}^{\infty} \frac{1}{i!} \left(\int_0^t X_\tau\, d\tau\right)^i = \\ = e^{\int_0^t X_\tau d\tau} \,.$$

For example, if $X_t \equiv X$ then $\overrightarrow{\exp}\int_0^t X_\tau\, d\tau = e^{tX}$ .

To demonstrate the flexibility of the obtained representation I shall derive formulas expressing two basic objects in the theory of ordinary differential equations - the perturbing flow of a given flow $F_t$ and the variation of $F_t$ .

4. <u>The perturbing flow</u>

Suppose the field $X_t$ and the corresponding flow $F_t = \overrightarrow{\exp}\int_0^t X_\tau\, d\tau$ fixed. Call an arbitrary field $Y_t$ a perturbing field for $X_t$ , the flow $\overrightarrow{\exp}\int_0^t (X_\tau + Y_\tau)d\tau$ - the corresponding perturbed flow.

<u>Problem.</u> Find a flow $C_t = C_t(Y_\tau)$ satisfying the equation

$$\overrightarrow{\exp}\int_0^t (X_\tau + Y_\tau)d\tau = C_t(Y_\tau) \circ \overrightarrow{\exp}\int_0^t X_\tau\, d\tau = C_t(Y_\tau) \circ F_t \,. \tag{8}$$

We call $C_t(Y_\tau)$ the perturbing flow for $F_t$ corresponding to $Y_t$.

The proposed solution coincides with the method of variation of constants and could be carried out as follows. According to (4) we can consider $\mathrm{Ad}\, F_t$ in the formula

$$(\mathrm{Ad}\, F_t)Z = F_t \circ Z \circ F_t^{-1}, \quad Z \in \mathscr{L}, \tag{9}$$

as a time-dependent linear transformation of $\mathscr{L}$. Differentiating we obtain the equation

$$\frac{d}{dt}\mathrm{Ad}\, F_t = \mathrm{Ad}\, F_t \circ \mathrm{ad}\, X_t ,$$

which suggests the notation

$$\mathrm{Ad}\, F_t = \overrightarrow{\exp}\int_0^t \mathrm{ad}\, X_\tau\, d\tau . \tag{10}$$

Differentiation of (8) yields

$$\frac{d}{dt} C_t(Y_\tau) = C_t(Y_\tau)\circ(\mathrm{Ad}\, F_t)Y_t ,$$

whence combining (8) and (10) we come to formulas

$$C_t(Y_\tau) = \overrightarrow{\exp}\int_0^t (\mathrm{Ad}\, F_\tau)Y_\tau\, d\tau = \tag{11}$$

$$= \overrightarrow{\exp}\int_0^t \Big(\overrightarrow{\exp}\int_0^\tau \mathrm{ad}\, X_s ds\Big)Y_\tau\, d\tau ,$$

$$\overrightarrow{\exp}\int_0^t (X_\tau + Y_\tau)d\tau = \overrightarrow{\exp}\int_0^t \Big(\overrightarrow{\exp}\int_0^\tau \mathrm{ad}\, X_s ds\Big)Y_\tau\, d\tau \circ$$

$$\overrightarrow{\exp}\int_0^t X_\tau\, d\tau ,$$

asserting two basic facts. 1) If $X_t, Y_t$ are analytic (in $t$ and $z$) and satisfy (2) then evaluating all chronological exponents on the right sides as the corresponding formal right Volterra series and performing the indicated operations we come to convergent (in appropriate regions) series defining the flows standing on the left sides. 2) For the general $C^\infty$-case the equalities (10)-(11) should be understood in the following asymptotic sense:

$$\Big\| \Big\{ \overrightarrow{\exp}\int_0^t \mathrm{ad}\, X_\tau\, d\tau - \Big(\mathrm{Id} + \sum_{i=1}^{k}\int_0^t d\tau_1 \int_0^{\tau_1} d\tau_2 \cdots \int_0^{\tau_{i-1}} d\tau_i\ \mathrm{ad}\, X_{\tau_i}\circ \cdots$$

$$\cdots \circ \mathrm{ad}\, X_{\tau_1}\Big)\Big\} Z \Big\|_{Q,o} \le C_{Q,k}\Big(\int_0^t \|X_\tau\|_{\hat{Q},k+1} d\tau\Big)^{k+1} \|Z\|_{\hat{Q},k+1} \quad \forall Z \in \mathscr{L},$$

$$k = 1, 2, \ldots ;$$

$$\Big\| \overrightarrow{\exp} \int_0^t (X_\tau + Y_\tau)d\tau - (\mathrm{Id} + \sum_{i=1}^{k} \int_0^t d\tau_1 \int_0^{\tau_1} d\tau_2 \ldots \int_0^{\tau_{i-1}} d\tau_i (\mathrm{Ad}\, F_{\tau_i})$$

$$\cdot Y_{\tau_i} \circ \ldots \circ (\mathrm{Ad}\, F_{\tau_1}) Y_{\tau_1}) \circ F_t \Big\|_{Q,0} \le C_{Q,k} \Big( \int_0^t \| F_\tau^{-1} \|_{\hat{Q},k+1} \| Y_\tau \|_{\hat{Q},k} d\tau \Big)^{k+1}$$

$\| F_t \|_{\hat{Q},k+1}$ . In case of time independent fields $X_t \equiv X$, $Y_t \equiv Y$ we have $e^{tX} \circ Z \circ e^{-tX} = e^{t\,\mathrm{ad}\,X} Z$ , $e^{t(X+Y)} = \overrightarrow{\exp} \int_0^t e^{\tau\,\mathrm{ad}\,X} Y d\tau \circ e^{tX}$ .

The second formula shows that even if the fields $X, Y$ are time--independent the corresponding perturbing flow is expressed through a chronological exponent.

Consider a flow $F_t = \overrightarrow{\exp} \int_0^t Y_\tau^{(1)} d\tau \circ \ldots \circ \overrightarrow{\exp} \int_0^t Y_\tau^{(m)} d\tau$ and suppose we have to define the field $Z_t$ that generates $F_t$ : $F_t = \overrightarrow{\exp} \int_0^t Z_\tau d\tau$ . We call $Z_t$ the right chronological logarithm of $F_t$ and denote $Z_t = \overrightarrow{\log} F_t$ . It is expressed by (see (9)-(10))

$$\overrightarrow{\log} \Big(\overrightarrow{\exp} \int_0^t Y_\tau^{(1)} d\tau \circ \ldots \circ \overrightarrow{\exp} \int_0^t Y_\tau^{(m)} d\tau \Big) \overset{\mathrm{def}}{=} \tag{12}$$

$$\overset{\mathrm{def}}{=} \lambda (Y_\tau^{(1)}, \ldots, Y_\tau^{(m)}) = F_t^{-1} \circ \frac{d}{dt} F_t =$$

$$= \Big(\overleftarrow{\exp} \int_0^t -\mathrm{ad}\, Y_\tau^{(m)} d\tau \circ \ldots \circ \overleftarrow{\exp} \int_0^t -\mathrm{ad}\, Y_\tau^{(2)} d\tau \Big) Y_t^{(1)} +$$

$$+ \Big(\overleftarrow{\exp} \int_0^t -\mathrm{ad}\, Y_\tau^{(m)} d\tau \circ \ldots \circ \overleftarrow{\exp} \int_0^t -\mathrm{ad}\, Y_\tau^{(3)} d\tau \Big) Y_t^{(2)} +$$

$$+ \ldots + \Big(\overleftarrow{\exp} \int_0^t -\mathrm{ad}\, Y_\tau^{(m)} d\tau \Big) Y_t^{(m-1)} + Y_t^{(m)} .$$

## 5. The variation of a flow

We start with the following problem. For a given flow $\hat{F}_t = \overrightarrow{\exp} \int_0^t Y_\tau d\tau$ determine a field $V_t(Y_\tau)$ which satisfies in an appropriate asymptotic sense (to be formulated precisely) the relation

$$\overrightarrow{\exp}\int_0^t Y_\tau\, d\tau \cong e^{V_t(Y_\tau)} = \mathrm{Id} + \sum_{i=1}^{\infty} \frac{1}{i!}\,(V_t(Y_\tau))^i .$$

It is natural to call $V_t(Y_\tau)$ the usual (not chronological) logarithm of the flow considered and to denote

$$V_t(Y_\tau) = \ln \overrightarrow{\exp}\int_0^t Y_\tau\, d\tau .$$

For a precise formulation we have to consider noncommutative nonassociative polynomials over $\mathbb{R}$, $P(Y_1,\dots,Y_k)$ in $k=1,2,\dots$ $\mathcal{L}$ - valued variables $Y_i$ with the Lie bracket multiplication; the P-s consequently will also be $\mathcal{L}$ - valued. A simple and explicit algorithm (see n° 6) prescribes a universal sequence of such polynomials

(13) $$P_1(Y_1),\quad P_2(Y_1,Y_2),\dots,\quad P_k(Y_1,\dots,Y_k),\dots ,$$

$P_k$ - homogeneous of degree $k$ in its variables, each variable having degree 1, for which the following theorem is valid.

<u>Theorem.</u> For every field $Y_t$ consider the formal series

(14) $$V_t(Y_\tau) = \sum_{i=1}^{\infty}\int_0^t d\tau_1\int_0^{\tau_1} d\tau_2 \dots \int_0^{\tau_{i-1}} d\tau_i P_i(Y_{\tau_1},\dots,Y_{\tau_i}) = \sum_{i=1}^{\infty} V_t^{(i)}(Y_\tau)$$

and call it the formal vector field associated with $Y_t$. Then the following asymptotics holds:

(15) $$\left\| \overrightarrow{\exp}\int_0^t Y_\tau\, d\tau - e^{\sum_{i=1}^{k} V_t^{(i)}(Y_\tau)} \right\|_{Q,o} \le C_{Q,k}\left(\int_0^t \|Y_\tau\|_{\hat{Q},k+1}\, d\tau\right)^{k+1}, \quad k=1,2,\dots ,$$

also expressed by either of the relations

(16) $$\ln \overrightarrow{\exp}\int_0^t Y_\tau\, d\tau \cong \sum_{i=1}^{\infty} V_t^{(i)}(Y_\tau) = V_t(Y_\tau) ,$$

$$\overrightarrow{\exp}\int_0^t Y_\tau\, d\tau \cong e^{\sum_{i=1}^{\infty} V_t^{(i)}(Y_\tau)} = e^{V_t(Y_\tau)} .$$

As an immediate consequence of (15) we obtain

$$(17)\qquad (V_t^{(i)}(Y_\tau))x = 0,\ i=1,\dots,k-1 \Longrightarrow (\overrightarrow{\exp}\int_0^t Y_\tau\, d\tau)x =$$
$$= x + (V_t^{(k)}(Y_\tau))x + \mathcal{O}(\int_0^t \|Y_\tau\|_{Q,k+1} d\tau)^{k+1}\ \forall x \in \mathbb{R}^n,$$
$$k=1,2,\dots .$$

Formulas (15), (17) justify the forthcoming terminology. Call the field $V_t^{(k)}(Y_\tau)$ the k-th variation of the identity flow $Id_t$ corresponding to $Y_t$ - the perturbing field of the zero field (which generates the identity flow), and denote $V_t^{(k)}(Y_\tau) = \delta^{(k)} Id_t(Y_\tau)$. Similarly, call the formal field (14) the (full) variation $\delta Id_t(Y_\tau)$ of the identity flow

$$(18)\qquad \delta Id_t(Y_\tau) = \sum_{i=1}^{\infty} \delta^{(i)} Id_t(Y_\tau),$$

and the formal series

$$(19)\qquad e^{V_t(Y_\tau)} = Id + \sum_{i=1}^{\infty} \frac{1}{i!} (V_t(Y_\tau))^i$$

- the formal flow corresponding to the formal field $V_t(Y_\tau)$. According to (14), (16) the following basic asymptotic expansion is valid:

$$(20)\qquad \overrightarrow{\exp}\int_0^t Y_\tau\, d\tau \cong e^{V_t(Y_\tau)} = e^{\delta Id_t(Y_\tau)} =$$
$$= Id + \sum_{i=1}^{\infty} \frac{1}{i!} (\delta Id_t(Y_\tau))^i,$$
$$\delta Id_t(Y_\tau) = \sum_{i=1}^{\infty} \delta^{(i)} Id_t(Y_\tau) =$$
$$= \sum_{i=1}^{\infty} \int_0^t d\tau_1 \int_0^{\tau_1} d\tau_2 \dots \int_0^{\tau_{i-1}} d\tau_i P_i(Y_{\tau_1},\dots,Y_{\tau_i}).$$

We call it the "Maclaurin series expansion" (around the zero field) of the flow $\overrightarrow{\exp}\int_0^t Y_\tau\, d\tau$ - the perturbing flow for $Id_t$ under $Y_t$, which can be also regarded as the corresponding perturbed flow. A composition rule in the set of formal flows (19) defined by (see (12))

$$e^{V_t(Y_\tau^{(1)})} \circ e^{V_t(Y_\tau^{(2)})} = e^{V_t(\lambda(Y_\tau^{(1)}, Y_\tau^{(2)}))}$$

turns it into a multiplicative group.

Unifying formulas (12), (20) (see also the construction of the $P_i$-s) we come to a generalization of the Campbell-Hausdorff formula

$$\text{(21)} \qquad \overrightarrow{\exp}\int_0^t Y_\tau^{(1)} d\tau \circ \ldots \circ \overrightarrow{\exp}\int_0^t Y_\tau^{(m)} d\tau \cong$$

$$\cong e^{\sum_{i=1}^{\infty} \int_0^t d\tau_1 \ldots \int_0^{\tau_{i-1}} d\tau_i P_i(Z_{\tau_1},\ldots,Z_{\tau_i})},$$

$$Z_t = (\overleftarrow{\exp}\int_0^t -\mathrm{ad}\, Y_\tau^{(m)} d\tau \circ \ldots \circ \overleftarrow{\exp}\int_0^t -\mathrm{ad}\, Y_\tau^{(2)} d\tau) Y_t^{(1)} +$$

$$+ \ldots + (\overleftarrow{\exp}\int_0^t -\mathrm{ad}\, Y_\tau^{(m)} d\tau) Y_t^{(m-1)} + Y_t^{(m)} .$$

The usual Dynkin form of this formula (when $Y_t^{(i)} \equiv Y$, $m=2$, $t=1$) seems to be unnecessarily complicated which results from the fact that it actually carries out all i-fold integrations indicated in (21). For analytic fields (both in $t$ and $z$) all formal series involved in (18)-(21) are convergent provided the appropriate norms of the $Y_t$-s are sufficiently small (I shall not go here into the details of precise formulation).

The crucial advantage of the introduced variations consists in the validity of the asymptotic relation (20) and in their invariant form - the $\delta^{(k)} Id_t$-s are vector fields and thus belong to the first tangent bundle of the underlying space (in our case of $\mathbb{R}^n$), consequently they act not only on $\Phi$ but also on $\mathbb{R}^n$ as "infinitesimal displacements" of $\mathbb{R}^n$. This permits to obtain by (18) the "formal infinitesimal displacement" of $\mathbb{R}^n$ - the full variation $\delta Id_t(Y_\tau)$ and finally, using the Maclaurin expansion (20) to come to the asymptotic evaluation of the perturbing flow $\overrightarrow{\exp}\int_0^t Y_\tau d\tau$ - the basic goal of many problems connected with ordinary differential equations, in particular of optimization problems.

"The usual variations" of the perturbing flow - the successive terms in the Volterra expansion

$$\overrightarrow{\exp}\int_0^t Y_\tau d\tau = Id + \int_0^t d\tau_1 Y_{\tau_1} + \int_0^t d\tau_1 \int_0^{\tau_1} d\tau_2 Y_{\tau_2} \circ Y_{\tau_1} + \ldots$$

have no invariant meaning starting from the quadratic term, therefore they act only on $\Phi$ but not on $\mathbb{R}^n$. Our actual achievement consists in extracting the "invariant variations" $\delta^{(k)} Id_t(Y_\tau)$ from the "usual ones". Their interrelations are established by (20), for example

$$\delta^{(2)}\mathrm{Id}_t(Y_\tau) = \int_0^t d\tau_1 \int_0^1 d\tau_2 Y_{\tau_2} \circ Y_{\tau_1} - \frac{1}{2}\delta^{(1)}\mathrm{Id}_t(Y_\tau)\circ$$
$$\circ\, \delta_t^{(1)}\mathrm{Id}_t(Y_\tau) .$$

Suppose an arbitrary flow $F_t = \overrightarrow{\exp}\int_0^t X_\tau\, d\tau$ rather than the identity flow is perturbed by $Y_t$. Then the corresponding variations $\delta^{(k)}F_t(Y_\tau)$ are defined by

$$\delta^{(k)}F_t(Y_\tau) = \int_0^t d\tau_1 \int_0^{\tau_1} d\tau_2 \ldots \int_0^{\tau_{k-1}} d\tau_k P_k((\overrightarrow{\exp}\int_0^{\tau_1} \mathrm{ad}\, X_s ds)Y_{\tau_1},$$
$$\ldots, (\overrightarrow{\exp}\int_0^{\tau_k} \mathrm{ad}\, X_s ds)Y_{\tau_k}) ,$$

the full variation $\delta F_t(Y_\tau) = \sum_{i=1}^{\infty} \delta^{(i)}F_t(Y_\tau)$, and for the perturbing flow we obtain the "Taylor series expansion around the initial field $X_t$":

$$\overrightarrow{\exp}\int_0^t (\overrightarrow{\exp}\int_0^\tau \mathrm{ad}\, X_s ds) Y_\tau\, d\tau \cong e^{\delta F_t(Y_\tau)} .$$

As a Taylor series expansion of the perturbed flow we may consider

$$\overrightarrow{\exp}\int_0^t (X_\tau + Y_\tau)d\tau \cong e^{\delta F_t(Y_\tau)} \circ e^{\delta \mathrm{Id}_t(X_\tau)} .$$

## 6. Construction of the polynomials (13)

Consider the free associative algebra $\mathrm{Ass}\,(\mathrm{ad}, Y_1, Y_2, \ldots)$ over $\mathbb{R}$ with (multiplicative) generators $\mathrm{ad}, Y_1, Y_2, \ldots$, and denote by $D(a)$, $a \in \mathrm{Ass}\,(\mathrm{ad}, Y_1, Y_2, \ldots)$, differentiation in the algebra defined on generators by $D(a) = a(\mathrm{ad})$, $D(a)Y_i = aY_i$, $i=1,2,\ldots$ .

Further, consider an arbitrary word from the algebra composed of generators $\mathrm{ad}, Y_1, Y_2, \ldots$ . To each $Y_k$ entering a given word $w$ we assign a nonnegative integer - the index of $Y_k$ in $w$ - by the following procedure. Represent $w = w_1 Y_k w_2$, where $w_1$ (may be an empty word) does not contain $Y_k$ and suppose $w_1 = v_1 \ldots . v_\ell$, where each of the $v_j$-s is one of the generators. Define a set $J \subset \{1, \ldots, \ell\}$ by the rule: $i \in J$ iff the following two conditions are satisfied: 1) the number of occurences of ad in the word $v_i v_{i+1} \ldots . v_\ell$ is equal to the number of occurences of the $Y_j$-s;

2) for every $i' > i$ the number of occurences of ad in the word $v_{i'} v_{i'+1} \dots v_{\ell}$ does not exceed the number of occurences of the $Y_j$-s . Then the index of $Y_k$ in $w$ is equal to the number of elements in $J$ (which may turn out to be empty).

Among all possible words composed of ad and the $Y_j$-s call regular words those which could be regarded (by appropriate distribution of parentheses) as elements of the free Lie algebra with the $Y_j$-s as generators and ad with its usual meaning (whenever possible this could be done only in a unique way).

Finally write down the sequence of real numbers $\beta_0 = 1$, $\beta_1 = \frac{1}{2}$ , $\beta_i = \frac{1}{i!} B_i$ , $i=2,3,\dots$ , where $B_i$ , $i \geq 2$ , is the i-th Bernoulli number: $B_3 = B_5 = B_7 = \dots = 0$ , $B_2 = \frac{1}{6}$ , $B_4 = -\frac{1}{30}$ , ... .

Now consider an element

$(D(\text{ad } Y_k) \circ \dots \circ D(\text{ad } Y_2))Y_1 \in \text{Ass}\,(\text{ad}, Y_1, Y_2, \dots)$, $k \geq 2$ ,

which is obtained from $Y_1$ by successive applications of the differentiation operators $D(\text{ad } Y_2), \dots, D(\text{ad } Y_k)$ and which is the sum of $(2k-3)!!$ regular words

$(D(\text{ad } Y_k) \circ \dots \circ D(\text{ad } Y_2))Y_1 = w_1 + \dots + w_{(2k-3)!!}$ , each of the symbols $Y_1, \dots, Y_k$ entering every word $w_j$ exactly once. Denote the index of $Y_i$ in $w_j$ by $N_{ij}$ and define $P_1(Y_1) = Y_1$ ,

$$P_k(Y_1, \dots, Y_k) = \beta_{N_{11}} \dots \beta_{N_{k1}} w_1 + \beta_{N_{12}} \dots \beta_{N_{k2}} w_2 + $$
$$+ \dots + \beta_{N_{1(2k-3)!!}} \dots \beta_{N_{k(2k-3)!!}} w_{(2k-3)!!} , \quad k \geq 2 .$$

Here are the first four polynomials:

$$P_1 = Y_1 , \quad P_2 = \frac{1}{2}[Y_2, Y_1] ,$$
$$P_3 = \frac{1}{6}[Y_3, [Y_2, Y_1]] + \frac{1}{6}[[Y_3, Y_2], Y_1] ,$$
$$P_4 = \frac{1}{2}([[Y_4, Y_3], [Y_2, Y_1]] + [[[Y_4, Y_3], Y_2], Y_1] +$$
$$+ [Y_4, [[Y_3, Y_2], Y_1]] + [Y_3, [[Y_4, Y_2], Y_1]] .$$

Author's address: Steklov Institute, 42 Vavilov Str.,
Moscow 117333, USSR

# THE RAYLEIGH AND VAN DER POL WAVE EQUATIONS, SOME GENERALIZATIONS*

W. S. Hall, Pittsburgh

Here are two interesting nonlinear partial differential equations. The first we call the Rayleigh wave equation,

$$y_{tt} - y_{xx} = \varepsilon(y_t - y_t^3) \qquad (1.1)$$
$$y(t,0) = y(t,\pi) = 0,$$

and the second is the wave equation of Van der Pol type,

$$y_{tt} - y_{xx} = \varepsilon(1 - y^2)y_t$$
$$y(t,0) = y(t,\pi) = 0 \qquad (1.2)$$

Each of these has been used to model physical phenomena, although they first appeared in the literature as curiosities. For example, in [3] we see (1.1) serving as a model for the large amplitude vibrations of wind-blown, ice-laden power transmission lines. Equation (1.2), on the other hand, can describe plane electromagnetic waves propagating between two parallel planes in a region where the conductivity varies quadratically with the electric field [5].

Just as their counterparts from ordinary differential equations can be transformed one to the other, (1.1) and (1.2) are related. As we shall see, solutions to each can be obtained by simple operations performed on the solution of a certain first order, nonlinear wave equation. In fact, the goal of this note is to show how certain aspects of this particular equation can be studied such as global existence, uniqueness, and the transient and steady state behavior for small $\varepsilon > 0$.

It is perhaps surprising that some second order equations can be solved as first order problems. However, this is strongly suggested by the form of (1.1) where $y$ itself is absent. Also, although two independent initial conditions are required for (1.1) and (1.2), each must be an odd, $2\pi$-periodic function of $x$. Two such odd functions can always be generated from an arbitrary $2\pi$-periodic function by separating it into its odd and even parts and integrating or differentiating the latter. Obviously, this

---

*Support for this work was provided in part by the Office of Research and in part by the Center for International Studies, University of Pittsburgh.

procedure can be reversed, and a single periodic function can be built from two odd functions. Thus it is possible for the initial value of an appropriate first order equation to carry the initial position and velocity for (1.1) and (1.2) in its odd and even parts.

Let us now derive the first order wave equation corresponding to (1.1). Let

$$u = y_t - y_x \tag{1.3}$$

and let P project a periodic function of x to its odd part. Since y and $y_t$ are odd in x, $y_t = Pu$. Hence, from (1.3),

$$u_t + u_x = y_{tt} - y_{xx} = \varepsilon(y_t - y_t^3) = \varepsilon(Pu - (Pu)^3) \tag{1.4}$$

Similarly, we can obtain (1.4) from (1.2) by the transformation,

$$y = \sqrt{3}z, \quad u = z - \int^t z_x \tag{1.5}$$

Of course, these derivations are formal, but they strongly suggest that once we have a solution to (1.4), then $\sqrt{3}$ Pu will solve (1.1) and $\int^t$ Pu will be a solution to (1.1). We shall not discuss here the question of whether these equations are equivalent. Rather we simply regard (1.4), or rather

$$u_t + u_x = \varepsilon(Pu - h(Pu) + f(t,x)), \tag{1.6}$$

where h is a suitable monotone increasing function and f is a $2\pi$-periodic forcing term depending on t and x, as the fundamental equation generalizing the previous examples. We mention, however, that the equivalency of (1.4) and (1.1) has been established in [7].

Some names associated with the study of (1.1) and (1.2) are Kurzweil [8], [9], Vejvoda and Štědrý [10], Chikwendu and Kevorkian [2], and Fink, Hall, and Hausrath [4], [5], [7]. Kurzweil's contribution is by far the most important, as he was able to prove the existence of exponentially asymptotically stable integral manifolds of periodic solutions for both (1.1) and a general form of (1.2). Vejvoda and Štědrý showed the existence of periodic solutions to these equations by elementary methods. In [2], a formal analysis of (1.1) appears, using a two-time method from the theory of ordinary differential equations. In [4] a convergent two-time method is developed, and in [7] a rather detailed analysis is made of the Rayleigh equation.

Let us begin the analysis of (1.6) by studying the question of global existence and uniqueness.

## 2. A Global Existence Theorem

Let H be the space of $2\pi$-periodic, square integrable functions of x with inner product $\langle u, v\rangle$ and norm $|u|$. Write (1.6) in the form,

$$\dot{u} - Au + \varepsilon Nu = \varepsilon u + \varepsilon f \tag{2.1}$$

where $Au = -du/dx$ and $Nu = h(Pu) + (I - P)u$. Assume $\varepsilon > 0$ and h is a continuous, monotone increasing, odd function defined for all real u. Then $Ph(Pu) = h(Pu)$, and so on the domain of h,

$$\langle h(Pu) - h(Pv), u - v\rangle = \langle h(Pu) - h(Pv), Pu - Pv\rangle \geq 0.$$

since $\langle u, Pv\rangle = \langle Pu, v\rangle$ for all u and v in H. Thus N is monotone on H. On its domain $H_1$ of elements having square integrable derivatives, $-A$ is trivially monotone. Hence $B = -A + \varepsilon N$ is monotone as well.

Now suppose B is maximal monotone and let f in $L_1(0, 2\pi; H)$. By Theorem 3.17 of Brezis [1], p. 105, for each $u_0$ in H, (1.6) has a unique weak solution on $[0, 2\pi]$. This means that there are sequences $u_n$ in $C(0, 2\pi; H)$ converging uniformly to u and $f_n \to f$ in $L_1(0, 2\pi; H)$ such that each pair $(u_n, f_n)$ strongly satisfies (1.6).

If f has some additional smoothness, then u is also differentiable in t. For example, if f is in $W_{1, 1}(0, 2\pi; H)$ and $u_0$ in $H_1$,

$$\left|\frac{du}{dt}(t)\right| \leq e^{\varepsilon t}\left|\varepsilon f(0+) - u_{0_x} + \varepsilon Pu_0 - \varepsilon h(Pu_0)\right| + \varepsilon\int_0^t (t-s)\left|\frac{df}{ds}(s)\right| ds$$

and when $f \equiv 0$, $u(t, u_0) = S(t)u_0$ is a semi-group on $[0, \infty)$ with expansion constant $e^{\varepsilon t}$.

The main condition that must be verified, therefore, is that B is maximal. First, note that it is enough to check that $-A + \varepsilon hP$ is maximal because then B is of the form maximal monotone plus monotone Lipschitzian defined on all of H. By [1], p. 34, B is also maximal monotone. According to a result of Minty [1], p. 23, $-A + \varepsilon hP$ is maximal monotone if it is monotone (already verified) and if the range of $I + \varepsilon hP - A$ is all of H. So we consider the equation,

$$\frac{du}{dx} + \varepsilon h(Pu) + u = f \tag{2.2}$$

where f is in H. Splitting u into its odd and even parts v, w we get the system

$$\frac{dw}{dx} + \varepsilon h(v) + v = g, \quad \frac{dv}{dx} + w = k \tag{2.3}$$

where $g = Pf$ and $k = (I - P)f$. But it is easy to see that a solution in $H_1$ to (2.3) exists if and only if $v$ is an odd function in $H_1$ satisfying

$$-Cv = \frac{d^2v}{dx^2} - \varepsilon h(v) - v = k_x - g \qquad (2.4)$$

where $k_x$ and $d^2v/dx^2$ are now in $H_{-1}$, the dual space of $H_1$ under $\langle\cdot,\cdot\rangle$. It is however, quite easy to verify that $C: H_1 \to H_{-1}$ is strongly monotone and hemicontinuous. Hence by a now classic result, $C$ is bijective and we are done.

## 3. A Perturbation Analysis Using Averaging

In this section we shall give some of the results of applying a modified method of averaging to (1.6). Suppose $u(t,\varepsilon)$ is one of its solutions. Let $v(\tau)$ solve the associated averaged equation,

$$\frac{dv}{d\tau} = F(v) = \frac{1}{2\pi}\int_0^{2\pi} e^{-At} F(t, e^{At}v)dt \qquad (3.1)$$

where $F(t,u) = Pu - h(Pu) + f(t,x)$ is the perturbing part of (1.6), and $\{e^{At}\}$ is the group generated by $A = -d/dx$. Let $X$ be a suitable space of initial values with norm $|\cdot|$. According to the theory we can expect the following:

(1) Suppose $\eta > 0$ is given and $u(0, \varepsilon) = v(0) = u_0$ is in $X$. Then there is a constant $L > 0$ such that for all $t$ in $[0, L/\varepsilon]$, $|u(t, \varepsilon) - e^{At}v(\varepsilon t)| < \eta$.

(2) Let $v_0$ be an equilibrium point for (3.1) and suppose $F'(v_0)$ has a bounded inverse on $X$. Then (1.6) has a $2\pi$-periodic solution given approximately by $e^{At}v_0$

(3) If the variational equation of (3.1) is exponentially asymptotically stable then so is the corresponding periodic solution to (1.6).

For a discussion of the averaging method and its application to various types of partial differential equations see [4] and [7]. A simplified explanation can be found in [6].

Statements (1), (2), and (3) above tell us simply that for times of order $\varepsilon^{-1}$, a solution to (3.1) for arbitrary initial value is asymptotic to the actual solution of (1.6), whereas the equilibrium points correspond to the periodic steady states. Thus we must analyze (3.1) for its behavior first as a differential equation with assigned initial value and secondly for the existence and nature of its constant solutions.

Before continuing in this direction, let us say something about the space X. The averaging method in its present form does not apply when the nonlinear term h is unbounded. As we would certainly like to admit polynomials, the $L_p$ spaces for $1 \leq p < \infty$ are ruled out. On the other hand, we need a set large enough to accommodate nonclassical steady states. For these reasons, we have taken X to be the $2\pi$-periodic essentially bounded functions of x with no mean value. This choice results in a series of interesting but sometimes bizarre consequences. The most notable is that the solution itself, which is of the form $u(t, \varepsilon) = e^{At}z(t, \varepsilon)$ (with z strongly continuous in t) is only weak* continuous since $\{e^{At}\}$ is simply right translation of the space variable x. Hence translations of solutions $u(t+h, \varepsilon)$, while certainly solutions in the autonomous case $f(t,x) \equiv 0$, are distinct and isolated from each other for each value of h in the norm topology of $L_\infty$. Thus even when (1.6) is autonomous, the periodic solutions can be exponentially asymptotically stable.

From (3.1) and the definitions of $e^{At}$ and $F(t,u)$,

$$\begin{aligned} F(v)(\tau,x) &= \frac{1}{2}v(\tau,x) - \frac{1}{2\pi}\int_0^{2\pi} h\left(\frac{v(\tau,x) - v(\tau,s)}{2}\right) ds + f_0(x) \\ f_0(x) &= \frac{1}{2\pi}\int_0^{2\pi} f(s, x+s)\, ds \end{aligned} \tag{3.2}$$

When $h(u) = u^3$, it is possible to solve (3.1) up to a quadrature depending only on the initial value $v_0$. The details are in [7]. When h is more general we can still say something about the behavior of the averaged equation. For example, if h is odd and monotone increasing, then we can prove solutions exist on $[0, \infty)$ but may not be bounded. In fact, if p is odd,

$$Nu = \frac{1}{2\pi}\int_0^{2\pi} h\left(\frac{u(x) - u(s)}{2}\right) ds,$$

and $\langle u,v\rangle$ is the usual inner product, then

$$\langle Nu,u^p\rangle = \frac{1}{2\pi}\int_0^{2\pi} ds \int_0^{2\pi} dx\, h\left(\frac{u(x) - u(s)}{2}\right) (u^p(x) - u^p(s))ds - \langle Nu, u^p\rangle$$

Thus, $2 \langle Nu, u^p\rangle \geq 0$ since $h(u)u \geq 0$.

Now let $z(\tau) = |v|_{p+1}$ be the $L_{p+1}$ norm of v.

From (3.1), (3.2), and the inequality on N,

$$\frac{dz}{d\tau} \leq \frac{1}{2}z + |f_0|,$$

and this gives $z(\tau) \leq \{|v_0| + 2|f_0|\}e^{\tau/2}$.

A better result can be obtained if we suppose $h(u)u \geq c_r u^{r+1}$ where $r > 1$ is odd, and if X restricted to the $\Pi$-antiperiodic elements $u(x+\pi) = -u(x)$. In this case we can show

$$\langle Nu, u^p \rangle \geq \frac{c_r}{2^r} |u|_{p+r}^{p+r},$$

and from the Holder inequality we get $\langle Nu, u^p \rangle \geq a_r |u|_{p+1}^{p+r}$ where $a_r$ is independent of p.

Now from the averaged equation we obtain

$$\frac{dz}{d\tau} \leq g(z) = \frac{1}{2}z + |f_0| - a_r z^r$$

Eventually, $g(x) \to -\infty$ as $\alpha \to +\infty$ monotonically. So for large $\alpha$, $g(\alpha) < -\delta$ and $\alpha_2 > \alpha_1$ implies $g(\alpha_2) < g(\alpha_1)$. For such an $\alpha$ let $|v(\tau)| > \alpha$. Since $|v(\tau)|_{p+1}$ is continuous in $\tau$ and converges monotonically to $|v(\tau)|$ we can apply Dini's Theorem to obtain a $P_0$ such that $|v(\tau)|_{p+1} > \alpha$ when $p > P_0$ uniformly for $\tau$ in $[0,L]$. Thus on $(0,L)$

$$\frac{dz}{d\tau} \leq g(z) < g(\alpha) < -\delta$$

and hence z, and therefore $|v(\tau)|$ itself is nonincreasing outside some large ball. Hence not only do we have global existence, but we also know all solutions are bounded. Of course, we cannot conclude the same is true of (1.6) since we are assured only that its solutions follow those of (3.1) on some long but finite time interval.

The constant solutions of (3.1) correspond to the periodic solutions of (1.6) provided the conditions of the implicit function theorem can be met. So let us examine the roots of $F(v)$. In most circumstances, the term $f_0(x)$ is absent even if f is not. In fact, unless f has a right traveling wave component, $f_0(x) \equiv 0$. So let us begin with this case.

If $f_0 = 0$ we expect that the most general solution to $F(v_0) = 0$ will be constant on subsets of $[0, 2\pi]$. Let $v_0$ be the $2\pi$-periodic extensions of

$$v_0(x) = \begin{cases} 0 & x \in A_0 \\ \alpha & \in A_1 \\ -\alpha & \in A_2 \end{cases} \qquad (3.3)$$

where $A_i$, $i = 0,1,2$ are disjoint measurable sets whose union is $[0, 2\pi)$. In order that $v_0$ have no mean value we need mes $A_1$ = mes $A_2$.

We thus obtain

$$F(v_0) = \begin{cases} -a_1 \; \{h(-a/2) + h(\alpha/2)\}, & x \in A_0 \\ F(\alpha) = \frac{1}{2}\alpha - a_1 h(\alpha) - a_0 h(\frac{\alpha}{2}), & x \in A_1 \cup A_2 \end{cases}$$

where $a_i$ = mes $A_i/2\pi$, $i = 0,1$. If h is odd, then F vanishes on $A_0$. Let us also suppose that $h'(0) = 0$, $h'(u) \to +\infty$ and $h''(u) > 0$ for $u > 0$. (For example, $h(u) = u^p$, $p \geq 3$ and odd, and $h(u) = \sinh u - u$ both satisfy this requirement). Then $F(\alpha)$ is concave on $\alpha > 0$, $F'(0) = \frac{1}{2}$ and $F'(\alpha)$ decreases steadily to $-\infty$ as $\alpha \to +\infty$. Hence there is a unique $\alpha > 0$ where $F(\alpha) = 0$.

When mes $A_0 = 0$, then $\alpha$ is simply the root of $\alpha = h(\alpha)$. Using techniques developed in [7] we can prove that $F'(v_0)$ is boundedly invertible when

$$\lambda = 1 - h'(\alpha) \neq 0 \qquad\qquad \mu = 1 - \tfrac{1}{2}h'(\alpha) \neq 0.$$

The solution to the variational equation is

$$\xi = e^{\lambda\tau}a + b(e^{\mu\tau} - 1)$$

where a and b are linear functionals of the initial value. Hence we have exponential decay and thus stability of the periodic solution if $\lambda$ and $\mu$ are both negative at the root of $F(\alpha)$.

It is interesting to note that the stability result above is independent of f if it has no right traveling wave component. Also, if mes $A_0 \neq 0$, then we can prove that $\xi$ is actually increasing exponentially for x in $A_0$. This indicates, (but does not prove, of course) that those equilibrium points with mes $A_0 \neq 0$ are unstable.

We shall conclude by looking at a particular case when $f_0(x) \neq 0$ Specifically, we take $h(u) = u^3$ and we restrict X to the $\pi$-anti-periodic elements so that the odd powers have no mean value. Then

$$F(v) = \tfrac{1}{2}v - \tfrac{1}{8}(v^3 + 3 < v^2 > v) + f_0(x)$$

$$<v^2> = \frac{1}{2\pi} \int_0^{2\pi} v^2(x)dx$$

Let us try to solve $F(v) = 0$ with $v = m \cos\theta$. This gives

$$m^3\cos^3\theta - (4 - 3m^2 < \cos^2\theta >) \, m\cos\theta - 8f_0(x) = 0 \qquad (3.4)$$

Replace $<\cos^2\theta>$ by $\beta$ and consider

$$m^3\cos^3\theta - (4 - 3m^2\beta)\, m \cos\theta - 8f_0(x) = 0. \quad (3.5)$$

Since $4\cos^3\theta - 3\cos\theta - \cos 3\theta = 0$ we will have a solution if

$$\frac{m^3}{4} = \frac{m(4-3m^2\beta)}{3} = \frac{8f_0(x)}{\cos 3\theta}$$

Hence

$$\cos 3\theta = 32f_0(x)/m^3 \text{ and } m = 4/\sqrt{3(1+4\beta)}\,,$$

and $\theta$ will be real if $-m^3 \le 32f_0(x) \le m^3$. Since $\beta \in [0,1]$, we can continue this procedure if

$$-2/3^{3/2} \le f_0(x) \le 2/15^{3/2} \qquad \text{a.e.}$$

Let $\theta_1(x,\beta)$ be a root of (3.5) in $0 \le \theta \le \pi$. Then $\theta_2 = \theta_1+2\pi/3$ and $\theta_0 = \theta_1 + 4\pi/3$ are also roots, and we can construct a candidate for an equilibrium point. Let $B_0$, $B_1$, $B_2$ be three mutually disjoint measurable sets whose union is $[0,\pi)$. Define

$$\tilde{v}_0(x,\beta) = \begin{cases} m\cos\theta_1(x,\beta) & x \in B_1 \\ m\cos\theta_2(x,\beta) & x \in B_2 \\ m\cos\theta_0(x,\beta) & x \in B_0 \end{cases}$$

and let $v_0$ be its $\pi$-antiperiodic extension. Then

$$<v^2> = \frac{1}{\pi}\int_0^{\pi} v_0^2(x)dx$$

and we have a solution to (3.4) if we can find $\beta$ in [0.1] such that

$$\beta = \frac{1}{\pi}\left\{\int_{B_1} \cos^2\theta_1(x,\beta)dx + \int_{B_2} \cos^2\theta_2(x,\beta)dx + \int_{B_3} \cos^2\theta_0(x,\beta)dx\right\}$$

But the right side of this equation is just a continuous map of the unit interval to itself. Hence we have at least one fixed point

It can be proved that if mes $B_0 = 0$, then the condition of the implicit function theorem and the requirement on the exponential decay of the variational equation are met. We remark that the sets $B_0$, $B_1$, $B_2$ play rolls analogous to those of $A_0$, $A_1$, and $A_2$ of the previous example and the solutions given here reduce to those given earlier as $f_0 \to 0$.

We conclude with the following observation. When $f \equiv 0$, the approximate steady states $e^{At}v_0$ are the actual solutions in a generalized sense. In particular, when $h(u) = u^3$, and $v_0(x) = 1$ on $A_1$, $-1$ on $A_2$ (where $A_1 \cup A_2 = [0, 2\pi)$ and mes $A_1$ = mes $A_2$) then a family of stable steady states for the Van der Pol wave equation is given by

$$\frac{\sqrt{3}}{2}\left\{ v_0(x-t) - v_0(-x-t) \right\} ,$$

and for the Rayleigh equation by

$$\frac{1}{2}\int^{t} \left\{ v_0(x-s) - v_0(-x-s) \right\} ds \quad .$$

References

[1] H.Brezis: Operateurs Maximaux Monotones, North Holland, Amsterdam, 1973

[2] S.C.Chikwendu and J.Kevorkian: A perturbation method for hyperbolic equations with small nonlinearities, SIAM J.Appl. Math. 22 (1972), 235-258

[3] A.D.Cooke, C.J.Myerscough and M.D.Rowbottom: The growth of full span galloping oscillations, Laboratory Note RD/L/N51/72, Central Electricity Research Laboratories,Leatherhead, Surrey, England

[4] J.P.Fink, W.S.Hall and A.R.Hausrath: A convergent two-time method for periodic differential equations, J.Differential Equations 15 (1974), 459-498

[5] J.P.Fink, A.R.Hausrath and W.S.Hall: Discontinuous periodic solutions for an autonomous nonlinear wave equation, Proc. Royal Irish Academy 75 A 16 (1975), 195-226

[6] W.S.Hall: Two timing for abstract differential equations, Lecture Notes in Mathematics 415, Ordinary and Partial Differential Equations, Springer, 1974, 368-372

[7] W.S.Hall: The Rayleigh wave equation - an analysis, J.Nonlinear Anal., Tech.Meth.Appl., to appear

[8] J.Kurzweil: Exponentially stable integral manifolds, averaging principle and continuous dependence on a parameter, Czech. Math.J. 16 (91), (1966), 380-423 and 463-492

[9] J.Kurzweil: Van der Pol perturbations of the equation for a vibrating string, Czech.Math.J. 17 (2), (1967), 588-608

[10] M.Štědrý and O.Vejvoda: Periodic solutions to weakly nonlinear autonomous wave equations, Czech:Math.J. 25 (100), (1975), 536-554

Author's address: Department of Mathematics/Statistics, University of Pittsburgh, PA 15260, U.S.A.

# THE DIRICHLET PROBLEM

W.Hansen, Bielefeld

Given a partial differential operator $L$ of second order on a relatively compact open subset $V$ of $\mathbb{R}^n$ and a continuous real function $f$ on $V^*$ the corresponding <u>Dirichlet problem</u> consists in finding a continuous real function $u$ on $\overline{V}$ such that $Lu = 0$ on $V$ and $u = f$ on $V^*$.

Since about twenty years ([1], [4])it is well known that a general treatment of this question is possible by using the concept of a harmonic space. We shall sketch how this is done and then discuss some recent developments.

## 1. Harmonic spaces

Let $X$ be a locally compact space with countable base. For every open $U$ in $X$ let $\mathcal{H}(U)$ be a linear space of continuous real functions on $U$, called <u>harmonic functions</u> on $U$, and suppose that $\mathcal{H} = \{\mathcal{H}(U) : U \text{ open in } X\}$ is a sheaf.

<u>Standard examples.</u> 1. Laplace equation. $X$ relatively compact open $\subset \mathbb{R}^n$, $\mathcal{H}(U) = \{u \in C^2(U) : \sum_{i=1}^{n} \frac{\partial^2 u}{\partial x_i^2} = 0\}$ 2. Heat equation. $X$ relatively compact open $\subset \mathbb{R}^{n+1}$, $\mathcal{H}(U) = \{u \in C^2(U) : \sum_{i=1}^{n} \frac{\partial^2 u}{\partial x_i^2} = \frac{\partial u}{\partial x_{n+1}}\}$.

A relatively compact open subset $V$ of $X$ is called <u>regular</u> if for every $f \in C(V^*)$ there exists a unique extension $H^V f$ on $\overline{V}$ which is harmonic on $V$ and positive if $f$ is positive.

Let us suppose that $(X,\mathcal{H})$ has the following properties:

I. The regular sets form a base of $X$.

II. For every open $U$ in $X$ and increasing sequence $(h_n)$ of harmonic functions on $U$ such that $h := \sup h_n$ is locally bounded the function $h$ is harmonic on $U$.

III. $1 \in \mathcal{H}(X)$, $\mathcal{H}^+(X)$ separates the points of $X$.

Then $(X,\mathcal{H})$ is a <u>harmonic space</u>.

<u>Remark.</u> We note that the general concept of a harmonic space in the sense of Constantinescu-Cornea [4] uses a slightly weaker form of property (I) and a separation property which is considerably weaker than our property (III). Accepting some technical modifications all the material we want to discuss can be presented in the more general situation (see [2], [3]). But probably the essential ideas become more clear in our setup.

Let $V$ be a regular set and $x \in V$. Then the mapping $f \longmapsto H^V f(x)$ is a positive linear form on $C(V^*)$, hence a positive Radon measure $\mu_x^V$ on $V^*$, called the harmonic measure (on $V$ at $x$).

## 2. The Dirichlet problem and the PWB-method

Let $U$ be a relatively compact open subset of $X$. Given a function $f \in C(U^*)$ the corresponding Dirichlet problem asks for a continuous extension of $f$ to a function $h \in C(\overline{U})$ which is harmonic in $U$. Therefore, one is interested in the linear space

$$H(U) := \{h \in C(\overline{U}) : h \text{ harmonic in } U\} .$$

If this Dirichlet problem is solvable for every $f \in C(U^*)$ then $U$ is regular, $H(U) \cong C(U^*)$, and vice versa. However, $U$ may be not regular and then there are functions $f \in C(U^*)$ for which the Dirichlet problem is not solvable.

But there is a method due to Perron, Wiener and Brelot (PWB-method) which yields a positive linear mapping $f \longmapsto H^U f$ such that $H^U f$ is harmonic on $U$ for every $f \in C(U^*)$ and such that $H^U f$ is the solution of the Dirichlet problem provided a solution exists.

The PWB-method of determining a so-called generalized solution of the Dirichlet problem uses hyperharmonic functions. A l.s.c. function $v : U \to ]-\infty, +\infty]$ is called hyperharmonic (on $U$) if $\mu_x^V(v) \leq v(x)$ for every regular $V$ such that $\overline{V} \subset U$ and for every $x \in V$.

Let ${}^*H(U) = \{v | v : \overline{U} \to ]-\infty, +\infty]$ l.s.c., $v$ hyperharmonic on $U\}$. We note that ${}^*H(U) \cap -{}^*H(U) = H(U)$. ${}^*H(U)$ is a convex cone satisfying the following boundary minimum principle: If $v \in {}^*H(U)$ and $v \geq 0$ on $U^*$ then $u \geq 0$ on $\overline{U}$.

Let $f \in C(U^*)$. Defining

$$\overline{H}^U f = \inf \{v \in {}^*H(U) : v \geq f \text{ on } U^*\} ,$$

$$\underline{H}^U f = \sup \{w \in -{}^*H(U) : w \leq f \text{ on } U^*\}$$

the boundary minimum principle yields $\underline{H}^U f \leq \overline{H}^U f$. If the Dirichlet problem for $f$ is solvable, i.e. if there exists a function $h \in H(U)$ such that $h = f$ on $U^*$ then evidently $h \leq \underline{H}^U f$ and $\overline{H}^U f \leq h$, hence $\underline{H}^U f = \overline{H}^U f = h$.

It can be shown that for every $f \in C(U^*)$

$$\overline{H}^U f = \underline{H}^U f =: H^U f$$

and furthermore $H^U f$ is harmonic on $U$, $H^U f = f$ on $U^*$.

A boundary point $z \in U^*$ is called <u>regular</u> if for all $f \in C(U^*)$ the generalized solution $H^U f$ is continuous at $z$. Evidently, $U$ is regular if and only if all boundary points of $U$ are regular. The generalized solution of the Dirichlet problem and a useful criterion for the regularity of boundary points can be obtained using balayage of measures.

## 3. Balayage

Let ${}^*H^+$ denote the set of all positive hyperharmonic functions on $X$. Given an arbitrary subset $A$ of $X$ and a function $u \in {}^*H^+$ one tries to find a smallest function $v \in {}^*H^+$ satisfying $v = u$ on $A$. The obvious candidate is the <u>pre-sweep</u> (or rêduite function)

$$R_u^A := \inf \{v \in {}^*H^+ : v = u \text{ on } A\} .$$

Since $R_u^A$ is not l.s.c. in general, one replaces $R_u^A$ by the greatest l.s.c. function $\leq R_u^A$. This is the <u>sweep</u> (or balayée function) of $u$ relatively to $A$:

$$\hat{R}_u^A(x) := \liminf_{y \to x} R_u^A(y) \qquad (x \in X) .$$

We have $\hat{R}_u^A \in {}^*H^+$ and obviously

$$0 \leq \hat{R}_u^A \leq R_u^A \leq u .$$

The initial interest leads then to the study of the base of $A$

$$b(A) := \bigcap_{u \in {}^*H^+} \{x \in X : \hat{R}_u^A(x) = u(x)\} .$$

It has the following fundamental properties:

$$\mathring{A} \subset b(A) \subset \overline{A} ,$$

$$b(A) = \{x \in X : \hat{R}_{u_0}^A(x) = u_0(x)\} \quad \text{for some} \quad u_0 \in {}^*H^+ \cap C ,$$

in particular, $b(A)$ is a $G_\delta$-set.

For every Radon measure $\mu \geq 0$ on $X$ with compact support there exists a unique Radon measure $\mu^A \geq 0$ on $X$ satisfying

$$\int u \, d\mu^A = \int \hat{R}_u^A \, d\mu \qquad \text{for all} \quad u \in {}^*H^+ .$$

$\mu^A$ is called the <u>swept out</u> of $\mu$ on $A$. It is carried by $\overline{A}$. By choosing for $\mu$ unit masses $\varepsilon_x$ at points $x \in X$ it follows that

$$b(A) = \{x \in X : \varepsilon_x^A = \varepsilon_x\} .$$

We are now able to express the solution of the generalized Dirichlet problem in terms of balayage:

For every relatively compact open set $U$ and every $f \in C(U^*)$ the solution $H^U f$ satisfies

$$H^U f(x) = \int f \, d\varepsilon_x^{\complement U} = \int f \, d\varepsilon_x^{U^*} \quad (x \in U) .$$

The set $U_r$ of regular boundary points is given by

$$U_r = b(\complement U) \cap \overline{U} .$$

## 4. The weak Dirichlet problem

Again let $U$ be a relatively compact open subset of $X$. The fact that a function $f \in C(U^*)$ may not admit an extension to a function $h \in H(U)$ led to the introduction of the generalized solution $H^U f$ which is a harmonic extension of $f$ but is not necessarily continuous at all points of the boundary $U^*$.

Another way of turning the problem is the following: Are there at least some subsets $B$ of the boundary such that every continuous function $f$ on $B$ admits a continuous extension to a function in $H(U)$ ? Because of a general minimum principle a natural candidate for such a set $B$ would be the Choquet boundary $Ch_{H(U)}\overline{U}$ of $\overline{U}$ with respect to $H(U)$.

The Choquet boundary $Ch_{H(U)}\overline{U}$ is the set

$$Ch_{H(U)}\overline{U} := \{x \in \overline{U} : M_x(U) = \{\varepsilon_x\}\}$$

where

$$M_x(U) := \{\mu : \mu(h) = h(x) \quad \text{for all} \quad h \in H(U)\}$$

denotes the set of all representing measures for $x$ (with respect to $H(U)$).

If for example $V$ is regular, $\overline{V} \subset U$ and $x \in V$ then $\mu_x^V$ is a representing measure for $x$. More generally, for every $x \in \overline{U}$ the swept-out $\varepsilon_x^{\complement U}$ of $\varepsilon_x$ on $\complement U$ is a representing measure for $x$. In particular, the Choquet boundary $Ch_{H(U)}\overline{U}$ is a subset of the set $U_r$ of regular points. For the Laplace equation these two sets coincide whereas for the heat equation the Choquet boundary may be a proper subset of $U_r$.

We have the following minimum principle: For every $h \in H(U)$ there exists a point $z \in Ch_{H(U)}\overline{U}$ such that $h \geq h(z)$. In particular, if $h_1, h_2 \in H(U)$ and $h_1 = h_2$ on $Ch_{H(U)}\overline{U}$ then $h_1 = h_2$.

Thus the following weak Dirichlet problem arises: Given a compact subset $K$ of $Ch_{H(U)}\overline{U}$ and a continuous function $f$ on $K$, is there a continuous extension to a function in $H(U)$ ?

The solution of this problem is obtained by the following result.

Theorem ([2]). For every $x \in \overline{U}$ there exists a unique measure $\mu_x \in M_x(U)$ which is carried by $Ch_{H(U)}\overline{U}$. For every $x \in \overline{U} \setminus Ch_{H(U)}\overline{U}$,

$$\mu_x = \varepsilon_x^{Ch_{H(U)}\overline{U}} .$$

A very general reasoning now yields the following consequence.

Corollary. 1. The weak Dirichlet problem is solvable.
2. $\{\rho \in H(U)^* : \rho \geq 0, \rho(1) = 1\}$ is a simplex.

Furthermore, a close study of the Choquet boundary yields a characterization of $Ch_{H(U)}\overline{U}$ which is similar to the one obtained for $U_r$ :

$$Ch_{H(U)}\overline{U} = \beta(\complement U) \cap \overline{U}$$

where $\beta(\complement U)$ is the greatest subset $C$ of $\complement U$ such that $b(C) = C$.

## 5. General PWB-method

We shall now see that for every $x \in U$ the measure $\varepsilon_x^{Ch_{H(U)}\overline{U}}$ and many other representing measures can be obtained by a procedure in the spirit of Perron-Wiener-Brelot.

For every compact subset $K$ of $U^*$ let ${}^*H_K(U)$ be the set of all functions $v$ which are limits of an increasing sequence $(v_n)$ of l.s.c. real functions $v_n$ on $\overline{U}$, hyperharmonic on $U$ and continuous on $\overline{U} \setminus K$. Then ${}^*H_K(U)$ is a convex cone such that $H(U) \subset {}^*H_K(U) \subset {}^*H_{U^*}(U) = {}^*H(U)$.

Furthermore

$$Ch_{H(U)}\overline{U} \subset Ch_{{}^*H_K(U)}\overline{U} \subset K \cup Ch_{H(U)}\overline{U}$$

where the last inclusion is a consequence of the local characterization of the Choquet boundary. Indeed, obviously $Ch_{{}^*H_K(U)}\overline{U} \subset U^*$. So let $x \in U^* \setminus (K \cup Ch_{H(U)}\overline{U})$. Then there exists an open neighborhood $V$ of $x$ such that $\overline{V} \cap K = \emptyset$. Defining $W = U \cap V$ we have $x \in V \cap \complement\beta(\complement U) \subset \complement\beta(\complement W)$ and hence $x \notin Ch_{H(W)}\overline{W}$. Thus $\varepsilon_x^{Ch_{H(W)}\overline{W}} \neq \varepsilon_x$ and being a representing measure of $x$ with respect to ${}^*H_\emptyset(W)$ the measure $\varepsilon_x^{Ch_{H(W)}\overline{W}}$ is a representing measure of $x$ with respect to ${}^*H_K(U)$.

Let $B$ be a Borel subset of $U^*$ containing $Ch_{H(U)}\overline{U}$. Defining

$${}^*H_B(U) = \bigcup_{K \text{ cp.} \subset B} {}^*H_K(U)$$

we thus have the following minimum principle: If $v \in {}^*H_B(U)$ and $v \geq 0$ on $B$ then $v \geq 0$ on $\overline{U}$.

Let $f \in C(U^*)$. Defining

$$\overline{H}_B^U f = \inf \{v \in {}^*H_B(U) : v \geq f \text{ on } B\},$$

$$\underline{H}_B^U f = \sup \{w \in -{}^*H_B(U) : w \leq f \text{ on } B\}$$

the minimum principle yields $\underline{H}_B^U f \leq \overline{H}_B^U f$. If there exists a function $h \in H(U)$ such that $h = f$ on $U^*$ then evidently $h \leq \underline{H}_B^U f$ and $\overline{H}_B^U f \leq h$, hence $\overline{H}_B^U f = \underline{H}_B^U f = h$. In the general situation we have the following result.

Proposition ([3]). For every $f \in C(U^*)$

$$\overline{H}_B^U f = \underline{H}_B^U f =: H_B^U f,$$

and $H_B^U f$ is harmonic on U. Furthermore, for every $x \in \overline{U} \setminus B$

$$H_B^U f(x) = \int f \, d\varepsilon_x^B .$$

Proof. It suffices to consider the case $f = v|_X$ for some continuous real $v \in {}^*H^+$. Then ${}^*H^+ \subset {}^*H_B(U)$ implies that

$$\overline{H}_B^U f|_{\overline{U}} \leq R_v^B .$$

Let K be a compact subset of B, $w = R_v^K|_{\overline{U}}$. Then $w \in -{}^*H_K(U), w \leq v$. Hence $R_v^K \leq \underline{H}_B^U f$ on $\overline{U}$. Taking the supremum of all $R_v^K$ we obtain

$$R_v^B|_{\overline{U}} \leq \underline{H}_B^U f .$$

References

[1] Bauer, H.: Harmonische Räume und ihre Potentialtheorie. Lecture Notes in Mathematics 22. Berlin-Heidelberg-New York: Springer 1966

[2] Bliedtner, J., Hansen, W.: Simplicial Cones in Potential Theory. Inventiones math. 29, 83-110 (1975)

[3] Bliedtner, J., Hansen, W.: Simplicial Cones in Potential Theory II (Approximation Theorems). Inventiones math. 46, 255-276 (1978)

[4] Constantinescu, C., Cornea, A.: Potential Theory on Harmonic Spaces. Berlin-Heidelberg-New York: Springer 1972

Author's address: Fakultät für Mathematik, Universität Bielefeld,
Postfach 8640, 4800 Bielefeld 1, Federal Republic of Germany

# MULTIPLE SOLUTIONS OF SOME ASYMPTOTICALLY LINEAR ELLIPTIC BOUNDARY VALUE PROBLEMS

P. Hess, Zürich

In this note we shall apply the additivity property of the Leray-Schauder topological degree in order to assure the existence of multiple solutions in two examples of nonlinear, asymptotically linear elliptic boundary value problems which both attracted remarkable interest in recent years:

I. Positive solutions of nonlinear eigenvalue problems,

II. Nonlinear perturbations of linear problems at resonance.

We do not bother here to state the results in the utmost generality possible. For detailed proofs the interested reader is referred to the forthcoming papers by Ambrosetti-Hess [1] (concerning example I) and Hess [2] (concerning example II).

## I. Positive solutions of nonlinear eigenvalue problems

We consider the question of existence of positive solutions of the problem

$$(1)\qquad \begin{cases} -\Delta u = \lambda f(u) & \text{in } \Omega \\ u = 0 & \text{on } \partial\Omega , \end{cases}$$

where $\Omega$ is a bounded domain in $\mathbb{R}^N$ $(N \geqq 1)$ with smooth boundary, $\lambda \geqq 0$ is a parameter, and $f : \mathbb{R}^+ \to \mathbb{R}$ is a continuous function satisfying the following conditions

(f1) $f(0) = 0$ ,

(f2) there exist $m_\infty > 0$ and a bounded function $g : \mathbb{R}^+ \to \mathbb{R}$ such that $f(s) = m_\infty s + g(s)$ $\forall s \geqq 0$ ,

(f3) the right-sided derivative $f'_+(0)$ at $s = 0$ exists, and $f'_+(0) < m_\infty$ .

Of course (1) admits always the trivial solution $u = 0$ . For $\lambda$ varying in a certain interval we shall prove the existence of positive (i.e. nonnegative and nontrivial) solutions. Let $\lambda_1 > 0$ denote the first eigenvalue of the Dirichlet problem

$$(2)\qquad \begin{cases} -\Delta u = \lambda u & \text{in } \Omega \\ u = 0 & \text{on } \partial\Omega ; \end{cases}$$

it is known that the corresponding eigenspace is 1-dimensional and spanned by an eigenfunction $\phi$ which we may choose to be positive in $\Omega$ . Let

$$\lambda_\infty := \frac{\lambda_1}{m_\infty} , \qquad \lambda_o := \begin{cases} \lambda_1/f'_+(0) & \text{if } f'_+(0) > 0 \\ +\infty & \text{otherwise.} \end{cases}$$

If $f$ is strictly monotone increasing, convex and $f'_+(0) > 0$ , Amann-Laetsch [3] prove the following:

For $\lambda \in [0,\lambda_\infty] \cup [\lambda_o,+\infty)$ , problem (1) has the trivial solution only. It has at least one positive solution for $\lambda \in (\lambda_\infty,\lambda_o)$ .

(For a uniqueness result cf. Amann [4], Ambrosetti [5].)

Their statement follows from an abstract theorem on order convex maps in general ordered Banach spaces and uses results from bifurcation theory (note that $\lambda_o$ is a bifurcation point from the trivial solution and $\lambda_\infty$ a bifurcation point from infinity). It seems that their method breaks down if $f$ is allowed to admit also negative values. We are interested in this case here.

Theorem 1. Let (f1) - (f3) be satisfied.

(i) If $\lambda \in (\lambda_\infty,\lambda_o)$ , there exists at least one positive solution of problem (1).

(ii) Suppose in addition

$$\liminf_{s\to+\infty} g(s) > 0 . \tag{3}$$

Then for all $\lambda \in [\lambda_\infty,\lambda_o)$ , problem (1) admits at least one positive solution. There exists $\varepsilon > 0$ such that for $\lambda \in (\lambda_\infty-\varepsilon,\lambda_\infty)$ , problem (1) has at least two positive solutions.

We briefly sketch the proof of Theorem 1, working in the Hilbert space $H = L^2(\Omega)$ . Let $L$ be the positive selfadjoint operator induced in $H$ by $-\Delta$ , with domain $D(L) = H^1_o(\Omega) \cap H^2(\Omega)$ . We extend the function $f$ to $\mathbb{R}$ by setting $f(s) = 0 \ \forall s \leqq 0$ , and denote by $F : H \to H$ the Nemytskii operator associated with $f$ : $(Fu)(x) = f(u(x)) \quad (x \in \Omega)$ for any function $u$ defined in $\Omega$ . By the maximum principle, any solution $u$ of $Lu = \lambda F(u)$ with $\lambda \geqq 0$ is nonnegative in $\Omega$ . Thus problem (1) is equivalent to the equation

$$u - \lambda L^{-1}F(u) = 0 \tag{4}$$

in $H$ . Note that $L^{-1} : H \to H$ is compact.

We now apply topological degree arguments to prove the existence of nontrivial solutions of (4). For $R > 0$ let $B_R$ denote the open ball in $H$ around $0$ , with radius $R$ .

Lemma 1. To each $0 \leqq \lambda < \lambda_o$ there exists $R_1 = R_1(\lambda) > 0$ such that $\deg(I-\lambda L^{-1}F, B_{R_1}, 0) = 1$.

Proof. The assumptions (f1) - (f3) imply the existence of a constant $c > 0$ such that $|f(s)| \leqq c|s|$ $\forall s \in \mathbb{R}$. Let $0 \leqq \lambda < \lambda_o$. By the homotopy invariance of the Leray-Schauder degree, Lemma 1 is proved if we can show that for some $R_1 > 0$,

$$u - t\lambda L^{-1}F(u) \neq 0 \quad \forall \|u\| = R_1 , \quad \forall t \in [0,1] .$$

This is established indirectly, using measure-theoretic arguments and the variational characterisation of the first eigenvalue $\lambda_1$ of $L$.

Lemma 2. Let $\lambda > \lambda_\infty$. Then $\deg(I-\lambda L^{-1}F, B_{R_2}, 0) = 0$ for all $R_2 = R_2(\lambda)$ sufficiently large.

In particular, if $\lambda_\infty < \lambda < \lambda_o$, we can choose $R_2(\lambda) > R_1(\lambda)$. Then, by the additivity property of the degree, $\deg(I-\lambda L^{-1}F, B_{R_2} \setminus \bar{B}_{R_1}, 0) = -1$, and consequently there is a (nontrivial) solution of (4) in $B_{R_2} \setminus \bar{B}_{R_1}$. This proves Theorem 1(i).

Idea of proof of Lemma 2. One shows, again indirectly, that to $\lambda > \lambda_\infty$, there exists $K = K(\lambda) > 0$ such that

$$(5) \qquad u - \lambda L^{-1}F(u) = \tau\phi , \text{ with } \tau \geqq 0 \Longrightarrow \|u\| < K .$$

Since $I - \lambda L^{-1}F$ is a bounded mapping, it then follows that there is a constant $a > 0$ having the property that

$$u - \lambda L^{-1}F(u) \neq a\phi \quad , \quad \forall u \in \bar{B}_K .$$

As, by (5),

$$u - \lambda L^{-1}F(u) \neq ta\phi \quad \forall \|u\| = K , \quad \forall t \in [0,1] ,$$

this implies that

$$\begin{aligned} 0 &= \deg(I-\lambda L^{-1}F-a\phi, B_K, 0) \\ &= \deg(I-\lambda L^{-1}F, B_K, 0) . \end{aligned}$$

The proof of Lemma 2 was inspired by a related argument used in Brown-Budin [6]; in our case however we are able to prove an a priori estimate of the form (5) only for this particular function $\phi$.

Lemma 3. If in addition (3) holds, the assertion of Lemma 2 remains valid also for $\lambda = \lambda_\infty$.

Since the degree is invariant in connected components of

$H \setminus (I-\lambda_\infty L^{-1}F)(\partial B_{R_2(\lambda_\infty)})$ , there exists $\varepsilon > 0$ such that

$\deg(I-\lambda L^{-1}F, B_{R_2(\lambda_\infty)}, 0) = 0$ , $\forall\ \lambda \in (\lambda_\infty - \varepsilon, \lambda_\infty)$ .

Lemma 4. For $0 < \lambda < \lambda_\infty$ we have
$\deg(I-\lambda L^{-1}F, B_{R_3}, 0) = 1$ , provided $R_3 = R_3(\lambda)$ is sufficiently large.

This is a simple consequence of the asymptotic behavior of the function $f$ .

Suppose now (3) holds, and let $\lambda \in (\lambda_\infty - \varepsilon, \lambda_\infty)$ . By Lemma 1, there is $R_1 > 0$ such that $\deg(I-\lambda L^{-1}F, B_{R_1}, 0) = 1$ . Further, Lemma 3 implies that $\deg(I-\lambda L^{-1}F, B_{R_2(\lambda_\infty)}, 0) = 0$ . Hence there exists a nontrivial solution of (4) in $B_{R_2(\lambda_\infty)} \setminus \bar{B}_{R_1}$ . Moreover, by Lemma 4 we find $R_3 > R_2(\lambda_\infty)$ such that

$\deg(I-\lambda L^{-1}F, B_{R_3}, 0) = 1$ . It follows that $\deg(I-\lambda L^{-1}F, B_{R_3} \setminus \bar{B}_{R_2(\lambda_\infty)}, 0) = 1$ and thus the existence of a second nontrivial solution is guaranteed. Theorem 1(ii) is proved.

II. Nonlinear perturbations of linear problems at resonance

The problem

$$\text{(6)} \qquad \begin{cases} (-\Delta-\lambda_k)u + g(u) = h & \text{in } \Omega \\ u = 0 & \text{on } \partial\Omega \end{cases} ,$$

is investigated. Here $\lambda_k$ denotes the k-th eigenvalue of the Dirichlet problem (2), $g : \mathbb{R} \to \mathbb{R}$ is a continuous function having limits $g_\pm := \lim_{s\to\pm\infty} g(s)$ $(\in\mathbb{R})$ and $h$ is a given element in $H = L^2(\Omega)$ . Let again $L$ be the realization in $H$ of the operator $-\Delta$ with Dirichlet boundary conditions, and denote by $G : H \to H$ the Nemytskii operator associated with $g$ . Supposing in addition that

$$\text{(7)} \qquad g_- < g(s) < g_+ \quad \forall s \in \mathbb{R} ,$$

it is well-known that condition

$$\text{(LL)} \qquad (h,v) < \int_\Omega (g_+v^+ - g_-v^-) \quad \forall v \in N(L-\lambda_k),\ v \neq 0 ,$$

is both necessary and sufficient for solvability of problem (6) (e.g. [7-12]). Moreover it follows that the range $R((L-\lambda_k)+G)$ is open in $H$ .

We consider here problem (6) under conditions which are in a certain sense opposite to hypothesis (7) and imply a closed range of $(L-\lambda_k) + G$ .

Let $H$ be decomposed as $H = N(L-\lambda_k) \oplus R(L-\lambda_k)$ . We set $H_1 := N(L-\lambda_k)$, $H_2 := R(L-\lambda_k)$ , and denote by $P_i$ $(i = 1,2)$ the orthogonal projection onto $H_i$. Let $h_i := P_i h \quad (i = 1,2)$.

We now suppose without loss of generality that $g_- \leqq 0 \leqq g_+$. Let $\mathcal{S} \subset H_1$ be the nonempty, closed, convex set defined by

$$\mathcal{S} := \{h_1 \in H_1 : (h_1, v) \leqq \int_\Omega (g_+ v^+ - g_- v^-) \quad \forall v \in H_1\} .$$

The following condition is imposed on $g$:

(g1) There exists $\delta \geqq 0$ such that

$$\begin{cases} g(s) \geqq g_+ & \forall s \geqq \delta \\ g(s) \leqq g_- & \forall s \leqq -\delta \end{cases} .$$

Let

$$\gamma_{\pm} := \liminf_{s \to \pm\infty} (g(s) - g_{\pm}) s \qquad (\geqq 0) .$$

<u>Theorem 2. Suppose that either</u>

(a) $k = 1$ <u>(perturbation in the first eigenvalue) and both</u> $\gamma_- > 0$, $\gamma_+ > 0$, <u>or</u>

(b) $k > 1$ <u>and at least one of</u> $\gamma_-$, $\gamma_+$ <u>is positive. Then to each</u> $h_2 \in H_2$ <u>there exists an open set</u> $\mathcal{S}_{h_2}$ in $H_1$, $\mathcal{S}_{h_2} \supset \mathcal{S}$, <u>such that</u>

(i) <u>if</u> $h_1 \in \mathcal{S}_{h_2}$, <u>then</u> (6) <u>admits at least one solution for</u> $h = h_1 + h_2$;

(ii) <u>if</u> $h_1 \in \mathcal{S}_{h_2} \setminus \mathcal{S}$, <u>then</u> (6) <u>admits at least two solutions for</u> $h = h_1 + h_2$.

Employing the strong maximum principle, assumption (a) can be slightly weakened (cf. Fučik-Hess [13]). Theorem 2 generalizes some results of Ambrosetti-Mancini [14,15]. A consequence is

<u>Theorem 3. Under either assumption of Theorem 2, the range of</u> $(L-\lambda_k) + G$ <u>is closed in</u> $H$.

It follows that assertion (i) of Theorem 2 remains valid for $h_1 \in \overline{\mathcal{S}_{h_2}}$. In order that $R((L-\lambda_k) + G)$ is closed in $H$, one can show by examples that in general conditions stronger than (g1) are needed.

<u>Idea of proof of Theorem 2.</u>

(i) Suppose $h_2 \in H_2$ is fixed, and let $h_1 \in \mathcal{S}$. Obviously the equation

$$(L-\lambda_k)u + G(u) = h$$

is equivalent to the equation

$$u + ((L-\lambda_k) + P_1)^{-1}(G(u) - P_1 u - h) = 0 .$$

Introduce the homotopy

$\mathcal{H}(t,u) := u + t((L-\lambda_k) + P_1)^{-1}(G(u) - P_1 u - h)$ , $t \in [0,1]$ , $u \in H$ .
It can be shown that there exists a rectangle $B = \{u \in H: \|P_1 u\| < c_1, \|P_2 u\| < c_2\}$ in $H$ such that $\mathcal{H}(t,u) \neq 0$ $\forall t \in [0,1]$, $\forall u \in \partial B$ and thus

$$\deg(\mathcal{H}(1,.), B, 0) = \deg(I, B, 0) = 1 .$$

Moreover to each $h_1 \in \mathcal{S}$ there is an open neighborhood $\mathcal{U}(h_1) \subset H_1$ such that the degree remains 1 for $h_1$ replaced by $\tilde{h}_1 \in \mathcal{U}(h_1)$ . Hence $\mathcal{S}_{h_2} := \bigcup_{h_1 \in \mathcal{S}} \mathcal{U}(h_1)$ suffices.

(ii) Let still $h_2 \in H_2$ be fixed, and suppose now $h_1 \in \mathcal{S}_{h_2} \setminus \mathcal{S}$ . Then $\exists \tilde{v} \in H_1$:

$$(h_1, \tilde{v}) > \int_\Omega (g_+ \tilde{v}^+ - g_- \tilde{v}^-) \quad (\geq 0)$$

and consequently $h_1 \neq 0$ . Since $R(G)$ is bounded in $H$ , $(1+K)h \notin R((L-\lambda_k) + G)$ for sufficiently large $K$ . Set

$$\mathcal{K}(t,u) := u + ((L-\lambda_k) + P_1)^{-1}(G(u) - P_1 u - (1+t)h),$$

$t \in [0,K]$, $u \in H$ . We can find a rectangle $C \supset \bar{B}$ (where $B$ is a rectangle as obtained in the proof of part (i)) such that $\mathcal{K}(t,u) \neq 0$ $\forall t \in [0,K]$ , $\forall u \in \partial C$ . Hence

$$\begin{aligned}\deg(\mathcal{H}(1,.),C,0) &= \deg(\mathcal{K}(0,.),C,0) \\ &= \deg(\mathcal{K}(K,.),C,0) = 0 .\end{aligned}$$

By the additivity property of the degree, the existence of a further solution in $C \setminus \bar{B}$ follows.

## References


[1] A. Ambrosetti, P. Hess: Publication to appear.

[2] P. Hess: Nonlinear perturbations of linear elliptic and parabolic problems at resonance: existence of multiple solutions. To appear in Ann. Scuola Norm. Sup. Pisa.

[3] H. Amann, T. Laetsch: Positive solutions of convex nonlinear eigenvalue problems. Indiana Univ. Math. J. 25, 259-270 (1976).

[4] H. Amann: Nonlinear eigenvalue problems having precisely two solutions. Math. Z. 150, 27-37 (1976).

[5] A. Ambrosetti: On the exact number of positive solutions of convex nonlinear problems. To appear.

[6] K.J. Brown, H. Budin: Multiple positive solutions for a class of nonlinear boundary value problems. J. Math. Anal. Appl. 60, 329-338 (1977).

[7] E.M. Landesman, A.C. Lazer: Nonlinear perturbations of linear elliptic boundary value problems at resonance. J. Math. Mech. 19, 609-623 (1970).

[8] S.A. Williams: A sharp sufficient condition for solution of a nonlinear elliptic boundary value problem. J. Differential Equations 8, 580-586 (1970).

[9] J. Nečas: On the range of nonlinear operators with linear asymptotes which are not invertible. Comment. Math. Univ. Carolinae 14, 63-72 (1973).

[10] P. Hess: On a theorem by Landesman and Lazer. Indiana Univ. Math. J. 23, 827-829 (1974).

[11] S. Fučik, M. Kučera, J. Nečas: Ranges of nonlinear asymptotically linear operators. J. Differential Equations 17, 375-394 (1975).

[12] H. Brézis, L. Nirenberg: Characterizations of the ranges of some nonlinear operators and applications to boundary value problems. To appear in Ann. Scuola Norm. Sup. Pisa.

[13] S. Fučik, P. Hess: Publication to appear.

[14] A. Ambrosetti, G. Mancini: Existence and multiplicity results for nonlinear elliptic problems with linear part at resonance. To appear in J. Differential Equations.

[15] A. Ambrosetti, G. Mancini: Theorems of existence and multiplicity for nonlinear elliptic problems with noninvertible linear part. To appear in Ann. Scuola Norm. Sup. Pisa.

Author's address: Mathematics Institute, University of Zürich,
Freiestrasse 36, 8032 Zürich, Switzerland.

DUAL FINITE ELEMENT ANALYSIS FOR SOME UNILATERAL BOUNDARY VALUE PROBLEMS

I.Hlaváček, Praha

A great number of papers of both technical and mathematical character has been devoted to the numerical solution of variational inequalities. For instance, in the book [1] more than 300 titles are quoted.

The boundary value problems with inequalities contain two important classes:

(i) problems with inequalities on the boundary of the domain under consideration,

(ii) problems with inequalities in the domain.

In general, the problems of both classes can be solved approximately by means of finite differences or finite elements. For problems of the class (i), however, where the inequalities are concentrated on the boundary, only the finite element method is applied in case of a general boundary. In the following, we restrict ourselves to the problems of the class (i) for elliptic equations. A survey of some recent results on the dual variational approach applied to several problems of the second order will be presented.

## 1. An equation of the second order

To point out the main idea, let us consider the following model problem in a bounded polygonal domain $G \subset R^2$:

$$-\Delta u = f \quad \text{in } G,$$
$$u = 0 \quad \text{on } \Gamma_u, \tag{1}$$
$$u \geq 0, \quad \frac{\partial u}{\partial n} \geq 0, \quad u\frac{\partial u}{\partial n} = 0 \qquad \text{on } \Gamma_a = \Gamma - \Gamma_u,$$

where $\Gamma_u$ and $\Gamma_a$ are parts of the boundary $\Gamma = \partial G$, $\partial u/\partial n$ denotes the derivative with respect to the outward normal $\underline{n}$.

We shall denote $H^k(G) = W_2^{(k)}(G)$ the Sobolev spaces with the usual norm $\|\cdot\|_k$ and $H^0(G) = L_2(G)$. The semi-norm consisting of all derivatives of the k-th order only, will be denoted by $|\cdot|_k$.

Assume that the right-hand side of the equation (1) $f \in L_2(G)$.

### 1.1. Primary variational formulation

Let us introduce the functional of the potential energy

$$L(v) = \frac{1}{2}\int_G |\operatorname{grad} v|^2 dx - \int_G fv dx$$

and the set of admissible functions

$$K = \{ v \in H^1(G) \mid \gamma v|_{\Gamma_u} = 0, \quad \gamma v|_{\Gamma_a} \geq 0 \},$$

(where $\gamma v$ denote the traces of the function $v$ ).

The problem (1) corresponds to the following variational (primary) problem:

(2) to find a function $u \in K$ such that
$L(u) \leq L(v) \; \forall \; v \in K$ .

The problem (2) has a unique solution. It is not difficult to prove that (i) any solution of the problem (1) satisfies the condition (2) and (ii) any solution of (2) satisfies the equation ($1_1$) in the sense of distributions and the boundary conditions ($1_3$) in a functional sense, i.e. in the space $H^{-1/2}(\Gamma)$ .

For the approximations to the primary problem - see e.g. [3],[5] .

1.2. Dual variational formulation

We often have problems when the gradient (or cogradient) of the solution $u$ is more interesting than the solution itself. In physical problems grad $u$ represents the vector of fluxes, in elasticity it corresponds to the stress tensor.

Therefore it may be useful to formulate the problem directly in terms of the unknown vector-function of the gradient. To this end let us introduce the set

$$Q = \{ q \in [L_2(G)]^2 \mid \text{div } q \in L_2(G) \} .$$

For $q \in Q$ we may define the functional (outward flux) $q.n \in H^{-1/2}(\Gamma)$ as follows

$$\langle q.n, w \rangle = \int_G (q.\text{grad } v + v \text{ div } q)dx \; \forall \; w \in H^{1/2}(\Gamma) ,$$

where $v \in H^1(G)$ is an extension of the function $w = \gamma v$ .

We write $s|_{\Gamma_a} \geq 0$ for a functional $s \in H^{-1/2}(\Gamma)$ if

$$\langle s, \gamma v \rangle \geq 0 \; \forall \; v \in K .$$

Let us introduce the set of admissible functions

$$U = \{ q \in Q \mid \text{div } q + f = 0 \quad \text{in } G, \; q.n|_{\Gamma_a} \geq 0 \} ,$$

the functional of complementary energy

$$S(q) = \frac{1}{2} \| q \|_0^2$$

and the dual variational problem:

to find $q^0 \in U$ such that

(3) $S(q^0) \leq S(q) \; \forall \; q \in U$ .

The problem (3) has a unique solution. Moreover, there is a connection between the solution $u$ of the primary problem and the solution $q^0$ of the dual problem as follows:

(4) $$q^0 = \text{grad } u ,$$

(5) $$L(u) + S(q^0) = 0 .$$

The proof of (4) and (5) can be based on the saddle point theory (cf. [5] ).

1.3. Approximations to the dual problem

Assume that we have a vector $\bar{q} \in Q$ such that $\text{div } \bar{q} + f = 0$ in the domain $G$ . (We can set e.g. $\bar{q} = (\bar{q}_1, 0)$, where

$$\bar{q}_1(x_1,x_2) = - \int_0^{x_1} f(t,x_2)dt \quad .)$$

Then it is readily seen that $q \in U$ if and only if $p = q - \bar{q} \in U_0$ , where

$$U_0 = \{ p \in Q \mid \text{div } p = 0, \quad (p.n + \bar{q}.n)|_{\Gamma_a} \geq 0 \} .$$

Thus we are led to the following problem, which is equivalent with the problem (3):

to find $p^0 \in U_0$ such that

(6) $$J(p^0) \leq J(p) \;\forall\; p \in U_0 ,$$

where

$$J(p) = \frac{1}{2} \| p \|_0^2 + (\bar{q}, p)_0 .$$

With respect to the definition of the set $U_0$ , instead of the standard finite element spaces $V_h$ we have to employ the spaces of the so called equilibrium (solenoidal) elements, which satisfy the equation $\text{div } p = 0$ in the domain $G$ at least in the sense of distributions. To this end we apply the spaces $N_h$ of piecewise linear triangular elements, which were proposed by Veubeke and Hogge in [6] and studied in the paper [7] . The latter study yields the following approximation property of the equilibrium spaces $N_h$ (on any regular family of triangulations):

(7) $$\forall\; p \in [H^2(G)]^2 , \text{ div } p = 0, \quad \exists\, \pi^h \in N_h :$$
$$\| p - \pi^h \|_0 \leq C\, h^2 |p|_2 .$$

In the following, we shall use the strongly regular family of triangulations $\{T_h\}$ (i.e., the minimal angle in $T_h$ is bounded from below and the ratio of any two sides in $T_h$ is bounded from above, the bounds being independent of the parameter $h$ - the maximal side in the triangulation $T_h$ ).

Assume that a function $F \in [H^2(G)]^2$ exists, such that

$$\operatorname{div} F = 0 \quad \text{in} \quad G, \quad F.n|_{\Gamma_a} = -\bar{q}.n|_{\Gamma_a} \quad .$$

Let us denote $-\bar{q}.n = g$ and construct a function $g_h \in L_2(\Gamma_a)$ such that its restriction $g_h|_{S_j}$ onto any side $S_j \subset \bar{\Gamma}_a$ of the triangulation $T_h$ coincides with the $L_2(S_j)$-projection of $g$ into the subspace of linear polynomials $P_1(S_j)$ .

Defining

$$U_{0h} = \{ p \in N_h \mid p.n|_{\Gamma_a} \geq g_h \} ,$$

we say that a vector $p^h \in U_{0h}$ is a finite element approximation to the dual problem, if

(8) $$J(p^h) \leq J(p) \; \forall \; p \in U_{0h} \quad .$$

(Note that $U_{0h} \not\subset U_0$ unless $g_h \geq g$ holds everywhere on $\Gamma_a$.)

Theorem 1. Let the boundary $\Gamma$ consist of a finite number of non-intersecting polygons $\partial G_j$ and let

$$\operatorname{meas} (\Gamma_u \cap \partial G_j) > 0 \; \forall \; j \; .$$

Assume that

$$p^0 \in [H^2(G)]^2 \; , \; (p^0 - F).n \in H^2(\Gamma_a \cap \Gamma_m)$$

holds for any side $\Gamma_m$ of the polygonal boundary.

Then for any strongly regular family of triangulations it holds

(9) $$\| p - p^h \|_0 \leq C(p^0) h^{3/2} \; .$$

Proof. Let us define the mapping $\Pi_T \in \mathcal{L}([H^1(T)]^2; [P_1(T)]^2)$ for all $T \in T_h$ by the following conditions: the $L_2(S_j)$-projection of the flux $q.n|_{S_j}$ is equal to the flux $(\Pi_T q).n|_{S_j}$ for each side $S_j$ of the triangle $T$ .

Define also the set

$$R(G) = \{ q \mid q \in [H^1(G)]^2 \; , \; \operatorname{div} q = 0 \}$$

and the mapping $r_h \in \mathcal{L}(R(G); N_h)$ by the conditions

$$r_h q|_T = \Pi_T q \; \forall \; T \in T_h \; .$$

(Then $\pi^h = r_h p$ can be taken in (7).)

Let us introduce the cones

$$C = \{ q \in Q \mid \operatorname{div} q = 0, \; q.n|_{\Gamma_a} \geq 0 \} \; , \quad C_h = C \cap N_h \; .$$

Lemma 1. Denote $U = p^0 - F \in C$ and assume that a $W_h \in C_h$ exists such that $2U - W_h \in C$ . Then it holds

(10) $$\| p^0 - p^h \|_0 \leq \| U - W_h \|_0 + \| F - r_h F \|_0 \; .$$

(For the proof of (10) one employs the variational inequalities characterizing $p^0$ and $p^h$ , respectively.)

Thus it suffices to find a suitable element $W_h \in C_h$ close to $U$. To this end we (i) construct a one-sided piecewise linear approximation $\psi_h$ of the boundary flux $U.n$ and (ii) define $W_h = r_h U$ in the interior elements $T \in T_h$, (iii) correct $r_h U$ in the boundary strip to obtain $W_h.n = \psi_h$ on $\Gamma$.

Let us present the approach in detail:

Lemma 2. If the assumptions of Theorem 1 hold, then there exists a piecewise linear function $\psi_h$ on $\Gamma$ (with the nodes determined by the triangulation $T_h$) and such that

$$\int_{\partial G_j} \psi_h ds = \int_{\partial G_j} (r_h U).n ds \quad \forall j, \tag{11}$$

$$0 \le \psi_h \le U.n \quad \text{on } \Gamma_a, \tag{12}$$

$$\|(r_h U).n - \psi_h\|_{0,\Gamma} \le c\, h^2 \sum_m |U.n|_{2,\Gamma_a \cap \Gamma_m}. \tag{13}$$

Proof. Denote $U.n = t$, $(r_h U).n = t_h$ and consider a side $S_i \subset \subset \overline{\Gamma}_a$. Let $t_I$ be the linear interpolate of $t$ over $S_i$. First we construct the function $\psi_h|_{S_i} \equiv \psi_h^i$.

$1^0$ If $t \ge t_I$ on $S_i$, setting $\psi_h^i = t_I$ we obtain

$$\|\psi_h^i - t_h\|_{0,S_i} \le \|t - t_I\|_{0,S_i} + \|t - t_h\|_{0,S_i} \le \le c\, h^2 |t|_{2,S_i}. \tag{14}$$

$2^0$ Let $t < t_I$ at some point $P \in \overline{S}_i$ such that the tangent to the graph of $t$ at $P$ lies under the graph of $t$ and, if $\psi_h^i$ corresponds to the tangent, $\psi_h^i \ge 0$ on $\overline{S}_i$. Then the estimate (14) holds. Thus we construct $\psi_h$ over $\Gamma_a$. On $\partial G_j - \Gamma_a$ we define

$$\psi_h = t_h + a_j, \tag{15}$$

where $a_j$ is a constant such that (11) takes place. The estimate (13) follows from (14) and (15).

Lemma 3. Let a piecewise linear (discontinuous, in general) function $\varphi$ on $\Gamma$ be given such that

$$\int_{\partial G_j} \varphi\, ds = 0 \quad \forall j. \tag{16}$$

Then there exists a vector-function $w^h \in N_h$ such that $w^h.n = \varphi$ on $\Gamma$ and

$$\|w^h\|_0 \le c\, h^{-1/2} \|\varphi\|_{0,\Gamma}. \tag{17}$$

Proof. Let $G_h$ be the union of all triangles $T \in T_h$ such that $T \cap \Gamma \neq \emptyset$ (a boundary strip of $G$ ). We determine $w^h \in N_h$ by means of properly chosen flux parameters on the sides of $T_h$ , such that $\mathrm{supp}\; w^h \subset G_h$ . In particular, we choose the flux parameters $\beta$ equal to the corresponding values of $\varphi$ on $\partial G_j$ and equal to zero on $\partial G_h^j - \partial G_j$ . As the sides connecting $\partial G_j$ and $\partial G_h^j - \partial G_j$ are concerned, we set $\beta = 0$ at the vertices of $\partial G_h^j - \partial G_j$ but the parameters at the vertices of $\partial G_j$ remain to be determined from the conditions (i) of the vanishing divergence and (ii) of the continuity of the fluxes along the interelementary boundaries. Using (16) the linear system for the remaining parameters can be solved and the solution $\beta$ estimated from above, making use of the strong regularity of the triangulations, as follows

$$|\beta_i| \leq C\, h^{-1} \|\varphi\|_{0,\Gamma} \; .$$

Then the estimate (17) follows easily.

To finish the proof of Theorem 1, let us set

$$\varphi = (r_h U).n - \psi_h \; ,$$

where $\psi_h$ is the one-sided approximation from Lemma 2, and consider the extension $w^h$ of $\varphi$ from Lemma 3. Then the function $W_h = r_h U - w^h$ satisfies the conditions of Lemma 1. The assertion (9) follows from (10), (7), (17) and (13).

Algorithm for the dual approximation

The problem (8) belongs to quadratic programming. It can be solved by various procedures, e.g. by the method of Uzawa (see [4]) or by a method of feasible directions (cf. [9]).

## 1.4. A posteriori error estimates and two-sided bounds for the energy

Having approximations of the primary and of the dual problem, we are able to calculate a posteriori error estimates and two-sided bounds for the energy of the solution $u$ , as follows.

Theorem 2. Let $u_h \in K$ and $\tilde{q}^h = \bar{q} + \tilde{p}^h \in U$ . Then it holds

$$|u - u_h|_1^2 \leq \|\tilde{q}^h - \mathrm{grad}\; u_h\|_0^2 + 2\int_{\Gamma_a} \tilde{q}^h.nu_h\, ds \equiv E \; ,$$

$$\|\tilde{q}^h - \mathrm{grad}\; u\|_0^2 \leq E \; ,$$

$$-2L(u_h) \leq |u|_1^2 = (f,u)_0 \leq 2S(\tilde{q}^h) \; .$$

Proof. One can easily obtain

$$|v - u|_1^2 \leq 2[L(v) - L(u)] \; \forall \; v \in K \; .$$

From (5) and (3) it follows that

$$-L(u) = S(q^0) \leq S(q) \; \forall \; q \in U \; .$$

Hence for any $v \in K$ and $q \in U$, $q - \bar{q} \equiv p \in U_{0h}$ we may write

$$|v - u|_1^2 \leq 2L(v) + 2S(q) = |v|_1^2 - 2(f,v)_0 + \|q\|_0^2 =$$

$$= \|q - q(v)\|_0^2 + 2(q,q(v)) - 2(f,v)_0 \; ,$$

where $q(v) = \text{grad } v$ . Moreover, we have

$$(q,q(v)) - (f,v)_0 = \int_G (q.\text{grad } v) + v \text{ div } q)dx =$$

$$= \langle q.n, \gamma v \rangle = \int_{\Gamma_a} q.nvds \; .$$

## 2. Some other unilateral problems

Recently, the following boundary value problems have been solved by the dual finite element method with analogous results as above:

(i) equations with an "absolute" term

$$(18) \qquad -\Delta u + u = f \quad \text{in } G$$

with the conditions $(1_3)$ on the whole boundary. Here the standard piecewise linear elements can be applied for both the primary and the dual approximations (see [5]);

(ii) problems with a non-homogeneous obstacle on the boundary, i.e. an equation (18) with the boundary conditions

$$u \geq g \; , \quad \frac{\partial u}{\partial n} \geq 0 \; , \quad (u - g)\frac{\partial u}{\partial n} = 0 \quad \text{on} \quad \Gamma,$$

where $g$ is a given function (see [10]);

(iii) semi-coercive problems of the type (1), i.e. the equation $(1_1)$ with the conditions $(1_3)$ on the whole boundary. The proof of convergence requires a different approach, because the one-sided approximations of the flux cannot be used (see [12]);

(iv) Signorini's problem in plane elastostatics and contact problems for two elastic bodies (see [14]). The triangular piecewise linear block-elements, which were proposed by Watwood and Hartz in [15] and studied in [16] , [17] , have been used for the dual approximations.

## References


[1] Glowinski R., Lions J.L., Trémolières R.: Analyse numérique des inéquations variationnelles. Dunod, Paris, 1976

[2] Falk R.S.: Error estimates for the approximation of a class of variational inequalities. Math.of Comp. 28, 1, (1974), 963-970

[3] Brezzi F., Hager W.W., Raviart P.A.: Error estimates for the finite element solution of variational inequalities. Numer.Math. 28, (1977), 431-443

[4] Céa J.: Optimisation, théorie et algorithmes. Dunod, Paris, 1971

[5] Hlaváček I.: Dual finite element analysis for unilateral boundary value problems. Apl.Mat. 22 (1977), 14-51

[6] Fraeijs de Veubeke B., Hogge M.: Dual analysis for heat conduction problems by finite elements. Inter.J.Numer. Meth.Eng. 5 (1972), 65-82

[7] Haslinger J., Hlaváček I.: Convergence of a finite element method based on the dual variational formulation. Apl. Mat. 21 (1976), 43-65

[8] Mosco U., Strang G.: One-sided approximations and variational inequalities. Bull.Amer.Math.Soc. 80 (1974), 308-312

[9] Zoutendijk G.: Methods of feasible directions. Elsevier, Amsterdam, 1960

[10] Hlaváček I.: Dual finite element analysis for elliptic problems with obstacles on the boundary. Apl.Mat. 22 (1977), 244-255

[11] Duvaut G., Lions J.L.: Les inéquations en mécanique et en physique. Dunod, Paris, 1972

[12] Hlaváček I.: Dual finite element analysis for semi-coercive unilateral boundary value problems. Apl.Mat. 23 (1978), 52-71

[13] Hlaváček I., Lovíšek J.: A finite element analysis for the Signorini problem in plane elastostatics. Apl.Mat. 22 (1977), 215-228

[14] Haslinger J., Hlaváček I.: Contact between elastic bodies, to appear

[15] Watwood V.B. Jr., Hartz B.J.: An equilibrium stress field model. Inter.J.Solids and Structures 4 (1968), 857-873

[16] Hlaváček I.: Convergence of an equilibrium finite element model for plane elastostatics, to appear

[17] Johnson C., Mercier B.: Some equilibrium finite element methods for two-dimensional elasticity problems, to appear


Author's address: Matematický ústav ČSAV, Žitná 25, 115 67 Praha 1, Czechoslovakia

GRADIENT ALTERNATING-DIRECTION METHODS

V. Il'in, Novosibirsk

## 1. Introduction

We shall be concerned with iterative methods for the solution of the system of equations

$$Ax = f, \tag{1}$$

where $A$ is a symmetric square matrix and $x$, $f$ are N-dimensional real vectors. We suppose that the eigenvalues $\lambda_k$ of the matrix $A$ are non-negative, $0 \leq \alpha \leq \lambda_1 \leq \lambda_2 \leq \dots \leq \lambda_N \leq \beta < \infty$, and that $A$ may be expressed as a sum of positive semi-definite matrices which are easily invertible, $A = \sum_{i=1}^{p} A_i$, $(A_i x, x) \geq 0$, $i = 1,2$. The alternating-direction iterative methods were introduced in the papers by Peaceman, Rachford and Douglas in 1955. These methods, which have passed extensive development since then, use the inversion of matrices of the form $I + \tau A_i$, where $I$ is the identity matrix and $0 < \tau < \infty$, in the intermediate stages [1]. Such methods may be understood to be based on the preliminary multiplication of the equation (1) by the matrix $H_\tau^{-1}$,

$$H_\tau = \frac{1}{\tau} (I + \tau A_1)(I + \tau A_2) \dots (I + \tau A_p), \tag{2}$$

where the iteration parameter $\tau$ is found in such a way that the condition number of the matrix $H_\tau^{-1} A$ be minimum. We shall not dwell on various versions of the algorithms studied in papers by Janenko, Kellogg, Samarskiǐ, Marčuk, D'jakonov, the author, and others (see e.g. the surveys in [2] - [4]), which differ in the ways of realization in the main. Instead, we give attention to the following form of the alternating-direction iterative methods:

$$x = x^{n-1} - \omega_n H_n^{-1} (Ax^{n-1} - f). \tag{3}$$

For $p = 2$ scheme (3) is equivalent to the Douglas-Rachford method if $w_n \equiv 1$ and to that of Peaceman-Rachford if $w_n \equiv 2$ [4].

The algorithms of alternating directions have found their principal application in the finite difference methods for solving elliptic equations with $p \geq 2$ independent variables. If, e.g., $A_i$ is a difference analogue of the operator of the second derivative with respect to one variable then $H_\tau$ is a product of easily invertible tridiagonal matrices (the so-called alternating-direction implicit methods, ADI). On the other hand, defining $A_1$ and $A_2$

as a lower and upper triangular matrix, respectively, (i.e. $A_1 + A_2 = A$, $A_1 = A_2'$), we obtain alternating-direction explicit methods (ADE) or point-triangular methods, studied by A.A. Samarskiĭ and the author. Some versions of these point-triangular methods coincide with particular realizations of the symmetric successive over-relaxation method (SSOR), see [2], [4] and the references quoted there.

The basic results on the optimization of iterative methods of the form (3) consist in the minimization of the spectral radius of the matrix

$$T_n = \sum_{s=1}^{n} (I - \omega_s H_{\tau_s}^{-1} A) \tag{4}$$

under the hypothesis that its spectrum is real. For example, if $\alpha > 0$, $p = 2$, $\omega_n \equiv 2$, and the matrices $A_1$, $A_2$ are commutative then the sequences $\tau_s$ of iteration parameters are known such that the number $n(\varepsilon)$ of iterations necessary for reducing the norm of the error $y^n = x - x^n$ $\varepsilon^{-1}$-times satisfies the inequality

$$n(\varepsilon) \leq C|\ln \varepsilon| . \ln \frac{\beta}{\alpha} , \tag{5}$$

where $C$ is a constant independent of the bounds of the spectrum of $A$. Another approach is connected with the use of constant $\tau_n \equiv \tau$ and the selection of $\omega_n$ according to the Chebyshev acceleration method. In this case, supposing $\alpha > 0$, $p = 2$, commutative $A_1$ and $A_2$, and the optimum value of $\tau$, we obtain the inequality [5]

$$n(\varepsilon) \leq \ln \frac{1-\sqrt{1-\varepsilon^2}}{\varepsilon} / \ln \frac{1-\sqrt{\gamma}}{1+\sqrt{\gamma}} , \tag{6}$$

where $\gamma = 2(\alpha/\beta)^{1/2}(1+\alpha/\beta)^{-1}$. For $\varepsilon << 1$, $\alpha/\beta << 1$ the inequality (6) may be written as

$$n(\varepsilon) \leq |\ln \varepsilon| / 2\sqrt{2\sqrt{\alpha/\beta}} . \tag{6a}$$

A number of papers present the convergence conditions for the iterative processes and estimates of $n(\varepsilon)$ for non-commutative matrices $A_i$ and $p > 2$. However, these estimates prove to be weaker than (5), (6) (cf. [1] - [4] and the references quoted there).

In the present paper we discuss the optimization of iteration parameters connected with minimizing certain functionals which characterize the suppression of errors in successive approximations. In other words, we shall investigate the application of the method of steepest descent, of the minimum residual method and the conjugate gradient method to the alternating-direction algorithms. The above methods have been introduced and studied by Kantorovič, Krasnosel'skiĭ, Kreĭn, Hestenes, Stiefel and others ( [6] - [9] ). We note that it seems that such an approach has been investigated for the first time in the papers by Godunov and Prokopov, Marčuk and Kuznecov, see [3], [10].

## 2. The method of steepest descent and the minimum residual method

Putting $\tau_n \equiv \tau$ in (3) and defining $\omega_n$ from the minimum condition of the functional

$$\Phi(x^{n+1}) = (Ax^{n+1},x^{n+1})-2(f,x^{n+1})=(Ay^{n+1},y^{n+1})-(f,x^{n+1}) \tag{7}$$

as

$$\omega_n = \frac{(r^n,H_\tau^{-1}r^n)}{(AH_\tau^{-1}r^n,H_\tau^{-1}r^n)}, \quad r_n = f - Ax^n, \tag{8}$$

we come to the method of steepest descent, for which the relation

$$\frac{\Phi(x^{n+1})}{\Phi(x^n)} = \frac{(Ay^{n+1},y^{n+1})}{(Ay^n,y^n)} = q_n = 1 - \frac{(r^n,H_\tau^{-1}r^n)^2}{(A^{-1}r^n,r^n)(AH_\tau^{-1}r^n,H_\tau^{-1}r^n)} \tag{9}$$

holds [3]. If the matrix $H^{1/2}AH^{-1/2}$ is positive definite and the bounds of its spectrum are $0<m<M<\infty$, then the inequality

$$q_n \leq q = \left(\frac{M-m}{M+m}\right)^2 \tag{10}$$

holds. In particular, if $p = 2$ and $A_i$ are symmetric, commutative, and satisfy the condition $\delta(x,x) \leq (A_ix,x) \leq \Delta(x,x)$, $i = 1,2$, $\delta > 0$ then we have

$$\begin{aligned} m &= 2\sqrt{\delta/\Delta}/(1+\sqrt{\delta/\Delta})^2, \\ M &= (1+\delta/\Delta)/(1+\sqrt{\delta/\Delta})^2, \\ q &= (1-\sqrt{\delta/\Delta})^4(1+\sqrt{\delta/\Delta})^4 \end{aligned} \tag{11}$$

for $\tau = (\delta\Delta)^{-1/2}$. This means that the method (3), (8) with a constant optimal value of $\tau$ converges no worse that the Peaceman--Rachford method (formula (5) with $\omega_n \equiv 2$).

Under the assumptions stated before (10) we can obtain an estimate of another kind for the method of steepest descent. Putting $k_n = (Ay^n,y^n)$ from (9) we have (see [5]):

$$k_n-k_{n+1} = \frac{(r^n,H_\tau^{-1}r^n)}{(AH_\tau^{-1}r^n,H^{-1}r^n)} \geq \frac{(r^n,H_\tau^{-1}r^n)}{M} \geq \frac{(r^n,y^n)^2}{M(H_\tau y^n,y^n)} \geq \tag{12}$$

$$\geq \frac{k_n^2}{Mq^n(H_\tau y^0, y^0)} \quad .$$

From this it follows immediately that

$$(13) \qquad (Ay^{n+1}, y^{n+1}) \leq (Ay^0, y^0) / [1 + (Ay^0, y^0) M^{-1} (Hy^0, y^0)^{-1} \sum_{s=1}^{n} q^{-s}].$$

The inequality (13) leads to the bound

$$(14) \qquad (Ay^{n+1}, y^{n+1}) \leq \frac{1-q^{-1}}{1-q^{-n}} M(H_\tau y^0, y^0)$$

for $q < 1$. If we do not require the positive definitness of $A$ (i.e. we admit $m = 0$, $q = 1$) then (13) implies the inequality

$$(15) \qquad (Ay^{n+1}, y^{n+1}) \leq M(H_\tau y^0, y^0)/n$$

obtained in [10] and for $H_\tau = I$ even earlier by Kantorovič [6]. In the practically most important case $q = 1 - \gamma$, $\gamma << 1$ the bound (14) gives a rather better result than (15):

$$(Ay^{n+1}, y^{n+1}) \leq M(H_\tau y^0, y^0)/n(1+n\gamma).$$

The minimum residual method is defined by (3) with $\tau_n = \tau$ and computing $\omega_n$ from the condition of minimum value of the functional $(r^{n+1}, r^{n+1})$:

$$(16) \qquad \omega_n = \frac{(AH_\tau^{-1} r^n, r^n)^2}{(AH_\tau^{-1} r^n, AH_\tau^{-1} r^n)} \quad .$$

In this case the residual $r^n$ satisfies the relation

$$(17) \qquad \frac{\| r^{n+1} \|^2}{\| r^n \|^2} = \frac{(r^{n+1}, r^{n+1})}{(r^n, r^n)} = \bar{q}_n = 1 - \frac{(AH_\tau^{-1} r^n, r^n)^2}{(r^n, r^n)(AH_\tau^{-1} r^n, AH_\tau^{-1} r^n)}$$

and bounds for $\bar{q}_n$ analogous to (10), (11) can be easily obtained. Since, under the above assumptions, the relation

$$\| r^n \| - \| r^{n+1} \| = \frac{(AH_\tau^{-1} r^n, r^n)^2}{\| AH_\tau^{-1} r^n \| ( \| r^n \| + \| r^{n+1} \| )} \geq \frac{(AH_\tau^{-1} r^n . r^n)}{2M \| r^n \|}$$

follows from (17) we obtain

$$(18) \qquad \| r^n \| - \| r^{n+1} \| \geq \frac{\| r^n \|^2}{2M \cdot \| H_\tau y^0 \| \bar{q}^{-n}}$$

with the help of the inequalities

$$(r^n,r^n) \leq (AH^{-1}r^n,r^n) \| H_\tau y^n \| . \| r^n \| , \| Hy^n \| \leq \bar{q}_n \| Hy^0 \| .$$

From this estimate

(19) $$\| r^{n+1} \| \leq \frac{1-\bar{q}^{-1}}{1-\bar{q}^{-n}} 2M \| H_\tau y^0 \|$$

follows for $\bar{q}<1$. Analogously to the previous case, the inequality

(20) $$(r^{n+1},r^{n+1}) \leq 4M^2(H_\tau y^0,H_\tau y^0)/n^2$$

is a consequence of (18) in the case of a singular matrix $A(\bar{q} \leq 1)$. The inequality (20) is obtained in [11].

Let us consider the problem of the choice of parameters $\tau$ in (3) for $p = 2$ based on the condition of minimum value of functionals. This problem is studied in [11]. In the method of steepest descent, calculating $\omega_n$ from the formula (8), we come to the nonlinear equation

(21) $$\tau_n = \left[\frac{(H_{\tau_n} r^n,r^{n+1})}{(H_{\tau_n}^{-1} A_1A_2H_{\tau_n}^{-1} r^n,r^{n+1})}\right]^{1/2},$$

which follows from the condition $\frac{d\,\Phi(x^{n+1})}{d\tau} = 0$. Similarly in the minimum residual method we obtain

(22) $$\tau_n = \left[\frac{(AH_{\tau_n}^{-1} r^n,r^{n+1})}{(H_{\tau_n}^{-1} A_1A_2H_{\tau_n}^{-1} r^n,r^{n+1})}\right]^{1/2}$$

from the equation $\frac{d(r^{n+1},r^{n+1})}{d\tau} = 0$. ($\omega_n$ in (22) is calculated from (16.) It can be easily seen that for these $\tau_n$ the functionals $\Phi(x^{n+1})$ and $(r^{n+1},r^{n+1})$ attain its minimum if $AH_{\tau_n}^{-1}$ is a symmetric positive definite matrix. Notice that if now $r^n$ possesses one dominant component in the expansion with respect to the eigenfunctions of the matrix $AH_{\tau_n}^{-1}$, then (21) and (22) become

(23) $$\tau_n \approx \left[\frac{(r^n,r^n)}{(A_1A_2r^n,r^n)}\right]^{1/2},$$

i.e., one approximate formula which under the assumptions made above is true even for $\omega_n \equiv 2$ particularly ( see [11] ).

It is possible to propose algorithms with the computation of iteration parameters using a posteriori information on the basis of the above considerations. We carry out the iterations by formula (3) with fixed values $\omega_n = 2$, $\tau_n = \tau_0$ (e.g., $\tau_0 = (\delta\Delta)^{-1/2}$ if $\delta$ and $\Delta$ are known) at first. If the condition

$$(24) \qquad \left| \frac{\| r^{n_k+1} \|}{\| r^{n_k} \|} - \frac{\| r^{n_k} \|}{\| r^{n_k-1} \|} \right| \leq \varepsilon$$

with a sufficiently small $\varepsilon$ is satisfied for some $n_k$ (this characterizes the isolation of the dominant harmonics in $r^{n_k}$) then we compute $\tau_n$ using formula (23) and $\omega_n$ using (8) or (16) in the next iteration. The following iterations are performed with $\omega_n = 2$ and $\tau_n = \tau_0$ again until the condition (24) is satisfied etc.

For the methods under consideration we present the results of experiments with alternating-direction implicit methods for the five-point approximation of the Laplace equation $\Delta u = 0$ on a square grid in a square domain with the boundary condition $u|_\Gamma = 1$. The iterations were performed on the grids $m \times m$ ($m = 10,20,40$) with $u^0_{ij} = 0$ till the condition $\max_{i,j} | u^{n+1}_{ij} - u^n_{ij} | \leq 10^{-5}$ was satisfied. For comparison we present the numbers of iterations for the Peaceman-Rachford method in Table 1 ($\omega_n = 2$). The first column corresponds to constant parameters close to their optimum values ($\tau = \tau_0 = 2.25$, $4.75$, $10$ for $m = 10,20,40$ respectively) and the second one to the optimum sequence $\tau_n$ "of Wachspress" [1]. The numbers of iterations with $\tau = \tau_0$ and $\omega_n$ calculated by the minimum residual method (16) are given in the third column while those with $\omega_n$ and $\tau_n$ calculated from (16), (23) and the condition (24) satisfied for $\varepsilon = 10^{-2}$ are presented in the fourth column. Finally the numbers of iterations in the fifth column correspond to $\omega_n$ and $\tau_n$ calculated from the formulae (16), (22) for each $n$. Although these last results seem to be best their character is purely illustrative since finding $\tau_n$ is here a very time-consuming process involving the solution of a nonlinear equation.

| m \ N | 1 | 2 | 3 | 4 | 5 |
|---|---|---|---|---|---|
| 10 | 17 | 9 | 17 | 13 | 9 |
| 20 | 31 | 13 | 28 | 15 | 11 |
| 40 | 60 | 16 | 54 | 18 | 13 |

Table 1.

The computations performed show that with $\omega_n$ calculated by the method of steepest descent (8) we obtain roughly the same results for the problems under consideration.

## 3. Conjugate gradient methods

Defining the conjugate gradient methods, as applied to the equation (1) multiplied by the matrix $H_\tau^{-1}$, in accord with [9], we obtain the following class of iteration algorithms [5]

$$(25)\qquad x_{n+1}=x_n+a_np_n,\quad p_{n+1}=Kq_{n+1}+b_np_n,\quad q_{n+1}=AH^{-1}Br_{n+1},$$
$$r_{n+1}=H^{-1}(f-Ax_{n+1}),\quad a_n=\frac{(q_n,p_n)}{(p_n,Rp_n)},\quad b_n=\frac{(q_{n+1},Kq_{n+1})}{(q_n,Kq_n)},$$
$$q_0=AH^{-1}Br_0,\quad r_0=H^{-1}(f-Ax_0),\quad p_0=Kq_0$$

where $R = AH_\tau^{-1}BH_\tau^{-1}A$, $B$ and $K$ are certain symmetric positive definite matrices, and $H_\tau$ is supposed to be symmetric for the sake of simplicity. The process given is optimal in the following sense: The vector $x_n$ minimizes the functional $\Phi(x) = (r,Br)$ on an $(n+1)$-dimensional hyperplane passing through the points $x_0,\dots,x_n$. As is shown in [8], [9], [5] it is valid that

$$(26)\qquad y_{n+1} = y_0 - TP_n(T)y_0\ ,$$

where $T = KR$ and $P_n(\nu)$ is a certain polynomial of degree $n$, holds for the error vectors. If $\nu_k$ and $z_k$, $k = 1,2,\dots,N$, are the eigenvalues and the corresponding orthogonal eigenvectors of the matrix $T$ and $\varphi(\nu_k)$ are the eigenvalues of the matrix $R$, which will be considered to be a function of $\nu_k$, then we arrive at

$$(27)\qquad \Phi(x_{n+1}) = \sum_{k=1}^{\nu} \varphi(\nu_k)\,[1-\nu_kP_n(\nu_k)]^2(y_0,z_k).$$

If $0<\nu_1 \leq \nu_k \leq \nu_N$ and $\varphi(\nu)$ is a polynomial it is possible to estimate the rate of the decrease of the functional $\Phi(x_{n+1})$ constructing a polynomial of the form

$$(28)\qquad t(\nu) = \varphi(\nu)\,[1-\nu P_n(\nu)]^2 = \varphi(\nu)F_{n+1}^2(\nu)$$

that satisfies the condition $F_{n+1}(0) = 1$ and possesses the least deviation from zero on the interval $[\nu_1,\nu_N]$.

Choosing the matrices $B$ and $K$ in various ways, we come to different versions of conjugate gradient methods. E.g., for a symmetric matrix $H_\tau A$ we obtain the following algorithms.

A. An analogue of the multistep method of steepest descent is obtained by putting $B = A^{-1}H_\tau$, $K = I$, $R = T = H_\tau^{-1}A$, $\Phi(x_n) = (A^{-1}H_\tau r_n, r_n) = (H_\tau^{-1}Ay_n, y_n)$:

(29) $$p_0 = r_0, \quad x_{n+1} = x_n + p_n(r_n,r_n)/(p_n, H_\tau^{-1}AP_n),$$

$$r_{n+1} = H^{-1}(f-Ax_{n+1}), \quad p_{n+1} = r_{n+1} - p_n(H_\tau^{-1}Ap_n, r_{n+1})/(p_n, H_\tau^{-1}Ap_n).$$

B. Choosing $B = I$, $K = A^{-1}H_\tau$, $R = AH_\tau^{-2}A$, $T = H_\tau^{-1}A$, $\Phi(x_n) = (r_n, r_n)$, we arrive at an analogue of the multistep minimum residual method:

(30) $$p_0 = r_0, \quad r_n = H_\tau^{-1}(f-Ax_n),$$

$$x_{n+1} = x_n + \frac{H_\tau^{-1}Ar_n, p_n)}{(H_\tau^{-1}Ap_n, H_\tau^{-1}Ap_n)} p_n, \quad p_{n+1} = r_{n+1} - \frac{(H_\tau^{-1}Ap_n, H_\tau^{-1}Ap_{n+1})}{(H_\tau^{-1}p_n, H_\tau^{-1}p_n)} .$$

If the matrix $H_\tau^{-1}A$ is not symmetric, analogous but somewhat more time-consuming algorithms can be constructed. We shall not study them here (see [5]).

Since the inequality

(31) $$\Phi(x_n \leq \max_{\nu \in [\nu_1, \nu_N]} [1-\nu P_{n-1}(\nu)]^2 \Phi(x_0)$$

follows immediately from (27), it is apparent that, on the assumptions made in deriving (6), (6a) for the Čebyšev acceleration method, these bounds remain true also for any of the conjugate gradient methods. Now $n(\varepsilon)$ denotes the number of iterations necessary for satisfying the condition $\Phi(x_n) \leq \varepsilon^2 \Phi(x_0)$.

Inequalities of another kind can be constructed with the help of special polynomials employed for estimating the functionals $\Phi(x_n)$. E.g. if $F_n(\nu)$ is the Lanczos polynomial

(32) $$F_n(\nu) = \frac{1-\cos[(n+1)\arccos(1-2\nu)]}{2(n+1)^2\nu}$$

we have

(33) $$|\nu F_n(\nu)| \leq (n+1)^{-2}$$

for $\nu \in [0,1]$. Analogously, using the polynomial

(34) $$\hat{F}_n(\nu) = \frac{(-1)^n \cos[(2n+1)\arccos(1-\sqrt{\lambda})]}{2(n+1)\sqrt{\lambda}}$$

of degree n (cf. [10]), we come to the inequality

$$(35)\qquad |\nu \hat{F}_n^2(\nu)| \leq (2n+1)^{-2}, \qquad \nu \in [0,1] .$$

Since $\varphi(\nu) = \nu$ in the method A and $\varphi(\lambda) = \lambda^2$ in the method B we immediately obtain from (33), (35) that the bounds

$$(36)\qquad (H_\tau^{-1}Ay_n, y_n) \leq \frac{(y_0,y_0)}{(2n+1)^2},$$

$$(37)\qquad (r_n,r_n) \leq \frac{(y_0,y_0)}{(n+1)^4}$$

hold for the multistep method of steepest descent (29) and for the multistep minimum residual method (30), respectively.

These bounds are obviously independent of the condition of the matrix $H_\tau^{-1}A$ and hold, in particular, also for A singular. The inequalities (36) and (37) give fast convergence for small n and a slower one for large n (this convergence is worse than ensured by (6) for $n \to \infty$ ).

It is possible to find estimates of the decrease of the functional $\Phi(x_n)$, which are - in a sense - a compromise. To this end we substitute a product of the Lanczos polynomial of degree k and the Čebyšev polynomial of the first kind of degree n - k, which possesses the least deviation from zero on the interval $[\nu_1, \nu_N]$, for $F_n(\lambda)$ instead of (32). Then we arrive at

$$(38)\qquad (r_n,r_n) \leq \min_k \left[\frac{2\gamma_0^{n-k}}{(k+1)^2(1+\gamma_0^{2(n-k)})}\right]^2 (y_0,y_0), \quad \gamma_0 = \frac{1-\gamma^{1/2}}{1+\gamma^{1/2}}$$

instead of (37). An analogous approach allows us to obtain the bound

$$(39)\qquad (Ay_n,y_n) \leq \min_k \left[\frac{2\gamma_0^{n-k}}{(2k+1)(1+\gamma_0^{2(n-k)})}\right]^2 (y_0,y_0)$$

instead of (36).

## 4. Conclusion

The gradient methods considered are not better than the algorithms based on the minimization of the spectral radius of the transition operator as far as the asymptotic rate of convergence is concerned. At the same time they are somewhat worse as concerns the economy of computation. On the other hand, our opinion is that the bounds and illustrative numerical experiments presented indicate that their efficiency is sufficiently good. Gradient methods possess suitable relaxation properties for small values of n (including also singular matrices). Apparently the nonlinear problem of the choice of parameters $\tau_n$ from variational considerations needs further investigation.

References

[1] Birkhof G., Varga R., Young D. Alternating direction implicit methods. Advances in computers, 1962, v.4, 140-274.

[2] Samarskiĭ A.A. Introduction into the Theory of Difference Schemes. Nauka, Moscow 1972. (Russian)

[3] Marčuk G.I., Kuznecov J.A. Iteration Methods and Quadratic Functionals. Nauka, Novosibirsk 1972. (Russian)

[4] Il'in V.P. Difference Methods for the Solution of Elliptic Equations. Izd. NGU, Novosibirsk 1970. (Russian)

[5] Il'in V.P. Some bounds for the conjugate gradient methods. Ž. Vyčisl. Mat. i Mat. Fiz. 16 (1976), 847-855. (Russian)

[6] Kantorovič L.V. On the method of steepest descent. Dokl. Akad. Nauk SSSR 3 (1947), 233-236. (Russian)

[7] Krasnosel'skiĭ M.A., Kreĭn S.G. An iteration process with minimum residuals. Mat. Sb. 31(73) (1952), 315-334. (Russian)

[8] Hestenes M., Stiefel E. Method of conjugate gradient for solving linear systems. J. Res. Nat. Bur. Standarta, 1952, 49, 409-436.

[9] Hestenes M. The conjugate gradient method for solving linear systems. Proc. Sympos. Appl. Math., v.6, New-York-Toronto-London, 1956, 83-102.

[10] Godunov S.K., Prokopov G.P. On the solution of the Laplace difference equation. Ž. Vyčisl. Mat. i Mat. Fiz. 9 (1969), 462-468. (Russian)

[11] Gorbenko N.I., Il'in V.P. On gradient alternating-direction methods. In: Nekotorye problemy vyčislitel'noĭ i prikladnoĭ matematiki, Nauka, Novosibirsk 1975, 207-214. (Russian)

Author's address: Vyčislitel'nyĭ centr SOAN SSSR,
630 090 Novosibirsk 90, USSR

NONLINEAR PARABOLIC BOUNDARY VALUE PROBLEMS WITH THE TIME DERIVATIVE IN THE BOUNDARY CONDITIONS

J.Kačur , Bratislava

The subject of this paper is motivated by the nonstationary, nonlinear and mixed boundary value problem for Schrödinger's equation considered in [2-4]. An approximate solution is constructed by solving a corresponding linearized boundary value problem. Construction of the approximate solution is convenient from the numerical point of view. Convergence and some properties of this approximate solution are investigated. Consider the equation

$$\frac{\partial u}{\partial t} + \sum_{i,j=1}^{N} \frac{\partial}{\partial x_i}\left( a_{ij}(x) \frac{\partial u}{\partial x_j}\right) + b_0(t,x,u,\nabla u) = 0 \tag{1}$$

for $(x,t)\in \Omega\times(0,T)$ where $T<\infty$ , $\Omega \subset E^N$ is a bounded domain with Lipschitzian boundary $\delta\Omega$ and $\nabla u\equiv\left(\frac{\partial u}{\partial x_1},\dots,\frac{\partial u}{\partial x_N}\right)$ . Let $\Gamma_1,\Gamma_2$ be open disjoint subsets of $\delta\Omega$ and $\Gamma_1\cup\Gamma_2\cup\Lambda = \delta\Omega$ where $\Lambda \subset \delta\Omega$, $mes_{N-1}\Lambda=0$. Together with (1) we consider

$$\begin{aligned} \frac{\partial u}{\partial t} &= -\frac{\partial u}{\partial \nu} - b_1(t,x,u) \qquad \text{for } (x,t)\in\Gamma_1\times(0,T) \\ 0 &= -\frac{\partial u}{\partial \nu} - b_2(t,x,u) \qquad \text{for } (x,t)\in\Gamma_2\times(0,T) \end{aligned} \tag{2}$$

where $\frac{\partial u}{\partial \nu} = \sum_{i,j=1}^{N} a_{ij}(x) \frac{\partial u}{\partial x_j} \cos(\nu,x_i)$ and $\nu$ is the outward normal to $\delta\Omega$ . The initial condition is

$$u(x,0) = \phi(x) \tag{3}$$

where $\phi$ is sufficiently smooth in $\overline{\Omega}$ .

Our concept of treating the problem (1) -(3) is based on Rothe's method developed in [5-9] .

Notation. We denote $W \equiv W_2^1(\Omega)$ (Sobolev space) , $(u,v)=\int_\Omega u\, v\, dx$ ,

$(u,v)_{\Gamma_j} = \int_{\Gamma_j} u\, v \, ds$ and $A[u,v] = \sum_{i,j=1}^{N} \int_{\Omega} a_{ij} \frac{\partial v}{\partial x_i}\frac{\partial u}{\partial x_j}\, dx$. By means of $A[u,v]$ for $u,v \in W$ we define the linear operator $A : W \to W^*$ (dual space to $W$). By $u_B(t)$ (from $L_2(\delta\Omega)$) we denote the trace of $u(t)$ from $W$ for fixed $t \in (0,T)$, by $\|\cdot\|$, $\|\cdot\|_W$, $\|\cdot\|_{\Gamma_1}$ and $\|\cdot\|_{\Gamma_2}$ the norms in the corresponding spaces $L_2(\Omega)$, $W$, $L_2(\Gamma_1)$ and $L_2(\Gamma_2)$. The letter $C$ will stand for any positive constant.

Assumptions. We assume $a_{ij} \in C^{0,1}(\bar{\Omega})$ for $i,j = 1,\ldots,N$ and

(4) $$\sum_{i,j=1}^{N} a_{ij}(x)\, \xi_i \xi_j \geq C_E\, |\xi|^2 ;$$

(5) $$|b_j(t,x,\xi) - b_j(t',x,\xi')| \leq C(|t-t'| + |t-t'|\,|\xi| + |\xi-\xi'|) \qquad j = 0,1,2 ;$$

(6) $$\phi \in W_2^2(\Omega) \text{ and } \frac{\partial \phi}{\partial \nu} = - b_2(0,x,\phi) \text{ for } x \in \Gamma_2 ;$$

(7) $$\left|\frac{\partial b_2(t,x,\xi)}{\partial \xi}\right| \leq C \leq \frac{C_E}{C_I^2} \quad \text{for } (x,t) \in \Gamma_2 \times (0,T),\ |\xi| < \infty,$$ where $C_I$ comes from the imbedding inequality $\|v\|_{L_2(\delta\Omega)} \leq C_I \|v\|_W$.

We shall be concerned with a weak solution of (1)-(3) which we define in a following way.

Definition. The function $u \in L_\infty(\langle 0,T\rangle, W) \cap C(\langle 0,T\rangle, L_2(\Omega))$ is a weak solution of (1)-(3) if

i) $u(0) = \phi$

ii) $\frac{du}{dt} \in L_\infty(\langle 0,T\rangle, L_2(\Omega))$ ; $\frac{du_B}{dt} \in L_\infty(\langle 0,T\rangle, L_2(\Gamma_2))$ ;

iii) the identity

(8) $$\left(\frac{du(t)}{dt}, v\right) + A[u(t), v] + \left(\frac{du_B(t)}{dt}, v\right)_{\Gamma_1} + (b_0(t,x,u(t),\nabla u(t)), v) + \sum_{j=1,2} (b_j(t,x,u_B(t)), v)_{\Gamma_j} = 0$$

holds for all $v \in W$ and a.e. $t \in (0,T)$.

Clearly, if a weak solution $u(t)$ is sufficiently smooth, then it satisfies (1)-(3) in the classical sense.

We define an approximate solution $u_n(t)$ (see (10)) of (1)-(3) in the following way. Let $n$ be a positive integer, $h = \frac{T}{n}$, $t_i = ih$ and $u_i \in W$ $i = 1,\ldots,N$ solutions of the linear elliptic problems

(1') $$\frac{u - u_{i-1}}{h} + A\,u + b_0(t_i,x,u_{i-1},\nabla u_{i-1}) = 0$$

$$u+h\frac{\partial u}{\partial \nu} = u_{i-1} - h\, b_1(t_i,x,u_{i-1}) \quad \text{on } \Gamma_1$$

(2 )

$$\frac{\partial u}{\partial \nu} = -\, b_2(t_i,x,u_{i-1}) \quad \text{on } \Gamma_2$$

where $u_0 = \phi$ . Precisely, successively for $i=1,\dots,N$ the elements $u_i \in W$ satisfy the identities

$$(9)\ \left(\frac{u_i-u_{i-1}}{h}, v\right) + A[u_i, v] + (b_0(t_i,x,u_{i-1},\nabla u_{i-1}), v) +$$

$$+\left(\frac{u_{i,B}-u_{i-1,B}}{h}, v\right)_{\Gamma_1} + \sum_{j=1,2} (b_j(t_i,x,u_{i-1,B}), v)_{\Gamma_j} = 0$$

for all $v \in W$. Existence and uniqueness of $u_i$ is well known. Now,we define

$$(10)\quad u_n(t) = u_{i-1} + (t-t_{i-1})h^{-1}(u_i-u_{i-1}) \quad \text{for } t_{i-1} \le t \le t_i\ ,\ i=1,\dots,N.$$

Theorem 1. Under the assumptions (4)-(7) there exists the unique weak solution $u \in L_\infty(\langle 0,T\rangle, W \cap W^2_{2,loc}(\Omega))$ of (1)-(3) and $u_n(t) \to u(t)$ in $L_2(\Omega)$ uniformly for $t \in \langle 0,T\rangle$.

Remark 1. Theorem 1 implies that $u(t)$ satisfies (1) for a. e. $(x,t)$ from $\Omega \times \langle 0,T\rangle$ in the classical sense.

Before proving Theorem 1, we prove some a priori estimates for $u_n(t)$.

Lemma 1. There exist $C_1$, $C_2$ and $n_0$ such that

$$(11)\quad \left\|\frac{u_i-u_{i-1}}{h}\right\|^2 + \left\|\frac{u_{i,B}-u_{i-1,B}}{h}\right\|^2_{\Gamma_1} + \frac{1}{h}\|u_i-u_{i-1}\|^2_W \le C_1 + C_2\, h \sum_{j=1}^{i} \|u_j\|^2_W$$

holds for all $n \ge n_0$ , $i=1,\dots,n$ .

The proof of (11) is based on the identity (9), suitable application of Young's inequality $(ab \le 2^{-1}(\varepsilon a)^2 + (2\varepsilon)^{-2} b^2)$ and the assumptions (4)-(7). We point out the basic steps of the proof.Subtracting(9)for $i=j$, $i=j-1$ and putting $v=u_j-u_{j-1}$ successively we obtain the recurrent inequality

$$(12)\quad (1-C_1h)\left(\left\|\frac{u_j-u_{j-1}}{h}\right\|^2 + \left\|\frac{u_{i,B}-u_{i-1,B}}{h}\right\|^2_{\Gamma_1} + \frac{C}{h}\|u_i-u_{i-1}\|^2_W\right) \le$$

$$\le (1+C_2h)\left(\left\|\frac{u_{j-1}-u_{j-2}}{h}\right\|^2 + \left\|\frac{u_{j-1,B}-u_{j-2,B}}{h}\right\|^2_{\Gamma_1} + \frac{C}{h}\|u_{i-1}-u_{i-2}\|^2_W\right) +$$

$$+ C_3h\sum_{i=1}^{j}\|u_i\|^2_W + C_4h$$

where (4), (6) and (7) has been used. Similarly, from (9) for $i=1$ and $v=\frac{u_1-\phi}{h}$ we obtain

$$(13)\quad \left\|\frac{u_1-\phi}{h}\right\|^2+\left\|\frac{u_{1,B}-\phi}{h}\right\|^2_{\Gamma_1}+\frac{C}{h}\|u_1-\phi\|^2_W \le C(\phi)$$

where (6) has been used. The constant $C(\phi)$ depends on $\|\phi\|_{W_2^2(\Omega)}$. There exist $\delta>0$, $K>0$ such that $(1-C_1h)^i \ge \delta$ and $(1+C_2h)^i \le K$ hold for all $n \ge n_0$ and $i=1,\dots,n$. Thus, from (12) and (13) Lemma 1 follows.

The estimate (11) implies

$$(14)\quad \|u_i\|^2 \le C_1+C_2h\sum_{j=1}^{i}\|u_j\|^2_W \ ,\ \|u_{i,B}\|^2_{\Gamma_1} \le C_1+C_2h\sum_{j=1}^{i}\|u_j\|^2_W$$

for all $n \ge n_0$ and $i=1,\dots,n$.

<u>Lemma 2.</u> Let $\varepsilon>0$. There exist $C_1(\varepsilon)$, $C_2(\varepsilon)$ such that

i) $A[u_i,u_i] \le C_1(\varepsilon)+C_2(\varepsilon)\sum_{j=1}^{i} h\|u_j\|^2_W + \varepsilon\|u_{i-1}\|^2_W$ ;

ii) $|(b_2(t_i,x,u_{i-1,B}),u_i)_{\Gamma_2}| \le C_1(\varepsilon)+C_2(\varepsilon)h\sum_{j=1}^{i}\|u_j\|^2_W + \varepsilon\|u_{i-1}\|^2_W$ .

From (9) for $v \in C_0^\infty(\Omega)$ and (5) we conclude

$$(15)\quad |A[u_i,v]| \le \left\|\frac{u_i-u_{i-1}}{h}\right\|\|v\| + C_1+C_2\|u_{i-1}\|_W\|v\| \ .$$

The estimate (15) takes place also for $v \in L_2(\Omega)$ and hence from (11), (14) and (15) for $v=u_i$ Assertion i) follows. Similarly, from (9), (11), (14) and Assertion i) we obtain Assertion ii).

<u>Lemma 3.</u> There exist $C$ and $n_0$ such that the estimates

i) $\left\|\frac{u_i-u_{i-1}}{h}\right\| \le C$ , $\left\|\frac{u_{i,B}-u_{i-1,B}}{h}\right\|_{\Gamma_1} \le C$ ;

ii) $\|u_i\|_W \le C$ ;

iii) $\|u_i-u_{i-1}\|^2_W \le \frac{C}{n}$

hold for all $n \ge n_0$, $i=1,\dots,n$.

Proof. From (9) for $v=u_i$, Lemma 1-2 and (4) we obtain

(16) $$C\,\|u_i\|_W^2 \le C_1(\varepsilon)+C_2(\varepsilon)\sum_{j=1}^{i} h\,\|u_j\|_W^2 + \varepsilon\,\|u_{i-1}\|_W^2 .$$

The estimate

(17) $$\|u_{i-1}\|_W^2 \le 2\,\|u_i\|_W^2 + 2\,\|u_i-u_{i-1}\|_W^2 \le 2\,\|u_i\|_W^2 + C_1 + C_2\sum_{j=1}^{i} h\,\|u_i\|_W^2$$

take place because of Lemma 1. The estimates (16) and (17) imply

$$\|u_i\|_W^2 \le C_1 + C_2\sum_{j=1}^{i} h\,\|u_j\|_W^2$$

from which we obtain (see e.g.[8]) $\|u_i\|_W \le C$ and hence Lemma 3 follows. From the regularity results (in the interior of the domain $\Omega$) for solutions of linear elliptic equations and Lemma 3 we obtain easily

(18) $$\|u_i\|_{W_2^2(\Omega^*)} \le C(\Omega^*) \qquad \text{for all} \quad n,\ i=1,\dots,n$$

where $\Omega^*$ is an arbitrary subdomain of $\Omega$, with $\overline{\Omega}^* \subset \Omega$.

Proof of Theorem 1. Lemma 3 implies

(19) $$\|u_n(t)-u_n(t^*)\| \le C\,|t-t^*|,\quad \|u_{n,B}(t)-u_{n,B}(t^*)\|_{\Gamma_1} \le C\,|t-t^*|$$

(20) $$\left\|\frac{d^-u_n(t)}{dt}\right\| \le C,\qquad \left\|\frac{d^-u_{n,B}\,t}{dt}\right\|_{\Gamma_1} \le C\,;$$

(21) $$\|u_n(t)\|_W \le C\,;$$

(22) $$\|u_n(t)\|_{W_2^2(\Omega^*)} \le C\,;$$

for all $n$, where $\frac{d^-}{dt}$ is the left hand derivative. From the compactness of the imbedding $W \to L_2(\Omega)$ and by the method of diagonalization we find outthat $u_n(t) \to u(t)$ in $L_2(\Omega)$ for all rational points $t$ of $\langle 0,T\rangle$ (here $\{u_n(t)\}$ is a suitable subsequence ofthe original $\{u_n(t)\}$). Hence using (19) we obtain that there exist $u : \langle 0,T\rangle \to L_2(\Omega)$ such that $u_n(t) \to u(t)$ for all $t \in \langle 0,T\rangle$. Using the Borel covering theorem we find out that this convergence is uniform in $\langle 0,T\rangle$. Reflexivity of $W$, (21) and (22) imply $u \in L_\infty(\langle 0,T\rangle, W \cap W^2_{2,loc}(\Omega))$. Then,similarly we conclude $u_{n,B} \to u_B$ in the norm of the space $C(\langle 0,T\rangle, L_2(\Gamma_1))$. Hence and from (19) we obtain

(23) $$\|u(t)-u(t^*)\| \le C\,|t-t^*|,\quad \|u_B(t)-u_B(t^*)\|_{\Gamma_1} \le C\,|t-t^*| .$$

Thus, applying the result of Y.Komura ( see e.g.[1]) from (23) we obtain $\frac{du}{dt} \in L_\infty(\langle 0,T\rangle, L_2(\Omega))$ and $\frac{du_B}{dt} \in L_\infty(\langle 0,T\rangle, L_2(\Gamma_1))$ . Let us denote

$x_n(t)=u_i$ for $t_{i-1}< t \le t_i$ , $i=1,\dots,n$ , $x_n(0)=u_0$,
$b_{j,n}(t,x,\xi)= b_j(t_i,x,\xi)$ for $t_{i-1} < t \le t_i$ , $i=1,\dots,n$ , $b_{j,n}(0,x,\xi)=$
$=b_j(0,x,\xi)$ where $\xi \in E^{N+1}$, $x\in\Omega$ for $j=0$ and $\xi\in E^1$, $x\in\Gamma_j$ for $j=1,2$.
Using our notation the identity (9) can be rewitten into the form

$$(24)\ \left(\frac{d^- u_n\, t}{dt}, v\right)+\left(\frac{d^- u_{n,B}\, t}{dt}, v\right)_{\Gamma_1}+ A\left[x_n(t), v\right] +$$

$$+\left(b_{o,n}(t,x,x_n(t-\tfrac{T}{n}), \nabla x_n(t-\tfrac{T}{n})), v\right) + \sum_{j=1,2}\left(b_{j,n}(t,x,x_{n,B}(t-\tfrac{T}{n})), v\right)_{\Gamma_j} = 0$$

for all $t\in(\frac{T}{n},T)$ and $n$ . Integrating (24) over $\langle\frac{T}{n}, t\rangle$ and taking limit for $n \to \infty$ we obtain

$$(25)\ (u(t), v)+(u_B(t), v)_{\Gamma_1} -(\phi, v) - (\phi, v)_{\Gamma_1} + \int_0^t \{ A[u(s), v] +$$

$$+(b_0(s,x,u(s), \nabla u(s)), v) + \sum_{j=1,2}(b_j(s,x,u_B(s)), v)\}\, ds = 0$$

for all $v\in W$ since we have the a priori estimates

$\|x_n(t)\|_W \le C$ , $\|x_n(t)\|_{W_2^2(\Omega^-)} \le C(\Omega^-)$, $\|x_n(t) - u_n(t)\| \le \frac{C}{n}$ and

$\|x_n(t-\frac{T}{n}) - x_n(t)\|_W^2 \le \frac{C}{n}$ . From (25) we find out that $u(t)$ is a weak solution of (1)-(3). If $u_1(t)$ , $u_2(t)$ are two solution of (1)-(3) then $u(t)= u_1(t) - u_2(t)$ satisfies

$$(26)\ \left(\frac{du}{dt}, v\right) + \left(\frac{du_B}{dt}, v\right)_{\Gamma_1}+ A\left[u, v\right] - C_1\|u\|\|v\| - C_2\|u\|_{\Gamma_1}\|v\|_{\Gamma_1} -$$

$$- C\|u\|_{\Gamma_2}\|v\|_{\Gamma_2} \le 0$$

for all $v\in W$ (C is from (7)). Substituing $u = v\exp(\lambda t)$ into (26) for sufficient large $\lambda$ and using (7) we obtain

$$\frac{d}{dt}\|v\|^2+ \frac{d}{dt}\|v_B\|^2_{\Gamma_1} \le 0$$

which implies uniqueness since $v(0)=v_B(0) = 0$ . From the uniqueness we conclude that the original sequence $\{u_n(t)\}$ converges to $u(t)$ in $C(<0,T>, L_2(\Omega))$ and the proof is complete.

Due to the a priori estimates for $u_n(t)$ and $x_n(t)$ the stronger regularity results for $u(t)$ can be proved.

Let $X$ be a reflexive Banach space with its dual space $X^*$ and the pairing $(. , .)$ . If $(w(t), v) \in C^1((0,T))$ and $\frac{d}{dt}(w(t), v) = (g(t), v)$ $(g(t) \in X)$ holds for all $v \in X^*$ then $w$ is weakly differentiable in $X$ (with respect to $t \in (0,T)$) and we denote $w^{\cdot}(t) = g(t)$.

Lemma 4. Let $u(t)$ be as in Theorem 1. Then

1) The function $u(.)$ is weakly continuous in $W$ and the estimates

$$\|u(t)\|_W \leq C \quad , \quad \|u(t)\|_{W_2^2(\Omega^{\cdot})} \leq C(\Omega^{\cdot})$$

hold for all $t \in (0,T)$ ;

2) The functions $A[u(t), v]$, $(b_0(t,x,u(t),\nabla u(t)), v)$ and $(b_j(t,x,u_B(t)), v)_{\Gamma_j}$ $(j=1,2)$ are continuous in $(0,T)$ for all $v \in W$;

3) The functions $u(t)$ , $u_B(t)$ are weakly derivable in $L_2(\Omega)$, $L_2(\Gamma_1)$ respectively and $u^{\cdot}(.)$ , $u_B^{\cdot}(.)$ are wekly continuous in the corresponding spaces. The estimate

$$(27) \quad \|u^{\cdot}(t)\| + \|u_B^{\cdot}(t)\|_{\Gamma_1} \leq C$$

holds for all $t \in (0,T)$ ;

4) $u^{\cdot} = \frac{du}{dt}$ , $u_B^{\cdot} = \frac{du_B}{dt}$ hold for a.e. $t \in (0,T)$ ($\frac{d}{dt}$ is the strong derivative) ;

5) The identity (8) (with $u^{\cdot}$, $u_B^{\cdot}$ instead of $\frac{du}{dt}$, $\frac{du_B}{dt}$) holds for all $t \in (0,T)$ .

Assertion 1) is a consequence of the a priori estimates for $u_n(t)$ , reflexivity of the spaces $W$, $W_2^2(\Omega^{\cdot})$ , (23) and the uniqueness of $u(t)$. Assertion 2) is a consequence of Assertion 1) .From Assertion 2) and (25) we conclude $(u(t), v) \in C^1((0,T))$ for all $v \in L_2(\Omega)$ and $(u_B(t),v)_{\Gamma_1} \in C^1((0,T))$ for all $v \in L_2(\Gamma_1)$ which implies the existence of $u^{\cdot}(t)$ , $u_B^{\cdot}(t)$ . Hence and from (25) Assertion 5) follows. Due to (8) and (24) we conclude $\left(\frac{d^- u_n(t)}{dt}, v\right) \to (u^{\cdot}(t), v)$ for all $v \in L_2(\Omega)$, $t \in (0,T)$ and $\left(\frac{d^- u_{n,B}(t)}{dt}, v\right) \to (u^{\cdot}(t), v)$ for all $v \in L_2(\Gamma_1)$, $t \in (0,T)$. Thus the a priori estimates of $\frac{d^- u_n(t)}{dt}$ , $\frac{d^- u_{n,B}(t)}{dt}$ imply (27). The identity (8) and Assertion 2) imply the weak continuity of $u^{\cdot}(.)$ and $u_B^{\cdot}(.)$ in

$L_2(\Omega)$, $L_2(\Gamma_1)$ respectively and Assertion 3) is proved.Asertion 4) is the well known result (see e.g. [6]) .

Using Theorem 1 ,Lemma 4 and a priori estimates for $u_n(t)$, $x_n(t)$ we can prove a stronger convergence results.

Theorem 2. Suppose (4)-(7).Let $u(t)$ , $u_n(t)$ and $x_n(t)$ be as in Theorem 1.Then

i) $x_n(t) \to u(t)$ in the norm of the space $W$ uniformly in $t \in \langle 0,T\rangle$;

ii) $u_n \to u$ in the norm of the space $C(\langle 0,T\rangle, W)$ ;

iii) $u(t)$ is a Hölder continuous function from $\langle 0,T\rangle \to W$. The estimate

$$\|u(t) - u(t')\|_W^2 \le C\,|t-t'|$$

holds for all $t,t' \in \langle 0,T\rangle$ .

For the proof we subtract (8) and (24) for $v = x_n(t) - u(t)$.Then, using Lemma 4 and a priori estimates for $x_n(t)$ , $u_n(t)$ and $u(t)$ we estimate

$$(28)\quad C_E\|x_n(t)-u(t)\|_W^2 \le C_1\|x_n(t)-u(t)\|+C_2\|x_{n,B}(t)-u_B(t)\|_{\Gamma_1}+$$
$$+C_3h\|x_{n,B}(t-\tfrac{T}{n})-u_B(t)\|_{\Gamma_2}+C\|x_{n,B}(t-\tfrac{T}{n})-u_B(t)\|_{\Gamma_2}\|x_{n,B}(t)-u_B(t)\|_{\Gamma_2}$$

where $C$ is from (7). Due to (7) and Lemma 3 (iii)) we obtain

$$C\|x_{n,B}(t-\tfrac{T}{n})-u_B(t)\|_{\Gamma_2}\|x_{n,B}(t)-u_B(t)\|_{\Gamma_2} \le C\,C_I^2\big(\|x_n(t)-u(t)\|_W^2+$$
$$+\|x_{n,B}(t-\tfrac{T}{n})-x_{n,B}(t)\|_W\|x_{n,B}(t)-u_B(t)\|_W\big) \le C\,C_I^2\big(\|x_n(t)-u(t)\|_W^2+$$
$$+C\sqrt{h}\big).$$

Hence, from Theorem 1,(7) and (28) Assertion i) follows. Assertion ii) follows from Assertion i) Lemma 3 (iii)) and the estimate

$$\|u_n(t)-u(t)\|_W \le 2\|x_n(t)-u(t)\|_W^2 \quad 2\|x_n(t)-u_n(t)\|_W^2 \le$$
$$\le 2\|x_n(t)-u(t)\|_W^2+C_1\sqrt{h}\,.$$

Finally, subtracting (8) for $t = t$ and $t = t'$ and putting $v = u(t) - u(t')$ we obtain

$$\|u(t)-u(t')\|_W^2 \le C_1|t-t'|+C_2\|u(t)\|_W|t-t'| + C\|u(t)-u(t')\|_{\Gamma_2}^2$$

where (23) and Lemma 4 has been used ($C$ is from (7)).Hence and from the estimate $C \|u(t)- u(t')\|^2_{\Gamma_2} \le C\, C_I^2 \|u(t) - u(t')\|^2_W$ Assertion iii) follows.

Remark 2. In [1] a similar result is proved for the case of A being a nonlinear, monotone and coercive operator and $b_j(t,x,\xi)$ $j=0,1,2$ being monotone in $\xi$ . Howewer in that case $u_i$ $(i= 1,\dots,n)$ are the solutions of a corresponding nonlinear elliptic boundary value problems.

Remark 3. All results hold true if either $\Gamma_1$ or $\Gamma_2$ is empty.

References

[1] J.Kačur : Nonlinear parabolic equations with the mixed nonlinear and nonstationary boundary conditions.Mathematica Slovaca, to appear.

[2] V.V.Barkovskij and V.L.Kulčickij : Generalized solutions of some mixed boundary value problems for Schrödinger's equation. Linear and nonlinear boundary value problems, AN USSR , Kiev, 1971 in Russian .

[3] V.L.Kulčickij : Regularity of solutions of some mixed boundary value problems for Schrödinger's equation.Linear and nonlinear boundary value problems, AN USSR,Kiev, 1971 in Russian .

[4] L.V.Ljubič : Solution of the heat conduction problem in a right dihedral angle with time derivatives in the boundary condition.Dokl.Akad.Nauk Ukrain.SSR Ser.A 1976,no.8,691-693 in Ukrainian .

[5] K.Rektorys : On application of direct variational methods to the solution of parabolic boundary value problems of arbitrary order in the space variables. Czech.Math.J., 21 96 ,1971, 318-339.

[6] J.Kačur : Method of Rothe and nonlinear parabolic equations of arbitrary order. Czech.Math.J.,to appear.

[7] J.Nečas : Application of Rothe's method to abstract parabolic equations.Czech.Math.J.,Vol.24 99 ,1974,No3,496-500 .

[8] J.Kačur : Application of Rothe's method to nonlinear evolution equations.Mat.Časopis Sloven.Akad.Vied,25,1975,No 1,63-81.

[9] J.Kačur and A.Wawruch : On an approximate solution for quasilinear parabolic equations. Czech.Math.J.,27, 102 1977,220-241.

Author's address : Institute of applied mathematics,Mlýnska dolina, 816 31 Bratislava, Czechoslovakia

# VARIATIONAL AND BOUNDARY VALUE PROBLEMS FOR DIFFERENTIAL EQUATIONS WITH DEVIATING ARGUMENT

G.A.Kamenskii and A.D.Myshkis, Moscow

The systems with after-effect that are described by differential equations with deviating arguments have the following characteristic property: for the estimation of the future in such a system it is necessary to know the past for the time equal to the time-lag. It means that the initial value space for such a system is a functional space S (with a given norm), and a natural analog to the simple variational problem is a problem of minimizing a functional with deviating argument on trajectories connecting two points of the space S. Such variational problems are named by us the infinite defect variational problems. In the same way, the boundary value problems for differential equations with deviting argument when the trajectories connect the points of the space S are called the infinite defect boundary value problems.

Various problems involving infinite defect are studied extensively now. N.N.Krasovskii [1] , § 45 has formulated and solved the problem of the quieting of a system with time-lag. For the quieting of a system without time-lag $x'(t) = Ax(t) + Bu(t)$ it is sufficient to find a control function $u(t)$ such that $x(t_1) = 0$ for a $t_1 > t_0$ and then put $u(t) = 0$ for $t \geq t_1$. In contrast, for the quieting of the system with time-lag

$$x'(t) = Ax(t) + Gx(t-\tau) + Bu(t)$$

a control function $u(t)$ such that $x(t) = 0$ for $t_1 \leq t \leq t_1 + \tau$ is needed. This problem, as is not difficult to see, is an infinite defect problem.

A.W. Krjashimskii and Yu.S. Osipov [2] studied a difference-differential game with a target set in a functional space. H.Banks and G.A.Kent derived Pontrjagin type maximum principle for control of neutral type difference-differential equations with a functional target set. (See[3]). The variational problems for functional with deviating argument were investigated by G.A. Kamenskii [4, 5] . All mentioned papers deal with infinite defect problems.

The boundary value problems for differential equations with deviating argument have been studied by A.Halanay [6], L.J. Grimm and K.Schmitt [7], G.A.Kamenskii and A.D.Myshkis [8] and others. In the last works was described the essential difference between the boundary value problems for equations with the deviations in the highest order derivatives and for equations without such deviations.

In this paper we investigate the variational problems for the functional with deviating argument of the more general type than in [4, 5]. We consider also analogous problems for functionals depending on functions of many arguments and finite difference method for solving the boundary value problem arising in the onedimentional case.

1. Variational problems for functionals with deviating argument

Consider the problem of the extremum of the functional

$$(1)\quad J(y) = \int_{\alpha}^{\beta} F(x, y(\omega_0(x)),\dots, y(\omega_m(x)), y'(\omega_0(x)),\dots, y'(\omega_m(x)))\,dx,$$

where $-\infty < \alpha \le \beta < \infty$, $F: [\alpha,\beta] \times (R^n)^{2m+2} \to R$, $m \ge 1$, in the class $\overset{\circ}{H}_p$ of the functions $y: R \to R^n$, $y(x) \equiv 0$ ( $x \bar{\in} [a,b]$ ), $\alpha \le a < b \le \beta$ ( a nonhomogeneous boundary value problem may be reduced to the homogeneous one by the standard change of variables), $y(x)$ is absolutely continuous, $y' \in L_p$, $1 \le p \le \infty$ with the natural norm. It is supposed that $F \in C_1$, $\omega_0(x) \equiv x$, all $\omega_j \in C_1[\alpha,\beta]$, $\omega_j'(x) \ne 0$, $\omega_j([\alpha,\beta]) \supseteq [a, b]$ and for $p < \infty$

$$(2)\quad |F(x, y_0,\dots,y_m,z_0,\dots,z_m)| + \sum_{j=0}^{m} \| F_{y_j}(\dots) \| \le K(\|y_0\|,\dots,\|y_m\|)\,(1 + \sum_{j=0}^{m} \|z_j\|^p)$$

$$(3)\quad \sum_{j=0}^{m} \| F_{z_j}(x, y_0,\dots,y_m,z_0,\dots,z_m) \| \le L(\|y_0\|,\dots,\|y_m\|)\,(1 + \sum_{j=0}^{m} \|z_j\|^{p-1})$$

$$(\forall\, x \in [\alpha,\beta],\quad y_0,\dots,y_m, z_0,\dots,z_m \in R^m)$$

with continuous $K$, $L$; $F_{y_j}$ is a n-dimentional vector.

Denote $\gamma_j = \omega_j([\alpha,\beta]) \to [\alpha,\beta]$ the inverse functions to the $\omega_j$. It is easy to prove that the functional $J(y)$ under our assumptions is differentiable and by simple changes of variables and by integrating by parts we can get the first variation of $J(y)$ in the form

$$\delta J = \int_b^a \Big[\sum_{j=0}^{m} \gamma_j'(x)\, F_{y_j}(\gamma_j(x), y(\omega_o(\gamma_j(x))),\dots,y(\omega_m(\gamma_j(x))),$$

$$y'(\omega_o(\gamma_j(x))),\dots,y'(\omega_m(\gamma_j(x)))) - \frac{d}{dx}\sum_{j=0}^{m} \gamma_j'(x) F_{z_j}(\dots)\Big]\delta y(x)\cdot$$

$$\cdot\, dx$$

( the subtrahend here is necessary to understand in the terms of the theory of distributions). By standard methods we get the proof of the following

Theorem 1. If the function $y$ is a stationary point for the functional (1) ( in particular, the point of extremum), then $y \in H_p$ satisfies almost everywhere on $[a, b]$ the equation

$$(4) \quad \sum_{j=0}^{m} \gamma_j'(x)\, F_{y_j}(\gamma_j(x),\ y(\omega_o(\gamma_j(x))),\dots,y(\omega_m(\gamma_j(x))),$$

$$y'(\omega_o(\gamma_j(x))),\dots,y'(\omega_m(\gamma_j(x)))) - \frac{d}{dx}\sum_{j=0}^{m} \gamma_j'(x) F_{z_j}(\dots) = 0.$$

It follows that the expression in (4) standing after the sign $\frac{d}{dx}$ has to be absolutely continuous. ( Mark that $y'(x)$ in general case does not belong to that class of functions).

Thus the $y(x)$ is the generalized solution of the equation (4) though the equation (4) is satisfied by $y(x)$ almost everywhere. Remind that you have to put in (4) $y(\omega_j(\gamma_\ell(x))) = y'(\omega_j(\gamma_\ell(x))) = 0$ every time when $\omega_j(\gamma_\ell(x)) \bar{\in} [a, b]$; and that $y(a) = y(b) = 0$. Suppose in addition that $F \in C_2$, $p \geq 2$ and for $p < \infty$ the matrices $F_{y_j y_1}$ satisfy (2), the matrices $F_{y_j z_1}$ satisfy (3) and the matices $F_{z_j z_1}$ - the analoguous inequality with the power p-2. Then by usual methods we may get the following representation of the increment of the functional (1)

$$(5) \qquad \Delta J = \delta J + \frac{1}{2}\delta^2 J + o(\|\delta y\|^2_{\overset{\circ}{H}_p}),$$

where

$$\delta^2 J = \int_\alpha^\beta \sum_{j,1=0}^{m} \Big[ (F_{y_j y_1}\delta y(\omega_j(x)))\cdot \delta y(\omega_\ell(x)) +$$

$$(F_{y_j z_1}\delta y(\omega_j(x))\cdot\delta y'(\omega_\ell(X)) + (F_{z_j z_1}\delta y'(\omega_j(x))\cdot\delta y'(\omega_\ell(x))\Big] dx.$$

Suppose also that mes $\{ x | \omega_j(x) = \omega_1(x) \} = 0$ $(\forall j, 1, j \neq 1)$. Then we may state the following analog to the necessary condition of Legendre.

<u>Theorem 2</u>. Suppose that the above mentioned conditions are satisfied and the functional (1) attains on $y$ the local minimum in the space $H_p$. Then for almost all $x \in [a, b]$ the matrix

$$\sum_{j=0}^{m} F_{z_j z_j}(\gamma_j(x), y(\omega_o(\gamma_j(x))), \ldots, y(\omega_m(\gamma_j(x))), y'(\omega_o(\gamma_j(x))), \ldots$$

$$\ldots, y'(\omega_m(\gamma_j(x))))\gamma'_j(x)$$

is non-negative.

For the proof it is necessary for any $x_0 \in ]a, b[$, $x_0 \bar{\in} \bigcup_{j \neq 1} \{ x | \omega_j(x) = \omega_\ell(x) \}$ to put $\delta y = \frac{1}{M} g(M(x-x_0))$, $g \in H_\infty$, $M \longrightarrow \infty$ and to use the arbitrarity of the finite function $g$.

2. <u>A variational problem for the quadratic functional depending on the functions of many deviating arguments</u>.

Let $S$ and $Q \subset S$ be non-empty open bounded sets in $R^n$ $(n \geq 2)$ and on $\bar{S}$ are given the functions $\omega_k: \bar{S} \to \omega_k(\bar{S}) \subset R^n$ having the inverse functions $\omega_k^{-1} = \gamma_k: \omega_k(\bar{S}) \to \bar{S}$ and $\omega_o(x) \equiv x$, $Q \subset \omega_k(S)$, $\omega_k \in C^2(\bar{S})$, $\gamma_k \in C^2(\omega_k(\bar{S}))$ $(k = 0, \ldots, m,\ m \geq 1)$.

Consider the problem of the minimum of the functional

$$J(u) = \iint\limits_S \Big[ \sum_{i,j=1}^{n} \sum_{k,l=0}^{m} a_{ijkl}(x) u_{x_i}(\omega_k(x)) u_{x_j}(\omega_1(x)) +$$

$$2 \sum_{i=1}^{n} \sum_{k,l=0}^{m} b_{ikl}(x) u_{x_i}(\omega_k(x)) u(\omega_1(x)) + \sum_{k,l=0}^{m} c_{kl}(x) u(\omega_k(x)) \cdot$$

$$\cdot u(\omega_1(x)) + 2 \sum_{i=1}^{n} \sum_{k=0}^{m} d_{ik}(x) u_{x_i}(\omega_k(x)) + 2 \sum_{k=0}^{m} e_k(x) u(\omega_k(x)) \Big] dx$$

in the subspace $H$ of the space $W_2^1(R^n)$ that is the closure of the set $H_0$ of the infinitely differentiable functions that are finite on $Q$. With other words we may say that $u$ belongs to the space $\overset{o}{W}{}_2^1(Q)$ and $u_{x_i}(\omega_k(X)) = u(\omega_k(x)) = 0$ by $\omega_k(x) \bar{\in} Q$. Here $a_{ijkl}$, $b_{ikl}$, $d_{ik} \in C^1(\bar{S})$, $c_{kl} \in C^0(\bar{S})$, $e_k \in L_2(S)$ $(i,j = 1, \ldots, n;$ $k,l = 0, \ldots, m)$. Without loss of generality we shall suppose that $a_{ijkl} = a_{jilk}$, $c_{kl} = c_{lk}$. Let $u$ be the extremal point for $J(u)$. Then for any $v \in H$

$$(6) \qquad \delta J(u, v) = 0.$$

By a change of variables in the integral representation of (6) we obtain

$$(7) \iint_Q \Big\{ \sum_l \Big[ \sum_{i,j,k} a_{ijkl}(\gamma_l(x)) u^{kl}_{x_i}(x) v_{x_j}(x) + \sum_{i,k} b_{ikl}(\gamma_l(x)) u^{kl}_{x_i}(x) \cdot$$

$$\cdot v(x) + \sum_{i,k} b_{ilk}(\gamma_l(x)) u^{kl}(x) v_{x_i}(x) + \sum_k c_{kl}(\gamma_l(x)) u^{kl}(x)\, v(x) +$$

$$+ \sum_i d_{il}(\gamma_l(x)) v_{x_i}(x) + e_l(\gamma_l(x))\, v(x) \Big] \Big| \gamma_l'(x) \Big| \Big\}\, dx = 0,$$

where $\bullet(\omega_k(\gamma_l(x))) = \bullet^{kl}(x)$, $\gamma_l'(x)$ - Jacobian $\dfrac{D(\gamma_{l1}, \ldots, \gamma_{ln})}{D(x_1, \ldots, x_n)}$. If the function $u \in H$ satisfies the equation (7) for any $v \in H$, we shall call $u$ the generalized solution of the differential equation

$$(8) -\sum_{i,j,k,l} \Big[ A_{ijkl}(x) u^{kl}_{x_i}(x) \Big]_{x_j} + \sum_{ikl} B_{ikl}(x)\, u^{kl}_{x_i}(x) +$$

$$+ \sum_{k,l} C_{kl}(x)\, u^{kl}(x) = F(x) \quad (x \in Q).$$

Here

$$A_{ijkl}(x) = a_{ijkl}(\gamma_l(x)) \big| \gamma_l'(x) \big|, \quad B_{ikl}(x) = b_{ikl}(\gamma_l(x)) \big| \gamma_l'(x) \big| -$$

$$- \sum_{r,s=1}^{n} b_{ikl}(\gamma_l(x)) \big| \gamma_l'(x) \big| \cdot (\omega_{kr})_{x_s}(\gamma_l(x)) \cdot (\gamma_{ls})_{x_i}(x),$$

$$C_{kl}(x) = - \sum_i \Big[ b_{ikl}(\gamma_l(x)) \big| \gamma_l'(x) \big| \Big]_{x_i} + c_{kl}(\gamma_l(x)) \big| \gamma_l'(x) \big|,$$

$$F(x) = \sum_{i,l} \Big[ d_{il}(\gamma_l(x)) \big| \gamma_l'(x) \big| \Big]_{x_i} - \sum_l e_l(\gamma_l(x)) \big| \gamma_l'(x) \big|,$$

$$\omega_k = (\omega_{k1}, \ldots, \omega_{kn}), \ \gamma_l = (\gamma_{l1}, \ldots, \gamma_{ln}).$$

We proved the following

Theorem 3. If the functional $J(u)$ attains on the function $u$ the extremum in the space $H$ then $u$ is the generalized solution of the equation (8).

It is easy to show on simple examples (not like for the equations without deviations of arguments) that any requirements on the smoothness of the right hand parts cannot guarantee the existence of twice differentiable solutions. Therefore it is necessary to use the above mentioned definition of the solution in all cases.

Consider now the boundary value problem for the equation (8) in the space H. The boundary condition has the form $u|_{\partial Q} = 0$ and $u^{kl}_{x_i}(x) = u^{kl}(x) = 0$ by $\omega_k(\gamma_l(x)) \overline{\in} Q$.

Define bounded operators

$$A: L_2^n(Q) \rightarrow L_2^n(Q),\ (Au)_i(x) = \sum_{j,k,l} A_{jikl}(x)\, u^{kl}_j(x);$$

$$R: H \rightarrow L_2(Q),\ (\ Ru\ )(X) = \sum_{i,k,l} B_{ikl}(x) u^{kl}_{x_i} + \sum_{k,l} C_{kl}(x) u^{kl}(x)$$

and adjoint operators

$$A^+: L_2^n(Q) \rightarrow L_2^n(Q), (A^+u)_i(x) = \sum_{j,k,l} A^{lk}_{ijkl}(x) \left|\omega_l'(\gamma_k(x))\gamma_k'(x)\right| u^{lk}_j(x)$$

$$R^+: H \rightarrow L_2(Q),\ (\ \overset{+}{R}u\ )(x) = - \sum_{i,k,l} \Big[\ B^{lk}_{ikl}(x) \left|\omega_l'(\gamma_k(x))\gamma_k'(x)\right| \cdot$$
$$\cdot u^{lk}(x)\Big]_{x_i} + \sum_{k,l} C_{kl}(x) \left|\omega_l'(\gamma_k(x)) \cdot \gamma_k'(x)\right| u^{lk}(x).$$

Denote by $(.,.)$ the scalar product in $L_2(Q)$ and by $(.,.)_n$ - - the scalar product in $L_2^n(Q)$. Suppose that for a $C > 0$

$$(9) \qquad (\ Au,\ u)_n \geq C(u,u)_n \qquad (\forall u \in L_2^n(Q)),$$

in this case it is natural to name the equation (8) elliptic. By definition the function $u \in H$ is a solution of the stated boundary value problem for the equation (8), if

$$(10) \quad (\ A\nabla u,\ \nabla v)_n + (\ Ru,\ v) = (\ F,\ v) \qquad (\forall\ v \in H\ ).$$

Consider also in the space H the homogeneous boundary value problem

$$(11) \qquad (\ A\nabla u, \nabla v)_n + (\ Ru,\ v\ ) = 0 \qquad (\forall\ v \in H\ )$$

and adjoint boundary value problem

$$(12) \qquad (\ A^+\nabla\ u, \nabla\ v\ )_n + (\ R^+u,\ v\ ) = 0\ \ (\ \forall v \in H\ ).$$

By means of reducing the equations (8) - (12) introduced above to the equations in the Hilbert space H and using the theory of compact operators in Hilbert spaces we obtain the following

Theorem 4. If the boundary value problem (11) has only zero solution, then the problem (10) has one and only one solution $u_F$ for any $F \in L_2(Q)$, and $\| u_F \|_H \leq C_1 \| F \|$.

If the boundary value problem (11) has non-zero solutions, then the problem (10) has solutions if and only if $(\ F,\ \tilde{u}\ ) = 0$, for all solutions $\tilde{u}$ of the problem (12). The dimensions of the solutions spaces of (11) and (12) are finite and equal.

In obtaining of the results of this section took part A.L. Skubachevskii.

3. The finite differences method of the numerical solution of the boundary value problem for the linear equtions with many senior members.

We describe here the finite difference method for linear equations with many senior members and with deviations commensurable with the length of the interval on wich we search for the solution. Such equations may be reduced to the equations with integer deviations of the form

$$(13)\quad \sum_{k=-m}^{m}\Big[ ( a_k(x)y'(x-k))' + b_k(x)y'(x-k) + c_k(x)y(x-k)\Big] = f(x),$$
$$0 \le x \le b \quad ( b \text{ - integer } ),$$

and boundary condutions

$$(14)\qquad y(x) = 0 \qquad \text{for} \qquad x \le 0 \quad \text{and} \quad x \ge b\,.$$

Suppose that all $a_k'$, $b_k$, $c_k$, $f \in C'[0, b]$ - the space of piecewise continuous functions with the possible jumps in the integers, The equation (13) may be written as the operator equation

$$(15)\qquad Ly = DQDy + RDy + Sy = f\,,$$

where

$$(Qz)(x) = \sum_{k=-m}^{m} a_k(x)z(x-k)\,,$$

$$(Rz)(x) = \sum_{k=-m}^{m} b_k(x)z(x-k),$$

$$(Sz)(x) = \sum_{k=-m}^{m} c_k(x)z(x-k)$$

with boundary conditions (14), D is the operator of differentiation, operators Q, R, S act in $L_2[0, b]$. We suppose that the operator Q has alwas the bounded inverse operator $Q^{-1}$. By $M_n$ we denote the space of functions defined on $T_n = \left\{0, \frac{1}{n}, \frac{2}{n}, \dots, b\right\}$. The operators $\Delta^+_n$, $\Delta^-_n$, $Q_n : M_n \to M_n$ are defined by the formulae:

$$(\Delta^+_n \xi)(s) = \begin{cases} n(\xi(s+h) - \xi(s)) & (s \in T_n \setminus \{b\}),\ h = \frac{1}{n}, \\ n(\xi(b) - \xi(b-h)) & (s=b)\,, \end{cases}$$

$$(\Delta_n^- \xi)(s) = \begin{cases} n(\xi(h) - \xi(0)) & (s = 0), \\ n(\xi(s) - \xi(s-h)) & (s \in T_n \setminus \{0\}), \end{cases}$$

$$(Q_n \xi)(s) = \sum_{k=-m}^{m} a_k(s)\, \xi(s-k) \quad (s \in T_n,\ \xi(s-k) = 0 \text{ for } s-k \bar{\in} T_n).$$

The operators $R_n$ and $S_n$ are definied similarly. Define operator $[\,\cdot\,]_n : C[0,b] \to M_n$ by equality $[y]_n(s) = y(s)$, $s \in T_n$ and define norms in $M_n$ by formulae

$$\|\xi\|_n^o = \Big(\sum_{\nu=1}^{bn-1} \xi^2(h\nu)\Big)^{\frac{1}{2}}, \quad \|\xi\|_n = \Big(\sum_{\nu=0}^{bn} \xi^2(h\nu)\Big)^{\frac{1}{2}},$$

$$\|\xi\|_n^C = \max_{s \in T_n} |\xi(s)|.$$

The approximate solution of the boundary value problem (13), (14) is a net function $\xi(s)$ satisfying the equation

$$(16) \quad (L_n \xi)(s) = (\Delta_n^- Q_n \Delta_n^+ + R_n \Delta_n^+ + S_n)\, \xi(s) = [f]_n(s), \quad s \in T_n \setminus \{0, b\};\ \xi(0) = \xi(b) = 0.$$

It is easy to prove that if $y \in H_\infty$, $y' \in C'[0,b]$, $y$ satisfies the condition (14), then

$$\lim_{n\to\infty} \| [Ry']_n - R_n \Delta^{\pm}[y]_n \|_n = 0.$$

If we put now $Ry = y$ and insert $Qy$ instead of $y$, we prove that if on every interval $]0,1[, \ldots, ]b-1, b[$ exist uniformly continuous $y'$ and $y''$ and $Qy' \in H_\infty$, $(Qy')' \in C'[0,b]$, then

$$\lim_{n\to\infty} \| [(Qy')']_n - \Delta_n^- Q_n \Delta_n^+ [y]_n \|_n^o = 0.$$

Thus we have prooved the following theorem of the approximation of the operator $L$ :

<u>Theorem 5.</u> If $Ly = f$, then

$$(17) \qquad \| L_n [y]_n - [f]_n \|_n^o \longrightarrow 0 \quad \text{for } n \longrightarrow \infty.$$

The following theorem states the stability of the finite difference scheme.

Theorem 6. For the injectivity of the operator $L$ it is necessary and sufficient that there exist $C > 0$ and $n_0$ such that

$$(18) \quad \| L_n \xi \|_n^o \geq C \{ \| \Delta_n^- Q_n \Delta_n^+ \xi \|_n + \| Q_n \Delta_n^+ \xi \|_n^C + \| \Delta_n^+ \xi \|_n + \| \xi \|_n^C \} \quad (n \geq n_0 ,\ \xi(0) = \xi(b) = 0).$$

The necessity is prooved by the assumption of the contrary by using the piecewise linear interpolation of the functions for which the expression in parenthesis in (18) is equal to 1. The sufficiency follows from the theorem 5.

From (18) it follows in particular that (16) has an exactly one solution for each $n \geq n_0$. If we put in (18) $\xi = [y]_n - [\xi]_n$ and apply (17), we prove the theorem of the approximation of the solution :

Theorem 7. If the operator $L$ is injective, $y$ is a solution of $Ly = f$ and $\xi_n$ is a solution of (16), then

$$\| [y]_n - \xi_n \|_n^o \longrightarrow 0 \quad \text{for} \quad n \longrightarrow \infty.$$

In obtaining the results of this section took part A.G. Kamenskii.

References

[1] Krasovskii N.N.: The theory of motion control, Nauka, Moscow 1968 (Russian)

[2] Krjazhimskii A.W., Osipov Yu.S.: A differential-difference game with functional target set, Prikl. mat. meh. 37 (1973), 3-13 (Russian)

[3] Banks H.T., Kent G.A. : Control of functional differential equations of retarded and neutral type to target sets in function space, SIAM J. Control, 1972, N.4, vol. 10

[4] Kamenskii G.A.: On extrema of functionals with deviating argument, Dokl. Akad. Nauk SSSR 224 (1975), No.6, 1952-1955 ( Russian)

[5] Kamenskii G.A.: Variational problems for functionals with deviating argument, Differencial'nye uravnenia s otklonyajuščimsa argumentom, Naukova dumka, Kiev 1977, 139-148 (Russian)

[6] Halanay A.: On a boundary value problem for linear systems with time-lag, J. of Diff. Equat. 2, N.1, 1966, 55-80.

[7] Grimm L.J., Schmitt K. : Boundary value problem for delay-differential equations, Bull. Amer. Math. Soc. 74, N.5, 1968, 997-1000

[8] Kamenskii G.A., Myshkis A.D.: Boundary value problem for quasilinear differential equations of divergent type of the second order with deviating argument, Differencial'nye uravnenia 10 (1974), N.12, 2137-2146 (Russian)

Authors' address : MIIT, Department of Applied Mathematics,
Obrazcova 15, Moscow - A-55, USSR

# ON A GENERAL CONCEPTION OF DUALITY IN OPTIMAL CONTROL

R. Klötzler, Leipzig

Many problems in the theory of differential equations and its applications can be formulated as problems of optimal control. For these problems again several conceptions of duality have been developed which are very useful from theoretical and numerical point of view. For example we all know in the theory of elasticity the important duality between the principle of Dirichlet and its dual problem as the principle of Castigliano.

In general, if we denote the original problem by

(1) $$F(x) \longrightarrow \text{Min}$$

subject to all $x \in X$ ,

then a dual problem is defined in general sense by any problem

(2) $$L(y) \longrightarrow \text{Max}$$

subject to all $y \in Y$ ,

with the property $F(x) \geqq L(y) \quad \forall\, x \in X ,\ y \in Y$ .

As a rule one aspires to construct such dual problems which satisfy the strong duality condition

$$\inf_X F \ (\text{ or } \min_X F\ ) = \sup_Y L\ (\text{ or } \max_Y L\ ) .$$

It is easily seen that such a conception of duality leads to both-side estimates of $\inf_X F$ and often also to corresponding error estimates with respect to an optimal solution $x_0$ .

For regular variational problems already K.O.Friedrichs [3] introduced dual variational problems in 1928. His theory requires besides assumptions of differentiability mainly convexity properties of the integrand. In the last decade by M.M.Cvetanov [9] , R.T.Rockafellar [8] and Ekeland/Temam [2] several investigations were stated, which may be viewed as an extension of the original conception of Friedrichs with respect to general problems of optimal control. In these papers the former assumptions of differentiability are essentially weakened, however convexity properties are again supposed and instead of Legendre transformation by Friedrichs now Fenchel's theory of conjugate functions is applied.

In the present paper we shall delineate a new conception of duality, which avoids any requirements on the convexity of the original problem. Simultaneously this treatment carries on relevant investigations on Bellman's differential equation and extensions of the classical theory of Hamilton and Jacobi by the author [5] - [7].

We consider problems of optimal control of the type

$$(3) \quad J(x,u) := \int_{\Omega} f(t,x,u)\,dt + \int_{\partial\Omega} l(t,x)\,do \longrightarrow \text{Min}$$

subject to all vector-valued state functions $x \in X$, control functions $u \in U(x)$, and constraints

$$(4) \quad x^i_{t_\alpha} = g^i_\alpha(t,x,u) \qquad (i = 1,\dots,n;\ \alpha = 1,\dots,m).$$

Here $\Omega$ is a strongly Lipschitz domain of $R^m$,

$$X = \left\{ x \in W^{1,n}_p(\Omega) \;\middle|\; (t,x(t)) \in \bar{G} \text{ on } \bar{\Omega},\ b(t,x(t)) = \sigma \text{ on } \partial\Omega \right\} \quad \text{with } p > m,$$

$$U(x) = \left\{ u \in L^r_p(\Omega) \;\middle|\; u(t) \in V(t,x(t)) \subset R^r \text{ a.e. on } \Omega \right\} \quad \text{for every } x \in X,$$

$G$ is an open set of $R^{n+m}$, and $V(.,.)$ is assumed to be a normal map from $\bar{G}$ into $R^r$ in the sense of Joffe/Tichomirov [4] p.338. Further we suppose $l$ and $b$ are real continuous functions on $\partial\Omega \times R^n$ and $f$ as well as $g^i_\alpha$ are real functions on $\bar{G} \times R^r$ satisfying the Carathéodory condition in the following meaning: they are (Lebesgue-) measurable functions with respect to the first argument $t$ and continuous functions for almost every fixed $t \in \Omega$. Therefore $f(.,x(.),u(.))$ and $g^i_\alpha(.,x(.),u(.))$ are measurable functions on $\Omega$ for every <u>process</u> $\langle x,u \rangle$, that means for every admissible pair $\langle x,u \rangle$ of problem (3). We denote the set of all processes by $\mathcal{P}$ and require the following additional assumptions:

$$(5a) \quad \mathcal{P} \neq \emptyset$$

(5b) $f(.,x(.),u(.))$ is minorized by a function $\psi \in L^1_1(\Omega)$ $\forall \langle x,u \rangle \in \mathcal{P}$.

In consequence of (5b) $f(.,x(.),u(.))$ is summable on $\Omega$ (in the broad sense) for every $\langle x,u\rangle \in \mathcal{P}$ and accordingly $J(x,u)$ is well-defined on $\mathcal{P}$.

Now we prepare the formulation of a corresponding duality principle. For this purpose we introduce the denotations

$H$ for the <u>Pontryagin function</u> defined by

(6) $$H(t,\xi,v,y) := -f(t,\xi,v) + y_i^\alpha g_\alpha^i(t,\xi,v) \quad ,$$

$\mathcal{H}$ for the <u>Hamiltonian function</u> defined by

(7) $$\mathcal{H}(t,\xi,y) := \sup_{v \in V(t,\xi)} H(t,\xi,v,y) \qquad \text{on } \bar{G} \times R^{nm} \quad ,$$

and $Q(t)$ for the following cuts of $\bar{G}$

(8) $$Q(t) := \begin{cases} \{\xi \in R^n \mid (t,\xi) \in \bar{G}\} & \forall\ t \in \Omega \\ \{\xi \in R^n \mid (t,\xi) \in \bar{G},\ b(t,\xi) = \sigma\} & \forall\ t \in \partial\Omega . \end{cases}$$

Moreover we select a subset $\mathcal{Y}$ from $W^{1,m}(G)$ consisting of all functions $S \in W_\infty^{1,m}(G)$ having the following properties:

(9a) each class of distribution derivatives of $S^\alpha$ $(\alpha = 1,\dots,m)$ contains a bounded representative $S_j^\alpha$ $(j = 1,\dots,n+m)$ ;

(9b) there are uniformly bounded sequences of functions $z_k^\alpha \in C^1(R^{n+m})$ and their derivatives satisfying pointwise the conditions

$$\lim_{k\to\infty} z_k^\alpha = S^\alpha \text{ and } \lim_{k\to\infty} (\partial z_k^\alpha / \partial \xi^j) = S_j^\alpha \quad \text{on } \bar{G} .$$

Obviously $\mathcal{Y}$ contains the set $C^{1,m}(\bar{G})$ of continuous differentiable (vector-valued) functions. By means of mollified functions of $S^\alpha$ we can easily see that also the set $D^{1,m}(\bar{G})$ of piecewise continuous differentiable functions belongs to $\mathcal{Y}$. In the following text we denote generally $S_i^\alpha = S_{t^i}^\alpha$ for $i = 1,\dots,m$ and $S_{i+m}^\alpha = S_{\xi^i}^\alpha$ for $i = 1,\dots,n$.

With these definitions we are starting from theorems on set-valued functions by Joffe/Tichomirov [4] ch. 8 and conclude the following lemmas.

<u>Lemma</u> 1 . $\mathcal{H}(.,.,.)$ is a measurable function on $\bar{G} \times R^{nm}$ . If dom $\mathcal{H}(t,\xi,.) \neq \emptyset$ ,then $\mathcal{H}(t,\xi,.)$ is a convex function .

Lemma 2 . Setting $y_i^\alpha = S^\alpha_{\xi^i}(t,\xi)$ in (7) for arbitrary functions $S \in \gamma$ we obtain $\mathcal{H}(.,.,.,S_\xi(.,.))$ as a measurable function on G .

Lemma 3 . For each $S \in \gamma$ and

$$\delta_S(t,\xi) := S^\alpha_{t^\alpha}(t,\xi) + \mathcal{H}(t,\xi,S_\xi(t,\xi)) \tag{10}$$

the function

$$\Lambda_S(.) := \sup_{\xi \in Q(.)} \delta_S(.,\xi) \tag{11}$$

is measurable on $\Omega$ .
In connection with the general equation of Hamilton-Jacobi

$$\phi^\alpha_{t^\alpha} + \mathcal{H}(t,\xi,\phi_\xi) = 0$$

it seems adequate to define this function $\delta_S$ as the "defect" of the Hamilton-Jacobi equation with respect to S .

Now we fix a process $\langle x,u \rangle$ and a function $S \in \gamma$. We obtain by using the expression (6) the equation

$$J(x,u) = \int_{\partial\Omega} l(t,x)\, do \; + $$
$$+ \int_\Omega \left\{ -H(t,x,u,S_\xi(t,x)) + S^\alpha_{\xi^i}(t,x)\, g^i_\alpha(t,x,u) \right\} dt \quad .$$

Furthermore, we insert in its second integrand $S^\alpha_{t^\alpha}(t,x) - S^\alpha_{t^\alpha}(t,x)$. Because of the fact that in consequence of (9) and (4)

$$\int_\Omega \left[ S^\alpha_{t^\alpha}(t,x) + S^\alpha_{\xi^i}(t,x)\, g^i_\alpha(t,x) \right] dt$$
$$= \int_\Omega \left[ S^\alpha_{t^\alpha}(t,x) + S^\alpha_{\xi^i}(t,x)\, x^i_{t^\alpha}(t) \right] dt$$
$$= \lim_{k\to\infty} \int_\Omega \left[ z^\alpha_{kt^\alpha}(t,x) + z^\alpha_{k\xi^i}(t,x)\, x^i_{t^\alpha}(t) \right] dt$$
$$= \lim_{k\to\infty} \int_\Omega \frac{d\, z^\alpha_k(t,x(t))}{dt^\alpha}\, dt$$
$$= \lim_{k\to\infty} \int_{\partial\Omega} z^\alpha_k(t,x)\, n_\alpha(t)\, do = \int_{\partial\Omega} S^\alpha(t,x)\, n_\alpha(t)\, do$$

holds, the following equality results:

$$J(x,u) = \int_{\Omega} \{-H(t,x,u,S_{\xi}(t,x)) + S^{\alpha}_{t^{\alpha}}(t,x)\} \; dt + \int_{\partial\Omega} [S^{\alpha}(t,x)\, n_{\alpha}(t) + l(t,x)] \; do \; .$$

Here the symbols $n_{\alpha}(t)$ ($\alpha = 1,\dots,m$) denote the components of the unit vector of the exterior normal on $\partial\Omega$ at the point $t$. If we observe that in consequence of (5),(7),Lemma 2 and $\mathcal{H}(t,x,S_{\xi}(t,x)) \geqq H(t,x,u,S_{\xi}(t,x)) \in L^1_1(\Omega)$ the function $\mathcal{H}(t,x,S_{\xi}(t,x))$ is summable too on $\Omega$ ,the last equation leads to the estimate

$$J(x,u) \geqq \int_{\Omega} \{-\mathcal{H}(t,x,u,S_{\xi}(t,x)) + S^{\alpha}_{t^{\alpha}}(t,x)\} \; dt + \int_{\partial\Omega} [S^{\alpha}(t,x)\, n_{\alpha}(t) + l(t,x)] \; do$$

and through Lemma 3 to

$$(12) \qquad J(x,u) \geqq L(S) := -\int_{\Omega} \Lambda_S(t)\, dt + \int_{\partial\Omega} \inf_{\xi \in Q(t)} [S^{\alpha}(t,\xi) n_{\alpha}(t) + l(t,\xi)] \; do \; .$$

From this development of formula (12) it is easy to notice in which cases there the equality occurs.We summarize our results in the following <u>duality theorem</u>.

<u>Theorem 1</u> . Let $\langle x,u \rangle$ be a process and $S \in \mathcal{Y}$. Then $J(x,u) \geqq L(S)$ in the sense of the detailed formulation of (12). Here the equality holds if and only if the following conditions are fulfilled:

$$(13a) \quad H(t,x,u,S_{\xi}(t,x)) = \mathcal{H}(t,x,S_{\xi}(t,x)) \text{ a.e. on } \Omega \; ,$$

$$(13b) \quad \delta_S(t,x(t)) = \Lambda_S(t) \quad \text{a.e. on } \Omega \; ,$$

$$(13c) \quad S^{\alpha}(t,x)n_{\alpha}(t) + l(t,x) = \inf_{\xi \in Q(t)} [S^{\alpha}(t,\xi)n_{\alpha}(t) + l(t,\xi)] \quad \text{a.e. on } \partial\Omega \; .$$

In virtue of ths Theorem 1 a duality is defined between the original problem (3) and its <u>dual problem</u> $L(S) \to \text{Max}$ on $\mathcal{Y}$ . This duality is a far-reaching generalization of several conceptions of duality which we cited above in the introduction.We can easily demonstrate that through a reduction of problem (3) to a Bolza problem the dual functional of Friedrichs and Rockafellar is generated by $L(S)$ under the special statement

$$S^{\alpha}(t,\xi) = y_0^{\alpha}(t) + y_i^{\alpha}(t)\,\xi^{i} \qquad (\alpha = 1,\ldots,m) \,. \tag{14}$$

Hence the duality of Friedrichs, Cvetanov, Rockafellar and Ekeland/Temam is formally included in our conception (12) by specialization on linear-affine functions $S$ with respect to $\xi$. From this fact it is obvious that in general the dual problem, restricted on the class $\mathcal{Y}_0 \subset \mathcal{Y}$ of functions (14), does not generate so good lower bounds of $\inf_{\mathcal{P}} J$ as $\sup L(S)$ on the whole $\mathcal{Y}$. An instructive comparison is supplied by the following example.

Example 1. It is to find in Euclidean metric the shortest way in the domain $\overline{G}_0 = \{\xi \in R^2 \mid r_1 \leq |\xi| \leq r_2\}$, $r_1 < r_2$, from an initial point $\xi_1 = (0,-r_1)$ to the endpoint $\xi_2 = (0,r_1)$. - Here we obtain $\inf_{\mathcal{P}} J = \pi r_1 = \sup_{\mathcal{Y}} L(S)$, attained by $S(\xi) = r_1 \arctan(\xi^2/|\xi^1|)$; but on the other hand $\sup_{\mathcal{Y}_0} L(S) = 2\,r_1$.

A further difference between these duality conceptions is the following. The duality of Rockafellar has for convex problems the advantage of being symmetric, as the double dual problem coincides with the original one. On the other hand, our duality in the sense of (12) leads to fundamental differences between the analytical structure of the functionals $J$ and $L$ so that this new duality is not symmetric.

As an application of Theorem 1 let us discuss the case in which for a given process $\langle x,u\rangle$ and $S \in \mathcal{Y}$ the equality $J(x,u) = L(S)$ is valid. Then the pair $(\langle x,u\rangle, S)$ is said to be a saddle point of the duality condition (12) and $\langle x,u\rangle, S$ are optimal solutions of (3) and of its dual problem respectively. Thus we can interpret the condition (13) equivalent to the saddle point property as a generalized form of Pontryagin's maximum principle. In this form it is especially a sufficient criterion for optimality of the process $\langle x,u\rangle$. In a recent paper [5] we proved that for problems (3) without state restrictions (disregarding boundary conditions) the condition (13) includes Pontryagin's maximum principle in the original form (for $m = 1$) and in the generalized form by L.Cesari [1] (for $m > 1$). The converse question is in general still unsolved: to what extent the condition (13) and the existence of the corresponding $S \in \mathcal{Y}$ is necessary for an optimal process $\langle x,u \rangle$ . Only for special classes of (3) with $m=1$ it is known that the Bellman function realizes this condition. For convex problems the stability theory of Rockafellar [8] answers this question.

Finally we mention two further results without giving their proofs, which are similar to the proof of Theorem 1.

Theorem 2 . The result of Theorem 1 holds even if we replace the set $\mathcal{Y}$ by $\mathcal{Y}_p := \{ S = S_1 + S_2 \mid S_1 \in \mathcal{Y}, \; S_2 \in W_p^{1,m}(\Omega) \}$.

Theorem 3 . Let $\langle x,u \rangle$ be a process and $S \in \mathcal{Y}_p$ ,restricted by the condition

$$(15) \qquad \delta_S(t,\xi) = S^{\alpha}_{t^\alpha}(t,\xi) + \mathcal{H}(t,\xi,S_\xi(t,\xi)) \leqq \sigma$$

for a.e. $t \in \Omega$ and every $\xi \in Q(t)$.

Then the inequality

$$(16) \qquad J(x,u) \geqq L_0(S) := \int_{\partial\Omega} \inf_{\xi \in Q(t)} \left[ S^{\alpha}(t,\xi) n_\alpha(t) + l(t,\xi) \right] do$$

is valid.Here the equality holds if and only if

$$(17a) \qquad H(t,x,u,S_\xi(t,x)) = \mathcal{H}(t,x,S_\xi(t,x)) \quad \text{a.e. on } \Omega ,$$

$$(17b) \qquad \delta_S(t,x(t)) = \sigma \quad \text{a.e. on } \Omega ,$$

$$(17c) \qquad S^{\alpha}(t,x) n_\alpha(t) + l(t,x) = \inf_{\xi \in Q(t)} \left[ S^{\alpha}(t,\xi) n_\alpha(t) + l(t,\xi) \right] \quad \text{a.e. on } \partial\Omega .$$

The estimate (16) induces a modified dual problem stated by the object

$$(18) \qquad L_0(S) \longrightarrow \text{Max} \quad \text{on} \quad \mathcal{Y}_p$$

under the constraint $\delta_S(t,.) \leqq \sigma$ for a.e. $t \in \Omega$.

In consequence of Lemma 1 this modified dual problem is a convex optimal problem on an infinite dimensional function space with a linear objective functional.If we denote the feasible set of (18) by $\mathfrak{A}$ and regard it as a subset of $W_p^{1,m}(G)$ , then formula (16) is true also on the closure $\overline{\mathfrak{A}}$ so that $\sup_{\overline{\mathfrak{A}}} L_0 \leqq \inf_{\mathcal{P}} J$ .

Example 2 (parametric variational problems) .

We consider simple integrals (m = 1)

$$J(x) = \int_{\sigma}^{T} f(x,\dot{x})dt \longrightarrow \text{Min} \quad \text{on} \quad W_p^{1,n}(\sigma,T)$$

under boundary conditions $x(\sigma) = x_o$ , $x(T) = x_T$ and state restrictions $x(t) \in \overline{G}_o \subset R^n \ \forall t \in [0,T]$ , where $G_o$ is a domain satisfying $\partial G_o \in C_1^o$ . Besides (5) we assume $f \geqq \sigma$ and $f(x,.)$ is a positive homogeneous function of the degree 1 . - Now we obtain by some here omitted computations under the additional assumption $S_t \equiv \sigma$ the result $L_o(S) = S(x_T) - S(x_o)$ and $\overline{\mathfrak{A}} = \{S \in W_p^{1,1}(G_o) \mid S_{\xi}(\xi) \in \mathfrak{F}(\xi)\}$ a.e. on $G_o$ , where $\mathfrak{F}(\xi)$ is the convex figuratrix set at the point $\xi$ in the sense of Carathéodory defined by

$$\mathfrak{F}(\xi) = \{z \in R^n \mid z_i v^i \leqq f(\xi,v) \ \forall \ v \in R^n\} .$$

## References


[1] L.Cesari, Optimization with partial differential equations in Dieudonné-Rashevsky form and conjugate problems, Arch.Rat. Mech.Anal. 33 (1969), 339-357

[2] I.Ekeland,R.Temam, Analyse convexe et problèmes variationnels, Gauthier-Villars, Paris 1974

[3] K.Friedrichs, Ein Verfahren der Variationsrechnung das Minimum eines Integrals als das Maximum eines anderen Ausdrucks darzustellen, Göttinger Nachr. 1929, 13-20

[4] A.D.Joffe,V.M.Tichomirov, Teoria ekstremalnych zadač, Nauka Moskva 1974

[5] R.Klötzler, On Pontryagin's maximum principle for multiple integrals, Beiträge z. Analysis 8 (1976), 67-75

[6] R.Klötzler, Einige neue Aspekte zur Bellmanschen Differentialgleichung, Materialy vsesoyuznogo simpoziuma po optimalnomu upravleniu i differencialnym igram, Tbilisi 1976, 146-154

[7] R.Klötzler, Weiterentwicklungen der Hamilton-Jacobischen Theorie, Sitzungsberichte der AdW der DDR (to appear)

[8] R.T.Rockafellar, Conjugate convex functions in optimal control and the calculus of variations, Journ.Math.Anal.Appl. 32 (1970), 174-222

[9] M.M.Cvetanov, O dvoystvennosti v zadačach variacionnogo isčislenia, Dokl. Bolgarskoj Akad. Nauk 21 (1968), 733-736



Author's address: Karl-Marx-Universität, Sektion Mathematik, Karl-Marx-Platz, 701 Leipzig, DDR

# BOUNDARY VALUE PROBLEMS
# FOR SYSTEMS OF NONLINEAR DIFFERENTIAL EQUATIONS

H.W.Knobloch, Würzburg

## 1. A General Existence Theorem.

The lecture is devoted to the study of two-point boundary-value problems (abbreviation: BVP) for second order vector differential eq. of the form

$$\ddot{x} = f(t,x). \tag{1.1}$$

Here $x = (x^1,\dots,x^n)^T$ is a n-dimensional column vector and the dot denotes differentiation with respect to the scalar variable $t$. We assume that $f$ and its partial derivatives with respect to $x$ are continuous functions of $(t,x)$ on some open bounded set $\mathcal{P}$ in the $(t,x)$-space. The boundary conditions are assumed to be of the form

$$x(0) = x_o, \quad x(1) = x_1. \tag{1.2}$$

To be more specific, we consider solutions $x(\cdot)$ of (1.1) on the interval $[0,1]$ which satisfy the condition

$$(t,x(t)) \in \mathcal{P} \quad , \quad 0 \le t \le 1 , \tag{1.3}$$

and assume the prescribed values $x_o$ and $x_1$ respectively for $t=0$ and $t=1$ respectively.

In the first part of this lecture we present a general existence theorem which seems to be new in case of dimension $n > 1$. In case $n = 1$ the hypothesis of the theorem essentially amounts to the existence of so called upper and lower solutions. These are (scalar) functions $\alpha(\cdot)$, $\beta(\cdot)$ of class $C^2$ which satisfy the inequalities

$$\begin{aligned} &\ddot{\alpha}(t) > f(t,\alpha(t)), \quad \ddot{\beta}(t) < f(t,\beta(t)), \ 0 < t < 1 . \\ &\alpha(t) < \beta(t), \quad 0 \le t \le 1 . \end{aligned} \tag{1.4}$$

It is well known that under these circumstances a solution $x(t)$ of the BVP (1.1), (1.2) exists, provided the boundary values $x_o$ and $x_1$ respectively are restricted to the intervals $[\alpha(0),\beta(0)]$, $[\alpha(1),\beta(1)]$ respectively. The existence of a solution is then established together with the a-priori estimate $\alpha(t) \le x(t) \le \beta(t)$ for $0 \le t \le 1$ (see e.g. [3]).

In order to find a generalization of the above result to higher dimensions we observe that the two first of the relations (1.4) admit a simple geometric interpretation. Let us consider the region $\Omega$ in the $(t,x)$-plane given by

(1.6) $\Omega = \{t,x : 0 < t < 1 , \alpha(t) < x < \beta(t)\}$ .

It is then easy to see that the said inequalities are equivalent with the following requirement:

(1.7) If a solution curve through a point $P_o = (t_o,x_o) \in \partial\Omega$ , where $0 < t_o < 1$, is tangent to $\partial\Omega$ then it touches the set $\Omega$ from the exterior.

At this point some explanations seem to be in order. By a solution curve we mean a curve in the $(t,x)$-space ($x$ need not be scalar from now on) which admits a parameteric representation $t \to (t,x(t))$, where $x(\cdot)$ is a solution of (1.1). "Tangent to $\partial\Omega$" means that the tangent to the curve at $P_o$ is in the tangent space to $\partial\Omega$ at $P_o$ . "Touching from the exterior" means that $(t,x(t)) \notin \bar{\Omega}$ if $t \neq t_o$ and $|t-t_o|$ sufficiently small.

In passing we note that the statement (1.6) can also be phrased in this way: The set $\{t,x,\dot{x}) : (t,x) \in \Omega\}$ is an isolating block for the first order (2n-dimensional) system which is equivalent with (1.1).

We next write down two further statements which are evident in case $n=1$ (and if $\Omega$ is defined according to (1.6)) but which are substantial requirements in the general situation. For notational convenience we will use from now on the symbol $\Omega_t$ in order to denote the cross section $\{x : (t,x) \in \Omega\}$ of a given set $\Omega$ . $\Omega_t$ is a subset of the x-space; its interior relative to this space will be denoted by $\mathring{\Omega}_t$ .

(1.8)
(i) $\Omega_t$ is convex, $\mathring{\Omega}_t$ is not empty.
(ii) There exists $q(\cdot) = (q^1(\cdot),\dots,q^n(\cdot))^T$, each $q^i$ being a function of class $C^2$ on $[0,1]$, such that $q(t) \in \mathring{\Omega}_t$ for every $t \in [0,1]$.

(1.9) $x_o \in \Omega_o$, $x_1 \in \Omega_1$ ($\Omega_o$, $\Omega_1 = \Omega_t$ for $t=0,1$).

The conditions (1.7)-(1.9) constitute the essential hypotheses of our general existence theorem. We add a further one which is of a more technical nature and can be relaxed somehow. It reduces the class of sets in the $(t,x)$-space which in the n-dimensional case will take the place of the special sets (1.6) to those which allow a simple analytic description.

(1.10) $\Omega$ is the intersection of finitely many regions of the form $\{t,x:\Phi(t,x)<0\}$. Each $\Phi$ is a scalar function of class $C^2$ on the whole $(t,x)$-space and satisfies
$k(t,x) \neq 0$, $H(t,x) > 0$ whenever $\Phi(t,x)=0$ and $(t,x)\in\partial\Omega$ .

Here $k$ and $H$ respectively denote the n-dimensional vector and the symmetric $n \times n$-matrix respectively which are given by

(1.11) $\quad k=(\Phi_{x^1},\ldots,\Phi_{x^n})^T, \; H=(\Phi_{x^i x^j})$ .

Note that the region $\Omega$ which is given (in case of $n=1$) by (1.6) falls in the category of sets which can be characterized in the form (1.10), (take $\Phi(t,x) = (x-\alpha(t))(x-\beta(t))$).

<u>Theorem 1.</u> Let $\Omega$ be an open subset of the $(t,x)$-space and let $\bar{\Omega} \subsetneq \mathcal{P}$ . Assume that (1.7)-(1.10) hold. Then there exists a solution $x(\cdot)$ of the BVP (1.1)-(1.3) with the property that $(t,x(t)) \in \bar{\Omega}$ for $0 \leq t \leq 1$ .

A proof of Theorem 1 - under slightly weaker hypotheses- can be found in the forthcoming paper [1] (cf. Theorem 5.2). It appears there in a setting which allows to treat by one and the same method various types of boundary conditions for differential eq. of the type (1.1) and also include certain cases where the right hande side of the differential eq. explicitely depends upon $\dot{x}$.

We conclude this section with a remark concerning the crucial hypothesis (1.7). Since it is of local nature one could expect that it can be replaced by conditions which do not involve a-priori knowledge of the solutions of (1.1). Indeed it is not difficult to convince oneself that the statement (1.7) is a consequence of the following requirement which has then to be met by every function $\Phi$ appearing in the analytic description (1.10) of $\Omega$ .

(1.12) $\quad \Phi(t_o,x_o) = 0$ and $\dot{\Phi}(t_o,x_o,\dot{x}) = 0 \Rightarrow \ddot{\Phi}(t_o,x_o,\dot{x}) > 0$ .

Here $\dot{\Phi}$, $\ddot{\Phi}$ have to be understood as formal first and second order derivatives of $\Phi$ with respect to eq. (1.1) . $\dot{\Phi}$ is a linear, $\ddot{\Phi}$ a quadratic polynomial in $\dot{x}$.

Various alternative versions of (1.12) have been developed in [1]. The following one is convenient for our purposes. (1.12) can be inferred from an inequality of the form

(1.13) $k^T f - \rho\Phi_t - l^T H l + \Phi_{tt} > 0$ ,

where the scalar $\rho$ and the vector $l$ are subject to the linear constraint

(1.4) $\quad 2k_t + 2Hl = \rho k$

(for the definition of $k$ and $H$ see (1.11)). The argument in $f,\Phi,H,k$ is $t_o,x_o$.

## 2. An Application of Theorem 1.

Let there be given a positive definite symmetric $n \times n$ - matrix $Q(t)$ which is elementwise of class $C^2$ on $[0,1]$ and let

(2.1) $\varphi(t,x) = x^T Q(t) x$

be the corresponding quadratic form. Furthermore let $\tilde{x}(\cdot)$ be a solution of the differential eq. (1.1) which exists on $[0,1]$ and satisfies condition (1.3) (but which need not satisfy the boundary condition (1.2)). $\tilde{x}(\cdot)$ has to be regarded as fixed throughout this section. We adopt the following notation

(2.2) $\tilde{\varphi}(t,x) = \varphi(t,x-\tilde{x}(t))$, $\Omega_\delta = \{t,x\colon 0 < t < 1,\ \tilde{\varphi}(t,x) < \delta^2\}$ .

Theorem 2. Let the matrix inequality

(2.3) $P(t,x) > 0$, $(t,x) \in \mathcal{P}$ , $0 \le t \le 1$;

hold where

(2.4) $P(t,x) = Q(t)F(t,x) + F(t,x)^T Q(t) + \ddot{Q}(t) - 2\dot{Q}(t)Q(t)^{-1}\dot{Q}(t)$

and $F(t,x) = f_x(t,x)$ is the Jacobian matrix of $f$ with respect to $x$. Furthermore let the positive number $\delta$ be chosen such that the inclusion

(2.5) $\bar{\Omega}_\delta \subseteq \mathcal{P}$

holds. Then the following statement is true. Whenever $x_o$, $x_1$ are such that

$$\tilde{\varphi}(0,x_o) \le \delta^2 \ , \quad \tilde{\varphi}(1,x_1) \le \delta^2$$

then the BVP (1.1) - (1.3) has a solution $x(\cdot)$ satisfying $\tilde{\varphi}(t,x(t)) \le \delta^2$ for all $t \in [0,1]$.

Proof. Most of the calculations which appear in the course of the proof are essentially the same as the ones used in the proof of Theorem 6.1 in [1]. Hence we skip here some details. On the other hand the procedure of the proof is simpler than in [1] and allows to dispose of the additional hypothesis $\tilde{\varphi}(0,x_o) = \tilde{\varphi}(1,x_1)$ which is required in [1] but which is superfluous. For the reader's convenience the proof is divided in two steps.

Step 1. We claim: For every compact subset $\mathcal{P}'$ of $\mathcal{P}$ there exists a positive number $\delta$ (depending upon $\mathcal{P}'$ only) such that the following statement holds true. Whenver $\Omega_\delta \subseteq \mathcal{P}'$ (regardless what $\tilde{x}(\cdot)$ is) then $\Omega_\delta$ satisfies all hypotheses of Theorem 1.

It is clear that one has to verify the isolating block property of $\Omega_\delta$ only. Since the boundary points $(t,x)$ of $\Omega_\delta$ with $0 < t < 1$ form the locus of the equation $\Phi(t,x) = \tilde{\varphi}(t,x) - \delta^2 = 0$ we may pursue the line described at the end of the previous section. Starting with this particular $\Phi$ we determine $H,k,\rho,l$ (cf. (1.11), (1.14)). It is easy to see by straightforward calculations that one arrives at the following result

$$(2.6)\qquad \begin{aligned} &k(t,x) = 2Q(x-\tilde{x}(t)),\ H(t,x) = 2Q,\\ &\rho = 0,\quad l=\dot{\tilde{x}}(t) - Q^{-1}\dot{Q}(x-\tilde{x}(t)) ,\end{aligned}$$

where $Q = Q(t)$. For this choice of $k$ and $H$ the quantity on the left hand side of (1.13) turns out to be

$$(2.7)\quad k^T(f(t,x)-\ddot{\tilde{x}}(t)) + (x-\tilde{x}(t))^T[\ddot{Q} - 2\dot{Q}Q^{-1}\dot{Q}](x-\tilde{x}(t)).$$

Let us now consider the function

$$p(t,x,\tilde{x}) = 2(x-\tilde{x})^TQ(t)[f(t,x)-f(t,\tilde{x})]$$

which is defined and of class $C^1$ on a neighborhood of the set

$$(2.8)\quad \{t,x,\tilde{x} : 0 \le t \le 1,\ (t,x) \in \mathcal{P}',\ (t,\tilde{x}) \in \mathcal{P}'\}.$$

The function $p$ and all its partial derivatives with respect to $x,\tilde{x}$ vanish whenever $x=\tilde{x}$. One easily recognizes that the second order term in the Taylor-expansion at $x=\tilde{x}$ can be expressed in terms of the Jacobian $F = f_x$ as

$$(2.9)\quad (x-\tilde{x})^T[Q(t)F(t,\tilde{x}) + F(t,\tilde{x})^TQ(t)](x-\tilde{x}).$$

It is then clear, by standard arguments, that the difference between $p$ and the above quadratic form is of order $\mathcal{O}(\|x-\tilde{x}\|^2)$. On the other hand it follows from (2.6) and (2.7) that the left hand side of (1.13) can be identified with

$$(2.10)\quad p(t,x,\tilde{x}) + (x-\tilde{x})^T[\ddot{Q} - 2\dot{Q}Q^{-1}\dot{Q}](x-\tilde{x}) \quad \text{for } \tilde{x} = \tilde{x}(t) .$$

Replacing $p$ in this formula by the quadratic form (2.9) turns (2.10) into the quadratic form $(x-\tilde{x})^TP(t,\tilde{x})(x-\tilde{x})$ (for the definition of $P$ see (2.4). It follows now from what was said in connection with (2.9) that the latter differs from the expression (2.10) by an error term of order $\mathcal{O}(\|x-\tilde{x}\|^2)$. Since, according to the hypothesis of our theorem, the matrix $P$ is positive on the compact set $\mathcal{P}'$ one can find $\delta > 0$ such that

$$p(t,x,\tilde{x}) + (x-\tilde{x})^T[\ddot{Q} - 2\dot{Q}Q^{-1}\dot{Q}](x-\tilde{x}) > 0$$

whenever $t,x,\tilde{x}$ belong to the set (2.8) and $\|x-\tilde{x}\| \le \delta$. For this choice of $\delta$ the sets $\Omega_\delta$ will then have all properties listed in Theorem 1 and hence the statement of Theorem 2 becomes an immediate consequence of what we found in Section 1.

<u>Step 2.</u> The general case - $\delta$ is now subject to the condition $\bar{\Omega}_\delta \subseteq \mathcal{P}$ only - can be handled in precisely the same way as in [1] (cf. the last subdivision of the proof of Theorem 6.1). We take $\bar{\Omega}_\delta$ as $\mathcal{P}'$ and choose a sequence of intermediate points $x_0^{(i)}$, $x_1^{(i)}$, $i=1,\dots,M-1$, respectively between $\tilde{x}(0),x_0$ and $\tilde{x}(1),x_1$ respectively such that the distance between two consecutive points is not

bigger than the $\delta$ which we elaborated in the first part of the proof. This allows us to solve for each $i=1,2,\dots,M$ the BVP

$$(2.11)_i \quad \ddot{x} = f(t,x), \quad x(0) = x_o^{(i)}, \quad x(1) = x_1^{(i)}$$

if we let the solution $x^{(i-1)}(\cdot)$ of the preceding problem $(2.11)_{i-1}$ play the role of $\tilde{x}(\cdot)$ and if we take the given $\tilde{x}(\cdot)$ as $x^{(o)}(\cdot)$. The solution of $(2.11)_M$ will then have all desired properties. Thereby the proof of the theorem is complete.

<u>Corollary</u>. The conclusion of Theorem 2 remains valid if one has instead of (2.3), (2.4) a matrix inequality of this form

$$(2.12) \quad Q(t)F(t,x) + F(t,x)^T Q(t) + \ddot{Q}(t) - 2R(t) > 0$$

where $R$ satisfies the following condition. One can find a positive symmetric matrix $\hat{Q}(t)$ of type $m' \times m'$, for some $m' \geq m$, such that

$$(2.13) \quad \hat{Q}(t) = \begin{pmatrix} Q(t) & * \\ * & * \end{pmatrix}, \quad \dot{\hat{Q}}(t)\hat{Q}(t)^{-1}\dot{\hat{Q}}(t) = \begin{pmatrix} R(t) & * \\ * & * \end{pmatrix}$$

where the asterisks denote submatrices of the types $m \times m'$, $m' \times m$ and $m \times m$ respectively. As before the inequality has to hold for all $(t,x) \in \mathcal{P}$ with $0 \leq t \leq 1$; one also has to assume that $\hat{Q}$ is elementwise of class $C^2$ on $[0,1]$.

<u>Proof.</u> We may assume without loss of generality that $m' > m$. Let $y$ be a new state variable which is of dimension $m'-m$ and let us consider the $m'$-dimensional system

$$(2.14) \quad \ddot{x}=f(t,x) - \mu Q^{-1}(t)Q_1(t)y, \quad \ddot{y} = \mu y .$$

where $\mu = \mu(t,x)$ is a scalar function and $Q_1$ is the submatrix in the right upper corner of $\hat{Q}$. We now treat (2.14) as a single system of the form $\ddot{\hat{x}} = \hat{f}(t,\hat{x})$ where $\hat{x}$ is the pair $(x,y)$. It is then not difficult to convince oneself that one can choose $\mu$ in such a way that the inequality

$$(2.15) \quad \hat{Q}\hat{F} + \hat{F}^T\hat{Q} + \ddot{\hat{Q}} - 2\dot{\hat{Q}}\hat{Q}^{-1}\dot{\hat{Q}} > 0$$

holds on the set $\{t,x,y : (t,x) \in \mathcal{P}, y=0\}$. This is a consequence of the hypothesis (2.12). Hence it is clear, in view of (2.5), that the inequality (2.15) will also hold on the closure of the set

$$(2.16) \quad \hat{\mathcal{P}} = \{t,x,y : (t,x) \in \Omega_\delta, \ \|y\| < \varepsilon\} ,$$

provided $\varepsilon > 0$ is sufficiently small. We wish to apply Theorem 2 to the diff. eq. (2.14) with $\hat{\mathcal{P}}$ and $\tilde{\hat{x}}(t) = (\tilde{x}(t),0)$ playing the roles of $\mathcal{P}$ and $\tilde{x}(t)$ respectively. It follows now by inspection that a solution $\hat{x}(\cdot)$ of (2.14) which assumes the boundary values

$$(2.17) \quad \hat{x}(0) = (x_o,0), \quad \hat{x}(1) = (x_1,0)$$

is necessarily of the form $\hat{x}(t) = (x(t),0)$, where $x(\cdot)$ is a solution of the BVP (1.1), (1.2). Therefore we need for the present situation not the full analogy of condition (2.5) but we can get along with the weaker requirement that the intersection of $\overline{\hat{\Omega}_\delta}$ with the set $\{t,\hat{x}=(x,y) : y = 0\}$ belongs to $\hat{\mathcal{P}}$. This however is true, in view of (2.16) and we can infer the existence of a solution $\hat{x}(t) = (x(t),0)$ of the eq. (2.14) which satisfies the boundary conditions (2.17) as well as the inequality

$$(\hat{x} - \tilde{\hat{x}})^T\hat{Q}(\hat{x} - \tilde{\hat{x}}) = (x - \tilde{x})^TQ(x - \tilde{x}) \le \delta^2$$

for all $t \in [0,1]$. Thereby the corollary is proved.

## 3. Uniqueness and Continuous Dependence.

In this section we will present a statement concerning uniqueness and continuous dependence of the solutions. It will bring out the importance of the conditions (2.12) and (2.13) for the study of the two-point boundary value problem. Related results have been established previously by Hartman ([2], Chapter XII, cf. in particular Theorem 4.3).

Theorem 3. Let the cross-sections $\mathcal{P}_t = \{x : (t,x) \in \mathcal{P}\}$ be convex, for $0 \le t \le 1$, and assume that matrix relations of the form (2.12), (2.13) hold for all $(t,x) \in \mathcal{P}$. Then for any two solutions $x(\cdot),\tilde{x}(\cdot)$ of eq. (1.1) which are such that the corresponding curves remain in $\mathcal{P}$ for $0 \le t \le 1$ the following statement holds true: The function

$$\rho(t) = (x(t) - \tilde{x}(t))^TQ(t)(x(t)-\tilde{x}(t)) = \tilde{\varphi}(t,x(t))$$

satisfies $\rho(t) \le \text{Max}\,(\rho(0),\rho(1))$ for $0 \le t \le 1$.

Proof. We first consider the linear case, i.e. we assume that $f(t,x)$ has the form $F(t)x$ and that we have

$$(3.1)\quad Q(t)F(t) + F(t)^TQ(t) + \ddot{Q}(t) - 2R(t) > 0$$

for all $t\in[0,1]$. It follows then from Theorem 2 and its corollary that there exists, for a r b i t r a r y choice of $x_0,x_1$, a solution $x(\cdot)$ of the differential equation $\ddot{x}=F(t)x$ which satisfies the boundary conditions (1.2) and which has the properties stated in the conclusion of the theorem with respect to an arbitrary solution $\tilde{x}(t)$ (take as $\mathcal{P}$ a sufficiently large region of the $(t,x)$-space and choose $\delta^2 = \text{Max}(\rho(0),\rho(1)))$. Solving the BVP (1.1), (1.2) in the linear case however amounts to solving n linear equations in n unknowns. Indeed one can determine the solution $x(\cdot)$ by setting up a system of linear equations for $\dot{x}(0)$. This system has the simple form $Ax_0 + B\dot{x}(0) = x_1$ and is clearly solvable for arbitrary $x_1$ if and only if $\det B \neq 0$. Hence for the linear BVP condition (3.1) guarantees uniqueness. This in turn implies that the conclusion of the theorem holds for an arbitrary pair of solutions of the linear differential eq. $\ddot{x} = F(t)x$, since $x(\cdot)$ then can be identified with a solution whose existence t o g e t h e r with the a-priori-estimate follows from Theorem 2.

For the proof in the non-linear case we use the same argument as Hartman (loc.cit.). Let $z(t) = x(t)-\tilde{x}(t)$, then $z(t)$ is solution of the linear differential eq. $\ddot{z} = \tilde{F}(t)z$, where

$$\tilde{F}(t) = \int_0^1 F(t,x(s,t))ds, \quad x(s,t) = sx(t) + (1-s)\tilde{x}(t).$$

Because of the convexity of $\mathcal{P}_t$ the relation (2.12) holds for $x = x(s,t)$ and $0 \leq s \leq 1$, $0 \leq t \leq 1$. If we integrate with respect to $s$ we obtain the relation (3.1) with $\tilde{F}$ instead of $F$. Hence the conclusion of the theorem follows from our previous considerations.

We add a further remark concerning the dependence of the solutions of the BVP (1.1) - 1.3) from the boundary data $x_0, x_1$. Under the hypothesis of Theorem 3 they are Lipschitz-continuous functions of $x_0, x_1$ as can be seen immediately from the statement of the theorem. It turns out however that they are even continuously differentiable functions of $x_0, x_1$ . This can easily be established from the following observation. As a consequence of (2.12) the variational eq. $\ddot{y} = F(t,x(t))y$ along a given solution $x(\cdot)$ falls into the category of linear differential eqs. satisfying condition (3.1). From what we found out in the course of the last proof the desired result follows then by a standard application of the implicit function theorem.

## References

[1] Knobloch, H.W. and Schmitt, K.: Non-linear boundary value problems for systems of differential equations. To appear in Proc. Roy.Soc.Edinburgh A, Vol. 78 (1977)

[2] Hartman, P.: Ordinary differential equations (New York: Interscience, 1964)

[3] Jackson, L.K. and Schrader, K.W.: Comparison theorems for non-linear differential equations. J.Differential Equations 3 (1967), 248-255

Author's address: Mathematisches Institut, Am Hubland, D-87 Würzburg F.R.G.

# BOUNDARY BEHAVIOR OF POTENTIALS

J. Král, Praha

Let L be an elliptic operator of the form

$$Lu = \sum_{i,k=1}^{m} \frac{\partial}{\partial x^k}\left(a_{ik} \frac{\partial}{\partial x^i} u\right) + \sum_{i=1}^{m} e_i \frac{\partial}{\partial x^i} u + cu$$

with sufficiently smooth coefficients in a domain $\Omega \subset R^m$ $(m > 2)$. It is well known that under certain conditions on L and $\Omega$ there exists a fundamental solution $G(x,y)$ on $\Omega \times \Omega$ which is smooth off the diagonal and has a specified singularity at points of the diagonal admitting locally uniform estimates of the type

$$G(x,y) = \mathcal{O}\left(\operatorname{dist}(x,y)^{2-m}\right), \tag{1}$$

$$|dG(x,y)| = \mathcal{O}\left(\operatorname{dist}(x,y)^{1-m}\right) \tag{2}$$

as $\operatorname{dist}(x,y) \to 0+$ (here dist... denotes the distance and d stands for the differential). For compactly supported finite signed Borel measures $\mu$ the potentials

$$G\mu(x) = \int_{\Omega} G(x,y)\, d\mu(y) \tag{3}$$

are locally integrable together with their derivatives and are often used to transform boundary value problems for L into integral equations. Various aspects of the method of potentials in the theory of partial differential equations together with ample references to the classical work of E.E.Levi, G.Giraud, M. Gevrey and others may be found in C.Miranda's monograph [1]. As pointed out by W.Feller [2], the leading part of the operator L can (possibly after multiplication by a suitable factor) conveniently be written in the form of the Laplace-Beltrami operator

$$\frac{1}{\sqrt{g}} \sum_{i,k=1}^{m} \frac{\partial}{\partial x^k}\left(g^{ik} \sqrt{g} \frac{\partial}{\partial x^i} \cdots\right)$$

corresponding to a Riemannian metric defined by the form

$$\sum_{i,k=1}^{m} g_{ik}\, dx^i\, dx^k \quad .$$

This permits a better insight in some properties of solutions of $Lu = 0$ ; in particular, the usual conormal derivative associated with $L$ reduces to the ordinary normal derivative corresponding to the Riemannian metric. We wish to indicate here that this point of view has useful applications in connection with investigation of boundary behavior of potentials and, in particular, their weak normal derivatives.

Instead of a domain in $R^m$ we shall thus consider an m-dimensional Riemannian manifold $\Omega$ (without boundary) which is smooth ( say, of class $C^\infty$ ) and oriented. On $\Omega$ we shall consider an operator $L$ of the form

$$Lu = *( d * du + du \wedge E + uC )$$

whose leading part is the Laplace-Beltrami operator on $\Omega$ ; here $*$ is the Hodge star operator mapping k-forms into (m-k)-forms , $d$ is the exterior derivative, $\wedge$ is the exterior product, $E$ is a differential (m-1)-form and $C$ is a differential m-form. The transpose of $L$ has the form

$$Mv = *( d * dv - dv \wedge E + v(C-dE) ) \; .$$

We shall suppose that we are given a function $G(x,y)$ on $\Omega \times \Omega$ which is smooth off the diagonal and satisfies in the weak sense the equations

$$L_x\, G(x,y) = \varepsilon_y \quad , \quad y \in \Omega \quad ,$$

$$M_y\, G(x,y) = \varepsilon_x \quad , \quad x \in \Omega \quad ,$$

where $\varepsilon_z$ denotes the Dirac measure concentrated at z; with the estimates of the form (1),(2) (where now the distance dist...

is derived from the Riemannian metric) we have then for each compactly supported signed Borel measure $\mu$ the potential (3) which together with $dG\mu$ is almost everywhere defined ( and locally summable ).

We shall fix an open set $Q \subset \Omega$ with a compact boundary $B \subset \Omega$ and denote by $C^*(B)$ the Banach space of all signed Borel measures with support contained in B ; the norm in $C^*(B)$ is given by total variation. $C_0^1$ will denote the class of all continuously differentiable functions with compact support on $\Omega$ . If $\mu \in C^*(B)$, then the weak normal derivative of $u = G\mu$ may be defined as the functional Nu over $C_0^1$ by the formula

$$\langle \varphi, Nu \rangle = \int_Q [d\varphi \wedge * du - \varphi du \wedge E - \varphi uC] .$$

( If the boundary B of Q is a properly oriented hypersurface, then $\langle \varphi, Nu \rangle = \int_B \varphi \wedge * du$ so that Nu is a reasonable weak characterization of the normal derivative .)

With the exception of the compactness requirement we make now no à priori restriction on the boundary B of Q and with each $\mu \in C^*(B)$ we associate the corresponding functional $NG\mu$. It is easily seen that the support of $NG\mu$ is contained in B ( in the sense that $\langle \varphi, NG\mu \rangle = 0$ whenever $\varphi \in C_0^1$ has support disjoint with B ). In general, $NG\mu$ need not be representable by a ( signed ) measure. On the other hand, if there is a representing measure $\nu$ for $NG\mu$, which means that

$$\langle \varphi, NG\mu \rangle = \int_\Omega \varphi \, d\nu$$

for all $\varphi \in C_0^1$ , then necessarily the support of $\nu$ is contained in B so that $\nu \in C^*(B)$ ; in this case we identify $NG\mu = \nu$, as usual.

We thus arrive naturally at the following

Question . What conditions on B guarantee that $NG\mu \in C^*(B)$ for every $\mu \in C^*(B)$ ?

In order to answer this question in geometric terms it appears useful to generalize the concept of a hit introduced in [3] , [4] in connection with investigation of Newtonian potentials in m-space. Let us denote by $H^1$ the length ( = 1-dimensional Hausdorff measure ) derived in the usual way from the metric in $\Omega$ . If $\Gamma$ is a simple arc and $P \subset \Omega$ is a Borel set, we call $\eta \in \Gamma$ a hit of $\Gamma$ on P ( and say that $\Gamma$ hits P at $\eta$ ) provided, for every neighborhood U of $\eta$ ,

$$H^1(U \cap \Gamma \cap P) > 0 \quad \text{and} \quad H^1((U \setminus P) \cap \Gamma ) > 0 .$$

Let us now fix a point $y \in \Omega$ and consider the tangent space $T\Omega_y$ of $\Omega$ at y ; let $S_y = \{ \theta \in T\Omega_y ; |\theta| = 1 \}$ denote the sphere of unit vectors and $d\sigma$ the element of the surface measure in $S_y$ ( induced by the metric in $T\Omega_y$ ), $A = \int_{S_y} d\sigma$ . If $r > 0$ is sufficiently small, then the exponential map at y

$$\exp_y : T\Omega_y \rightarrow \Omega$$

is well defined and 1-1 on the set

$$\{ \rho\theta ; \theta \in S_y , 0 \leq \rho < r \}$$

and we may consider the geodesic arcs

$$\Gamma_r(y,\theta) = \{ \exp_y \rho\theta ; 0 < \rho < r \} , \qquad \theta \in S_y .$$

We shall denote by $n_r^Q(y,\theta)$ the total number of all hits of $\Gamma_r(y,\theta)$ on Q ( $0 \leq n_r^Q(y,\theta) \leq +\infty$). It can be shown that the function

$$\theta \mapsto n_r^Q(y,\theta)$$

is Borel measurable so that we may define

$$v_r^Q(y) = \frac{1}{A} \int_{S_y} n_r^Q(y,\theta)\, d\,\sigma(\theta) .$$

Thus $v_r^Q(y)$ is just the average number of points at which the open geodesic arcs of length r starting at y hit Q. It is also useful to adopt the following notation. Let $K \subset \Omega$ be a compact set. Then, for sufficiently small $r > 0$, $v_r^Q(y)$ is defined for all $y \in K$ and we put

$$V_0^Q(K) = \lim_{r \downarrow 0} \sup\left\{ v_r^Q(y);\ y \in K \right\} .$$

With this notation we have the following answer to the above question.

Theorem 1 . If $NG\mu \in C^*(B)$ for every $\mu \in C^*(B)$, then necessarily

$$V_0^Q(B) < +\infty . \tag{4}$$

Conversely, if (4) holds, then $NG\mu \in C^*(B)$ whenever $\mu \in C^*(B)$ and the operator

$$NG : \mu \longmapsto NG\mu \tag{5}$$

is bounded on $C^*(B)$ .

The basic ideas of the proof of this theorem are similar to those employed in section 1 in [3] .

If we assume (4) and denote by $C(B)$ the Banach space of all continuous functions on B ( equipped with the maximum norm ),then

$$Wf(y) = \langle f,\ NG\,\varepsilon_y \rangle$$

represents a continuous function of the variable $y \in B$ for every $f \in C(B)$ and the operator (5) is dual to the operator

$$W : f \longmapsto Wf \tag{6}$$

acting on $C(B)$ .

The operator W, which is closely connected with the classical double layer potentials, admits various concrete integral representations analoguous to those obtained in section 2 in [3] for Newtonian potentials. They are partly based on the fact that (4) implies that Q has finite perimeter

$$P(Q) = \sup \left\{ \int_Q d\psi \; ; \; |*\psi| \leq 1 \right\},$$

where $\psi$ ranges over differential (m-1)-forms with compact support in $\Omega$, and on some results concerning sets with finite perimeter ( compare [5]-[7] ).

The operator (6) is more easily treated than (5) and its analytic properties are closely tied with geometric structure of B. As an illustration we shall evaluate the quantity

$$\omega(\alpha) = \inf_T \| W + \alpha I - T \| ,$$

where T ranges over all compact operators on C(B), $\alpha \in R^1$ and I is the identity operator. For simplicity we shall state the formula under a mild simplifying restriction requiring $\mathrm{vol}(U_y \setminus Q) > 0$ for every neighborhood $U_y$ of any $y \in B$ ( vol... denotes the volume in $\Omega$ ). We have

Theorem 2 . If (4) holds, then the density

$$D_Q(y) = \lim_{r \downarrow 0} \frac{\mathrm{vol}(\{ z \in Q; \mathrm{dist}(z,y) < r \})}{\mathrm{vol}(\{ z \in \Omega \; ; \mathrm{dist}(z,y) < r \})}$$

exists for all $y \in B$ and the following equality holds for any $\alpha \in R^1$

$$\omega(\alpha) = \lim_{r \downarrow 0} \sup_{y \in B} ( |\alpha - D_Q(y)| + v_r^Q(y) ).$$

Moreover,

$$\min \left\{ \frac{\omega(\alpha)}{\alpha} \; ; \; \alpha \in R^1 \right\} = 2\,\omega(\tfrac{1}{2}) = 2\, V_0^Q(B) .$$

Results analoguous to theorems 1,2 were originally established for logarithmic and Newtonian potentials and proved to be useful in connection with the Radon scheme [8] for treating the Dirichlet and the Neumann problem as well as related problems in potential theory ( compare [3],[4],[9] - [11] including further references ). The above results permit similar applications in a more general setting . In distinction to local results we have described here, however, some of these applications depend on global behavior of the kernel G. These considerations remain beyond the scope of the present lecture.

Finally we wish to mention that the quantity $v_r^Q(.)$ permits also to obtain necessary and sufficient conditions for the existence of angular limits of potentials analoguous to those known for logarithmic or Newtonian potentials ( cf. [12],[13] ) and admits further generalizations useful in various investigations ( cf.[14] - [17] ).

## R e f e r e n c e s


[1] C.Miranda : Partial differential equations of elliptic type, Springer-Verlag 1970.

[2] W.Feller : Über die Lösungen der linearen partiellen Differentialgleichungen zweiter Ordnung vom elliptischen Typus, Math. Ann. 102( 1930 ), 633-649.

[3] J.Král : The Fredholm method in potential theory, Trans. Amer. Math. Soc. 125 ( 1966 ), 511 - 547 .

[4] J.Král : On the Neumann problem in potential theory, Comment. Math. Univ. Carolinae 7 ( 1966 ), 485 - 493.

[5] E. De Giorgi : Nuovi teoremi relativi alle misure (r-1)-dimensionali in uno spazio ad r dimensioni, Ricerche Mat. 4 (1955), 95 - 113 .

[6] J.Mařík : The surface integral, Czechoslovak Math. J. 6 (1956), 522 - 558 .

[7] H.Federer : Geometric measure theory, Springer-Verlag 1969 .

[8] J.Radon : Über die Randwertaufgaben beim logarithmischen Potential, Sitzungsber. Öster. Akad. Wiss., Math.-naturwiss. Kl. Abt. IIa, Bd 128 (1919), 1123-1167.

[9] Yu.D.Burago, V.G. Mazja : Nekotoryje voprosy teorii potenciala i teorii funkcij dlja oblastej s nereguljarnymi granicami, Zap. Naučn. Sem. Leningrad. Otdel. Mat. Inst. Steklov. t. 3, Leningrad 1967 .

[10] I.Netuka : The third boundary value problem in potential theory, Czechoslovak Math. J. 22 ( 1972 ), 554 - 580 .

[11] R.Kleinman, W.Wendland : On Neumann's method for the exterior Neumann problem for the Helmholtz equation, J. Math. Anal. Appl. 57 ( 1977 ) , 170 - 202 .

[12] J.Veselý : On the limits of the potential of the double distribution, Comment. Math. Univ. Carolinae 10 (1969),189-194.

[13] M.Dont : Non-tangential limits of the double distribution, Časopis pěst. mat. 97(1972), 231-258.

[14] J.E.Brothers : A characterization of integral currents, Trans. Amer. Math. Soc. 150(1970), 301-325.

[15] J.Král, J.Lukeš : Integrals of the Cauchy type, Czechoslovak Math. J. 22(1972), 663-682.

[16] M.Dont : On a boundary value problem for the heat equation, Czechoslovak Math. J. 25(1975), 110-133 .

[17] J.Veselý : On a generalized heat potential, Czechoslovak Math. J. 25(1975), 404-423 .

Author's address: Mathematical Institute of the Czechoslovak Academy of Sciences, Žitná 25, 115 67 Praha 1, Czechoslovakia

# SOME MODIFICATIONS OF SOBOLEV SPACES AND NON-LINEAR BOUNDARY VALUE PROBLEMS

A. Kufner, Praha

This paper deals with applications of two types of function spaces of Sobolev type to (generally non-linear) elliptic partial differential equations. Especially, the following spaces are considered:

(i) Anisotropic Sobolev spaces denoted by

$$W^{E,p}(\Omega) \ ;$$

(ii) Sobolev weight spaces denoted by

$$W^{k,p}(\Omega; \sigma) \ .$$

Here $\Omega$ is a bounded domain in the Euclidean space $R^N$ with a boundary $\partial\Omega$ which can be locally described by Lipschitzian functions and $p$ is a real number, $p > 1$. The spaces under consideration are defined as follows:

(i) The space $W^{E,p}(\Omega)$ with $E$ an arbitrary but fixed set of N-dimensional multiindices is defined as the set of all functions $u \in L_{1,loc}(\Omega)$ such that their distributional derivatives $D^\alpha u$ with $\alpha \in E$ belong to the space $L_p(\Omega)$. The space $W^{E,p}(\Omega)$ is a separable reflexive Banach space under the norm

$$\|u\|_{E,p} = \Big( \sum_{\alpha \in E} \int_\Omega |D^\alpha u(x)|^p \, dx \Big)^{1/p}$$

provided the multiindex set $E$ contains the zero multiindex $\theta =$ $= (0,0,\dots,0)$.

(ii) The space $W^{k,p}(\Omega; \sigma)$ with $k$ a positive integer and $\sigma = \sigma(x)$ an almost everywhere positive function defined on $\Omega$ (and called the weight function) is defined as the set of all functions $u \in L_{1,loc}(\Omega)$ such that their distributional derivatives $D^\alpha u$ of order $|\alpha| \leq k$ ( $\alpha$ denotes the usual N-dimensional multiindex, $|\alpha|$ its length) have the property

$$\int_\Omega |D^\alpha u(x)|^p \, \sigma(x) \, dx < \infty$$

for all $\alpha$ with $|\alpha| \leq k$ . The space $W^{k,p}(\Omega;\sigma)$ is a separable reflexive Banach space under the norm

$$\|u\|_{k,p,\sigma} = \Big( \sum_{|\alpha| \leq k} \int_{\Omega} |D^{\alpha}u(x)|^{p} \, \sigma(x) \, dx \Big)^{1/p} .$$

The aim of this paper is to extend some results concerning the solvability of linear partial differential equations in the above mentioned spaces and in some cases the uniqueness of the solution (which is treated in the *weak* sense here) also to the case of *non-linear* partial differential equations.

## 1. Anisotropic Sobolev spaces

1.1. In [1], S. M. NIKOL'SKIǏ investigated linear partial differential equations on $\Omega$ of the form

$$(1) \qquad \sum_{\alpha,\beta \in E} (-1)^{|\alpha|} D^{\alpha}\big(a_{\alpha\beta}(x) \, D^{\beta}u(x)\big) = f(x) , \quad x \in \Omega ,$$

where $E$ is a given set of N-dimensional multiindices and $a_{\alpha\beta}(x)$ are given functions from $L_{\infty}(\Omega)$ . Provided that the differential operator on the left-hand side in (1) is conditionally elliptic which means that a constant $c_0 > 0$ exists such that

$$\sum_{\alpha,\beta \in E} a_{\alpha\beta}(x) \xi_{\alpha} \xi_{\beta} \geq c_0 \sum_{\gamma \in E} |\xi_{\gamma}|^2$$

for all $x \in \Omega$ and $\xi_{\alpha} \in R^1$ , the concept of the weak solution of a boundary value problem for (1) can be introduced to be a certain function $u \in W^{E,2}(\Omega)$ , and existence theorems can be proved.

1.2. Remark. Equations of the form (1) occur in applications. E. g., the fourth order equations appear in elasticity, namely, in the theory of plates. Analogous problems are often formulated also in terms of non-linear equations.

1.3. Here, we shall deal with *non-linear* analoga of the equation (1), i.e. with equations of the form

$$(2) \qquad \sum_{\alpha \in E} (-1)^{|\alpha|} D^{\alpha} a_{\alpha}(x; \delta_E u(x)) = f(x) , \quad x \in \Omega .$$

The symbol $\delta_E u$ denotes the vector function

$$\{ D^{\beta}u \; ; \; \beta \in E \} ;$$

let us denote by $\varkappa(E)$ the number of components of this vector and assume that the "coefficients" $a_\alpha(x;\xi)$ in the equation (2) are defined for $x \in \Omega$ and $\xi = \{\xi_\beta\} \in R^{\varkappa(E)}$, satisfy the Carathéodory condition and the growth conditions

$$|a_\alpha(x;\xi)| \leqq g_\alpha(x) + c_\alpha \sum_{\beta \in E} |\xi_\beta|^{p-1} \tag{3}$$

with $p > 1$ for $\alpha \in E$, $x \in \Omega$, $\xi \in R^{\varkappa(E)}$ and with constants $c_\alpha > 0$ and functions $g_\alpha \in L_q(\Omega)$, $q = p/(p-1)$.

1.4. The multiindex set $E$ and the parameter $p$ in (3) allow now to introduce the (anisotropic) space $W^{E,p}(\Omega)$. Before defining the weak solution of a boundary value problem for the equation (3), let us summarize various assumptions and notions:

1.5. Assumptions. (I) Let us assume that the set $E$ fulfils the following conditions:

(i) $E$ is a convex set;

(ii) if $\beta \in E$ and $\gamma$ is such a multiindex that $\gamma \leqq \beta$ (i.e., $\gamma_i \leqq \beta_i$ for $i = 1,\dots,N$) then $\gamma \in E$, too.

(II) Let us introduce the space

$$W_0^{E,p}(\Omega)$$

as the closure of the space $C_0^\infty(\Omega)$ of infinitely differentiable functions with compact support in $\Omega$ with respect to the norm $\|\cdot\|_{E,p}$, and a space $V$ such that

$$W_0^{E,p}(\Omega) \subset V \subset W^{E,p}(\Omega);$$

both space $W_0^{E,p}(\Omega)$ and $V$ are again normed by the expression $\|\cdot\|_{E,p}$.

(III) Let us introduce a Banach space $Q$ such that $V \subset Q$, a constant $c_1$ exists with

$$\|v\|_Q \leqq c_1 \|v\|_{E,p} \qquad \text{for every } v \in V,$$

and $C_0^\infty(\Omega)$ is dense in $V$.

(IV) Let a function $\varphi \in W^{E,p}(\Omega)$ and functionals $f \in Q^*$ and $g \in V^*$ be given such that

$$\langle g, v \rangle = 0 \quad \text{for } v \in W_0^{E,p}(\Omega). \tag{4}$$

1.6. Definition. Let the coefficients $a_\alpha$ from (2) fulfil the growth conditions (3) and let the assumptions (I) - (IV) from 1.5 be satisfied. A function $u \in W^{E,p}(\Omega)$ is said to be a *weak solution of the boundary value problem* for the equation (2) - we shall denote this boundary value problem by $(\{a_\alpha\}; W^{E,p}(\Omega); V, Q)$ - if the following conditions are satisfied:

(i) $u - \varphi \in V$,

(ii) the identity

$$\sum_{\alpha \in E} \int_\Omega a_\alpha(x; \delta_E u(x))\, D^\alpha v(x)\, dx = \langle f, v \rangle + \langle g, v \rangle \tag{5}$$

holds for every $v \in V$.

*1.7. Notation.* Let $B$ be a subset of $E$ such that

(i) $E$ is the convex hull of $B \cup \theta$;

(ii) for every $\beta \in E - B$ there exist multiindices $\alpha^{(1)}, \ldots, \alpha^{(N)} \in B$ such that

$$\beta \leqq \alpha^{(i)}, \quad \beta_i < \alpha_i^{(i)} \quad \text{for } i = 1, \ldots, N.$$

Such a set $B$ is called a *complete basis of $E$*.

Let us write

$$a_\alpha(x; \xi) = a_\alpha(x; \omega, \eta)$$

where $\omega \in R^{\varkappa(E-B)}$ corresponds to the elements of $E - B$ and $\eta \in R^{\varkappa(B)}$ corresponds to the elements of $B$, i.e.

$$a_\alpha(x; \delta_E u(x)) = a_\alpha(x; \delta_{E-B} u(x), \delta_B u(x)).$$

Now, we are able to formulate the following existence theorem:

*1.8. Theorem.* Let the assumptions of Definition 1.6 be fulfilled. Further, let the coefficients $a_\alpha$ fulfil the *monotonicity condition*

$$\sum_{\alpha \in B} (a_\alpha(x; \omega, \xi) - a\,(x; \omega, \eta))(\xi_\alpha - \eta_\alpha) \geqq 0 \tag{6}$$

for every $\omega \in R^{\varkappa(E-B)}$ and $\xi, \eta \in R^{\varkappa(B)}$, the condition

$$\frac{\sum_{\alpha \in B} \xi_\alpha a_\alpha(x; \xi)}{\sum_{\alpha \in E} |\xi_\alpha| + \sum_{\alpha \in E} |\xi_\alpha|^{p-1}} \to \infty \quad \text{if} \quad \sum_{\alpha \in E} |\xi_\alpha| \to \infty \tag{7}$$

and the *coerciveness condition*

(8) $$\sum_{\alpha \in E} \xi_\alpha a_\alpha(x;\xi) \geqq c_1 \sum_{\alpha \in E} |\xi_\alpha|^p - c_2$$

for every $\xi \in R^{\varkappa(E)}$ with $c_1 > 0$, $c_2 \geqq 0$.

Then there exists at least one weak solution $u \in W^{E,p}(\Omega)$ of the boundary value problem $(\{a_\alpha\}; W^{E,p}(\Omega); V,Q)$.

P r o o f : We use the Leray-Lions theory concerning general equations in monotone operators (see [2]) and the properties of the anisotropic Sobolev spaces $W^{E,p}(\Omega)$ developed in [3].

Let us define an operator $A(u,v)$ on $V \times V$ by

$$A(u,v) = A_1(u,v) + A_2(u),$$

where

$$\langle A_1(u,v),w\rangle = \sum_{\alpha \in B} \int_\Omega a_\alpha(x; \delta_{E-B}(v(x) + \varphi(x)), \delta_B(u(x) + \varphi(x)))\, D^\alpha w(x)\, dx,$$

$$\langle A_2(u),w\rangle = \sum_{\alpha \in E-B} \int_\Omega a_\alpha(x; \delta_E(u(x) + \varphi(x)))\, D^\alpha w(x)\, dx$$

for $w \in V$, and let the operator $\mathcal{A}$ on $V$ be defined by

$$\mathcal{A}(u) = A(u,u).$$

The growth conditions (3) imply that $A$ maps $V \times V$ into $V^*$ and that $A$ is bounded. From conditions (6) and (7) and from the compactness of the imbedding

$$W^{E,p}(\Omega) \hookrightarrow W^{E-B+\Theta}(\Omega)$$

proved in [3] it follows that the operator $\mathcal{A}$ is pseudomonotone (in the sense of [2], Chap. 2, Prop. 2.6). The coerciveness condition (8) implies that $\mathcal{A}$ is coercive. Therefore, the equation

$$\mathcal{A}w = F$$

has at least one solution $w \in V$ for every $F \in V^*$ by [2], Chap. 2, Theorem 2.7.

Since the identity (5) can be rewritten as

$$\langle \mathcal{A}(u-\varphi), v\rangle = \langle F, v\rangle$$

with $\langle F,v\rangle = \langle f,v\rangle + \langle g,v\rangle$, we conclude that the element

$$u = w + \varphi \in W^{E,p}(\Omega)$$

is a weak solution of the boundary value problem $(\{a_\alpha\}; W^{E,p}(\Omega); V,Q)$.

<u>1.9. Remark.</u> It is also possible to introduce a "variational formulation" of the boundary value problem $(\{a_\alpha\}; W^{E,p}(\Omega); V,Q)$:

assuming in addition that the coefficients $a_\alpha$ are symmetric in the following sense:

$$\frac{\partial a_\alpha}{\partial \xi_\beta} = \frac{\partial a_\beta}{\partial \xi_\alpha} \quad \text{for} \quad \alpha, \beta \in E$$

in the sense of distributions in $\xi \in R^{\varkappa(E)}$ for almost every $x \in \Omega$, we can show that the functional

$$\Phi(v) = \int_0^1 \Big( \sum_{\alpha \in E} \int_\Omega a_\alpha(x;\ t\,\delta_E v(x) + \delta_E \varphi(x))\ D^\alpha v(x)dx\Big)dt$$
$$- \langle f,v\rangle - \langle g,v\rangle$$

defined on $V$ attains (under the assumptions of Theorem 2.4) its minimum on $V$ at a certain "point" $u_0$ and that this element $u_0$ determines the weak solution $u$ of the boundary value problem $(\{a_\alpha\};\ W^{E,p}(\Omega);\ V,Q)$: it is $u = u_0 + \varphi$.

1.10. At the first sight, it is not clear how the boundary value problem $(\{a_\alpha\};\ W^{E,p}(\Omega);\ V,Q)$ is to be interpreted, i.e., to which "classical" boundary value problem it corresponds. Therefore we shall give here one example which illustrates the difference between the isotropic and anisotropic cases:

If the set $E$ is defined by

$$E = \{\alpha\ ;\ |\alpha| \leq k\}$$

with $k$ a positive integer, then the corresponding space $W^{E,p}(\Omega)$ is the "usual" Sobolev space $W^{k,p}(\Omega)$. In this case, the choice

$$V = W_0^{E,p}(\Omega) \quad (\text{i.e.,} \quad V = W_0^{k,p}(\Omega))$$

corresponds to the Dirichlet problem for the equation (2). Now, let us show what the Dirichlet problem means in the anisotropic case.

1.11. Example. Let $N = 2$, let $\Omega$ be the square $]0,1[\times]0,1[$ and $E$ the set $\{(2,0),\ (1,1),\ (1,0),\ (0,1),\ (0,0)\}$. The equation (2) then assumes the form

$$(9) \quad \frac{\partial^2}{\partial x^2} a_{(2,0)}(x,y;\ \delta_E u) + \frac{\partial^2}{\partial x\, \partial y} a_{(1,1)}(x,y;\ \delta_E u) -$$
$$- \frac{\partial}{\partial x} a_{(1,0)}(x,y;\ \delta_E u) - \frac{\partial}{\partial y} a_{(0,1)}(x,y;\ \delta_E u) +$$
$$+ a_{(0,0)}(x,y;\ \delta_E u) = f(x,y)$$

where $\delta_E u = \left\{ \frac{\partial^2 u}{\partial x^2}, \frac{\partial^2 u}{\partial x \partial y}, \frac{\partial u}{\partial x}, \frac{\partial u}{\partial y}, u \right\}$. Let us choose $V = W_0^{E,p}(\Omega)$, $Q = L_r(\Omega)$ with a suitable value $r$, $\varphi \equiv 0$ (which corresponds to homogeneous boundary conditions) and let $f \in Q^*$ be defined by the function $f(x,y)$ in (9); in view of the choice of $V$ and of the condition (4), it is not necessary to choose $g$. If the smoothness of the weak solution and of the data of our boundary value problem allow to introduce the concept of a classical solution, then it can be shown that the "abstract" boundary value problem $(\{a_\alpha\}; W^{E,p}(\Omega); V,Q)$ corresponds to the following "Dirichlet problem for the equation (9)" : The solution $u$ has to satisfy equation (9) on $\Omega$ and the following boundary conditions on $\partial\Omega$ :

$$u\big|_{\partial\Omega} = 0 , \quad \frac{\partial u}{\partial x}\Big|_{\partial\Omega} = 0 , \quad \frac{\partial u}{\partial y}\Big|_{\Gamma} = 0$$

where $\Gamma$ is the part of $\partial\Omega$ described by the conditions $\{x = 0$ or $x = 1\}$. In other words, the "Dirichlet problem" (with homogeneous boundary conditions) means that $u = 0$ on the whole boundary $\partial\Omega$ while the normal derivative $\frac{\partial u}{\partial n} = 0$ only on $\Gamma$ and <u>no</u> values for $\frac{\partial u}{\partial n}$ are prescribed on $\partial\Omega - \Gamma$.

For a comparison, let us note that if we add the multiindex (0,2) to the set $E$, we obtain the space $W^{2,p}(\Omega)$ and the boundary conditions corresponding to the Dirichlet problem for this choice of multiindices are the usual ones:

$$u = 0 \text{ and } \frac{\partial u}{\partial n} = 0 \text{ on the } \underline{\text{whole}} \text{ boundary } \partial\Omega .$$

## 2. Sobolev weight spaces

2.1. Spaces of this type are useful for the investigation of uniformly elliptic as well as degenerate elliptic equations. In the case of a degenerate equation, the weight function $\sigma$ is prescribed by the degeneration; in the case of a uniformly elliptic equation, the application of a Sobolev weight space is motivated by the desire of having a possibility of extending the class of solvable boundary value problems, e.g., by extending the class of admissible right-hand sides of the equation or the class of boundary conditions.

We shall deal here with the latter case. Then the question arises for <u>what</u> type of weight functions one can obtain assertions about existence and uniqueness of a (weak) solution of a boundary value problem for an equation of order $2k$ in the space $W^{k,p}(\Omega;\sigma)$. This question was partially answered by some authors in the case of l i n e a r differential equations (i.e., $p = 2$) and of the D i r i c h l e t problem and the m i x e d problem: In [4], [5], [6] weight functions of the type

$$\sigma(x) = [\text{dist}\,(x,M)]^{\varepsilon} \tag{10}$$

with $M \subset \partial\Omega$ and $\varepsilon$ a real number were investigated and it was shown that there exist positive numbers $c_1$, $c_2$ such that if $-c_1 < \varepsilon < c_2$ then the Dirichlet problem (or the mixed problem) is uniquely solvable in the space $W^{k,2}(\Omega;\sigma)$ with weight functions $\sigma$ from (10). Further, in [7] weight functions of the type

$$\sigma(x) = s(\text{dist}\,(x,M)) \tag{11}$$

with $s = s(t)$ defined for $t \geqq 0$ were investigated and conditions on $s$ were given which guarantee again the existence and uniqueness of the solution of the (linear) Dirichlet problem in the space $W^{k,2}(\Omega;\sigma)$.

<u>2.2.</u> Here, we shall deal with n o n - l i n e a r equations of the type

$$\sum_{|\alpha| \leqq k} (-1)^{|\alpha|} D^{\alpha} a_{\alpha}(x;\, \delta_k u(x)) = f(x)\,, \quad x \in \Omega\,, \tag{12}$$

and with weight functions of the type (10). The symbol $\delta_k u$ denotes the vector function

$$\{D^{\beta}u;\ |\beta| \leqq k\}\,,$$

$\varkappa$ is the number of components of this vector and it is assumed that the "coefficients" $a_{\alpha}(x;\xi)$ in the equation (3) are defined for $x \in \Omega$ and $\xi \in R^{\varkappa}$, satisfy the <u>Carathéodory condition,</u> the <u>growth conditions</u>

$$|a_{\alpha}(x;\xi)| \leqq c_{\alpha}(1 + \sum_{|\beta| \leqq k} |\xi_{\beta}|^{p-1}) \tag{13}$$

for $|\alpha| \leqq k$, $x \in \Omega$, $\xi \in R^{\varkappa}$ with constants $c_{\alpha} > 0$, the <u>monotonicity condition</u>

$$\sum_{|\alpha| \leqq k} (a_{\alpha}(x;\xi) - a_{\alpha}(x;\eta))(\xi_{\alpha} - \eta_{\alpha}) > 0 \tag{14}$$

for every choice of $\xi, \eta \in R^{\varkappa}$, $\xi \neq \eta$, and the <u>coerciveness condition</u>

$$(15) \qquad \sum_{|\alpha| \leqq k} \xi_\alpha a_\alpha(x;\xi) \geqq c_1 \sum_{|\beta|=k} |\xi_\beta|^p + c_2 |\xi_\theta|^p - c_3$$

for every $\xi \in R^{\varkappa}$ with constants $c_1 > 0$, $c_2 \geqq 0$, $c_3 \geqq 0$.

<u>2.3.</u> In the following, we shall investigate the Dirichlet problem only; therefore it is necessary to introduce the space

$$W_0^{k,p}(\Omega;\sigma)$$

as the closure of the space $C_0^\infty(\Omega)$ with respect to the norm $\|\cdot\|_{k,p,\sigma}$. – Further, we shall consider weight functions $\sigma$ of the type (10) where $M$ is a subset of the boundary $\partial\Omega$ (usually it is $M = \partial\Omega$ or $M = \{x_0\}$ with $x_0 \in \partial\Omega$).

<u>2.4. Definition.</u> Let $\varphi$ be a given function from $W^{k,p}(\Omega;\sigma)$, $f$ a given functional from the dual space $(W_0^{k,p}(\Omega;\sigma))^*$. A function $u \in W^{k,p}(\Omega;\sigma)$ is said to be a <u>weak solution of the Dirichlet problem</u> for the equation (12) if the following conditions are satisfied:

(i) $u - \varphi \in W_0^{k,p}(\Omega;\sigma)$,

(ii) the identity

$$(16) \qquad \sum_{|\alpha| \leqq k} \int_\Omega a_\alpha(x; \delta_k u(x)) D^\alpha v(x)\,dx = \langle f, v \rangle$$

holds for every $v \in C_0^\infty(\Omega)$.

Now we are able to formulate the following existence and uniqueness theorem:

<u>2.5. Theorem.</u> Let the assumptions in 2.2, 2.3 and 2.4 be fulfilled. Then there are positive numbers $d_1$, $d_2$ such that if we consider the space $W^{k,p}(\Omega;\sigma)$ where $\sigma$ is a weight function of the type (10) with $\varepsilon \in ]-d_1, d_2[$, then there exists one and only one weak solution $u \in W^{k,p}(\Omega;\sigma)$ of the Dirichlet problem for the equation (12). Further, a constant $c > 0$ exists such that

$$\|u\|_{k,p,\sigma} \leqq c(\|\varphi\|_{k,p,\sigma} + \|f\|).$$

P r o o f : Analogously as in the proof of Theorem 1.8, we use the Leray-Lions theory (see [2]) and the properties of the Sobolev weight spaces $W^{k,p}(\Omega;\sigma)$ with weights of the type (10) (see e.g. [4], [8]): If we define the operator $A$ on $V = W_0^{k,p}(\Omega;\sigma)$ by

$$\langle Au,w\rangle = \sum_{\alpha \leq k} \int_\Omega a_\alpha(x;\ \delta_k(u(x) + \varphi(x)))D^\alpha w(x)dx$$

for $w \in C_0^\infty(\Omega)$, then the growth, monotonicity and coerciveness conditions (13), (14) and (15) imply that $A$ is a bounded, hemi-continuous, monotone and coercive operator from $V$ into $V^*$. Therefore, the equation

$$Aw = f$$

has one and only one solution $w \in V$ for every $f \in V^*$ by [2], Chap. 2, Theorems 2.1 and 2.2. Consequently, the element $u = w + \varphi \in W^{k,p}(\Omega;\sigma)$ is the (unique) weak solution of the Dirichlet problem.

<u>2.6. Remark.</u> The foregoing theorem states only the <u>existence</u> of admissible values $\varepsilon$ such that the Dirichlet problem is uniquely solvable in the space $W^{k,p}(\Omega;\sigma)$ with $\sigma$ of the type (1) without explicit bounds for these values of $\varepsilon$. A computation of the values $d_1$, $d_2$ which determine the admissible interval $]-d_1,d_2[$ for $\varepsilon$ shows that they depend mainly on the geometrical properties of $\partial\Omega$, more precisely on the properties of the subset $M \subset \partial\Omega$ (of course, in addition to their dependence on the other data of the boundary value problem).

In the conclusion, it should be noted that the results of Section 1 were obtained together with J. RÁKOSNÍK.

<u>References:</u>

[1] NIKOL'SKIǏ, S. M.: Some properties of differentiable functions defined on an n-dimensional open set. Izv. Akad. Nauk SSSR Ser. Mat. 23(1959), 213-242 (Russian).

[2] LIONS, J.-L.: Quelques méthodes de résolution des problèmes aux limites non linéaires. Dunod, Gauthier-Villars, Paris 1969.

[3] RÁKOSNÍK, J.: Some remarks to anisotropic Sobolev spaces. Beiträge zur Analysis 13 (to appear).

[4] NEČAS, J.: Les méthodes directes en théorie des équations elliptiques. Academia, Prague 1967.

[5] KUFNER, A.: Lösungen des Dirichletschen Problems für elliptische Differentialgleichungen in Räumen mit Belegungsfunktionen. Czech. Math. J. 15(90) (1965), 621-633.

[6] KADLEC, J.; KUFNER, A.: On the solution of the mixed problem. Comment. Math. Univ. Carolinae 7 (1966), 1, 75-84.

[7] KUFNER, A.; OPIC, B.: Solution of the Dirichlet problem in Sobolev space with weight of a general typ. Trudy Sem. S.L. Soboleva, No 2, Novosibirsk 1976, 35-48 (Russian).

[8] KUFNER, A.: Einige Eigenschaften der Sobolevschen Räume mit Belegungsfunktionen. Czech.Math.J. 15(90) (1965), 597-620.

Author's address: Mathematical Institute of the Czechoslovak Academy of Sciences, Žitná 25, 115 67 Praha 1, Czechoslovakia.

## SOME PROBLEMS IN NEUTRON TRANSPORT THEORY [+)]

J.Kyncl, I.Marek, Praha

### 1. Introduction. Formulation of the problem.

Today when many nuclear power stations produce a significant portion of the world's electricity, it may be assumed that most problems in reactor physics have been solved. Indeed, there are mathematical models which function well in the sense that they are in good agreement with experimental models of both qualitative and quantitative level. However, despite this fact, there remain some unsolved problems of both theoretical and practical nature. We will show this claim on two basic problems which frequently occur in reactor calculations.

Problem 1. To determine the critical value $\gamma_0 \in \Gamma \subset [0,+\infty)$ of the fuel enrichment and the corresponding neutron density $N$ in reactor body $G$ ; to investigate the numerical aspects.

Problem 2. To examine the time behaviour of the neutron density in particular for $t \to \infty$ .

By the reactor we mean a convex body $G$ where absorption, scattering and creation of neutron take place. The neutron density $N$ is a function of velocity $\underline{v} = (v_x, v_y, v_z)$ , space variables $\underline{r} = (x,y,z)$ and time $t$ .

Since any interaction between neutrons themselves is negligible, the models of reactor physics must be linear. Obviously, such models apply in other fields too when the particles under consideration do not interact as in the case of neutrons.

From physical point of view the most rigorous model is defined by the Boltzmann linearized operator, that is $N$ satisfies

$$(1.1) \qquad \frac{d}{dt} N = LN + SN + FN$$

and

$$(1.2) \qquad N(\underline{r},\underline{v},t) = 0 \quad \text{for} \quad \underline{r} \in \partial G \quad \text{and} \quad (\underline{n},\underline{v}) < 0$$

where $\underline{n}$ is the outer normal, and

$$(1.3) \qquad LN \equiv -\underline{v} \operatorname{grad} N - v \Sigma(\underline{r},\underline{v}) N ,$$

$$(1.4) \qquad SN \equiv \int d\underline{v}' v \, \Sigma_s(\underline{v}' \to \underline{v},\underline{r}) \, N(\underline{r},\underline{v},t) ,$$

---

+) Delivered by the second author.

$$(1.5) \qquad FN \equiv \int d\underline{v}' v' \, \nu(\underline{v}') \, \Sigma_f(v') \, \chi(\underline{v}' \to \underline{v}) N(\underline{r},\underline{v},t)$$

and where $\Sigma$, $\Sigma_s$, $\Sigma_f$, $\nu$ and $\chi$ are characteristics of the reactor medium with respect to collisions with neutrons; $v$ denotes the length of the vector $\underline{v}$.

Problem 1 is connected with the stationary case of (1.1) in which the characteristics $\Sigma, \Sigma_s$, etc. have to be arranged in such a way that the problem

$$(1.6) \qquad 0 = (L + S + F)N_0, \quad N_0(\underline{r},\underline{v}) = 0$$
$$\text{for } \underline{r} \in \partial G \text{ and } (\underline{n},\underline{v}) < 0 ,$$

has a positive solution for fixed geometrical configuration of $G$. More precisely, the stationary state depends on some (criticality) parameter: $L = L(\gamma)$, $S = S(\gamma)$, $F=F(\gamma)$ and one has to find a $\gamma_0$ such that the problem

$$(1.7) \qquad \left[L(\gamma) + S(\gamma) + \frac{1}{\tau(\gamma)} F(\gamma)\right] N_0 = 0 ,$$
$$N_0(\underline{r},\underline{v}) = 0 \quad \text{for } \underline{r} \in \partial G, \ (\underline{v},\underline{n}) < 0$$

has a nonnegative nontrivial solution $N_0$ corresponding to $\tau(\gamma_0) = 1$ (criticality condition). From theoretical point of view the Problem 1 was more or less completely solved (see [2],[3],[5],[20]). Concerning the calculations there are difficulties connected with the fact that the data of most of the models change rapidly with respect to space variables. A real reactor situation is such that domain $G$ consists of periodically changing subdomains $G_p$ each of which contains cells with fuel kernel and several shells having various purposes. The cells are rather small in the sense that when any variant of the finite element method is used the majority of elements are then determined by discontinuous data. However, in practice we are not interested in the detailed behaviour of the differential neutron density; only some global characteristics of the solution are needed such as integral neutron flux, parameter of criticality etc. In order to be able to obtain such characteristics some appropriate procedures and simplifications have to be used. A very efficient procedure how to overcome difficulties with rapidly oscillating data a homogenization method has been invented. The homogenization procedure is an approximate method by which the original problem is approximated by another one (usually with respect to space variables) in which the data do not change so rapidly as the original data; the data are in a special way averaged, or, as we say homogenized.

Some mathematical and numerical aspects of the problem of criticality and homogenization will be considered in a subsequent paper. In this paper Problem 2 will be considered in detail.

## 2. Fundamental decay mode and asymptotic behaviour in time.

In order to investigate the asymptotic time-behaviour of any particle distribution $N$, we examine in more detail the Boltzmann operator $A = L + S + F$ defined by (1.2) - (1.5). We remark that $F = 0$ if the medium under consideration is nonmultiplying (a moderator). Let $Y$ be any $L^p(G \times R^3, w)$ for $p \in (1, +\infty)$, where $w \geqq 0$ is a suitable weight function.

It is known that $L$ is an infinitesimal generator of a semigroup $T(t;L)$ of class $(C_0)$ and that

$$\|T(t;L)\| \leqq e^{-\lambda^* t},$$

where

$$\lambda^* = \inf\{v\,\Sigma(\underline{r},\underline{v}) : \underline{v} \in R^3,\ \underline{r} \in R^3,\ v = |\underline{v}|\}.$$

Since both of the operators $S$ and $F$ are bounded, we see that $A$ is an infinitesimal generator of a semigroup of operators $T(t;A)$ of class $(C_0)$ ([6 , p.403, Th.13.2.1]) . The semigroup $T(t;L)$ can be written explicitly by integrating the corresponding first-order differential equation, and we can conclude that $T(t;L)$ is a semigroup of positive operators, that is, $T(t;L)$ leaves invariant the cone of elements in $Y$ with nonnegative representatives. Since the operators $S$ and $F$ are defined by nonnegative kernels, the semigroup $T(t;A)$ is also nonnegative ([6 , p.418, Corollary 4]).

Moreover, since the kernel of $S$ is positive almost everywhere, we may conclude that the semigroup $T(t;A)$ is <u>primitive</u> for sufficiently large $t > 0$, that is, for every nonnegative $u \in Y$, $u \neq 0$, there is a $\tau_u > 0$ such that $v = T(t;A)u$ is positive almost everywhere for $t > \tau_u$.

The spectrum $\sigma(A)$ has the following structure: Every $\lambda$ for which $\operatorname{Re}\lambda \leqq -\lambda^* \leqq 0$ belongs to the continuous spectrum, i.e. $C\sigma(A) \supset \{\lambda : \operatorname{Re}\lambda \leqq -\lambda^*\}$. On the other hand, $\lambda$ for which $\operatorname{Re}\lambda > -\lambda^* + \|S + F\|$ belongs to the resolvent set $\rho(A)$.

If the body $G$ is sufficiently small, there are no further points in $\sigma(A)$ except those in $\{\lambda : \operatorname{Re}\lambda \leqq -\lambda^*\}$ ([1]). Hence, we must assume that the strip $-\lambda^* < \operatorname{Re}\lambda \leqq -\lambda^* + \|S + F\|$ has a nonempty intersection with $\sigma(A)$. Let $\lambda_0$ be such that any $\lambda$ for which $\mathrm{Re}\,\lambda > \lambda_0$ belongs to the resolvent set $\rho(A)$, while there exists a $\lambda_1 \in \sigma(A)$ with $\mathrm{Re}\,\lambda_1 = \lambda_0$.

It is known that $N(t) = T(t;A)N_0$, where $N_0 \in Y$, is a unique solution of the problem ([6, p.359, Theorem 11.5.3])

$$\frac{d}{dt} N = AN \,,\; N(0) = N_0 \,.$$

We assume that $N_0$ is nonnegative almost everywhere and now investigate the behaviour of $N$ as $T \to +\infty$.

A standard procedure [19, p.210-213] consists in estimating the semigroup operator by using the resolvent inversion formula [6, Theorem 11.6.1, p.363]

$$T(t;A)N_0 = \lim_{\omega \to \infty} \frac{1}{2\pi i} \int_{\alpha - i\omega}^{\alpha + i\omega} e^{\lambda t} R(\lambda, A) N_0 d\lambda, \; \alpha > \max(0, \lambda_0) \,,$$

where $\lambda_0 = \mathrm{Re}\,\lambda_0$ is such that $\mathrm{Re}\,\lambda > \lambda_0$ implies $\lambda \in \rho(A)$. For such a procedure we must have complete information about that part of the spectrum of $A$ in the region $\mathrm{Re}\,\lambda > -\lambda^*$.

We propose a more direct and much simpler approach. We formulate it in an abstract way.

Let $Y$ be a real Banach space, $X = Y \oplus iY$ its complexification. Let $K \subset Y$ be a generating and normal cone. We say that a linear bounded operator $T \in B(Y) = (Y \to Y)$ is K-positive if $Tx \in K$ whenever $x \in K$. We also have a partial ordering in $Y$ defined as follows $x \leqq y \Longleftrightarrow y - y \in K$. Similarly $T \leqq S \Longleftrightarrow (S-T)K \subset K$. We call an element $y \in K$ quasiinterior if $x'(y) > 0$ for all linear functionals $0 \neq x' \in Y'$ such that $x'(x) \geqq 0$ for all $x \in K$; here $Y'$ is the dual space of $Y$.

If the cone $K$ is such that the partial ordering of $Y$ generated by $K$ is a lattice order, that is, $\sup\{x,y\}$ and $\inf\{x,y\}$ exist for every pair of elements $x$ and $y$ in $Y$, we call $Y$ a Banach lattice.

In the following theorems (Theorem 1-3) we assume that $Y$ is a Banach lattice generated by a cone $K$. We shall apply some deep results due to F.Niiro and I.Sawashima [16] and H.H.Schaefer [18, p.328-333].

Theorem 1. Let $A$ be an infinitesimal generator of a semigroup of operators $T(t;A)$ of class $(C_0)$. Let $\lambda_0$ be such that

$$\lambda \in \sigma(A) \Longrightarrow \mathrm{Re}\,\lambda \leqq \lambda_0 = \mathrm{Re}\,\lambda_0 \,, \tag{2.1}$$

and let $\lambda_o$ be a Fredholm eigenvalue, i.e. an isolated pole of $R(\mu,A)$ to which there corresponds a finite-dimensional eigenspace

(2.2) $\mathcal{H}_o = \{u \in Y : (A - \lambda_o I)^k u = 0$ for some $k = 1,2,\dots\}$.

Let the semigroup $T(t;A)$ be K-positive for $t \geqq 0$.

Then

(2.3) $\sigma(A) \cap \{\lambda : \mathrm{Re}\,\lambda = \lambda_o\} = \{\lambda_o,\dots,\lambda_s\}$

and

(2.4) $T(t;A) = \sum_{j=o}^{s} e^{\lambda_j t}[B_j + Z_j(t)] + W(t)$,

where

(2.5) $B_j B_k = B_k B_j = \delta_{jk} B_j$, $B_j Z_j(t) = Z_j(t) B_j = Z_j(t)$,

(2.6) $B_j W(t) = W(t) B_j = \Theta$, $j,k = 0,\dots,s$

and

(2.7) $\lim_{t\to\infty} e^{-\lambda_o t} \|W(t)\| = 0$, $\lim_{t\to\infty} t^{-q+1} \|Z_j(t)\| = 0$,

where $q$ is the order of $\lambda_o$ as a pole of $R(\mu,A)$.

Moreover, $B_{o,q} = \lim_{\rho\to\lambda_o} (\rho - \lambda_o)^q R(\rho,A)$ is K-positive and hence, if $u_o \in K$, then $v_o = B_{oq} u_o \neq 0$ is an eigenvector of $A$ in $K : Av_o = \lambda_o v_o$.

<u>Theorem 2</u>. If the semigroup $T(t;A)$ in Theorem 1 is such that for every $u \in K$, $u \neq 0$, there exists $\tau_o(u) > 0$ and a positive integer $p = p(u)$ such that $[T(t;A)]^p u$ is quasi-interior with respect to $K$ for $t > \tau_o(u)$, then

(a) $s = 0$ in (2.3), $B_{o,q} = B_o$;

(b) $\mathcal{H}_o = B_o Y$ with $\dim \mathcal{H}_o = 1$,

and $B_o v$ is quasi-interior whenever $v \in K$, $v \neq 0$; if $y \in K$ is any eigenvector of $A$ then $y = cu_o = B_o v_o$, $v_o \in K$, $v_o \neq 0$. Furthermore,

(c) $T(t;A) = e^{\lambda_o t} B_o + W(t)$

with

(2.8) $\lim_{t\to\infty} e^{-\lambda_o t} \|W(t)\| = 0$.

Under the hypotheses of Theorem 2, we consider the Cauchy problem

$$(2.9) \qquad \frac{d}{dt}u(t) = Au \ , \quad u(0) = u_0 \in K \ .$$

By Theorem 2 we have the following representation of the solution

$$u(t) = T(t;A)u_0 = \\ = e^{\lambda_0 t} B_0 u_0 + W(t)u_0 \ .$$

It follows from (2.8) that

$$(2.10) \qquad \lim_{t\to\infty} e^{-\lambda_0 t} u(t) = B_0 u_0 \ .$$

Thus we have

Theorem 3. The asymptotic behaviour of any solution $u(t)$ of (2.9) is non-oscillatory.

Remark. Let $\mu_0$ be an eigenvalue of $A$, and $M_0$ the corresponding eigenvector. We see that any $M(t)$ of the form $e^{\lambda_0 t} M_0$ is always a solution of (2.9) with $u(0) = M_0$. Such a solution is called a decay mode; a decay mode is called fundamental if $M_0 \in K$, $M_0 \neq 0$.

It is easy to see that the normalized fundamental decay mode is unique if $T(t;A)$ fulfils the hypotheses of Theorem 2.

To apply our previous theory we have to show only that the point $\lambda_0$, the bound of the spectrum $\sigma(A)$, is an isolated pole of the resolvent operator $R(\mu,A)$. We emphasize this fact because a complete analysis of the existence of decay modes and the uniqueness of the fundamental decay mode can be made without any further information about the spectrum of the operator $A$. This makes our approach different from the sort of analysis proposed by others. On the other hand, we describe only the peripheral part of the spectrum of the semigroup $T(t;A)$. If we make assumptions involving compactness about $ST(t;L)S$ [22] or other closely related assumptions, we can give a complete description of $\sigma(A)$. Actually, under certain assumptions concerning compactness of $T(t_1;L)S \ldots T(t_k;L)S$ it has been shown that every $\mu \in \sigma(T(t;A))$ for which $|\mu| > e^{-\lambda^* t}$ has the form $\mu = e^{\lambda t}$, where $\lambda$ is an isolated pole of $R(\mu,A)$ with finite-dimensional invariant subspace $\mathcal{H}(\lambda) = \{ u : (A - \lambda I)^k u = 0 \text{ for some } k = 1,2,\ldots \}$, [21],[22], [19]. However, these assumptions are not fulfilled in general, e.g. for some models including the case of inelastic scattering in the high-energy range [11].

On the other hand, our theory does not cover the model excluding the up-scattering. With some minor modifications this case can also be considered by our method and the main results, such as the final Theorem, remain valid in general.

We already know that the semigroup $T(t;A)$ is K-positive in $Y$, where $K$ is the cone of elements of $Y = L^p(G\times R^3, w)$, $1 < p < +\infty$, with nonnegative representatives, $w \geqq 0$. It follows that $R(a,A)$ is also K-positive, where $a > \max(0, \lambda_o)$ ([6, Theorem 11.7.2]).

Let us write $A$ in the form $A = L + S_1 + S_2 + F$, in which $S_1$ includes the elastic scattering and the inelastic scattering in high-energy range and $S_2$ the inelastic scattering in low-energy range. Since $S_1$ is bounded and $S_2$ and $F$ are compact operators, we have that

$$R(a,A) = R(a,L+S_1) + R(a,A)(S_2+F)R(a,L+S_1) .$$

A crucial assumption for the applicability of our theory is the fulfilment of the strict inequality

$$(2.11) \qquad r = r(R(a,A)) > r(R(a,L+S_1)) = r_1 ,$$

the relation $r \geqq r_1$ being trivial.

Actually, we have

<u>Lemma</u>. Under the assumption (3.11) for some $a > \max(0, -\lambda^* + \|S+F\|)$ the point $\lambda_o$ is a pole of the resolvent operator $R(\mu, A)$.

<u>Remark</u>. We note that the validity of (2.11) follows from the compactness of $(S+F)T(t;L)(S+F)$ and similar other assumptions, as we have mentioned above. The converse is obviously not necessarily true, as we have mentioned, in the case of inelastic scattering in the high-energy range.

<u>Proof of the Lemma</u>. The operator $R(a,A)$ is an operator of Radon-Nikolskii type [14] whence it follows that its peripheral spectrum consists of a finite set of Fredholm eigenvalues $\nu_o, \ldots, \nu_s$. Obviously,

$$\nu_j = \frac{1}{a-\lambda_j}, \text{ where } \lambda_j \in \sigma(A), \; j = 0, \ldots, s,$$

$|\nu_j| = r(R(a,A))$. We identify $\lambda_o$ by setting $r(R(a,A)) = \dfrac{1}{a-\lambda_o}$.

Let $R(a,A)y_j = \nu_j y_j$ , $y_j \neq 0$ and let $b > a$ . We see that

$$R(b,A)y_j = \frac{1}{b-\lambda_j} y_j .$$

Thus, for all $b > a$ we have that

$$\left| \frac{1}{b-\lambda} \right| \leqq \left| \frac{1}{b-\lambda_o} \right|$$

for every $\lambda \in \sigma(A)$ . Since $b$ can be arbitrarily large, we conclude that $\lambda \in \sigma(A)$ implies that $\mathrm{Re}\,\lambda \leqq \lambda_o$ . Because $r = r(R(a,A)) \in \sigma(R(a,A))$ , the spectral mapping theorem shows that $\lambda_o$ is a Fredholm eigenvalue of $A$ since $r$ has this property with respect to $R(a,A)$ . This completes the proof of the Lemma.

The conclusion of the Lemma implies that Theorems 1-3 apply to those cases of neutron transport where the assumption (3.11) holds. In our opinion, this is the case in most of the models used until now.

As a consequence we have the following final result.

__Theorem.__ If $\sigma(A) \cap \{\lambda : \mathrm{Re}\,\lambda > -\lambda^*\} \neq \emptyset$ , then there exists exactly one normalized fundamental decay mode $(\lambda_o, M_o)$ and we have that for every solution $N$ of

$$\frac{d}{dt} N = AN , \quad N(0) = N_o \geqq 0 ,$$

$$\lim_{t\to\infty} \| e^{-\lambda_o t} N(t) - cM_o \| = 0 ,$$

where $c > 0$ is a constant independent of $t$ .

More precisely, $cM_o = PN_o$ , where $P$ is the residue of the Laurent expansion of $R(\mu,A)$ about the point $\lambda_o$ .

We remark that this last theorem gives a solution to Problem 10 of Kaper's Collection of problems in [7].

__References__

[1] Albertoni S., Montagnini B.: On the spectrum of neutron transport equation in finite bodies. J.Math.Anal.Appl.Vol.13(1966), 19-48.

[2] Birkhoff G.: Reactor criticality in transport theory. Proc.Nat. Acad.Sci. USA, 45(1958), 567-569.

[3] Birkhoff G., Varga R.S.: Reactor criticality and nonnegative matrices. SIAM J.Appl.Math.Vol.6(1958), 354-377.

[4] Borysiewicz M., Mika J.: Time behavior of thermal neutrons in moderating media. J.Math.Anal.Appl.Vol.26(1969), 461-478.

[5] Habetler G.J., Martino M.A.: The multigroup diffusion equations of reactor physics. Report KAPL-1886, July 1958.

[6] Hille E., Phillips R.S.: Functional Analysis and Semigroups. Revised Edition. Providence 1957. Russian translation Izd.Inostrannoj Lit.Moscow 1962.

[7] Kaper H.G.: A collection of problems in transport theory. Transport Theory and Stat.Physics 4(3)(1975), 125-134.

[8] Kato T.: Perturbation Theory for Linear Operators. Springer-Verlag, Berlin-Heidelberg-New York 1965.

[9] Krein M.G., Rutman M.A.: Linear operators leaving a cone invariant in a Banach space. Uspekhi mat.nauk III:1 (1948), 3-95. (Russian); Amer.Math.Soc.Translations no. 26(1950), 128 pp.

[10] Kyncl J., Marek I.: Relaxation lengths and nonnegative solutions in neutron transport. Apl.mat.22(1977), 1-13.

[11] Larsen E.W., Zweifel P.F.: On the spectrum of the linear transport operator. J.Math.Phys.Vol.15(1974), 1987-1997.

[12] Marek I.: Frobenius theory of positive operators. Comparison theorems. SIAM J. Appl.Math. 19(1970), 607-628.

[13] Marek I.: On Fredholm points of compactly perturbed bounded linear operators. Acta Univ.Carol.-Math.Phys.Vol.17(1976), No.1, 65-72.

[14] Marek I.: On some spectral properties of Radon-Nikolskii operators and their generalizations. Comment.Math.Univ.Carol. 3:1 (1962), 20-30.

[15] Mika J.: Neutron transport with anisotropic scattering. Nucl.Sci. Engi.11(1961), 415-427.

[16] Niiro F., Sawashima I.: On spectral properties of positive irreducible operators in an arbitrary Banach lattice and problems of H.H.Schaefer. Sci Papers College General Education Univ.Tokyo 16(1966), 145-183.

[17] Sawashima I.: Spectral properties of some positive operators. Natur.Sci.Rep.Ochanomizu, Univ.15(1964), 55-64.

[18] Schaefer H.H.: Banach Lattices and Positive Operators. Springer Verlag Berlin-Heidelberg-New York 1974.

[19] Shikhov S.B.: Lectures in Mathematical Theory of Reactors. I. Linear Theory. Atomizdat, Moscow 1973. (Russian)

[20] Varga R.S., Martino M.A.: The theory for numerical solution of time-dependent and time-independent multigroup diffusion equations. Proc.of the Second Intern.Conf. in the Peaceful Uses of A.E.XVI, 570-577. Ref.P/154, Geneve 1958.

[21] Vidav I.: Existence and uniqueness of nonnegative eigenfunctions of the Boltzmann operator. J.Math.Anal.Appl.22(1968), 144-155.

[22] Vidav I.: Spectra of perturbed semigroups with applications to transport theory. J.Math.Anal.Appl.30(1970), 264-279.

Authors' address: Nuclear Research Institute, 25068 Řež u Prahy, Czechoslovakia
Caroline University, Faculty of Mathematics and Physics, Malostranské nám.25, 11800 Praha 1, Czechoslovakia

# ON FORMULATION AND SOLVABILITY OF BOUNDARY VALUE PROBLEMS FOR VISCOUS INCOMPRESSIBLE FLUIDS IN DOMAINS WITH NON-COMPACT BOUNDARIES

O.A.Ladyženskaja, Leningrad

Studying boundary value problems for viscous incompressible fluids I have introduced two function spaces, namely $\overset{\circ}{J}(\Omega)$ and $H(\Omega)$. The former is the closure in the norm $L_2(\Omega)$ of the family $J^{\infty}(\Omega)$ of all infinitely differentiable solenoidal vector functions $\vec{u}(x)$ with compact supports which belong to the domain $\Omega$ of the Euclidean space $R^n$ $(n=2,3)$. The latter is the closure of the same family $J^{\infty}(\Omega)$ in the norm of Dirichlet integral. Let us denote by $\overset{\circ}{D}(\Omega)$ the Hilbert space which is the closure in the norm of Dirichlet integral of the family $\dot{C}^{\infty}(\Omega)$ of all infinitely differentiable vector-functions $\vec{u}(x)$ with compact supports which belong to $\Omega$. The scalar product in $\overset{\circ}{D}(\Omega)$ is defined by

$$(1)\qquad [\vec{u},\vec{v}] = \int_{\Omega} \vec{u}_x \vec{v}_x dx = \int_{\Omega} \sum_{i,k=1}^{n} u_{ix_k} v_{ix_k} dx .$$

We introduce the same scalar product in $H(\Omega)$. $H(\Omega)$ is a proper subspace of the space $\overset{\circ}{D}(\Omega)$. We shall regard $\overset{\circ}{J}(\Omega)$ as a subspace of the Hilbert space $L_2(\Omega)$ and introduce a scalar product in both of them by

$$(2)\qquad (\vec{u},\vec{v}) = \int_{\Omega} \vec{u}\vec{v}dx; \quad (\vec{u},\vec{u})^{1/2} \equiv \|\vec{u}\|_{2,\Omega} .$$

Let us give a motivation for introducing these spaces and show why they proved useful and suitable for the study of Navier-Stokes equations. First let us consider the Stokes problem

$$(3)\qquad -\nu\Delta\vec{v} = -\nabla\rho + \vec{f}(x) ,$$

$$(4)\qquad \operatorname{div}\vec{v} = 0 , \quad \vec{v}|_{\partial\Omega} = 0 ,$$

restricting ourselves to the case of homogeneous boundary conditions. If $\Omega$ is an unbounded domain in $R^3$ then $\vec{v}$ has to satisfy an additional condition

$$(5_1)\qquad \vec{v}(x) \to 0 \quad \text{for} \quad |x| \to \infty ,$$

while for $\Omega \subset R^2$ it must satisfy

$$(5)\qquad \vec{v}(x) \to 0 \quad \text{for} \quad |x| \to \infty .$$

In this most current formulation, $\rho$ is subjected to no boundary conditions. Therefore I wanted to "get rid" of $\rho$ and to obtain such a system of relations for $\vec{v}$ which would enable us to determine uniquely $\vec{v}$ and then to find $\rho$ from $\vec{v}$. At the same time, I did not want to put any restrictions on the behavior or $\rho$ near to $\partial\Omega$ and infinity lest I should have to verify them when determining $\rho$ from $\vec{v}$. To this aim I formed the scalar product of (3) with $\vec{\eta} \in \dot{J}^{\infty}(\Omega)$, integrated over $\Omega$ and transformed the resulting equation into the form

$$(6) \qquad [\vec{v}, \vec{\eta}] = (\vec{f}, \vec{\eta})$$

using the integration by parts formula, the equation (4) and the fact that $\rho \in L_{2,loc}$. Provided $\vec{f}$ is not too bad, namely

$$(7) \qquad |(\vec{f}, \vec{\eta})| \le c_f \|\vec{\eta}\|_H$$

for all $\vec{\eta} \in H(\Omega)$, then $\vec{v} \in H(\Omega)$ is found uniquely from the identity (6) [1, Chap.II]. With regard to all this I introduced the following definition of a (generalized) solution of the problem (3) - (5):

A function $\vec{v}$ is called a solution of the problem (3)-(5) if it belongs to $H(\Omega)$ and satisfies the equality (6) for all $\vec{\eta} \in \dot{J}^{\infty}(\Omega)$.

If $\vec{f} \in L_{2,loc}$ then it is relatively easy to prove that $\vec{v} \in W^2_{2,loc}$ and satisfies the system (3) with a certain function $\rho \in W^1_{2,loc}$. The function $\rho$ is determined uniquely provided it is normed, say, by the condition $\int_{\Omega_1} \rho \, dx = 0$, $\bar{\Omega}' \subset \Omega$.

Such an approach to the problem (3)-(5) is attractive for its simplicity and generality: it permits to include simultaneously arbitrary domains from $R^2$ and $R^3$ not only for the Stokes system but for the complete nonlinear Navier-Stokes system as well (see [1, Chap.IV]). It accounts also for the Stokes paradox: for unbounded domains $\Omega \subset R^3$ the solution from $H(\Omega)$ converges for $|x| \to \infty$ to zero while for $\Omega \subset R^2$ it converges to a constant, generally non--zero. Thus the suggested re-formulation of the problem (3)-(5) proved to be successful from the mathematical point of view: we have satisfied all the requirements of the problem (3)-(5) proving at the same time its unique solvability for a wide class of right hand side terms $\vec{f}$. Nevertheless, to obtain uniqueness I had to consider $\vec{v}$ in the space $H(\Omega)$. This assumption has not been included in the classical formulation of the problem (3)-(5) and the question whether the suggested specification of the problem (3)-(5) is the only possible is essential. First of all, $\vec{v} \in H(\Omega)$ implies finiteness of the Dirichlet integral for $\vec{v}$. We know quite a number of problems in which the solution, interesting from the physical point of view, does not possess this property. However, to omit it (in the case of nonlinear Navier-Stokes equations and general type of domains) does not seem possible at the moment, and therefore we restrict ourselves by considering only such $\vec{v}$'s for which $\|\vec{v}_x\|_{2,\Omega} < \infty$. This assumption together with zero boundary conditions means that $\vec{v}$ has to be an element of $\overset{\circ}{D}(\Omega)$. Moreover, taking into account, the equality div $\vec{v} = 0$ we conclude that $\vec{v}$ belongs to the space $\hat{H}(\Omega)$ which consists of all elements of $\overset{\circ}{D}(\Omega)$ which have zero divergence.

It is clear that

$$H(\Omega) \subset \hat{H}(\Omega) \subset \overset{\circ}{D}(\Omega),$$

which raises a question about the dimension of the quotient space $\hat{H}(\Omega)|_{H(\Omega)}$. Its investigation was initiated by J.Heywood [2]. He proved that $\hat{H}(\Omega) = H(\Omega)$ for domains $\Omega$ (bounded or not) with compact smooth boundaries of the class $C^2$. Moreover, he indicated domains for which $\hat{H}(\Omega)$ is wider than $H(\Omega)$. In the three-dimensional case this holds for the whole space $R^3$ divided by the plane $\{x : x_1 = 0\}$ with "holes" cut in it. For such $\Omega$ we have dim $\hat{H}|_H = 1$ and the elements of $\hat{H}|_H$ may be characterized either by the quantity $\alpha$ ($\alpha \in R^1$) of the total flow through all the holes (their number is assumed finite and they must be bounded two-dimensional domains with smooth boundaries) or by the difference of the limit values of $\rho$ for $x_1 \to \pm\infty$.

In accordance with this, for such domains the system (3)-(5) has a unique solution $\vec{v}$ from $H(\Omega)$ which has a prescribed total flow through the holes. The solution $\vec{v}$ determined above (i.e. $\vec{v}$ from $H(\Omega)$) corresponds to the value of $\alpha$ equal to zero.

Together with V.A.Solonnikov we have carried out a more detailed analysis of the cases $\hat{H} = H$ and dim $\hat{H}|_H \geq 1$. Furthermore, we have investigated problems of formulation and solvability of boundary value problems for general nonlinear Navier-Stokes equations in the space $\hat{H}$ when $\hat{H}$ is wider than $H$. The results obtained have been published in [3], [4]. They have been continued in the thesis of K.Pileckas and in a joint paper [5] by V.A.Solonnikov and K.Pileckas. Let us mention the results of [3], [4] without presenting the precise formulations. First, we proved that $\hat{H}$ coincides with $H$ for domains (bounded or not) with compact "not too bad" boundaries (e.g. Lipschitzian). To this aim we had to consider two auxiliary problems:

(8) $\quad \operatorname{div} \vec{u} = \varphi \ , \quad \vec{u} \in \overset{\circ}{D}(\Omega)$

with $\varphi \in L_2(\Omega)$ , $\int\limits_{\Omega} \varphi dx = 0$ , and

(9) $\quad \rho_{x_i} = \sum\limits_{k=1}^{n} (R_{ik})_{x_k} + f_i \ , \quad \int\limits_{\Omega_1} \rho dx = 0 \ , \quad i=1,\dots,n \ ,$

with $\rho \in W^1_{2,loc}$ and $R_{ik} \in L_2(\Omega)\ W^1_{2,loc}$ , $f_i \in L_2(\Omega)$ .
For (8) we found a solution $\vec{u}$ which satisfies an inequality $\|\vec{u}_x\|_{2,\Omega} \le c_{\Omega}\|\varphi\|_{2,\Omega}$ with a constant $c_{\Omega}$ which is invariant with respect to similarity mapping of the domain $\Omega$ . For $\rho$ satisfying (9) we proved an estimate

$$\|\rho\|_{2,\Omega} \le c_{\Omega,\Omega'}\Big( \sum_{i,k=1}^{n} \|R_{ik}\|_{2,\Omega} + \sum_{i=1}^{n} \|f_i\|_{2,\Omega}\Big) .$$

Non-smoothness of the boundary precluded us from using the theory of hydrodynamic potentials. And it is this type of boundaries that we have to deal with even if the boundary $\partial\Omega$ of the original domain $\Omega$ is smooth but not compact.

The above presented auxiliary results are useful not only for the problems just considered. They have been applied to deal with problems with free surfaces which meet non-smoothly a rigid wall [6] . They can be used also in the case of the problem (3), (4) on a bounded domain $\Omega$ to prove $\rho \in L_2(\Omega)$ for all $\vec{f}$ satisfying the condition (7).

However, let us come back to the problem whether $\hat{H}$ and $H$ coincide or not. We have proved that $\hat{H} = H$ provided $\Omega$ has one exit to infinity. If $\Omega$ has $m$ exits to infinity, $m>1$ and each of them includes a circular cone (an angle in the case $\Omega \subset R^2$) then $\dim \hat{H}|_H = m-1$ . The elements $\vec{v}$ of the quotient space $\hat{H}|_H$ can be characterized by choices of numbers $\alpha_k$ , $k=1,\dots,m-1$ which indicate the flows $\vec{v}$ through $m-1$ exits (as $\vec{v} \in \hat{H}$, the flow through the last exit equals $\alpha_m = - \sum\limits_{k=1}^{m-1} \alpha_k$). For elements $\vec{v}$ from $H$ all $\alpha_k$ are equal to zero. In accordance with this, the problem (3)-(5) for such $\Omega$ allows the following more precise formulation:

to find a vector function $\vec{v}$ from $\hat{H}$ for which the flows through $m-1$ exits are equal to $\alpha_k$ , $k=1,\dots,m-1$ and which satisfies the identity (6) for all $\vec{\eta} \in J^{\infty}(\Omega)$ (or, which is the same, for all $\vec{\eta} \in H$).

Its unique solvability follows from the above proved solvability of the problem (3)-(5) in the space $H$. Indeed, let $\vec{a}$ be an element of the space $\hat{H}$ with given flows $\alpha_k$, $k=1,\dots,m-1$ and let us seek $\bar{v}$ in the form $\vec{u}+\vec{a}$, $\vec{u} \in H$. For $\vec{u}$ we obtain the problem

$$-\nu\Delta\vec{u} = -\nabla\rho + \nu\Delta\vec{a} + \vec{f}, \quad \vec{u} \in H$$

whose unique solvability was proved in [1, Chap.II].

The nonlinear problem

$$(10) \qquad -\nu\Delta\vec{v} + \sum_{k=1}^{n} v_k\vec{v}_{x_k} = -\nabla\rho + \vec{f}(x),$$

$$\operatorname{div}\vec{v} = 0, \quad \vec{v}\big|_{\partial\Omega} = 0,$$

on domains $\Omega$ with $m$ exits to infinity which extend "sufficiently quickly" (e.g. they may contain cones (angles)) allows an analogous formulation:

to find $\vec{v}$ from $\hat{H}$ with prescribed flows $\alpha_k$, $k=1,\dots,m-1$ through $m-1$ exits and satisfying the identity

$$\nu[\vec{v},\vec{\eta}] - (v_k\vec{v}, \vec{\eta}_{x_k}) = (\vec{f},\vec{\eta})$$

for all $\vec{\eta} \in \overset{\circ}{J}{}^{\infty}(\Omega)$.

The solvability of this problem follows also from the results which I proved about the solvability of the system (10) in $H$ provided at least one of the representants $\vec{a}$ of the element of $\hat{H}|_H$ which corresponds to the prescribed values $\alpha_k$, $k=1,\dots,m-1$ possesses the following property:

$$(11) \qquad \int_\Omega \vec{a}^2\vec{\eta}^2 dx \le \nu_1 \int_\Omega \vec{\eta}_x^2 dx, \qquad \nu_1 \in (0,\nu)$$

for all $\vec{\eta} \in H(\Omega)$. In the paper [4] such $\vec{a}$'s will be constructed for "almost" all the class of domains $\Omega$ for which we proved $\dim \hat{H}|_H = m-1$ in [3]. Here $\alpha_k$, $\nu^{-1}$ as well as the other data of the problem are subjected to no smallness requirements.

If the domain $\Omega$ has $m$ "sufficiently quickly" extending exits to infinity and $r$ "insufficiently quickly" extending ones then the prescribed values of $\alpha_k$, $k=1,\dots,m-1$ of flows through the exits of the first kind are added to the equations (3)-(5) and (10) provided $m>1$. The dimension $\dim \hat{H}|_H$ is then equal to $m-1$.

The words "insufficiently quickly" extending exit indicate the fact that for any element $\vec{v}$ from $\hat{H}(\Omega)$ the flow through this exit is equal to zero. It is not difficult to obtain sufficient conditions guaranteeing this property of an exit. For example, let it have the form $B = \{x : x_1 > a , (x_2,x_3) \in S(x_1)\}$ , where $S(x_1)$ is a family of two-dimensional domains with meas $S(x_1) > 0$ . If $\vec{v} \in \overset{o}{D}(\Omega)$ then it is well known that for almost all $x_1$ the following inequalities hold:

$$|j(x_1)|^2 \equiv \Big( \int_{S(x_1)} |\vec{v}(x_1,x_2,x_3)| dx_2 dx_3 \Big)^2 \leq \text{meas } S(x_1) ,$$

$$\int_{S(x_1)} \vec{v}^2(x) dx_2 dx_3 \leq c \text{ meas}^2 S(x_1) \int_{S(x_1)} \vec{v}_x^2(x) dx_2 dx_3 .$$

This together with the finiteness of the Dirichlet integral implies that $j(x_1^k) \to 0$ for a certain sequence $x_1^k$ , k=1,2,... of values of $x_1$ tending to infinity, provided

$$\int_a^\infty \lambda_1(x_1) \text{ meas}^{-1} S(x_1) dx_1 = \infty . \tag{12}$$

On the other hand, the flow $\int_{S(x_1)} v_3(x_1,x_2,x_3) dx_2 dx_3$ is independent of $x_1 > a$ . Consequently, it is equal to zero.

It turned out that the convergence of the integral (12) together with the Lipschitz condition on $\partial B$ already guarantee the property called by myself "sufficiently quick" extension of the exit $B$ . In the paper [5] we have weakened our original assumption for an exit to contain a cone in the sense just described.

Let us pass now to non-stationary problems

$$\vec{v}_t - \nu \Delta \vec{v} + \sum_{k=1}^{n} v_k \vec{v}_{x_k} = - \nabla \rho + \vec{f}(x,t) , \tag{13}$$

$$\text{div } \vec{v} = 0 , \quad \vec{v}\big|_{\partial\Omega} = 0 , \quad \vec{v}\big|_{t=0} = \vec{\varphi}(x) .$$

In my papers, and actually in papers of other authors as well it is assumed that $v \in L_2((0,T) , \overset{o}{J}{}_2^1(\Omega))$ . The space $\overset{o}{J}{}_2^1(\Omega)$ is the closure of the space $\overset{\to}{J}{}^\infty(\Omega)$ in the norm of $W_2^1(\Omega)$ . The scalar product in both $\overset{o}{J}{}_2^1(\Omega)$ and $\overset{o}{\vec{W}}{}_2^1(\Omega)$ is given by

$$(\vec{u},\vec{v})^{(1)} = \int_\Omega (\vec{u}\vec{v} + \vec{u}_x v_x) dx . \tag{14}$$

All elements $\vec{v}$ of the space $\overset{o}{\vec{W}}{}_2^1(\Omega)$ which satisfy the equality $\text{div } \vec{v} = 0$ form a subspace which will be denoted by $\hat{\overset{o}{J}}{}_2^1(\Omega)$ . It is easily seen that

$$\overset{o}{J}{}_2^1(\Omega) \subset \hat{\overset{o}{J}}{}_2^1(\Omega) \subset \hat{\overset{o}{W}}{}_2^1(\Omega) .$$

J.Heywood proved that $\hat{\mathring{J}}{}^1_2(\Omega)$ coincides with $\hat{\mathring{J}}{}^1_2(\Omega)$ for domains $\Omega$ with compact boundaries of the class $C^2$ as well as for the domain from $R^3$ which was described above in connection with his results. For an analogous domain in the plane (i.e. the whole $R^2$ except a straight line with some open segments cut in it) he raised the problem of finding $\dim \hat{\mathring{J}}{}^1_2 \big| \mathring{J}{}^1_2$ . Together with V.A.Solonnikov we have shown that for all three-dimensional domains of the above considered types we have $\dim \hat{\mathring{J}}{}^1_2 \big| \mathring{J}{}^1_2 = \dim \hat{H} \big|_H$ while in the planar case

$\hat{\mathring{J}}{}^1_2 \big| \mathring{J}{}^1_2$ . According to this result we can formulate the problem (13) for three-dimensional domains with $\dim \hat{\mathring{J}}{}^1_2 \big| \mathring{J}{}^1_2 = m-1 > 0$ in the following way:

to find a vector function $\vec{v}$ from the space $L_2((0,T); \hat{\mathring{J}}{}^1_2(\Omega))$ with the prescribed flows $\alpha_k(t)$ , $k=1,\dots,m-1$ through the "sufficiently quickly" extending exits and satisfying the integral identity corresponding to the Navier-Stokes system.

We omit here the integral identity which replaces the Navier--Stokes system. It is known that it can be written in various, nonetheless equivalent forms. The analogue of the condition (12) for the space $\hat{\mathring{J}}{}^1_2$ is $\int_a^\infty \text{meas}^{-1} S(x_1)dx_1 = \infty$ (this is clear from the above argument for elements of $\hat{H}$ and from the fact that the elements of $\hat{\mathring{J}}{}^1_2$ have a finite $L_2(\Omega)$-norm). It guarantees that the flow through an exit $B$ for any element from $\hat{\mathring{J}}{}^1_2(\Omega)$ is equal to zero. The divergence of the integral implies a "sufficiently quick" extension of the exit $B$ for the couple of spaces $\hat{\mathring{J}}{}^1_2$ and $\mathring{J}{}^1_2$ . The question of solvability of the problem (13) in the given formulation reduces by means of a substitution $\vec{v}(x,t) = \vec{u}(x,t) + \vec{a}(x,t)$ to the problem of finding the function $\vec{u}$ from the space $L_2((0,T); \mathring{J}{}^1_2(\Omega))$ . For the function $\vec{a}(x,t)$ we can take any element of $\hat{\mathring{J}}{}^1_2(\Omega)$ with the prescribed flows $\alpha_k$ , $k=1,\dots,m-1$ which are constructed in [3] . If $\alpha_k(t)$ possesses some smoothness properties in $t$ (e.g. $\alpha_k \in C^1([0,T])$) then the existence of $\vec{u}$ is established in essentially the same way as for the problem (13) in the space $L_2((0,T); \mathring{J}{}^1_2(\Omega))$ . In general, the problem (13) in the new formulation allows an application of the results by E.Hopf as well as their modifications and a majority of results proved by the author for the problem (13) in the original formulation (i.e. for $\alpha_k(t) \equiv 0$) . They are given together with their proofs in [1 , Chap.VI] .

## References

(All references except [2] in Russian)

[1] Ladyženskaja O.A.: Mathematical problems of the dynamics of viscous incompressible fluid. 1st edition Fizmatgiz 1961, 2nd edition Nauka 1970

[2] Heywood J.G.: On uniqueness questions in the theory of viscous flow. Acta Mathematica 136 (1976), 61-102

[3] Ladyženskaja O.A., Solonnikov V.A.: On some problems of vector analysis and generalized formulations of boundary value problems for Navier-Stokes equations. Zapiski nauč.sem. Leningr.otd.Matem.Inst.Akad.Nauk SSSR 59 (1976), 81-116

[4] Ladyženskaja O.A., Solonnikov V.A.: On the solvability of boundary and initial-boundary value problems for the Navier--Stokes equations on domains with non-compact boundaries. Vestnik Leningr.gosud.univ. 13 (1977), 39-47

[5] Pileckas K., Solonnikov V.A.: On some spaces of solenoidal vectors and on solvability of the boundary value problem for the system of Navier-Stokes equations on domains with non--compact boundaries. Zapiski nauč.sem.Leningr.otd.Matem. inst.Akad.nauk SSSR 1978

[6] Solonnikov V.A.: Solvability of the problem of planar motion of a viscous incompressible capillary fluid in a non-closed vessel. Preprint LOMI, R-5-77, Leningrad 1977

Author's address: Institut matematiki, Krasnoputilovskaja d. 2, 198152 Leningrad, USSR

# BOUNDARY VALUE PROBLEMS AT RESONANCE FOR VECTOR SECOND ORDER NONLINEAR ORDINARY DIFFERENTIAL EQUATIONS

J. Mawhin, Louvain-la-Neuve

## 1. INTRODUCTION

The aim of this paper is to use a version of the Leray-Schauder continuation theorem (see section 2) to prove the existence of solutions for second order vector ordinary differential equations with linear boundary conditions. The type of sharp sufficient conditions given in section 3 are motivated by the recent work of Kazdan and Warner [4], de Figueiredo and Gossez [2], and Brézis and Nirenberg [1] for semilinear scalar elliptic equations of the form

$$Lu = f(x, u),$$

and dealing with resonance problems at the first eigenvalue of $L$. We consider linear boundary value problems for systems of ordinary nonlinear differential of the form

$$x'' + f(t, x, x') = 0$$

with f verifying the Caratheodory conditions, allowing for f a dependence in x'. Moreover the technique of proof differs from the ones used in the papers quoted above and could be successfully applied to some of the problems considered there, as illustrated in [5]. In the special case of a system

$$x'' + f(t, x) = 0 \tag{1.1}$$

with boundary conditions

$$x(o) = x(\pi) = 0$$

$$(1.2) \qquad (\text{resp. } x(o) - x(2\pi) = x'(o) - x'(2\pi) = 0$$

the obtained conditions for existence of a solution are

$$(1.3) \quad F(t) := \overline{\lim_{|x|\to\infty}} \; |x|^{-2}(x|f(t,x) \leqslant 1 \;,\; \int_0^{\pi} F(t)\sin^2 t\,dt < \frac{\pi}{2}$$

$$(1.4) \qquad (\text{resp. } F(t) \leqslant 0 \;, \int_0^{2\pi} F(t)dt < 0)$$

(with $(\,|\,)$ the inner product in $\mathbb{R}^n$) so that $F(t)$ can be equal to the first eigenvalue 1 (resp. 0) of the associated linear problem

$$x'' + \mu x = 0$$

$$x(o) = x(\pi) = 0 \quad (\text{resp. } x(o) - x(2\pi) = x'(o) - x'(2\pi) = 0)$$

on any subset of $[o, \pi]$ (resp. $[o, 2\pi]$) of positive measure such that the second condition in (1.3) (resp. (1.4)) is verified. In the case of a first order vector equation

$$x' + f(t, x) = 0$$

it has been shown in [5] that the first condition in (1.4) could be replaced by the mere existence of $F(t)$ and the second one replaced by

$$\int_0^{2\pi} F(t)dt \neq 0$$

It is an open question to know if (1.4) for (1.1) - (1.2) could be generalized by

$$F(t) \leqslant a \;,\quad \int_0^{2\pi} F(t)dt < 0$$

for same $a > 0$. Let us finally notice that Theorem 2 gives existence conditions distinct of the ones of Hartman's type described for example in [3], Chapter V.

## 2. A USEFUL VERSION OF THE LERAY-SCHAUDER CONTINUATION THEOREM

Let $X$, $Z$ be real normed spaces, $L : \text{dom}\, L \subset X \to Z$ a linear Fredholm mapping of index zero and $N : X \to Z$ a (not necessary linear) mapping which is $L$-compact on bounded subsets of $X$ (shortly $L$-completely continuous on $X$). See [ 3 ] p. 12-13 for the corresponding definitions. We shall be interested in proving the existence of a solution for the abstract problem

$$Lx = Nx \tag{2.1}$$

Theorem 1. *Assume that there exist an open bounded set* $\Omega \subset X$ *and a linear* $L$*-completely continuous mapping* $A : X \to Z$ *such that* :

(i) $o \in \Omega$

(ii) $\ker (L - A) = \{0\}$

(iii) $Lx \neq (1 - \lambda) Ax + \lambda Nx$ *for all* $x \in \text{dom}\, L \cap \partial\Omega$ *and all* $\lambda \in ]0,1[$ .

*Then (2.1) has at least one solution in* $\text{dom}\, L \cap \overline{\Omega}$.

For a proof of this theorem, special cases of which have been used by many authors, see [5 ]. If $L$ is invertible, then a natural choice for $A$ is $A = 0$ and one gets the usual Leray-Schauder condition for mappings of the form $I - L^{-1}N$ and $L^{-1}N$ completely continuous.
If $L$ is not invertible, it is shown in [5 ] how to use this theorem to give more simple proofs of some results of Cesari and Kannan, De Figueiredo, Brézis and Nirenberg for abstract equations. One could get similarly the results of De Figueiredo and Gossez on elliptic problems.

## 3. AN EXISTENCE THEOREM FOR VECTOR SECOND ORDER DIFFERENTIAL EQUATIONS WITH LINEAR BOUNDARY CONDITIONS

Let $I = [a,b]$ be a compact internal of $\mathbb{R}$, $f : I \times \mathbb{R}^n \times \mathbb{R}^n \to \mathbb{R}^n$ a mapping satisfying the Caratheodory conditions.

Let (1.1) denote the Euclidian norm in $\mathbb{R}^n$ and $(u|v)$ the inner product of u and v.

Let now

$$\Lambda : C^1(I, \mathbb{R}^n) \to \mathbb{R}^n$$

be a linear continuous mapping such that

(A) For each $x \in \ker \Lambda$, $(x'(b) \mid x(b)) - (x'(a) \mid x(a)) = 0$

(B) The problem

$$x'' + \mu x = 0, \quad \Lambda(x) = 0$$

has a nontrivial solution if and only if $\mu = \mu_i$ $(i = 1, 2, \ldots)$ with

$$0 \leqslant \mu_1 < \mu_2 < \ \ldots.$$

of finite multiplicity and such that the corresponding eigenfunctions form a complete orthogonal set in $L^2(I, \mathbb{R}^n)$ with the inner product

$$< y,z > = \int_I (y(t) \mid z(t)) \, dt$$

(C) There exists $\gamma > 0$ such that for each $x \in \ker\Lambda$ such that

$$< x,u > = 0$$

for all u belonging to the eigenspace relative to $\mu_1$, one has

$$|x|_\infty \leqslant \gamma |x'|_2$$

where

$$|x|_\infty = \max_{t \in [a,b]} |x(t)| , \quad x' = dx/dt \quad \text{and}$$

$$|y|_2 = \left( \int_I |y(t)|^2 dt \right)^{1/2}$$

Important special cases of mappings $\Lambda$ satisfying (A) - (B) - (C) are given by

$\Lambda x = (x(a), x(b))$ (Dirichlet or Picard boundary conditions)

$\Lambda x = (x'(a), x'(b))$ (Neumann boundary conditions)

$\Lambda x = (x(a) - x(b), x'(a) - x'(b))$ (periodic boundary conditions)

$\Lambda x = (x(a), x'(b)), \ldots$

We shall be interested in proving the existence of solutions for the boundary value problem

$$x'' + f(t, x, x') = 0 \tag{3.1}$$

$$\Lambda(x) = 0, \tag{3.2}$$

i.e. the existence of mapping $x : I \to \mathbb{R}^n$ having an absolutely continuous first derivative and verifying (3.1) a.e. in $I$ and (3.2).

By taking $X = C^1(I, \mathbb{R}^n)$ with the norm

$$\|x\|_\infty = \max(|x|_\infty, |x'|_\infty),$$

$Z = L^1(I, \mathbb{R}^n)$ with the norm $|x|_1 = \int_I |x(t)| \, dt$, dom $L = \{x \in X : x'$ is absolutely continuous on $I$ and $\Lambda(x) = 0\}$, $L : \text{dom } L \subset X \to Z$, $x \mapsto -x''$, $N : X \to Z$, $x \mapsto f(., x, x')$, it is routine to check that $L$ is Fredholm of index zero and closed and that $N$ is $L$-completely continuous. The problem is then equivalent to an abstract one of type (2.1) and we shall prove the following existence theorem.

Theorem 2. *Let f be like above and let $\Lambda$ satisfy conditions (A), (B) and (C). Assume that the following conditions hold :*

$$F(t) := \overline{\lim_{|x| \to \infty}} \left( \sup_{u \in \mathbb{R}^n} \frac{(x \mid f(t, x, u))}{|x|^2} \right) \leqslant \mu_1 \tag{D}$$

*uniformly in* $t \in I$, *i.e. for each* $\varepsilon > 0$, *there exists* $\rho_\varepsilon > 0$ *and* $\delta_\varepsilon \in L^1(I, \mathbb{R}_+)$ *such that for a.e.* $t \in I$, *all* $x$ *with* $|x| \geqslant \rho_\varepsilon$ *and all* $u \in \mathbb{R}^n$, *one has*

$$(x \mid f(t, x, u)) \leqslant (\mu_1 + \varepsilon) |x|^2 + \delta_\varepsilon |x|$$

(E) For each $\varphi$ belonging to the eigenspace relative to $\mu_1$, one has

$$\int_I F(t)|\varphi(t)|^2 dt < \mu_1 \int_I |\varphi(t)|^2 dt$$

(F) For each $R > 0$ there exists $\varphi_R \in C^0(\mathbb{R}_+, \mathbb{R}_+^*)$ nondecreasing with $\lim_{s \to \infty} s^{-2}\varphi_R(s) = 0$ and such that for a.e. $t \in I$ and all $x$ with $|x| \leq R$, one has

$$|f(t, x, u)| \leq \varphi_R(|u|).$$

Then problem (3.1-3.2) has at least one solution.

Proof

We shall apply Theorem 1 with $Ax = ax$ and

$$a < \mu_1 \tag{3.3}$$

which clearly satisfies the $L$-complete continuity requirement and condition (ii). The inequality in (iii) becomes here

$$x'' + (1 - \lambda) ax + \lambda f(t, x, x') \neq 0 \tag{3.4}$$

and we shall first show that there exists $R > 0$ such that for any $\lambda \in ]0, 1[$ and any $x$ verifying $\Lambda(x) = 0$ and

$$|x|_\infty \geq R$$

(3.4) is satisfied. If it is not the case, there will exist a sequence $(\lambda_n)$ with $\lambda_n \in ]0, 1[$ $(n \in \mathbb{N})$ and a sequence $(x_n)$ of elements of dom $L$ such that

$$|x_n|_\infty \geq n$$

and

$$x_n'' + (1 - \lambda_n) ax_n + \lambda_n f((t, x_n, x'_n) = 0 \tag{3.5}$$

Using the assumption (A) (3.5) implies

$$|x'_n|_2^2 = (1 - \lambda_n)a|x_n|_2^2 + \lambda_n < x_n, f(., x_n, x'_n)> . \tag{3.6}$$

Let now $\varepsilon > 0$ ; Caratheodory conditions on f and assumption (D) insure the existence of $\tilde{\delta}_\varepsilon \in L^1(I, \mathbb{R}_+)$ such that for a.e. $t \in I$

$$(x_n(t) \mid f(t, x_n(t), x'_n(t)) \leq (\mu_1 + \varepsilon)|x_n(t)|^2 + \tilde{\delta}_\varepsilon(t)|x_n(t)|$$

which together with (3.3) and (3.6) gives

$$|x'_n|_2^2 \leq (\mu_1 + \varepsilon)\, |x_n|_2^2 + |\tilde{\delta}_\varepsilon|_1 |x_n|_\infty \tag{3.7}$$

Now let us write

$$x_n = u_n + v_n \tag{3.8}$$

where $u_n$ (resp. $v_n$) belongs (resp. is orthogonal) to the eigenspace associated to $\mu_1$. Then

$$|x'_n|_2^2 - \mu_1 |x_n|_2^2 = |v'_n|_2^2 - \mu_1 |v_n|_2^2$$

and we have the Wirtinger's inequality (which follows from assumption (B))

$$\mu_2 |v_n|_2^2 \leq |v'_n|_2^2 .$$

Those results together with (3.7) and assumption (C) imply that

$$|v_n|_\infty \leq \gamma\, \mu_2\, (\mu_2 - \mu_1)^{-1}\, [\, \varepsilon(b-a)^2 |x_n|_\infty^2 + |\tilde{\delta}_\varepsilon|_1 |x_n|_\infty \,]$$

and hence that

$$|x_n|_\infty^{-1}\, |v_n|_\infty \to 0 \quad \text{if} \quad n \to \infty. \tag{3.9}$$

Let now

$$y_n = x_n / |x_n|_\infty \quad , \quad n = 1, 2, \ldots .$$

Using (3.9), the finite dimension of the eigenspace associated to $\mu_1$ and going to subsequences, one can assume that (for the norm $|.|_\infty$),

$$y_n \to y \quad , \quad \lambda_n \to \lambda_0$$

with y in the eigenspace associated to $\mu_1$, $|y|_\infty = 1$ and $\lambda_0 \in [0,1]$.

From (3.6) and Wirtinger's inequality we then obtain, if $X_n$ denotes the characteristic function of the set

$$I_n = \{t \in I : |x_n(t)| \neq 0\} = \{t \in I : |y_n(t)| \neq 0\},$$

$$\lambda_n \int_I X_n(t)|y_n(t)|^2 \frac{\big(x_n(t)|f(t, x_n(t), x'_n(t)\big)}{|x_n(t)|^2} dt$$

$$\geq [\mu_1 - (1-\lambda_n)a]\,|y_n|_2^2$$

Hence, by assumption (D) and by Fatou's lemma,

$$\lambda_o \int_I \overline{\lim_{n\to\infty}} \left[ X_n(t)\;|y_n(t)|^2 \frac{\big(x_n(t)\mid f(t, x_n(t), x'_n(t)\big)}{|x_n(t)|^2} \right] dt + (1-\lambda_o)a|y|_2^2$$

(3.10)

$$\geq \mu_1 |y|_2^2$$

But, on $I_n$, $|x_n(t)| = |y_n(t)|\;|x_n|_\infty \to \infty$ if $n \to \infty$ and, as $y_n \to y$, $X_n \to 1$ a.e. in $I$, which implies together with (3.10) and assumption (D) that

$$\lambda_o \int_I F(t)\;|y(t)|^2\,dt + (1-\lambda_o)\,a \int_I |y(t)|^2\,dt$$

$$\geq \mu_1 \int_I |y(t)|^2\,dt,$$

a contradiction with assumption (E). Thus, if $x \in \operatorname{dom} L$ satisfies

(3.11) $$x'' + (1-\lambda)\,a\,x + \lambda f(t, x, x') = 0$$

for some $\lambda \in ]0,1[$, one has

$$|x|_\infty < R$$

and hence, for a.e. $t \in I$, using assumption (F),

$$|x''(t)| \leq |a|\,R + \varphi_R(|x'(t)|) = \tilde{\varphi}_R(|x'(t)|)$$

with $\tilde{\varphi}_R \in C^0(\mathbb{R}_+, \mathbb{R}_+^*)$ nondecreasing and $\lim_{s\to\infty} s^{-2}\,\tilde{\varphi}_R(s) = 0$.

By a result of Schmitt (see e.g. [ 3 ] p. 69) this implies the existence of $S = S(R) > 0$ such that any solution $x \in \operatorname{dom} L$ of (3.11) verifies

$$|x'|_\infty < S.$$

Therefore, if

$$\Omega = \{x \in C^1(I, \mathbb{R}^n) : |x|_\infty < R , |x'|_\infty < S\}$$

conditions (i) and (iii) of Theorem 1 are satisfied and the proof is complete.

## REFERENCES


1. H. BREZIS and L. NIRENBERG, Characterizations of the ranges of some nonlinear operators and applications to boundary value problems, Annali Scuola Norm. Sup., to appear.

2. D.G. de FIGUEIREDO and J.P. GOSSEZ, Perturbation non linéaire d'un problème elliptique linéaire près de sa première valeur propre, C.R. Acad. Sci. Paris (A) 284 (1977) 163-166.

3. R.E. GAINES and J. MAWHIN, "Coincidence Degree and Nonlinear Differential Equations", Lecture Notes in Math. vol. 568, Springer, Berlin, 1977.

4. J.L. KAZDAN and F.W. WARNER, Remarks on some quasilinear elliptic equations, Comm. Pure Applied Math. 28 (1975) 567-597.

5. J. MAWHIN, Landesman - Lazer's type problems for nonlinear equations, Conf. Semin. Mat. Univ. Bari n° 147, 1977.


Author's address : Université de Louvain, Institut Mathématique, B-1348 Louvain-la-Neuve, Belgique.

BEHAVIOUR OF SOLUTIONS TO THE DIRICHLET PROBLEM
FOR THE BIHARMONIC OPERATOR AT A BOUNDARY POINT

V.G. Maz'ya, Leningrad

1°. Introduction. According to the classical result by Wiener [1], [2] the regularity of a boundary point 0 for the Laplace equation in a domain $\Omega \subset R^n$, $n > 2$ is equivalent to the divergence of the series

$$\sum_{k=1}^{\infty} 2^{k(n-2)} \operatorname{cap}(C_{2^{-k}} \setminus \Omega)$$

where $C_\rho = \{x \in R^n : \rho/2 \leq |x| \leq \rho\}$ and cap is the harmonic capacity. Wiener's theorem was extended (sometimes only with respect to sufficiency) to different classes of linear and quasilinear second order partial differential equations ([3] - [11] and others). However, results of this type for higher order equations seem to be unknown.

In the present paper we study the behaviour near a boundary point of solutions to the Dirichlet problem with zero boundary data for the equation $\Delta^2 u = f$, $f \in C_0^\infty(\Omega)$, $\Omega \subset R^n$. The proof covers only dimensions $n = 4,5,6,7$ (the case $n < 4$ is not interesting). We show in particular that the condition

$$\sum_{k=1}^{\infty} 2^{k(n-4)} \operatorname{cap}_2(C_{2^{-k}} \setminus \Omega) = \infty \ , \quad n = 5,6,7,$$

where $\operatorname{cap}_2$ is the so called biharmonic capacity, guarantees the continuity of the solution at the point 0. This result follows from an estimate of the modulus of continuity. Such estimates, formulated in terms of the rate of divergence of Wiener's series were known only for second order equations ([12], [7], [9], [13]).

In the last section we obtain some pointwise estimates for the Green function $G(x,y)$ of the Dirichlet problem for $\Delta^2$ valid without any restrictions on the boundary $\partial\Omega$. In particular it is proved that $|G(x,y)| \leq c|x-y|^{4-n}$ where $n = 5,6,7$ and $c$ is a positive constant depending only on $n$.

The author takes pleasure in thanking E.M. Landis for stimulating discussions.

2°. Preliminaries and definitions. Let $\Omega$ denote an open subset of Euclidean space $R^n$ with a compact closure $\bar{\Omega}$ and a boundary $\partial\Omega$. Let 0 be a point of $\partial\Omega$ and $B_\rho = \{x : |x| < \rho\}$, $C_\rho = B_\rho \setminus \bar{B}_{\rho/2}$. We denote by $c, c_1, \dots$ positive constants depending only on $n$ and write $\nabla_\ell = \{\partial^\ell / \partial x_1^{\alpha_1} \dots \partial x_n^{\alpha_n}\}$, $\nabla_1 = \nabla$. We consider only real functions.

Let $\overset{\circ}{W}{}_2^2(\Omega)$ be the closure of the space $C_0^\infty(\Omega)$ in the norm $\|\nabla_2 u\|_{L_2(\Omega)}$.

We introduce the biharmonic capacity of a compact $e$ with respect to an open domain $G$, $G \supset e$:

$$\mathrm{cap}_2(e;G) = \inf\Big\{ \int_G |\nabla_2 u|^2\, dx : u \in C_0^\infty(G),$$

$$u = 1 \text{ in a neighbourhood of } e \Big\}.$$

We write $\mathrm{cap}_2(e)$ instead of $\mathrm{cap}_2(e;R^n)$.

Let $\Gamma$ denote the fundamental solution for the biharmonic operator, i.e.

$$\Gamma(x) = \frac{|x|^{4-n}}{2(n-4)(n-2)\omega_n} \quad \text{if } n>4, \tag{1}$$

$$\Gamma(x) = (4\omega_4)^{-1} \log \frac{d}{|x|} \quad \text{if } n = 4,$$

where $\omega_n = \mathrm{mes}_{n-1}\, \partial B_1$ and $d$ is a constant.

3°. "Weighted" positivity of $\Delta^2$.

Lemma 1. Let $u \in \overset{\circ}{W}{}_2^2(\Omega) \cap C^\infty(\Omega)$ and $4 \le n \le 7$. Then for every point $p \in \Omega$ (and in the case $n = 4$ for any $d$ satisfying $d \ge \mathrm{diam}\,(\mathrm{supp}\, u)$) we have

$$u(p)^2 + c\int_\Omega \Big[(\nabla_2 u(x))^2 + \frac{(\nabla u(x))^2}{|p-x|^2}\Big]\, \Gamma(x-p)\, dx \le \tag{2}$$

$$\le 2\int_\Omega \Delta u(x).\Delta(u(x)\,\Gamma(x-p))\, dx.$$

Proof. Let $(r,\omega)$ be the spherical coordinates with the center $p$ and let $G$ denote the image of $\Omega$ under the mapping $x \to (t,\omega)$ where $t = -\log r$. Since

$$r^2\Delta u = r^{2-n}(r\,\partial/\partial r)\,[r^{n-2}(r\,\partial/\partial r)u] + \delta_\omega u$$

where $\delta_\omega$ is the Beltrami operator on the unit sphere $S^{n-1}$ we get for the function $v(t,\omega) = u(x)$

$$r^2\Delta u = v_{tt} - (n-2)v_t + \delta_\omega v = Lv.$$

Consider first the case $n>4$. By a simple computation

$$c(n)\int_\Omega \Delta u(x).\Delta(u(x)\,\Gamma(x-p))\, dx = \int_G e^{(4-n)t} Lv.L(ve^{(n-4)t})\, dt\, d\omega = \tag{3}$$

$$= \int_G (v_{tt}-(n-2)v_t+\delta_\omega v)(v_{tt}+(n-6)v_t-2(n-4)v+\delta_\omega v)\, dt d\omega$$

where $c(n) = 2(n-2)(n-4)\omega_n$. We remark that

$$(4) \qquad 2\int_G v_t v \, dt\, d\omega = \int_{S^{n-1}} v(\infty,\omega)^2 \, d\omega = \omega_n u(p)^2.$$

The following identities are also obvious:

$$(5) \qquad \int_G v_t \delta_\omega v \, dt\, d\omega = 0, \qquad \int_G v_t v_{tt} \, dt\, d\omega = 0.$$

Thus the last integral in (3) becomes

$$(6) \qquad \int_G [v_{tt}^2-(n-2)(n-6)v_t^2-2(n-4)v_{tt}v+2v_{tt}\delta_\omega v+(\delta_\omega v)^2- \\ - 2(n-4)v\delta v]\, dt\, d\omega + \frac{c(n)}{2} u(p)^2.$$

After integrating by parts we rewrite (6) as

$$(7) \qquad \int_G \{v_{tt}^2+(\delta_\omega v)^2+2v_t(-\delta_\omega v_t)+2(n-4)v(-\delta_\omega v) + \\ + [5-(n-5)^2]v_t^2\}\, dt\, d\omega + \frac{c(n)}{2} u(p)^2.$$

Using the former variables $(r,\omega)$ we obtain

$$\int_\Omega [u_{rr}^2+ \frac{2}{r^2}(\nabla_\omega u_r)^2+2\frac{n-4}{r^4}(\nabla_\omega u)^2+\frac{(7-n)(n-3)}{r^2}u_r^2]\frac{dx}{r^{n-4}} + \frac{c(n)}{2}u(p)^2.$$

This completes the proof of (2) for $n = 5,6$. In the case $n = 7$ one can use the inequality

$$\int_\Omega u_{rr}^2 \frac{dx}{r^{n-4}} \geq \int_\Omega u_r^2 \frac{dx}{r^{n-2}}$$

which is a corollary of the one-dimensional inequality

$$\int_0^\infty w(r)^2 r\, dr \leq \int_0^\infty w'(r)^2 r^{(3)}\, dr.$$

Now let $n = 4$. We have

$$\int_\Omega 4\omega_4 \; \Delta u(x).\Delta(u(x)\,\Gamma(x-p))\, dx = \int_\Omega \Delta u(x)\Delta(u(x)\log\frac{d}{|x-p|})\, dx = \\ = \int_G Lv.L((\ell+t)v)\, dt\, d\omega$$

where $\ell = \log d$. The last integral is equal to

$$(8) \qquad \int_G (\ell + t)(Lv)^2 \, dt \, d\omega + 2 \int_G (v_t - v) Lv \, dt \, d\omega .$$

Applying (4) and (5) we rewrite (8) in the form

$$(9) \quad \int_G (\ell + t)(Lv)^2 \, dt \, d\omega + 2 \int_G [(\nabla_\omega v)^2 - v_t^2] \, dt \, d\omega + 2\omega_4 u(p)^2 .$$

For the first integral in (9) we have

$$\int_G (\ell + t)(Lv)^2 \, dt \, d\omega = \int_G [v_{tt}^2 + 4v_t^2 + (\delta_\omega v)^2](\ell + t) \, dt \, d\omega +$$

$$+ 2 \int_G (v_{tt} \delta_\omega v - 2v_t \delta_\omega v - 2v_{tt} v_t)(\ell + t) \, dt \, d\omega ,$$

and integrating by parts, we get

$$\int_G (\ell + t)(Lv)^2 \, dt \, d\omega = \int_G [v_{tt}^2 + 4v_t^2 + (\delta_\omega v)^2 + 2(\nabla_\omega v_t)^2](\ell + t) \, dt \, d\omega -$$

$$- 2 \int_G [(\nabla_\omega v)^2 - v_t^2] \, dt d\omega \quad .$$

Therefore

$$4\omega_4 \int_\Omega \Delta u \circ \Delta(u \Gamma) \, dx = \int_G [v_{tt}^2 + 4v_t^2 + (\delta_\omega v)^2 +$$

$$+ 2(\nabla_\omega v_t)^2] (\ell + t) \, dt \, d\omega + 2\omega_4 u(p)^2 .$$

This identity together with the following easily checked one

$$\int_{S^{n-1}} (\delta_\omega v)^2 \, d\omega \geq (n-1) \int_{S^{n-1}} (\nabla_\omega v)^2 \, d\omega$$

implies

$$2 \int_\Omega \Delta u \circ \Delta(u \Gamma) \, dx \geq c \int_G [(\nabla_2 v)^2 + (\nabla v)^2] (\ell + t) \, dt \, d\omega +$$

$$+ u(p)^2 \geq c \int_\Omega [(\nabla_2 u)^2 + \frac{(\nabla u)^2}{|x-p|^2}] \log \frac{d}{|x-p|} \, dx + u(p)^2 .$$

The proof is complete.

Lemma 1 fails for $n \geq 8$. Indeed, let the function $u \in C_0^\infty(\Omega \setminus p)$ depend only on $r = |x-p|$. Then (see [7])

$$c(n)\int_{\Omega} \Delta u(x)\circ\Delta(u(x)\,\Gamma(x-p))\,dx = \omega_n\int_{-\infty}^{+\infty} v_{tt}^2\,dt - c\int_{-\infty}^{+\infty} v_t^2\,dt$$

where $v(t) = u(e^{-t})$. Therefore rhe estimate (2) is impossible.

$4^{\circ}$. Local estimates. In the next lemma and henceforth we use the notation:

$$M_{\rho}(u) = \rho^{-n}\int_{\Omega\cap C_{2\rho}} u^2\,dx,$$

$$N_{\rho}(u) = \int_{\Omega\cap B_{2\rho}} [(\nabla_2 u)^2 + \frac{(\nabla u)^2}{|x-p|^2}]\,\Gamma dx$$

where $\Gamma = \Gamma(x-p)$ and we set $d = 3\rho$ for the case $n = 4$ in the definition of $\Gamma$.

Lemma 2. Let $\eta \in C_0^{\infty}(B_{2\rho})$, $\eta = 1$ in a neighbourhood of the ball $B_{\rho}$; $u \in \overset{0}{W}{}_2^2(\Omega)\cap C_0^{\infty}(\Omega)$. Then for any point $p \in B_{\rho/2}$

$$\int_{\Omega} \Delta(\eta^2 u)\Delta(\eta^2 u\,\Gamma)\,dx \le \int_{\Omega} \Delta u.\Delta(\eta^4 u\,\Gamma)\,dx + \tag{10}$$

$$+ c\,M_{\rho}(u)^{1/2}\,N_{\rho}(\eta^2 u)^{1/2} + c\,M_{\rho}(u).$$

Proof. Since

$$\Delta(\eta^2 u)\Delta(\eta^2 u\,\Gamma) - \Delta u.\Delta(\eta^4 u\,\Gamma) =$$
$$= [\Delta,\eta^2]u.\Delta(\eta^2 u\,\Gamma) - \Delta u.[\Delta,\eta^2]\eta^2 u\,\Gamma =$$
$$= [\Delta,\eta^2]u.[\Delta,\eta^2\,\Gamma] - \Delta u.\,[[\Delta,\eta^2].\eta^2\,\Gamma]u$$

(the square brackets denote the commutator of operators), we must estimate the difference of the integrals

$$i_1 = \int_{\Omega}[\Delta,\eta^2]u.[\Delta,\eta^2\,\Gamma]u\,dx, \quad i_2 = \int_{\Omega}\Delta u.[[\Delta,\eta^2],\eta^2\,\Gamma]u\,dx.$$

We begin with the estimate of $i_2$. Clearly

$$[[\Delta,\eta^2],\eta^2\Gamma]u = 2u\nabla\eta^2\nabla(\eta^2\,\Gamma) = 4u\eta^2(2\,\Gamma(\nabla\eta)^2 + \eta\,\nabla\eta\nabla\Gamma).$$

Hence

$$i_2 = \int_{\Omega} u\,\Delta(\varphi_2\eta^2 u)\,dx, \tag{11}$$

where $\varphi_2 = 4(2\,\Gamma(\nabla\eta)^2 + \eta\,\nabla\eta\cdot\nabla\,\Gamma)$. In general, we denote further by $\varphi_i$ the functions from $C_0^{\infty}(B_{2\rho}\setminus\bar{B}_{\rho})$ satisfying

$$|\nabla_k \varphi_i| \leq c\rho^{i-n-k}, \qquad k = 0,1, \dots$$

The inequality

$$|i_2| \leq c\, M_\rho(u)^{1/2} N_\rho(\eta^2 u)^{1/2} + c M_\rho(u)$$

is a straightforward consequence of (11). Now we pass to the estimate of $i_1$. Since

$$[\Delta, \eta^2]u \circ [\Delta, \eta^2 \Gamma]u =$$

$$= (4\eta\nabla\eta\circ\nabla u + u\Delta\eta^2)(2\nabla u\circ\nabla(\eta^2\Gamma) + u\Delta(\eta^2\Gamma)),$$

we have

$$(12) \qquad i_1 = 8\int_\Omega (\nabla u.\nabla\eta)\eta(\nabla(\eta^2\Gamma).\nabla u)\, dx + \int_\Omega \varphi_0 u^2\, dx,$$

where $\varphi_0 = \Delta\eta^2.\Delta(\eta^2\Gamma) - \operatorname{div}(\Delta\eta^2\circ\nabla(\eta^2\Gamma)) - 2\operatorname{div}(\Delta(\eta^2\Gamma).\eta\nabla\eta)$. The first term on the right hand side of (12) can be written in the form

$$i_1 = 8\int_\Omega (\nabla u\circ\nabla\eta)(2\Gamma\nabla\eta + \eta\nabla\Gamma).\nabla(\eta^2 u)\, dx +$$

$$+ 8\int_\Omega u^2 \operatorname{div}\{(\nabla\eta.\nabla(\eta^2\Gamma))\nabla\eta\}\, dx =$$

$$= \int_\Omega u \operatorname{div}(\varphi_2\nabla(\eta^2 u))\, dx + \int_\Omega u^2\varphi_0\, dx.$$

Hence

$$|i_i| \leq c\, M_\rho(u)^{1/2} N_\rho(\eta^2 u)^{1/2} + c M_\rho(u),$$

which completes the proof.

Using Lemmas 1 and 2 we get

Corollary 1. Let $4 \leq n \leq 7$, $u \in \mathring{W}_2^2(\Omega)$, $\Delta^2 u = 0$ in $\Omega \cap B_{2\rho}$. Then for all points $p \in B_{\rho/2}$

$$(13) \quad u(p)^2 + \int_{\Omega\cap B_\rho} ((\nabla_2 u)^2 + |x-p|^{-2}(\nabla u)^2)\Gamma(x-p)\, dx \leq c\, M_\rho(u).$$

Corollary 2. Let $4 \leq n \leq 7$ and let the function $u \in \mathring{W}_2^2(\Omega)$ satisfy the equation $\Delta^2 u = 0$ in $\Omega\setminus B_\rho$. Then for all points $p \in \Omega\setminus B_{2\rho}$,

$$(14) \qquad |u(p)| \leq c\left(\frac{\rho}{|p|}\right)^{n-4} M_\rho(u)^{1/2}.$$

Proof. Let $G$ be the image of $\Omega$ under the inversion $p \to p|p|^{-2}$. We make use of the Kelvin transform $U(q) = |q|^{4-n} u(q|q|^{-2})$

which maps $u$ into a biharmonic function in $G \cap B_{\rho^{-1}}$. One can easily see that the Kelvin transform preserves the class $\overset{\circ}{W}{}_2^2$. By the inequality (13) for all points $q \in G \cap B_{(2\rho)^{-1}}$

$$U(q)^2 \leq c\rho^n \int_{B_{2\rho^{-1}} \setminus B_{\rho^{-1}}} U(y)^2\, dy$$

or which is the same,

$$|q|^{2(4-n)} u(q|q|^{-2})^2 \leq c\rho^n \int_{B_{2\rho^{-1}} \setminus B_{\rho^{-1}}} |y|^{2(4-n)} u(y|y|^{-2})^2\, dy.$$

Setting here $p = q|q|^{-2}$, $x = y|y|^{-2}$ we obtain the estimate (14).

$5^{\circ}$ Local estimates in terms of capacity.

Lemma 3. Let $4 \leq n \leq 7$ and let the function $u \in \overset{\circ}{W}{}_2^2$ satisfy the equation $\Delta^2 u = 0$ in $\Omega \cap B_{2\rho}$. Then for all points $p \in B_{\rho/2}$

$$u(p)^2 + \int_{\Omega \cap B_\rho} ((\nabla_2 u)^2 + |x-p|^{-2}(\nabla u)^2)\,\Gamma(x-p)\, dx \leq \tag{15}$$

$$\leq \frac{c}{\gamma(\rho)} \int_{\Omega \cap C_{2\rho}} ((\nabla_2 u)^2 + |x-p|^{-2}(\nabla u)^2)\,\Gamma(x-p)\, dx.$$

where $\gamma(\rho) = \rho^{4-n} \mathrm{cap}_2(C_{2\rho} \setminus \Omega)$ for $n > 4$ and $\gamma(\rho) = \mathrm{cap}_2(C_{2\rho} \setminus \Omega; B_{2\rho})$ for $n = 4$; in the case $n = 4$ we set $d = 3\rho$ in the definition of the fundamental solution.

Proof. The results of [14], [15] imply

$$\int_{\Omega \cap C_{2\rho}} u^2\, dx \leq \frac{c\rho^4}{\gamma(\rho)} \int_{\Omega \cap C_{2\rho}} ((\nabla_2 u)^2 + \rho^{-2}(\nabla u)^2)\, dx.$$

Noting that $\rho \geq c|x-p|$, $\Gamma(x-p) \geq c\rho^{4-n}$ for $x \in C_{2\rho}$, $p \in B_{\rho/2}$ and using Corollary 1 we complete the proof.

Lemma 4. Under the conditions of Lemma 3 for $2r < \rho$ it holds

$$\int_{\Omega \cap B_r} [(\nabla_2 u)^2 + |x|^{-2}(\nabla u)^2] \frac{dx}{|x|^{n-4}} \leq c\, M_\rho(u) \exp\Big(-c \int_r^\rho \gamma(\tau) \frac{d\tau}{\tau}\Big). \tag{16}$$

Proof. By (15), for sufficiently small $\varepsilon > 0$ and $r \leq \rho$

$$\int_{\Omega \cap (B_r \setminus B_\varepsilon)} ((\nabla_2 u)^2 + |x-p|^{-2}(\nabla u)^2)\,\Gamma(x-p)\, dx \leq$$

$$\leq \frac{c}{\gamma(r)} \int_{\Omega\cap C_{2r}} ((\nabla_2 u)^2 + |x-p|^{-2}(\nabla u)^2)\Gamma(x-p)\, dx.$$

Taking limits with $p \to 0$ and then with $\varepsilon \to +0$ we get

$$\int_{\Omega\cap B_r} ((\nabla_2 u)^2 + |x|^{-2}(\nabla u)^2)|x|^{4-n}\, dx \leq$$

$$\leq \frac{c_1}{\gamma(r)} \int_{\Omega\cap C_{2r}} ((\nabla u)^2 + |x|^{-2}(\nabla u)^2)|x|^{4-n}\, dx.$$

We denote the left hand side of this inequality by $\psi(r)$ and set $r = 2^{-k}$. Then

$$(1+c_2\gamma(2^{-k}))\psi(2^{-k}) \leq \psi(2^{1-k}). \tag{17}$$

Since $\gamma$ is a bounded function, the estimate (17) is equivalent to

$$\psi(2^{-k}) \leq \exp[-c_3\gamma(2^{-k})]\,\psi(2^{1-k}).$$

So for $m > \ell$

$$\psi(2^{-m}) \leq \exp[-c_3 \sum_{j=m}^{\ell-1} \gamma(2^{-j})]\ \psi(2^{-\ell}). \tag{18}$$

Let numbers $m$ and $\ell$ satisfy the inequalities $2^{-m-1} \leq r \leq 2^{-m}$ and $2^{-\ell} \leq \rho \leq 2^{1-\ell}$. Then (18) and (13) yield

$$\psi(r) \leq c \exp\left[-c_3 \sum_{j=m}^{\ell-1} \gamma(2^{-j})\right] M_\rho(u).$$

Using simple properties of the biharmonic capacity (see for example [15]) we obtain (16) from the last estimate.

$6^\circ$. Regularity of a boundary point. We say that a point $0 \in \partial\Omega$ is regular for the biharmonic operator if the solution $u \in \overset{\circ}{W}{}_2^2(\Omega)$ of the equation $\Delta^2 u = f$ with an arbitrary right hand side from $C_0^\infty(\Omega)$ is continuous at $0$.

Theorem 1. Let $4 \leq n \leq 7$ and

$$\int_0 \gamma(\tau)\,\frac{d\tau}{\tau} = \infty \tag{19}$$

where $\gamma$ is the function introduced in Lemma 3. Then the point $0$ is regular for $\Delta^2$. Moreover if $u \in \overset{\circ}{W}{}_2^2(\Omega)$ and $\Delta^2 u = 0$ in $\Omega \cap B_{2\rho}$ for some $\rho > 0$ then there exists a constant $c$ such that

$$\text{(20)} \qquad \lim_{r \to 0} \exp\Big(c \int_r^{\rho} \gamma(\tau) \frac{d\tau}{\tau}\Big) \sup_{|p|<r} |u(p)| = 0.$$

Proof. According to (15) we have for all $p \in B_{r/2}$ with $r \leq \rho$

$$\text{(21)} \qquad u(p)^2 \leq \frac{c}{\gamma(r)} \int_{\Omega \cap C_{2r}} ((\nabla_2 u)^2 + |x|^{-2}(\nabla u)^2)|x|^{4-n}\, dx.$$

Let $S(r) = \sup\{u(p)^2 : p \in B_{r/2}\}$. From (21) it follows that

$$\int_0^{r/2} S(\tau)\gamma(\tau)\frac{d\tau}{\tau} \leq c \int_0^{r} \frac{d\tau}{\tau} \int_{\Omega \cap C_{2\tau}} ((\nabla_2 u)^2 + |x|^{-2}(\nabla u)^2)|x|^{4-n} dx =$$

$$= c \int_0^{r/2} \frac{d\tau}{\tau} \int_{\tau}^{2\tau} R^3\, dR \int_{S^{n-1}} ((\nabla_2 u)^2 + R^{-2}(\nabla u)^2)\, d\omega$$

which by the change of integration order becomes

$$\int_0^{r/2} S(\tau)\gamma(\tau)\frac{d\tau}{\tau} \leq c \int_{\Omega \cap B_r} ((\nabla_2 u)^2 + |x|^{-2}(\nabla u)^2)|x|^{4-n}\, dx.$$

Using this estimate and Lemma 4 we obtain

$$\text{(22)} \qquad \int_0^{r/2} S(\tau)\gamma(\tau)\frac{d\tau}{\tau} \leq c\, M_{\rho}(u) \exp\Big(-c \int_r^{\rho} \gamma(\tau)\frac{d\tau}{\tau}\Big).$$

Let

$$\xi(\tau) = \int_{\tau}^{\rho} \gamma(t)\frac{dt}{t}.$$

The inequality (22) assumes the form

$$\int_{\xi(r/2)}^{\infty} S(\tau(\xi))\, d\xi \leq c\, M_{\rho}(u)\exp(-c\,\xi(r)).$$

Since the function $\xi \to S(\tau(\xi))$ decreases and $\xi(r) \geq \xi(r/2) - c$, $c > 0$ we conclude

$$\xi(r/2) S(\xi^{-1}(2\xi(r/2))) \leq \int_{\xi(r/2)}^{2\xi(r)} S(\tau(\xi))d\xi \leq c\, M_{\rho}(u)\exp(-c\,\xi(r/2)),$$

where $\xi^{-1}$ is the inverse function to $\xi(\tau)$. We set $R = \xi^{-1}(2\xi(r/2))$. Then

$$\xi(R)\exp\Big(\frac{c}{2}\xi(R)\Big)S(R) \leq 2c\, M_{\rho}(u)$$

for all $R \leq \xi^{-1}(2\xi(\rho/4))$. Therefore

$$\lim_{R \to 0} \exp\left(\frac{c}{4}\xi(R)\right)S(R) = 0.$$

The result follows.

An immediate consequence of Theorem 1 is

Corollary 3. If $4 \leq n \leq 7$ and

$$\lim_{r \to 0} \frac{1}{\log \frac{1}{r}} \int_r^{\rho} \gamma(\tau) \frac{d\tau}{\tau} > 0$$

then the solution $u \in \mathring{W}_2^2(\Omega)$ of the equation $\Delta^2 u = f$ with $f \in C_0^\infty(\Omega)$ satisfies the inequality $|u(x)| \leq c|x|^\alpha$, $\alpha > 0$ in a neighbourhood of $0$.

<u>$7^0$. Examples of regular points for $\Delta^2$.</u> The proof of the following assertions can be performed in the same way as the proofs of analogous facts for $(p,1)$-capacity in [9], p. 53-55.

If $n = 4$ and the point $0$ belongs to a continuum which is a part of $R^n \setminus \Omega$ then $\gamma(\tau) \geq \text{const} > 0$ and consequently the condition of Corollary 3 holds.

Let the exterior of $\Omega$ in a neighbourhood of the point $0$ contain the domain $\{x: 0 < x_n < 1,\ x_1^2 + \dots + x_{n-1}^2 < f(x_n)^2\}$, where $f(t)$ is an increasing positive continuous function on $(0,1)$ such that $f(0) = f'(0) = 0$. Then $\gamma(\tau) \geq c|\log f(\tau)|^{-1}$ for $n = 5$ and $\gamma(\tau) \geq c[\tau^{-1}f(\tau)]^{n-5}$ for $n > 5$.

Hence the point $0$ is regular for $\Delta^2$, if

$$\int_0 |\log f(\tau)|^{-1}\tau^{-1}\, d\tau = \infty \qquad \text{for } n = 5,$$

$$\int_0 [\tau^{-1}f(\tau)]^{n-5}\tau^{-1}\, d\tau = \infty \qquad \text{for } n = 6,7.$$

<u>$8^0$. Estimates for the Green function.</u> Let $G(x,y)$ be the Green function of the Dirichlet problem for the biharmonic operator.

<u>Theorem 2.</u> Let $5 \leq n \leq 7$ and $d_y = \text{dist}(y, \partial\Omega)$. Then

$$|G(x,y) - \Gamma(x-y)| \leq c\, d_y^{4-n} \qquad \text{if } |x-y| \leq d_y, \tag{23}$$

$$|G(x,y)| \leq c|x-y|^{4-n} \qquad \text{if } |x-y| > d_y,$$

and consequently $|G(x,y)| \leq c|x-y|^{4-n}$ for all $x \in \Omega$, $y \in \Omega$.

Proof. Let $B(y) = \{x: |x-y| < d_y\}$ and $aB(y) = \{x: |x-y| < ad_y\}$.

We denote by $\eta$ a function from $C_0^\infty[0,1)$ equal to unity on the segment $[0,1/2)$ and set

$$H(x,y) = G(x,y) - \eta\left(\frac{|x-y|}{d_y}\right)\Gamma(x-y).$$

Obviously the function $x \to H(x,y)$ belongs to the class $\overset{\circ}{W}{}_2^2(\Omega) \cap C^\infty(\Omega)$, the support of the function $x \to \Delta_x^2 H(x,y)$ lies in $B(y) \setminus \frac{1}{2} B(y)$ and $|\Delta_x^2 H(x,y)| \le d_y^{-n}$. Applying Lemma 1 to the function $x \to H(x,y)$ we get

$$H(p,y)^2 \le 2 \int_{B(y)\cap\Omega} \Delta_x^2 H(x,y) . H(x,y)\,\Gamma(x-p)\,dx.$$

Therefore

$$\sup_{p\in 2B(y)\cap\Omega} H(p,y)^2 \le \tag{24}$$

$$\le \sup_{x\in B(y)\cap\Omega} |H(x,y)| \sup_{p\in 2B(y)\cap\Omega} \int_{B(y)\cap\Omega} |\Delta_x^2 H(x,y)|\,\Gamma(x-p)\,dx,$$

and hence

$$\sup_{p\in 2B(y)\cap\Omega} |H(p,y)| \le c d_y^{-n} \sup_{p\in 2B(y)\cap\Omega} \int_{B(y)\cap\Omega} \Gamma(x-p)\,dx \le c\, d_y^{4-n} . \tag{25}$$

Since $\Delta_p^2 H(p,y) = 0$ for $p \bar{\in} B(y)$ we obtain from (25) and Corollary 2 (in which $O$ must be substituted by $p$) for $p \bar{\in} 2B(y)$

$$|H(p,y)| \le c \left(\frac{d_y}{|p-y|}\right)^{n-4} \sup_{x\in 2B(y)\cap\Omega} |H(x,y)| \le c|p-y|^{4-n}.$$

The result follows.

Theorem 3. Let $n = 4$, $d_y = \operatorname{dist}(y,\partial\Omega)$, let $\Omega$ be a domain with a diameter $\mathcal{D}$ and

$$\Gamma(x-y) = (4\omega_4)^{-1}\log\frac{\mathcal{D}}{|x-y|} .$$

Then

$$|G(x,y) - \Gamma(x-y)| \le c_1 \log\frac{\mathcal{D}}{d_y} + c_2 \qquad \text{if } |x-y| \le d_y,$$

$$|G(x,y)| \le c_3 \log\frac{\mathcal{D}}{d_y} + c_4 \qquad \text{if } |x-y| > d_y.$$

Proof. Proceeding in the same way as in the proof of Theorem 2 we come to (24). Hence

$$\sup_{p \in 2B(y)\cap\Omega} |H(p,y)| \leq c d_y^{-4} \sup_{p \in 2B(y)\cap\Omega} \int_{B(y)\cap\Omega} \Gamma(x-p)\, dx \leq$$

$$\leq c_1 \log \frac{\mathcal{D}}{d_y} + c_2$$

which together with Corollary 2 gives for $p \overline{\in} 2B(y)$

$$|H(p,y)| \leq c \sup_{x \in 2B(y)\cap\Omega} |H(p,y)| \leq c(c_1 \log \frac{\mathcal{D}}{d_y} + c_2).$$

Since $G(p,y) = H(p,y)$ for $p \overline{\in} 2B(y)$ the result follows.

References


[1] N. Wiener, Certain notions in potential theory, J. Math. and Phys. 3 (1924), 24-51.

[2] N. Wiener, The Dirichlet problem, J. Math. and Phys., 3 (1924), 127-146.

[3] O.A. Oleynik, On Dirichlet problem for elliptic equations, Matem. Sb., 24, N 1 (1949), 3-14 (Russian).

[4] R.M. Hervé, Recherches axiomatiques sur la theorie des fonctions surharmoniques et du potentiel, Ann. Inst. Fourier, 12 (1962), 415-517.

[5] N.V. Krylov, On the first boundary value problem for elliptic equations, Diff. Uravneniya 3, N 2 (1967), 315-325 (Russian).

[6] W. Littman, G. Stampacchia, H.F. Weinberger, Regular points for elliptic equations with discontinuous coefficients. Ann. Scuola Norm. Sup. Pisa, 17 (1963), 43-77.

[7] E.M. Landis, The second order elliptic and parabolic equations, "Nauka", Moscow, 1971 (Russian).

[8] A.A. Novruzov, On regularity of boundary points for elliptic equations with continuous coefficients, Vestn. Moskovskogo Universiteta, N 6 (1971), 18-25 (Russian).

[9] V.G. Maz'ja, On continuity at a boundary point of the solution of quasilinear elliptic equations. Vestnik Leningradskogo Universiteta, 25 (1970), 42-55 (Russian).

[10] R. Gariepy, W. Ziemer, A gradient estimate at the boundary for solutions of quasilinear elliptic equations, Bull. Amer. Math. Soc., 82 (1976), 629-631.

[11] A.A. Novruzov, On some questions of qualitative theory of linear and quasilinear elliptic equations of second order. Doklady AN SSSR, 204, N 5 (1972), 1053-1056. (Russian).

[12] V.G. Maz'ja, On the behavior near the boundary of the Dirichlet problem for second order elliptic equations in divergence form. Matem. Zametki, 2, N 2 (1967), 209-220 (Russian).

[13] A.A. Novruzov, On the modulus of continuity of the Dirichlet problem at a regular boundary point. Matem. Zametki, 12, N 1 (1972), 62-72 (Russian).

[14] V.G. Maz'ja, On the Dirichlet problem for arbitrary order elliptic equations in unbounded domains. Doklady AN SSSR, 150, N 6 (1963), 1221-1224 (Russian).

[15] V.G. Maz'ja, On $(p, \ell)$-capacity, embedding theorems and the spectrum of a selfadjoint elliptic operator. Izv. AN SSSR, ser. matem., 37 (1973), 356-385 (Russian).

Author's address: Marata 19, kv. 7, 191025 Leningrad, USSR

ASYMPTOTIC METHODS FOR SINGULARLY PERTURBED
LINEAR DIFFERENTIAL EQUATIONS IN BANACH SPACES

Janusz Mika, Świerk-Otwock

## Introduction

Take a Banach space $\mathfrak{X}$ with the norm $\|\cdot\|$ and a singularly perturbed differential equation

$$\frac{dz(t)}{dt} = T(t)z(t) + m(t) \tag{1}$$

where $T(t)$ is a time-dependent linear operator, $m(t)$ a given function, and $\epsilon$ a small positive parameter. Such equations were analyzed by Krein [1] who proved that the zero order asymptotic solution consisting of three parts: inner, outer, and intermediate, converges uniformly to the exact solution. In [2] the author found the uniformly convergent asymptotic solution of any fixed order containing only outer and inner parts matched together by a proper choice of initial conditions. Such a procedure was already used by Vasil'eva and Butuzov [3] for systems of ordinary differential equations.

From the practical point of view, much more interesting are systems of singularly perturbed differential equations

$$(2)\qquad \begin{aligned} \epsilon\frac{dx(t)}{dt} &= A(t)x(t) + P(t)y(t) + q(t) ; \\ \frac{dy(t)}{dt} &= Q(t)x(t) + B(t)y(t) + r(t) , \end{aligned}$$

which were analyzed by the author in the zero order approximation in [4] and in any fixed order approximation in [5] . The results can be also applied to a single differential equation in a Banach space

$$\frac{dz(t)}{dt} = \frac{1}{\epsilon} A\, z(t) + Bz(t) + m(t) \tag{3}$$

such that the singularly perturbed operator A has an eigenvalue at the origin contrary to the conditions which have to be satisfied for (1) . If A and B are bounded operators then the Banach space can be split into a direct sum of two subspaces and (3) written as a system of differential equations which can be treated similarly as (2) . In particular, one can prove the uniform convergence of the asymptotic solution to the exact one for the initial value problem for the equation (3) . The detailed analysis will be published elsewhere.

The asymptotic expansion method for differential equations with singularly perturbed operators having an eigenvalue at the origin was first applied by Hilbert to the Boltzmann equation in kinetic theory. The presented results give a rigorous justification for the Hilbert approach in case of bounded operators.

As an example, the linear Boltzmann equation for neutrons in a discretized form is considered and the asymptotic expansion method used to derive the diffusion equation.

This work was supported in part by the International Atomic Energy Agency under the Research Contract No. 1236/RB.

## Hilbert asymptotic expansion method

Let A be a bounded linear operator from $\mathfrak{X}$ into itself with zero as its semisimple isolated eigenvalue. The corresponding finite-dimensional eigenspace $\mathcal{N}$ consists only of such elements $y \in \mathfrak{X}$ that $Ay = 0$. The space $\mathfrak{X}$ can be represented as a direct sum

$$\mathfrak{X} = \mathcal{N} + \mathcal{M}$$

where both $\mathcal{N}$ and $\mathcal{M}$ are invariant subspaces of A and A is one-to-one from $\mathcal{M}$ into itself [8, Section 148] . If P is the projector from $\mathfrak{X}$ onto $\mathcal{N}$ and $Q = I - P$ then

$$P\mathfrak{X} = \mathcal{N} \quad ; \quad Q\mathfrak{X} = \mathcal{M}.$$

The spectrum of A, except the point at the origin, is assumed to be located in the left half-plane and bounded away from the imaginary axis so that

$$\alpha = \inf_{\lambda \in SpA;\, \lambda \neq 0} Re\, \lambda < 0 \,, \tag{4}$$

Thus the uniformly continuous semigroup $G(t)$ generated by QAQ taken as an operator from $\mathcal{M}$ into itself satisfies the inequality

$$\|G(t)\| \leqslant M \cdot \exp(\alpha t) \quad ; \quad 0 \leqslant t < \infty$$

where M is a constant.

If B is a bounded operator in $\mathfrak{X}$ and $m(t)$ a function with values from $\mathfrak{X}$ n times continuously differentiable on $[0, t_o]$ , then with the assumed properties of A, B the equation (3) has on $[0, t_o]$ a unique, strongly differentiable solution $z(t)$ for the initial condition

$$z(0) = \theta$$

where $\theta$ is an arbitrary element of $\mathfrak{X}$ .

Define new functions

$$Qz(t) = v(t) \;; \quad Pz(t) = w(t) \;;$$
$$Qm(t) = q(t) \;; \quad Pm(t) = r(t) \;;$$
$$Q\theta = \mu \;; \quad P\theta = \eta$$

and transform (3) into an equivalent system of equations

$$(5)\quad \begin{aligned} \frac{dv(t)}{dt} &= QAQv(t) + QBQv(t) + QBPw(t) + q(t) ; \\ \frac{dw(t)}{dt} &= PBQv(t) + PBPw(t) + r(t) \end{aligned}$$

with the corresponding initial condition

$$(6)\qquad v(0) = \mu \; ; \quad w(0) = \eta .$$

The functions $v(t)$ and $q(t)$ have values from $\mathcal{M}$ and $w(t)$ and $r(t)$ from $\mathcal{N}$ . The operators in (5) are defined accordingly so that, for instance, QBP is an operator from $\mathcal{N}$ into $\mathcal{M}$ . In deriving (5) it was taken into account that for any y

$$PAy = APy \equiv 0.$$

The system of equations (5) is analogous to (2) so that an approach similar to that presented in [5] can be applied to obtain the asymptotic solution of any fixed order to (5) with the initial condition (6) .

Let the outer asymptotic solution of order n be defined as

$$\bar{v}^{(n)}(t) = \sum_{k=0}^{n} \epsilon^k \bar{v}_k(t) \quad ; \quad \bar{w}^{(n)}(t) = \sum_{k=0}^{n} \epsilon^k \bar{w}_k(t) \; ;$$

then $\bar{v}_k(t)$ and $\bar{w}_k(t)$ satisfy the equations

$$(7)\quad \begin{aligned} QAQ\bar{v}_k(t) + QBQ\bar{v}_{k-1}(t) + QBP\bar{w}_{k-1}(t) + \delta_{1k} q(t) &= \frac{d\bar{v}_{k-1}(t)}{dt} \; ; \\ \frac{d\bar{w}_k(t)}{dt} = PBQ\bar{v}_k(t) + PBP\bar{w}_k(t) + \delta_{ok}\, r(t) \quad &; \; k = 0,1,\dots n, \\ \bar{v}_{-1}(t) = \bar{w}_{-1}(t) &\equiv 0. \end{aligned}$$

Similarly, if the inner asymptotic solution of order n is defined as

$$\tilde{v}^{(n)}(\tau) = \sum_{k=0}^{n} \epsilon^k \tilde{v}_k(\tau) \quad ; \quad \tilde{w}^{(n)}(\tau) = \sum_{k=0}^{n} \epsilon^k \tilde{w}_k(\tau) \quad ; \quad \tau = t/\epsilon \; ;$$

then $\tilde{v}_k(\tau)$ and $\tilde{w}_k(\tau)$ satisfy the equations

$$(8)\quad \begin{aligned} \frac{d\tilde{v}_k(\tau)}{d\tau} &= QAQ\tilde{v}_k(\tau) + QBQ\tilde{v}_{k-1}(\tau) + QBP\tilde{w}_{k-1}(\tau) \; ; \\ \frac{d\tilde{w}_k(\tau)}{d} &= PBQ\tilde{v}_{k-1}(\tau) + PBP\tilde{w}_{k-1}(\tau) \quad ; \; k = 0,1,\dots n; \\ \tilde{v}_{-1}(\tau) &= \tilde{w}_{-1}(\tau) \equiv 0. \end{aligned}$$

The algorithm of solving (7) and (8) consists of the following steps:

(i) $\tilde{w}_k(\tau)$ is found by solving the second equation in (8) and using the condition that $\lim_{\tau\to\infty} \tilde{w}_k(\tau) = 0$; for k = 0 it gives simply $\tilde{w}_0(\tau) \equiv 0$;

(ii) $\bar{v}_k(t)$ is eliminated from the equations (7) and the resulting equation for $\bar{w}_k(t)$ is solved with the initial condition

$$\bar{w}_k(0) = \delta_{ok} - \tilde{w}_k(0) ;$$

(iii) $\bar{v}_k(t)$ is calculated from $\bar{w}_k(t)$ with the first equation in (7) ;

(iv) the first equation in (8) is solved with

$$\tilde{v}_k(0) = \delta_{ok} - \bar{v}_k(0) .$$

The functions $\tilde{v}_k(\tau)$ and $\tilde{w}_k(\tau)$ decay in the norm exponentially with time as $\exp(\alpha\tau)$ where $\alpha$ is defined in (4) .

If the asymptotic solution of order n is taken as

$$v^{(n)}(t) = \bar{v}^{(n)}(t) + \tilde{v}^{(n)}\left(\frac{t}{\epsilon}\right) \quad ; \quad w^{(n)}(t) = \bar{w}^{(n)}(t) + \tilde{w}^{(n)}\left(\frac{t}{\epsilon}\right) ,$$

then $\{v^{(n)}(t), w^{(n)}(t)\}$ tends uniformly on $[0, t_0]$ to the exact solution $\{v(t), w(t)\}$ of (5) and (6) faster than $\epsilon^n$.

## Application of the Hilbert method in neutron transport theory

A singularly perturbed differential equation containing the operator with an eigenvalue at the origin was first considered by Hilbert in the kinetic theory (see e.g. [6]). The Hilbert expansion of the Boltzmann equation, later modified by Chapman and Enskog, has played an important role in statistical physics. However, it is based essentially on intuitive grounds. Only in the case of the linearized Boltzmann equation for special initial conditions it was rigorously analyzed by Grad [7] .

The results obtained by the author and presented in previous section indicate the convergence of the Hilbert expansion for differential equations in Banach spaces but an obvious disadvantage of the analysis is that the operators have to be bounded. Nevertheless, the obtained results can be applied to unbounded operators provided a mollifying procedure is used. An example of such an approach will be given in this section.

The behavior of neutrons in reactor systems is described by the linear Boltzmann equation. In practical situations, this equation is by far too complicated to be treated directly so that various approximate models have to be applied. One of the most widely used is the diffusion approximation. In the literature there are several ways of deriving the neutron diffusion equation from the Boltzmann equation but so far the Hilbert expansion method was not utilized.

Take a simplest possible reactor system consisting of an infinite slab of thickness a and assume that all neutrons have the same speed which will be taken as equal to unity. The Boltzmann equation for such a system is

$$\begin{aligned}(9)\quad \frac{\partial z}{\partial t} &= -_1\xi \frac{\partial z}{\partial \varrho} - (\beta_a(\varrho) + \beta_s(\varrho)\ z(\varrho, \xi, t) \\ &+ \beta_s(\varrho) \int_{-1} d\xi\ h_s(\varrho; \xi, \xi')\ z(\varrho, \xi', t) + m(\varrho, \xi, t) \quad ; \ 0 \leqslant \varrho \leqslant a; \\ & -1 \leqslant \xi \leqslant 1 \ ; \ 0 \leqslant t < \infty .\end{aligned}$$

Here $z(\varrho, \xi, t)$ is the neutron distribution function dependent on the position variable $\varrho$, angular variable $\xi$, and time t; $\beta_a(\varrho)$ and $\beta_s(\varrho)$ are absorption and scattering frequencies, respectively; $h_s(\varrho; \xi, \xi')$ is the scattering kernel assumed to have the form

$$(10)\quad h_s(\varrho; \xi, \xi') = \sum_{l=0}^{\infty} \frac{2l+1}{2}\, b_l(\varrho)\, p_l(\xi)\, p_l(\xi'),$$

where $p_l(\xi)$ are Legendre polynomials and

$$(11)\quad b_o(\varrho) \equiv 1 \ ; \ b_l(\varrho) < 1 \ ; \ 0 \leqslant \varrho \leqslant a \ ; \ l = 1,2,\ldots.$$

The solution to (9) is assumed to be periodic so that the boundary condition to be satisfied by $z(\varrho, \xi, t)$ is

$$(12)\quad z(0, \xi, t) = z(a, \xi, t) \quad ; \ -1 \leqslant \xi \leqslant 1; \ 0 \leqslant t < \infty .$$

The equation (9) is supplemented by the initial condition

$$(13)\quad z(\varrho, \xi, 0) = \theta(\varrho, \xi); \ 0 \leqslant \varrho \leqslant a \ ; \ -1 \leqslant \xi \leqslant 1.$$

It can be shown [8] that the equation (9) with the conditions (12) and (13) has a unique solution in the Hilbert space of square integrable functions.

The unbounded differential operator in (9) can be replaced by its finite difference counterpart if the interval $[0,a]$ is covered by the mesh of points $0 = \varrho_0 < \varrho_1 < \ldots < \varrho_J = a$ and all the functions of $\varrho$ are replaced by vectors or matrices whose components are values of the relevant functions taken at $\varrho_1 \ldots \varrho_J$. The boundary condition (12) is accounted for by identifying the values of $z(\varrho, \xi, t)$ at $\varrho_0$ and $\varrho_J$.

The derivative in (16) is replaced by the matrix

$$D = \begin{vmatrix} 0 & \varkappa_1 & 0 & 0 & \cdot & \cdot & \cdot & 0 & 0 & -\varkappa_1 \\ -\varkappa_2 & 0 & \varkappa_2 & 0 & \cdot & \cdot & \cdot & 0 & 0 & 0 \\ 0 & -\varkappa_3 & 0 & \varkappa_3 & \cdot & \cdot & \cdot & 0 & 0 & 0 \\ \cdot & \cdot & \cdot & \cdot & \cdot & \cdot & \cdot & \cdot & \cdot & \cdot \\ 0 & 0 & 0 & 0 & \cdot & \cdot & \cdot & -\varkappa_{J-1} & 0 & \varkappa_{J-1} \\ \varkappa_J & 0 & 0 & 0 & \cdot & \cdot & \cdot & 0 & -\varkappa_J & 0 \end{vmatrix}$$

where $\varkappa_1 = (\varrho_2 - \varrho_J)^{-1}$, $\varkappa_J = (\varrho_1 - \varrho_{J-1})^{-1}$, and $\varkappa_j = (\varrho_{j+1} - \varrho_{j-1})^{-1}$ ; j = 2, .... J-1. All the remaining operators in (9) are in their discretized form diagonal.

In the matrix notation one can write (9) as

$$\frac{dz}{dt} = -\xi Dz(\xi,t) - (\beta_a + \beta_s)z(\xi,t) + \beta_s\int_{-1}^{1} d\xi\, h_s(\xi,\xi')\,z(\xi',t) + m(\xi,t) \; ; \tag{14}$$

and the initial condition (13) as $z(\xi,0) = \Theta(\xi)$.

As the space $\mathfrak{X}$ we shall take the product $\mathfrak{X} = \mathcal{L} \times \ldots\ldots \times \mathcal{L}$ of Hilbert spaces $\mathcal{L}$ of functions square integrable over $[-1,1]$.

Define the operator A as

$$(Ax)(\xi) = \beta_s\left(\int_{-1}^{1} d\xi'\, h(\xi,\xi')\,x(\xi') - x(\xi)\right) \; ; \; x \in \mathfrak{X}$$

with

$$h(\xi,\xi') = h_s(\xi,\xi') + \tfrac{3}{2}(I - b_1)\; p_1(\xi)\,p_1(\xi')$$

and B as

$$(Bx)(\xi) = -\xi Dx(\xi) - \beta_a x(\xi) - \tfrac{3}{2}\beta_s(I-b_1)\,p_1(\xi)\int_{-1}^{1} d\xi'\, p_1(\xi')\,x(\xi') \; ; \quad x \in \mathfrak{X},$$

where I is the unit matrix.

With the above definitions, introducing the coefficient $\frac{1}{\epsilon}$ in front of A one can write (14) in the form (3) and apply the Hilbert expansion method. The null space $\mathcal{N}$ is now the sum $\mathcal{N} = \mathcal{N}_0 + \mathcal{N}_1$, where $\mathcal{N}_0 = \mathcal{N}_0^{(0)} \times \ldots\ldots \times \mathcal{N}_0^{(0)}$ ; $\mathcal{N}_1 = \mathcal{N}_1^{(0)} \times \ldots\ldots \times \mathcal{N}_1^{(0)}$ ; and $\mathcal{N}_0^{(0)}$ and $\mathcal{N}_1^{(0)}$ are linear manifolds in $\mathcal{L}$ spanned by the elements $\frac{1}{2}p_0(\xi)$ and $\frac{3}{2}p_1(\xi)$, respectively. The projectors corresponding to $\mathcal{N}_0$ are given by the formulas

$$(P_0x)(\xi) = \tfrac{1}{2}\int_{-1}^{1} d\xi'\, p_0(\xi')\,x(\xi') \; ; (P_1x)(\xi) = \tfrac{3}{2}\,p_1(\xi)\int_{-1}^{1} d\xi'\, p_1(\xi')\,x(\xi').$$

From (11) it follows that the spectrum of A excluding the point at origin satisfies the requirement (4) .

Defining new functions

$$P_0 z(t) = \varphi(t) \; ; \quad P_1 z(t) = \xi\, \chi(t) \; ;$$
$$P_0 m(t) = m_0(t) \; ; \quad P_1 m(t) = \xi\, m_1(t) \; ;$$
$$P_0 \Theta = \Theta_0 \; ; \quad P_1 \Theta = \xi\, \Theta_1$$

and applying the procedure of the previous section one gets in the zero order approximation the following outer asymptotic equations

$$\frac{d\,\overline{\varphi}_0(t)}{dt} = -\beta_a\,\overline{\varphi}_0(t) - \tfrac{1}{3} D\,\chi_0(t) + m_0(t) \; ; \tag{15}$$

$$\frac{d\overline{\chi}_o(t)}{dt} = -D\,\overline{\varphi}_o(t) - \beta_t\,\overline{\chi}_o(t) + m_1(t)$$

and the corresponding initial conditions $\overline{\varphi}_o(0) = \theta_o$ ; $\overline{\chi}_o(0) = \theta_1$. The matrix $\beta_t$ is defined as $\beta_t = \beta_a + \beta_s(I - b_1)$.

The equations (15) are a discretized representation of the equations obtained from the first order spherical harmonics approximation as applied to (9) with $\overline{\varphi}_o(t)$ as the neutron density and $\overline{\chi}_o(t)$ as the neutron current. Obviously, they could be improved with higher order outer solutions together with inner solutions according to the general scheme developed in the previous section.

The equations (15) have a different structure from the diffusion equation. In fact, if $\beta_a$ and $\beta_t$ are constant matrices, then by eliminating $\overline{\chi}_o(t)$, (23) can be reduced to the second order differential equation being a discretized version of the telegraph equation.

To obtain the discretized diffusion equation from (15) one has to introduce another small parameter, say, $\delta$ in front of the derivative $\frac{d\overline{\chi}_o(t)}{dt}$ and apply the asymptotic expansion method as for the equation (2) . As the result in the zero order approximation one gets the outer asymptotic equations with respect to both small parameters $\epsilon$ and $\delta$ which after eliminating the current $\overline{\overline{\chi}}_{oo}(t)$ lead to the discretized diffusion equation

$$\frac{d\overline{\overline{\varphi}}_{oo}(t)}{dt} = \frac{1}{3} D\,\beta_t^{-1} D\,\overline{\overline{\varphi}}_{oo}(t) - \beta_a\,\overline{\overline{\varphi}}_{oo}(t) + \overline{\overline{m}}(t)$$

with the initial condition

$$\overline{\overline{\varphi}}_{oo}(0) = \theta_o$$

and

$$\overline{\overline{m}}(t) = m_o(t) - \frac{1}{3}\,\beta_t^{-1}\,Dm_1(t) .$$

Thus it is seen that the repeated application of the asymptotic expansion method to the Boltzmann equation with respect to two different small parameters $\epsilon$ and $\delta$ leads in the zero order outer approximation to the diffusion equation, at least in the discretized representation. However, the physical justification for introducing both small parameters into (14) and (15) is not so convincing as in the case of the Boltzmann equation for gases. The more detailed discussion of this problem will be performed elsewhere.

## References.

[1] S.G.Krein: "Linear Differential Equations in Banach space", Amer. Math.Soc., Providence, R.I., 1971.

[2] J.Mika: Higher order singular perturbation method for linear differential equations in Banach spaces, Ann.Polon.Math., 32, 43-57 /1976/.

[3] A.B.Vasil'eva and V.F.Butuzov: "Asymptotic Expansions of Solutions to Singularly Perturbed Equations", Nauka, Moscow, 1973 /Russian/.

[4] J.Mika: Singular perturbation method in neutron transport theory, J.Math.Phys., 15, 872-879 /1974/.

[5] J.Mika: Singularly perturbed evolution equations in Banach spaces, J.Math.Anal.Appl., 58, 189-201 /1977/.

[6] C.Cercignani: "Mathematical Methods in Kinetic Theory", Macmillan, London, 1969.

[7] H.Grad: Asymptotic theory of the Boltzmann equation, Phys.Fluids, 6, 147-181 /1963/.

[8] A.Belleni-Morante: Neutron transport in a nonuniform slab with generalized boundary conditions, J.Math.Phys., 11, 1553-1558 /1970/.

Author's address: Computing Center CYFRONET, Institute of Nuclear Research, Świerk, 05-400 Otwock, Poland.

# NON LINEAR QUASI VARIATIONAL INEQUALITIES AND STOCHASTIC IMPULSE CONTROL THEORY

Umberto Mosco, Roma

1. It has been shown by A. Bensoussan and J.L.Lions [1] [2] that the Hamilton-Jacobi function related to a stochastic optimal control problem with both continuous and impulse control can be obtained as the strong regular solution of a quasi-variational inequality involving a second order semi-linear partial differential operator.

The aim of this paper is to prove the existence of such a regular solution, for the stationary case and Dirichlet boundary condition, under the assumption that the higher order coefficients of the partial differential operator are constants. We shall also prove that the solution, which is unique, is the limit of monotone iterative algorithms converging from above and from below.

2. The inequality (Q.V.I.) we are interested in can be stated as follows:

$$(1)\qquad \begin{cases} u \in H^1_0(\mathcal{O}) \cap C(\bar{\mathcal{O}}), \quad Lu \in L^2(\mathcal{O}) \\ u(x) \leq M(u)(x) \quad \text{for all } x \in \bar{\mathcal{O}} \\ Lu + G(u) \leq f \quad \text{a.e. in } \mathcal{O} \\ (u - M(u))(Lu + G(u) - f) = 0 \quad \text{a.e. in } \mathcal{O} \end{cases}$$

where $M(u)$ is defined by

$$(2)\qquad M(u)(x) = 1 + \inf_{\substack{\xi \in \mathbb{R}^N_+ \\ \xi \geq 0}} u(x+\xi)$$

and the data of the problem are:

(i) $\mathcal{O}$, a bounded open subset of $\mathbb{R}^N$, with a smooth boundary $\Gamma$, say $\Gamma$ of class $C^2$;

(ii) $L$, a second order linear uniformly elliptic partial differential operator in divergence form,

$$Lu = - \sum_{i,j=1}^{N} \frac{\partial}{\partial x_i}\left(a_{ij} \frac{\partial u}{\partial x_j}\right) + \sum_{j=1}^{N} a_j \frac{\partial u}{\partial x_j} + a_o u ,$$

whose coefficients satisfy the conditions

$$a_{ij} \in C^1(\bar{\mathcal{O}}) , \; a_j, a_o \in L^\infty(\mathcal{O}) , \; a_o \geq 0 \text{ a.e., } \forall i,j = 1,\dots,N ;$$

$$\sum_{i,j=1}^{N} a_{ij}\xi_i\xi_j \geq \beta \sum_{i=1}^{N} \xi_i^2 \quad \text{a.e. in } \mathcal{O}, \text{ for all } \xi \in \mathbb{R}^N ,$$

where $\beta > 0$ is some given constant;

(iii) G, a first order nonlinear partial differential operator of the form

$$G(u)(x) = -H(x,Du(x))$$

where $Du = \left\{\frac{\partial u}{\partial x_1},\dots, \frac{\partial u}{\partial x_N}\right\}$, the (real valued) function

$$H(x, p) \quad , \quad x \in \mathcal{O} , \; p \in \mathbb{R}^N$$

being concave in p for a.e. $x \in \mathcal{O}$ and satisfying in addition the conditions:

$$|H(x,p)| \leq h(x) + c_o|p| \qquad \text{a.e. } x \in \mathcal{O} , \quad \forall p \in \mathbb{R}^N$$

$$|H(x,p') - H(x,p'')| \leq c_o|p'-p''| \quad \text{a.e. } x \in \mathcal{O}, \; \forall p',p'' \in \mathbb{R}^N$$

for some constant $c_o \geq 0$ and some $h \in L^\infty(\mathcal{O})$;

(iv) f, a given function of $L^\infty(\mathcal{O})$.

We also assume that the following conditions are verified:

(v) $\inf_{\mathcal{O}} a_o \geq (2\beta)^{-1}\left\{\sum_{j=1}^{N} \|a_j\|_\infty^2 + 2c_o^2\right\}$

(vi) $\inf_{\mathcal{O}} f - \sup_{\mathcal{O}} h \geq \inf_{\mathcal{O}} a_o$.

We shall come back to the last assumption (vi) in the following Remark 2 and Remark 3.

Our first result is the following (see U.Mosco [3]).

THEOREM 1. Let us suppose, in addition to what required in (i) ...

(vi) above, that the coefficients $a_{ij}$ of the operator L are constants in $\mathcal{O}$. Then, the solution u of problem (1) exists and is unique. Moreover, it satisfies the additional regularity properties

$$(3) \qquad u \in W^{2,p}(\mathcal{O}, \quad \forall p \geq 2 \quad , \quad Lu \in L^{\infty}(\mathcal{O}),$$

in particular,

$$u \in C^{1,\alpha}(\bar{\mathcal{O}}) \quad \text{for all} \quad 0 < \alpha < 1. \qquad \blacksquare$$

REMARK 1. The specific function H that appears in the QVI arising from the stochastic impulse control theory is the so called *Hamiltonian function*, defined by

$$(4) \qquad H(x,p) = \min_{d \in U} \left[ g_0(x,d) + p \cdot g_1(x,d) \right] ,$$

where U is a subset of some $\mathbb{R}^Q$, $Q \geq 1$, $g_0 : \bar{\mathcal{O}} \times U \to \mathbb{R}$ and $g_1 : \bar{\mathcal{O}} \times U \to \mathbb{R}^N$. If, for instance, we assume that U is compact and that $g_0$ and $g_1$ are continuous in $d \in U$ for fixed x a.e. $\in \mathcal{O}$ and continuous in $x \in \bar{\mathcal{O}}$ uniformly with respect to $d \in U$, then the Hamiltonian (4) verifies the properties required in (iii) above, with $c_0 \geq 0$ any constant and $h \in L^{\infty}(\mathcal{O})$ any function such that

$$|g_0(x,d)| \leq h(x) \qquad \forall x \text{ a.e.} \in \mathcal{O} \qquad \forall d \in U ,$$

$$|g_1(x,d)| \leq c_0 \qquad \forall x \in \bar{\mathcal{O}} , \quad \forall d \in U ;$$

see also A.Bensoussan, J.L.Lions, *loc. cit.* ■

3. Before stating further results, let us introduce the notion of *weak solution* of the QVI considered above.

We introduce the bilinear form

$$(5) \quad a(v,w) = \sum_{i,j=1}^{N} a_{ij} \frac{\partial v}{\partial x_i} \frac{\partial w}{\partial x_j} dx + \sum_{j=1}^{N} \int_{\mathcal{O}} a_j \frac{\partial v}{\partial x_j} w \, dx + \int_{\mathcal{O}} a_0 vw dx$$

which is well defined for any functions v,w in the Sobolev space $H^1(\mathcal{O})$, and we denote by $(\cdot,\cdot)$ the $L^2(\mathcal{O})$ inner product. If $\langle \cdot,\cdot \rangle$ denotes the duality pairing between the space $H^1_0(\mathcal{O})$ and its dual $H^{-1}(\mathcal{O})$, the identity

(6) $$\langle Lu, w\rangle = a(u,w) \quad ,$$

when $u \in H^1(\mathcal{O})$, $w \in H^1_0(\mathcal{O})$, defines L as an operator from the space $H^1(\mathcal{O})$ to the dual $H^{-1}(\mathcal{O})$ of $H^1_0(\mathcal{O})$, and the identity

(7) $$\langle A(u),w\rangle = a(u,w) + (G(u),w)$$

also defines the non linear operator

(8) $$A = L + G$$

from the space $H^1(\mathcal{O})$ to $H^{-1}(\mathcal{O})$.
For any given function

(9) $$\psi : \mathcal{O} \to \mathbb{R} \text{ measurable, with } \psi \geq 0 \text{ a.e. on } \Gamma$$

we denote by $\sigma_f^A(\psi)$ the (unique) solution v of the following *variational inequality* (V.I.)

(10) $$\begin{cases} v \in H^1_0(\mathcal{O}), \quad v \leq \psi \quad \text{a.e. in } \mathcal{O} \\ \langle A(v), v-w\rangle \leq (f,v-w) \\ \forall w \in H^1_0(\mathcal{O}) \;, \; w \leq \psi \quad \text{a.e. in } \mathcal{O} \end{cases}$$

where A is the operator (8).
For any function

(11) $$u \in L^\infty(\mathcal{O} \text{ , with } u \geq -1 \text{ a.e. in } \mathcal{O},$$

the function $\psi = M(u)$ - with $M(u)$ defined by (2), where the inf is now taken as the ess inf in the space $L^\infty(\mathcal{O})$ - clearly verifies assumption (9)
Therefore, we can consider the function

(12) $$(\sigma_f^A \circ M)(u) = \sigma_f^A(M(u)) \quad :$$

according to the definition of $\sigma_f^A$ just given, the function (12) is the solution of the following V.I.

$$(13)\qquad \begin{cases} v \in H^1_0(\mathcal{O})\ , \qquad v \leq M(u) \quad \text{a.e. in } \mathcal{O} \\ \langle A(v), v-w \rangle \leq (v, v-w) \\ \forall w \in H^1_0(\mathcal{O}), \qquad w \leq M(u)\ . \end{cases}$$

We say that a function u defined in $\mathcal{O}$ is a (weak) *subsolution* of problem (1) if u satisfies the condition (11) and, moreover,

$$(14)\qquad u \leq (\sigma^A_f \circ M)(u) \qquad \text{a.e. in } \mathcal{O}.$$

The function u is said a *supersolution* of (1) if in addition to (11) we have

$$(15)\qquad u \geq (\sigma^A_f \circ M)(u) \qquad \text{a.e. in } \mathcal{O}.$$

We say that u is a *weak solution* of problrm (1) if u verifies (11) and moreover

$$(16)\qquad u = (\sigma^A_f \circ M)(u) \qquad \text{a.e. in } \mathcal{O},$$

that is, u is a *fixed point* of the mapping $\sigma^A_f \circ M$. Let us remark, incidentally, that in corresponce of the $L^\infty$-estimates for problem (13), the map $\sigma^A_f \circ M$ carries the subset of $L^\infty(\mathcal{O})$ defined by (11) into a subset of $H^1_0(\mathcal{O}) \cap L^\infty(\mathcal{O})$.

According to the definition above, a weak solution of problem (1) is thus any function u that satisfies the conditions

$$(17)\qquad \begin{cases} u \in H^1_0(\mathcal{O}) \cap L^\infty(\mathcal{O}), \quad u \geq -1 \quad \text{a.e. in } \mathcal{O} \\ u \leq M(u) \quad \text{a.e. in } \mathcal{O} \\ a(u,u-w) + (\ (u), u-w) \leq (f, u-w) \\ \forall w \in H^1_0(\mathcal{O})\ , \ w \leq M(u) \ \text{a.e. in } \mathcal{O} \end{cases}$$

where a(u,v) is the form (5). It is quite easy to check that if u is the solution of (1) (and we then refer to u as to the *strong* solution of problem (1)), then u is in particular a weak solution in the sense just defined.

REMARK 2. Under the assumptions of Theorem 1, in particular as a consequence of (vi), the function $z \equiv -1$ in $\mathcal{O}$ is a subsolution of problem (1). It follows, in fact, from the $L^\infty$ estimates already mentioned, that if $\psi = 1$ and (vi) holds, then the solution $v=\sigma_f^A(1)$ of (10) satisfies the condition $-1 \leq v$ a.e. in $\mathcal{O}$, which is to say, due to the fact that $z \equiv -1$ implies $M(z) \equiv 1$, $z \leq (\sigma_f^A \circ M)(z)$ a.e. in $\mathcal{O}$. ■

REMARK 3. In Theorem 1 it suffices to assume, in place of (vi), the weaker hypothesis

(vii) It exists a subsolution z of problem (1) ($z \geq -1$ a.e. in $\mathcal{O}$).

The hypothesis is satisfied provided $f - h$ is not too negative in $\mathcal{O}$, which is indeed also a *necessary* condition for the existence of the solution u of (1), as it can be easily checked directly on (1) taking into account the comparison theorems and the fact that u vanishes on $\Gamma$. ■

THEOREM 2. Let us suppose that the hypothesis of Theorem 1 are satisfied, with (vi) possibly replaced by the weaker assumption (vii). Let z be a function verifying (vii) and let $u^o$ be any given function such that

$$(18) \qquad u^o \in H_0^1(\mathcal{O}), \quad A(u^o) \in L^\infty(\mathcal{O}), \quad u^o \geq z \text{ a.e. in } \mathcal{O}.$$

Then, the solution $u^n$ of the V.I.

$$(19) \qquad u^n = (\sigma_f^A \circ M)(u^{n-1})$$

defined iteratively for $n = 1,2,\ldots$, exists for every n and converges as $n \to \infty$ to the (strong) solution u of problem (1) in the weak topology of $W^{2,p}(\mathcal{O})$ for any $p \geq 2$, the sequence $\{A(u^n)\}$ being bounded in $L^\infty(\mathcal{O})$. Furthermore, if $u^o$ is a supersolution [subsolution, resp.] of problem (1), then the sequence $\{u^n\}$ is non-increasing [non-decreasing, resp.] in $\mathcal{O}$. ■

Theorem 1 is clearly a consequence of Theorem 2, so we shall prove Theorem 2 in what follows.

Let us remark that when $G = 0$, that is, $H \equiv 0$, and *all* coefficients of L are constants, then the existence results stated above were proved by J.L.JOLY, U.MOSCO and G.M.TROIANIELLO [1] [2], who also obtained a dual pointwise estimate of the solution.

The existence of a *weak* solution of problem (1) has been proved by A.BENSOUSSAN and J.LIONS, *loc.cit.*, by using the theory of *monotone operators*.

Under the assumptions (i)-(vi) above, the weak solution of (1) is also unique. This can be proved by extending to the case at hand the uniqueness result due to T.LAETSCH [1] for the linear case $A = L$ and $f \geq 0$, in a modified form communicated to the author by J.L.JOLY, allowing arbitrary $f \in L^\infty(\mathcal{O})$. It should be also remarked that the uniqueness of the *strong* solution of problem (1), when H is given by (4), also follows *via* the interpretation of u as the Hamilton-Jacobi function of the related control problem, as shown in A.BENSOUSSAN and J.L.LIONS, *loc.cit.*.

When all coefficients of L are constants and the function $H(x,p)$ is of the form $H(x,p) = H_0(x) + H_1(p)$, then a pointwise dual estimate of $A(u)$ a.e. in $\mathcal{O}$, can be also obtained, see U.MOSCO [2].

4. Let $\xi \in \mathbb{R}^N_+$. We define the map

$$\pi_\xi \circ \tau_\xi : H^1_0(\mathcal{O}) \cap L^\infty(\mathcal{O}) \to H^1(\mathcal{O}) \cap L^\infty(\mathcal{O}) \tag{20}$$

by setting

$$\begin{cases} \pi_\xi \circ \tau_\xi(v)(x) = \tau_{-\xi}(v)(x) = v(x+\xi) & \text{if } x \in \mathcal{O}'_\xi \\ \pi_\xi \circ \tau_\xi(v)(x) = 0 & \text{if } x \in \mathcal{O}-\mathcal{O}'_\xi \end{cases}$$

where $\mathcal{O}'_\xi$ is the, possibly empty, open subset of all $\xi \in \mathcal{O}$ such that $x + \xi \in \mathcal{O}$.

We define the map

$$\pi'_\xi \circ A'_\xi \circ \tau'_\xi : H^1_0(\mathcal{O}) \cap H^2(\mathcal{O}) \to L^2(\mathcal{O}) \ , \tag{21}$$

by setting

$$\begin{cases} \pi'_\xi \circ A'_\xi \circ \tau'_\xi(v)(x) = A'_\xi(\tau'_{-\xi}(v))(x) & \text{if } x \in \mathcal{O}'_\xi \\ \pi'_\xi \circ A'_\xi \circ \tau'_\xi(v)(x) = A(0)(x) = -H(x,0) & \text{if } x \in \mathcal{O}-\mathcal{O}'_\xi \ , \end{cases}$$

where $A'_\xi$ denotes the restriction of the operator A to the open set $\mathcal{O}'_\xi$ and $\tau'_{-\xi}(v)$ the restriction to $\mathcal{O}'_\xi$ of the translated function $\tau_{-\xi}(v)(x) = v(x + \xi)$. We have indeed

$$\pi'_\xi \circ A'_\xi \circ \tau'_\xi : H^1_0(\mathcal{O}) \cap W^{2,p}(\mathcal{O}) \to L^p(\mathcal{O}) \tag{22}$$

for every $p \geq 2$.

Let $\mathcal{F}$ denote the family of all *finite* subsets F of $\mathbb{R}^N_+$, such that $0 \in F$. For each fixed $F \in \mathcal{F}$, we define the map

$$M_F : H^1_o(\mathcal{O}) \cap L^\infty(\mathcal{O}) \to H^1(\mathcal{O}) \cap L^\infty(\mathcal{O}) \tag{23}$$

by setting

$$M_F(v) = 1 + \inf_{\xi \in F} \pi_\xi \circ \tau_\xi(v) \wedge 0 ,$$

where both *inf* and $\wedge$ denote the usual a.e. lattice operation in $L^\infty(\mathcal{O})$.

The main point in the proof of Theorem 2 consists in proving the uniform dual estimate stated in the following

PROPOSITION 1. For every fixed $F \in \mathcal{F}$ let us consider the sequence $\{u^n_F\}$ defined recursively by

$$u^o_F = u^o \quad , \quad u^n_F = (\sigma^A_f \circ M_F)(u^{n-1}_F) \tag{24}$$

where $(\sigma^A_f \circ M_F)(u^{n-1}_F) = \sigma^A_f(M_F(u^{n-1}_F))$ is obtained for each n=1,2,... as the solution of the V.I. (10) for $\psi = M_F(u^{n-1}_F)$. Then all solutions $u^n_F$'s exist and they verify the following dual estimate

$$\|A(u^n_F)\|_{L^\infty(\mathcal{O})} \leq c \tag{25}$$

with a constant c which is *independent on* $n \geq 1$ *and* $F \in \mathcal{F}$. ∎

The proof of the Proposition is done in four steps, the first two of which rely only on *potential* theoretic properties of the non-linear operator A.

The first step consists in proving that for any function $v \in H^1_o(\mathcal{O})$, such that $A(v) \in L^p(\mathcal{O})$ with $p \geq 2$, the distribution $A(M_F(v))$ of $H^{-1}(\mathcal{O})$ is actually a *measure* that can be estimated from below, in the sense of measures, by a $L^p$ function (depending on F), namely

$$A(M_F(v)) \geq \inf_{\xi \in F} \pi'_\xi \circ A'_\xi \circ \tau'_\xi(v) \wedge A(0). \tag{26}$$

For more details we refer to U.MOSCO [2].

The second step relies on the *dual estimates* for nonlinear V.I., see Th. 4.1 in U.Mosco [1].By assuming that $\psi \in H^1(\mathcal{O})$ and $A(\psi)$ is a measure in $\mathcal{O}$, the solution v of the V.I. (10) is shown to satisfy in

$\mathcal{O}$ the estimate

$$(27) \qquad f \geq A(v) \geq f \wedge A(\psi)$$

in the sense of measures.

We now consider the sequence of iterates (24) for a given $F \in \mathcal{F}$. Since the initial function $u^{o}$ is assumed to satisfy (18), we may suppose that $u_F^{n-1}$ verifies

$$(28) \qquad u_F^{n-1} \in H_o^1(\mathcal{O}) \quad , \quad A(u_F^{n-1}) \in L^p(\mathcal{O}), \text{ with } p \geq 2.$$

As in the first step above, it follows that $A(M_F(u_F^{n-1}))$ is a measure, and the estimate (26) holds, with $v = u_F^{n-1}$. According to the second step above, the estimate (27) also holds, with $v = u_F^{n-1}$. It follows then that $u_F^n$ satisfies the inequalities

$$(29) \qquad f \geq A(u_F^n) \geq f \wedge_{\xi \in F} \pi_\xi^{\prime} \cdot A_\xi^{\prime} \circ \tau_\xi^{\prime}(u_F^{n-1}) \wedge A(0), \text{ a.e. in } \mathcal{O}.$$

By well known regularity theorems, it also follows from (28) that

$$u_F^{n-1} \in W^{2,p}(\mathcal{O}) \ ,$$

hence, by (22),

$$\pi_\xi^{\prime} \circ A_\xi^{\prime} \circ \tau_\xi^{\prime}(u_F^{n-1}) \in L^p(\mathcal{O})$$

for every $\xi \in \mathbb{R}_+^N$. Since $f \in L^p(\mathcal{O})$ and $A(0) = -H(x,0) \in L^p(\mathcal{O})$ by our assumptions (iii) and (iv), it follows then from (29) that $u_F^n$ too verifies the properties (28). Thus the estimate (29) holds *for all* $n \geq 1$.

Let us also remark that the following uniform estimate holds

$$(30) \qquad \|u_F^n\|_{L^\infty(\mathcal{O})} \leq 1 \qquad \forall n \geq 1 \ , \ \forall F \in \mathcal{F}.$$

In fact, since $u^o \geq z \geq -1$, we may assume, for given F, that $u_F^{n-1} \geq z \geq -1$, hence

$$1 \geq M_F(u_F^{n-1}) \geq M_F(z)$$

It follows, by well know comparison theorems, that

$$\sigma_f^A(M_F(u_F^{n-1})) \geq \sigma_f^A(M_F(z)) ,$$

therefore

$$u_F^n \geq z$$

since z is a subsolution; on the other hand,

$$1 \geq M_F(u_F^n) \geq u_F^n .$$

Thus

$$1 \geq u_F^n \geq z \geq -1 \quad \text{a.e. in } \mathcal{O} ,$$

for all $n \geq 1$.

Let us note incidentally that the last property assures that the sequence of iterates $\{u_F^n\}_{n \geq 1}$ actually exists, each $u_F^n$ being the solution of the V.I. (13), where M(u) is replaced by $M_F(u_F^{n-1})$; in fact, $M_F(u_F^{n-1}) \geq 0$ a.e. in $\mathcal{O}$.

Up to now no use has been made of the assumption that the coefficients $a_{ij}$'s of the operator L are constants. This property of L, however, will now be used in the following argument.

Let us denote by $L_o$ the leading part of the operator L, i.e.

$$L_o v = - \sum_{i,j=1}^{n} \frac{\partial}{\partial x_i}(a_{ij} \frac{\partial v}{\partial x_j}) ,$$

and for any $\xi \in \mathbb{R}_+^N$ such that $\mathcal{O}'_\xi \neq \emptyset$ , according to our previous notation for the operator $A'_\xi$, let us also denote by $(L_o)'_\xi$ the restriction of the operator $L_o$ to the open set $\mathcal{O}'_\xi$ and by $\pi'_\xi \circ (L_o)'_\xi \cdot \tau'_\xi$ the operator (21), with $(L_o)'_\xi$ playing the role of $A'_\xi$.

Since $L_o$ commutes with the translations, we have

$$\| (L_o)'_\xi (\tau'_{-\xi}(v)) \|_{L^\infty(\mathcal{O}'_\xi)} \leq \| L_o(v) \|_{L^\infty(\mathcal{O})}$$

for all $\xi \in \mathbb{R}_+^N$ such that $\mathcal{O}'_\xi \neq \emptyset$. Hence also, by our assumption (iii) on the function H,

$$\text{(31)} \quad \| \pi'_\xi \circ (L_o)'_\xi \circ \tau'_\xi(v) \|_{L^\infty(\mathcal{O})} \leq \| L_o(v) \|_{L^\infty(\mathcal{O})} , \quad \forall \xi \in \mathbb{R}_+^N .$$

Let now v be any function satisfying

$$(32)\qquad v \in H^1_o(\mathcal{O}) \quad , \quad A(v) \in L^\infty(\mathcal{O}) \ .$$

We then have

$$(33)\qquad v \ , \ Dv \ , \ L_o v \in L^\infty(\mathcal{O})$$

and by simple interpolation estimates it also follows from (31) and the hypotheses (ii) and (iii) that

$$(34)\qquad \pi'_\xi \circ A'_\xi \circ \tau'_\xi(v) \in L^\infty(\mathcal{O}) \qquad \forall \xi \in \mathbb{R}^N_+$$

and

$$(35)\qquad \| \pi'_\xi \circ A'_\xi \circ \tau'_\xi(v) \|_{L^\infty(\mathcal{O})} \leq c$$

with $c > 0$ a constant independent on $\xi$.

We are now in a position to make the third step of the proof, which consists in proving that

$$(36)\qquad A(u^n_F) \in L^\infty(\mathcal{O}) \quad , \quad \forall n \geq 0 \ , \ \forall F \in \mathcal{F}.$$

Since $A(u^o_F) = A(u^o) \in L^\infty(\mathcal{O})$ by (18), in order to prove (36) it suffices to show that for every $F \in \mathcal{F}$ and $n \geq 1$,

$$(37)\qquad A(u^{n-1}_F) \in L^\infty(\mathcal{O})$$

implies

$$(38)\qquad A(u^n_F) \in L^\infty(\mathcal{O}) \ .$$

In fact, by (37) the function $v = u^{n-1}_F$ satisfies the hypothesis (32), hence (34) holds, that is,

$$\pi'_\xi \circ A'_\xi \circ \tau'_\xi(u^{n-1}_F) \in L^\infty(\mathcal{O}) \ ;$$

moreover, by the assumptions (iii) and (iv), we also have $A(0) = -H(x,0) \in L^\infty(\mathcal{O})$ and $f \in L^\infty(\mathcal{O})$. Therefore (38) is an immediate consequence of the recursive estimate (29).

The last step of the proof of Proposition 1 consists in proving that the uniform estimate (25) holds. Again by trivial interpolation estimates, it suffices prove that

(39) $$\|L_o u_F^n\|_{L^\infty(\mathcal{O})} \le c \qquad \forall n \ge 0 \quad \forall F \in \mathcal{F},$$

with c some constant independent on n and F.

Let us remark first that for every $F \in \mathcal{F}$ each iterate $u_F^n$, $n \ge 1$, can be regarded - with notation taken from above - as the solution of the VI

(40) $$u_F^n = \sigma_{f_{n,F}}^{L_o}(M_F(u_F^{n-1}))$$

where

(41) $$f_{n,F} = f - A_1(u_F^n)$$

and

$$A_1(u_F^n) = A(u_F^n) - L_o(u_F^n) = \sum_{j=1}^n a_j \frac{\partial u_F^n}{\partial x_j} + a_o u_F^n + G(u_F^n).$$

By the assumptions (ii) (iii) and (iv), we have

$$\|f_{n,F}\|_{L^\infty(\mathcal{O})} \le c\|Du_F^n\|_{L^\infty(\mathcal{O})} + c\|u_F^n\|_{L^\infty(\mathcal{O})} + c$$

for every $n \ge 1$ and $F \in \mathcal{F}$, hence also, by (30),

(42) $$\|f_{n,F}\|_{L^\infty(\mathcal{O})} \le c\|Du_F^n\|_{L^\infty(\mathcal{O})} + c$$

for some constants c independent on n and F.

As a consequence of (36), we have

(43) $$L_o(u_F^n) \in L^\infty(\mathcal{O})$$

moreover, $Du_F^n \in L^\infty(\mathcal{O})$, hence also, from (42)

(44) $$f_{n,F} \in L^\infty(\mathcal{O}) \quad , \quad \forall n \ge 1 \quad \forall F \in \mathcal{F}.$$

We are thus in a position, for every $F \in \mathcal{F}$ and each fixed $n \ge 1$, to make use of estimate (29), with

$$A = L_o \qquad \text{and} \qquad f = f_{n,F}\ .$$

We thus have

$$(45) \qquad f_{n,F} \geq L_o u_F^n \geq f_{n,F} \bigwedge_{\xi\in F} \pi'_\xi \circ (L_o)'_\xi \circ \tau'_\xi (u_F^{n-1}) \wedge 0 \quad \text{a.e. in } \mathcal{O},$$

for every $F \in \mathcal{F}$ and all $n \geq 1$.

By taking (31) and (44) into account, it follows from (45) that

$$(46) \qquad \|L_o u_F^n\|_{L^\infty(\mathcal{O})} \leq \max\left\{\|f_{n,F}\|_{L^\infty(\mathcal{O})} \; , \; \|L_o(u_F^{n-1})\|_{L^\infty(\mathcal{O})}\right\}$$

for every $F \in \mathcal{F}$ and all $n \geq 1$.

On the other hand, by (42),(30) and classical interpolation estimates, we have

$$(47) \qquad \|f_{n,F}\|_{L^\infty(\mathcal{O})} \leq c\|L_o u_F^n\|^\delta_{L^\infty(\mathcal{O})} + c \; ,$$

with $\frac{1}{2} < \delta < 1$ and c some constants independent on $n \geq 1$ and $F \in \mathcal{F}$.

By taking (43) into account, the uniform estimate (39) is then an immediate consequence of (46) and (47).

The proof of Proposition 1 has thus been completed.

5. Let us now consider the sequence of iterates (19). The existence of $u^n$, for every $n \geq 1$, is shown as before the existence of the iterates $u_F^n$.

We then prove that the sequence $\{u^n\}_{n\geq 1}$ verifies the uniform estimate

$$(48) \qquad \|A(u^n)\|_{L^\infty(\mathcal{O})} \leq c$$

with a constant c independent on n.

In fact, let for each $k=1,2,\ldots$, $F_k \in \mathcal{F}$ be such that

$$(49) \qquad d_{F_k,\mathcal{O}} \leq \frac{1}{k}$$

where for each $F \in \mathcal{F}$ we define

$$d_{F,\mathcal{O}} = \sup_{\substack{\xi \in \mathbb{R}^N_+ \\ |\xi| \leq \operatorname{diam} \mathcal{O}}} \operatorname{dist}(\xi,F)$$

In consequence of the estimate (25), (a subsequence of) the $\{u^n_{F_k}\}_{k\geq 1}$ converges to a function

$$\tilde{u}^n \in H^1_o(\mathcal{O}) \cap W^{2,p}(\mathcal{O}) \quad , \quad p \geq 2 \; ,$$

weakly in $W^{2,p}(\mathcal{O})$ for all $p \geq 2$ and strongly in $H^1_0(\mathcal{O})$. Therefore, again by (25), the sequence $\{A(u^n_{F_k})\}_{k\geq 1}$ converges to $A(\tilde{u}^n)$ in the weak$^*$topology of $L^\infty(\mathcal{O})$ and

(50) $$\|A(\tilde{u}^n)\|_{L^\infty(\mathcal{O})} \leq c,$$

with c a constant independent on n.

We then prove that for every $F \in \mathcal{F}$ and every $n \geq 1$, we have

(51) $$\|u^n - u^n_F\|_{L^\infty(\mathcal{O})} \leq c\, n\, d_{F,\mathcal{O}}$$

with c independent on $n \geq 1$ and $F \in \mathcal{F}$, hence $\tilde{u}^n = u^n$ for all $n \geq 1$ and (48) follows from (50). For more details on (51) we refer to U.Mosco [2].

Let us now consider the sequence $\{u^n\}$, beginning with the case when $u^0$ is a super-solution or a sub-solution. Then, as a consequence of comparison theorems, $\{u^n\}$ is monotone; by taking (48) into account, we can conclude that the sequence $\{u^n\}$ converges to some function $\tilde{u}$ weakly in $W^{2,p}(\mathcal{O})$ and strongly in $H^1_0(\Omega)$ and $L^\infty(\mathcal{O})$. On the other hand, since $\tilde{u} \geq -1$ in $\mathcal{O}$, the solution $\sigma^A_f \circ M(\tilde{u})$ exists and we have for every $n \geq 1$

$$\|u^n - \sigma^A_f \circ M(\tilde{u})\|_{L^\infty(\mathcal{O})} = \|\sigma^A_f \circ M(u^{n-1}) - \sigma^A_f \circ M(\tilde{u})\|_{L^\infty(\mathcal{O})} \leq \|u^{n-1} - \tilde{u}\|_{L^\infty(\mathcal{O})},$$

since the map $\sigma^A_f \circ M$ is non-expansive for the $L^\infty$ norm. Therefore, $u^n$ converges to $\sigma^A_f \circ M(\tilde{u})$ in $L^\infty(\mathcal{O})$ as $n \to \infty$, hence

$$\tilde{u} = \sigma^A_f \circ M(\tilde{u}) \quad \text{a.e. in } \mathcal{O}$$

which is to say, the function (52) coincides a.e. in $\mathcal{O}$ with the (unique) solution u of problem (1).

As for $\{u^n\}$ constructed from $u^0$ as in the general case, let us remark that, by comparison theorems, $u^n$ verifies

$$(\sigma^A_f \circ M)^n z \leq u^n \leq (\sigma^A_f \circ M)^{n-1} \bar{u},$$

and that, from what we have already proved, both sequences $\{(\sigma^A_f \circ M)^n z\}$ and $\{(\sigma^A_f \circ M)^n \bar{u}\}$ converge to u in $L^\infty(\Omega)$. This suffices, by taking (48) again into account, to complete the proof of the theorem.

The author wishes to thank Haïm Brezis and Jean Pierre Puel for some stimulating discussion on the subject of this paper.

REFERENCES

A.BENSOUSSAN and J.L.LIONS, [1] C.R.Acad. Sc. Paris, t.278 série A, 1974, p.675; [2], Optimal Impulse and Continuous Control: Methos of non linear quasi-variational inequalities, Trudy Matěmaticěskovo Institut Imeni Stěklova, to appear.

J.L.JOLY, U.MOSCO and G.M.TROIANIELLO, [1] C.R. Acad. Sc. Paris, t.279, Série A, 1974, p.937; [2], On the regular solution of a quasi-variational inequality connected to a problem of stochastic impulse control, J.Math.Anal.and Appl.61, 2, 1977.

T.LAETSCH, [1] J.F. Analysis 18, 1975, p. 286-287.

U.MOSCO, [1] Implicit variational problems and quasi-variational inequalities, in Nonlinear Operators and the Calculus of Variations, Bruxelles 1975, Ed. L.WAELBROECK, Springer Lecture Notes in Mathematics, 543, 1976, p. 83-156; [2] Sur l'existence de la solution regulière de l'inéquation quasi variationnelle non linéaire du contrôle optimale impulsionnel et continue, Journées d'Analyse Non linéaire, Besançon 1977, Springer Lecture Notes in Mathematics, to appear; [3] Régularité forte de la fonction d'Hamilton-Jacobi du contrôle stochastique impulsionnel et continu, C.R. Acad. Sc. Paris, t. 286, Série A, 1978, p. 211.

Author's address: Umberto Mosco, Istituto dell'Università di Roma, Italy, and
CEREMADE, Université Paris IX Dauphine, Paris, France

ON THE REGULARITY OF WEAK SOLUTIONS TO VARIATIONAL EQUATIONS AND INEQUALITIES FOR NONLINEAR SECOND ORDER ELLIPTIC SYSTEMS

J. Nečas, Praha

The history of the solution of the 19th Hilbert's problem, i.e. in fact of the problem of regularity, is described in the book by O.A. Ladyženskaja, N.N. Uralceva [1] and in the paper by Ch.B. Morrey [2]. The weak solution is a vector function from the Sobolev space $W^{1,2}(\Omega)$ satisfying the equations in the sense of distributions and it is regular if it belongs to $C^{(1)}(\Omega)$ (interior regularity) or $C^{(1)}(\bar{\Omega})$ (regularity up to the boundary).

The problem of the regularity for the dimension $n=2$ was solved very soon (1937) by Ch.B. Morrey [3], also for systems. The generalization of this result for higher dimensions, but only for a single second-order equation, was done by E. De Giorgi [4] in 1957 and his method, based in fact on the maximum principle, was further developed by several authors, see J. Moser [5], G. Stampacchia [6], the book [1] and others.

If we consider a vector function $u = (u_1, u_2, \dots u_m)$ from $[W^{1,2}(\Omega)]^m$, satisfying the system

$$(1) \qquad -\frac{\partial}{\partial x_i}\left[a_i^r(x,u,\nabla u)\right] + a_0^r(x,u,\nabla u) = f_r(x), \qquad r=1,2,\dots m,$$

if $f_r \in L_2(\Omega)$ and if, for example, $u = u^0$ on $\partial\Omega$, where $u^0 \in [W^{1,2}(\Omega)]^m$ is a prescribed function, then the existence of a weak solution as well as its uniqueness can be proved relatively easily, see for example the book by J.L. Lions [7], under the standard assumptions:

$$(2) \qquad \left|\frac{\partial a_i^r}{\partial \frac{\partial u_s}{\partial x_j}}\right| + \left|\frac{\partial a_i^r}{\partial u_s}\right| + \left|\frac{\partial a_0^r}{\partial \frac{\partial u_s}{\partial x_j}}\right| + \left|\frac{\partial a_0^r}{\partial u_s}\right| \le c,$$

$$(3) \qquad \frac{\partial a_i^r}{\partial \frac{\partial u_s}{\partial x_j}} \eta_i^r \eta_j^s \ge \alpha \eta_i^r \eta_i^r, \qquad \alpha > 0,$$

$$(4) \qquad a_i^r(x,\eta_s,\eta_j^s)\eta_i^r + a_0^r(x,\eta_s,\eta_j^s)\eta_r \ge \beta \eta_i^r \eta_i^r, \qquad \beta > 0,$$

provided that the derivatives of $a_i^r$ in (2), (3) satisfy the Caratheodory condition.

Instead of equations, we can study inequalities, if for example on $\partial\Omega$ (or on some part of $\partial\Omega$) a unilateral condition of

Signorini's type

$$(5)\qquad b_{rs}u_s \geq \psi_r, \qquad r=1,2,\dots,k \leq m,$$

is given. Writing

$$(6)\qquad K \equiv \{v \in [W^{1,2}(\Omega)]^m \mid b_{rs}v_s \geq \psi_r \text{ on } \partial\Omega\},$$

we look for $u \in K$ such that, $\forall v \in K$,

$$(7)\qquad \int_\Omega a_i^r(x,u,\nabla u)\left(\frac{\partial v_r}{\partial x_i} - \frac{\partial u_r}{\partial x_i}\right) dx \geq \int_\Omega f_r(v_r-u_r)\,dx.$$

For the existence and other questions, see [7]. The conditions (2) and (3) guarantee the first step to the interior regularity, i.e., the proof of the inclusion $u \in [W^{2,2}(\Omega')]^m$, $\bar{\Omega}' \subset \Omega$. If $\Omega' = \Omega$, we get the first step to the regularity up to the boundary. For the idea of the proof of this step, see also [7]. If, for simplicity, we restrict ourselves in the following to the case $a_i^r(x,u,\nabla u) = a_i^r(\nabla u)$, $a_0^r = 0$, then we can immediately see that this first step leeds to an equation in variations obtained through integration by parts of the equation

$$(8)\qquad \int_\Omega a_i^r(\nabla u)\frac{\partial \varphi_r}{\partial x_i}\,dx = \int_\Omega f_r \varphi_r\,dx, \qquad \varphi_r \in \mathcal{D}(\Omega);$$

if we denote by $u'$ some derivative, then we get from (8), substituting here $\varphi'$:

$$(9)\qquad \int_\Omega a_{ij}^{rs}\frac{\partial u_s'}{\partial x_j}\frac{\partial \varphi_r}{\partial x_i}\,dx = \int_\Omega f_r'\varphi_r\,dx,$$

where $a_{ij}^{rs} = \dfrac{\partial a_i^r}{\partial \frac{\partial u_s}{\partial x_j}}$. (9) is a linear system in $u'$ with, in general, only measurable, bounded coefficients $a_{ij}^{rs}$.

Let us mention the known fact that once being $u \in [C^{(1)}(\Omega)]^m$, we get arbitrary higher regularity of the solution, provided that the coefficients and right-hand sides are regular enough.

The significance of the problem of regularity is underlined by the fact that the regularity up to the boundary, provided that the coefficients are analytic, implies that in the potential case the set of critical values is a sequence, tending to zero, see S. Fučík, J. Nečas, J. Souček, V. Souček [8]. Also the Newton's type methods are convergent only in the space of regular solutions.

For more general systems than discussed in the paper [2], J.

Stará proved the regularity in [9], also for $r=2$, using the method of the papers by J. Nečas [10], [11] concerning higher order single equations.

E. Giusti, M. Miranda constructed in the paper [12], for $n\geq 3$, a regular functional, whose critical point is $u = \frac{x}{|x|}$. This functional is continuous on $[W^{1,2}(\Omega)]^m$, but not differentiable. In the same work, a system with coefficients $a_i^r = A_{ij}^{rs}(u)\frac{\partial u_s}{\partial x_j}$, $s = 1,2,\dots,n$, is constructed with the ellipticity condition

$$A_{ij}^{rs}\eta_i^r\eta_j^s \geq \alpha|\eta|^2 \tag{10}$$

and the same solution $\frac{x}{|x|}$. Some variations of this type of example can be found in the paper by S.A. Arakčejev [13].

Let us start with a more detailed description of the results of the paper by J. Nečas [14] with small complements.

The easiest example of a fourth order system with a non-regular solution is

$$\Delta^2 u_i + \frac{1}{(n+1)^2(n-2)}\frac{\partial^2}{\partial x_j \partial x_k}[\Delta u_i \Delta u_j \Delta u_k] = 0, \tag{11}$$

provided that this system is defined on the set of $u$'s such that $\Delta u_i \Delta u_i \leq (n+1)^2$. The solution of this system is $u_i = x_i|x|$ and the corresponding conditions (3), (4) are satisfied for $n\geq 6$.

$K$ being the unit ball $|x|<1$, let us consider the system (in the weak formulation),

$$\begin{aligned}\int_K \frac{\partial u_{ij}}{\partial x_k}\frac{\partial \varphi_{ij}}{\partial x_k}dx + \lambda_2\int_K \frac{\partial u_{kk}}{\partial x_i}\frac{\partial \varphi_{\ell\ell}}{\partial x_i}dx + \\ +\lambda_3\int_K \frac{\partial u_{\alpha j}}{\partial x_\alpha}\frac{\partial \varphi_{\beta j}}{\partial x_\beta}dx + \lambda_4\int_K \frac{\partial u_{\alpha\beta}}{\partial x_\alpha}\frac{\partial \varphi_{ii}}{\partial x_\beta}dx + \\ +\lambda_1\int_K \frac{\partial u_{\gamma i}}{\partial x_\gamma}\frac{\partial u_{\alpha j}}{\partial x_\alpha}\frac{\partial u_{\beta\ell}}{\partial x_\beta}\frac{\partial \varphi_{ij}}{\partial x_\ell}dx = 0,\end{aligned} \tag{12}$$

$i,j=1,2,\dots n$, where $\lambda_3 = \frac{2(n-1)^3-9n}{9(n-1)}$ for $n\geq 5$, $\lambda_3 = \frac{2(n-1)^3-n}{n-1}$ for $3\leq n\leq 4$, $\lambda_1 = \frac{n+\lambda_3(n-1)}{(n-1)^4(n+1)^2}n^2$, $\lambda_4 = -\frac{1+\lambda_3(n-1)}{(n-1)^2}$ and $\lambda_2$ is large enough. If we put $u_{ij} = \frac{x_i x_j}{|x|} - \frac{1}{n}\delta_{ij}|x|$, then $u_{ii}=0$ and $u_{ij}$ satisfy (12). The coefficients of (12) are defined only for such "$\xi_k^{ij}$" where

$\frac{\partial u_{\alpha i}}{\partial x_\alpha} \frac{\partial u_{\beta i}}{\partial x_\beta} \leq \left( \frac{n^2-1}{n} + \delta \right)^2$, $\delta > 0$ and small enough.

For $n \geq 5$, we have (3) and (4), for $3 \leq n \leq 4$, we have (4). If we replace the nonlinear term in (12) by

$$(13) \qquad \lambda_1 \left( \varepsilon + \frac{(n^2-1)^2}{n^2} \right) \int_K \frac{\partial u_{\gamma i}}{\partial x_\gamma} \frac{\partial u_{\alpha j}}{\partial x_\alpha} \frac{\partial u_{\beta \ell}}{\partial x_\beta} \cdot$$

$$\cdot \left( \varepsilon + \frac{\partial u_{ab}}{\partial x_a} \frac{\partial u_{cb}}{\partial x_c} \right)^{-1} \frac{\partial \varphi_{ij}}{\partial x_\ell} \, dx$$

with $\varepsilon > 0$ small enough, we get the same result with coefficients defined everywhere.

If we consider the functional of the total potential energy in finite elasticity under the incompressibility constraint then there exists a universal, isotropic body, see C. Truesdell [16], and its deformation from the so-called 5th class, a critical point of the functional under the constraint, which is not regular. But the set of irregular points is a segment, so it is not possible to get in this way immediately an example of an irregular solution without a constraint because the Hausdorff measure of irregular points must be less than 1, see E. Giusti [17].

The example (12), (13) is a vector function $|x| f\left(\frac{x}{|x|}\right)$ and the functions $f_i(\xi)$ are not linear in $\xi$. If we write such an example in polar coordinates $r, \vartheta_1, \vartheta_2, \ldots \vartheta_{n-1}$, $0 < r < \varepsilon$, $0 < \vartheta_j < \pi$, $j=1,2,\ldots,n-2$, $0 < \vartheta_{n-1} < 2\pi$, putting

$$(14) \qquad \begin{aligned} x_1 &= r \cos \vartheta_1, \; x_2 = r \sin \vartheta_1 \cos \vartheta_2, \ldots, x_{n-1} = \\ &= r \sin \vartheta_1 \sin \vartheta_2 \ldots \sin \vartheta_{n-2} \cos \vartheta_{n-1}, \\ x_n &= r \sin \vartheta_1 \sin \vartheta_2 \ldots \sin \vartheta_{n-2} \sin \vartheta_{n-1}, \end{aligned}$$

$$(15) \qquad \partial_1 v = \frac{\partial v}{\partial r}, \; \frac{1}{r} \frac{\partial v}{\partial \vartheta_1} = \partial_2 v, \; \frac{1}{r \sin \vartheta_1} \frac{\partial v}{\partial \vartheta_2} = \partial_3 v, \ldots,$$

$$\frac{1}{r \sin \vartheta_1 \ldots \sin \vartheta_{n-2}} \frac{\partial v}{\partial \vartheta_{n-1}} = \partial_n v,$$

we first get $\partial_i v = a_{ij} \frac{\partial v}{\partial x_j}$, where $a_{ij}$ is an orthonormal matrix. Let us define the elementary differential operators

$$(16) \qquad \bar{\partial}_2 h = \frac{\partial h}{\partial \vartheta_1}, \; \bar{\partial}_3 h = \frac{1}{\sin \vartheta_1} \frac{\partial h}{\partial \vartheta_2}, \; \ldots, \; \bar{\partial} h = (\bar{\partial}_2 h, \ldots \bar{\partial}_n h).$$

We introduce the space $W^{1,2}(S)$, $S$ being the unit sphere, as the closure of infinitely differentiable functions in the norm

$$(17)\qquad \Big(\int_S [f^2 + \bar{\partial}_{j'} f \bar{\partial}_{j'} f]\, dS\Big)^{1/2}$$

where the indices with primes are summed from 2 up to $n$. Starting from the system (8), we get for $f$ another system on the unit sphere

$$(18)\qquad \int_S [-(n+1)A_1^r(\vartheta, f, \bar{\partial} f)\tilde{f}_r + A_{j'}^r(\vartheta, f, \bar{\partial} f)\bar{\partial}_{j'}\tilde{f}_r]\, dS = 0,$$

where $f, \tilde{f} \in [W^{1,2}(S)]^m$.

We get immediately

$$(19)\qquad \left|\frac{\partial A_j^r}{\partial f_s}\right| + \left|\frac{\partial A_j^r}{\partial\, \partial_{j'} f_s}\right| \leq c,$$

$$(20)\qquad \frac{\partial A_j^r}{\partial(\bar{\partial}_{i'} f_s)}\xi_{j'}^r \xi_{i'}^s \geq \alpha\, \xi_{j'}^r \xi_{j'}^r, \qquad \alpha > 0.$$

Let $J$ be the kernel of (18), i.e., the set of all the solutions from $[W^{1,2}(S)]^m$. We introduce $J_0 \subset J$, the trivial subset of $J$, consisting of the linear combinations of the coordinate functions $\cos\vartheta_1$, $\sin\vartheta_1\cos\vartheta_2$, ... $\sin\vartheta_1 \sin\vartheta_2 \ldots \sin\vartheta_{n-2}\sin\vartheta_{n-1}$.

Let us consider a weak solution to the system $(K_\varepsilon \equiv \{|x| \leq \varepsilon\})$

$$(21)\qquad \int_{K_\varepsilon} a_i^r(\nabla u)\frac{\partial \varphi_r}{\partial x_i}\, dx = 0.$$

We easily get

Theorem 1. The necessary condition for the regularity of every weak solution to (21) is $J = J_0$.

Proof: Let us suppose the contrary and let us take $f \in J \setminus J_0$. Put $u = rf(\vartheta)$; $u$ satisfies (21) and so

$$\frac{\partial u_r}{\partial x_i}(0) = \lim_{r\to 0}\frac{\partial u_r}{\partial x_i}(x) = \lim_{r\to 0}[a_{1i}(\vartheta)f_r(\vartheta) + a_{j,i}(\vartheta)\bar{\partial}_{j'}f_r(\vartheta)] = $$

$$= \frac{\partial u_r}{\partial x_i}(x);$$ hence $u_r(x)$ is a linear function and, therefore, $f \in J_0$, which is impossible, q.e.d.

So the study of the kernel $J$ leads to the construction of an irregular solution in 3 and 4 dimensions. If $J = J_0$, we can hope that this condition is sufficient for the regularity of every weak solution of this equation.

Let us consider some sufficient conditions for the regularity in 3 dimensions in more detail. We refer to the papers [14] and [15].

Let us consider the Euler equation

$$(22)\qquad \int_\Omega \frac{\partial F}{\partial \xi_i^r}(\nabla u)\frac{\partial \varphi_r}{\partial x_i}\,dx = \int_\Omega f_i^r \frac{\partial \varphi_r}{\partial x_i}\,dx,$$

where the Lagrangian $F(\xi)$ is defined and continuous together with its 4 derivatives in the cube $\overline{K}_a = \{\xi \mid |\xi_i^r| \le a\}$. Let $\partial\Omega$ be smooth enough, let $f_i^r \in W^{2,2}(\Omega)$, $u_r^0 \in W^{3,2}(\Omega)$, and let us look for a solution $u$ of (22) such that $u \in [W^{3,2}(\Omega)]^m$, $u = u^0$ on $\partial\Omega$, $\|u\|_{1,\infty} = \max_{r,i}\left(\max_{x\in\bar\Omega}\left|\frac{\partial u_r}{\partial x_i}(x)\right|\right) < a$. We shall suppose the ellipticity condition

$$(23)\qquad \frac{\partial^2 F}{\partial \xi_i^r \partial \xi_j^s}(\xi)\eta_i^r\eta_j^s \ge c_1|\eta|^2, \qquad c_1 > 0$$

and the regularity condition

$$(24)\qquad c_1 - 3a^2 T > 0,$$

where

$$(25)\qquad \frac{\partial^4 F}{\partial \xi_i^r \partial \xi_j^s \partial \xi_k^t \partial \xi_e^v}(\xi)\eta_i^r\eta_j^s\eta_k^t\eta_e^v \le T\sum_{r,i}(\eta_i^r)^4.$$

<u>Theorem 2.</u> (A priori estimate.) Let (23), (24) be satisfied. If $u$ is the solution in question, then

$$(26)\qquad \|u\|_{3,2} \le c(1 + \|f\|_{2,2}^2 + \|u^0\|_{3,2}^2).$$

For the half space $R_3^+$ we get

<u>Theorem 3.</u> Let $\Omega = R_3^+$, $u^0 = 0$, let (23), (24) be satisfied. Then

$$(27)\qquad \|u\|'_{3,2} \le c(\|f\|_{2,2} + \|f\|_{2,2}^2)$$

$$\left(\|u\|'_{k,2} \equiv \left(\int_{R_3^+} \sum_{1\le\alpha\le k}(D^\alpha u)^2\,dx\right)^{1/2}\right).$$

<u>Main idea of the proof:</u> Let $'$ be the derivative $\frac{\partial}{\partial x_1}, \frac{\partial}{\partial x_2}$.

We have

$$(28) \quad \int_{R_3^+} (f_i^r)'' \frac{\partial \varphi_r''}{\partial x_i}\, dx = \int_{R_3^+} \frac{\partial^2 F}{\partial \xi_i^r \partial \xi_j^s} \frac{\partial u_s''}{\partial x_j} \frac{\partial \varphi_r''}{\partial x_i}\, dx +$$

$$+ \int_{R_3^+} \frac{\partial^3 F}{\partial \xi_i^r \partial \xi_j^s \partial \xi_k^t} \frac{\partial u_s'}{\partial x_j} \frac{\partial u_t'}{\partial x_k} \frac{\partial \varphi_r''}{\partial x_i}\, dx \equiv I_1 + I_2 .$$

Substituting the function $u$ for $\varphi$ in (28), we get

$$(29) \quad I_2 = -\frac{1}{3} \int_{R_3^+} \frac{\partial^4 F}{\partial \xi_i^r \partial \xi_j^s \partial \xi_k^t \partial \xi_\ell^v} \frac{\partial u_r'}{\partial x_i} \frac{\partial u_s'}{\partial x_j} \frac{\partial u_t'}{\partial x_k} \frac{\partial u_v'}{\partial x_\ell}\, dx .$$

Through integration by parts, we obtain $\forall\, v \in \mathcal{D}(R_1)$ :

$$(30) \quad \int_{-\infty}^{\infty} (v')^4\, dx \leq 9 \max_{x \in R_1} [v(x)]^2 \cdot \int_{-\infty}^{\infty} (v'')^2\, dx .$$

From (28) - (30) we obtain

$$(31) \quad (c_1 - 3\, T\, a^2) \| u'' \|'_{1,2} \leq \| f \|_{1,2} .$$

(29), (30) imply also

$$(32) \quad \int_{R_3^+} \sum_{r,i} \left( \frac{\partial u_r'}{\partial x_i} \right)^4 dx \leq \frac{9a^2}{(c_1 - 3Ta^2)^2} \| f \|_{2,2}^2 .$$

Because the derivatives $\dfrac{\partial^2 u_s}{\partial x_3^2}$ can be calculated from the elliptic system

$$(33) \quad - \frac{\partial^2 F}{\partial \xi_i^r \partial \xi_j^s} \frac{\partial^2 u_s}{\partial x_i \partial x_j} = - \frac{\partial f_i^r}{\partial x_i}, \qquad r = 1,2,\dots m,$$

we get from (32) and (33)

$$(34) \quad \| u \|_{2,4} \leq c_2 c_1^{-1} [c_1 - 3Ta^2]^{-1/2} [\| f \|_{2,2}^{1/2} + \| f \|_{2,2}] .$$

Differentiating (33) first with respect to $x_1$ and $x_2$, we get

$$(35) \quad \| u' \|_{2,2} \leq c_3 c_1^{-2} [c_1 - 3Ta^2]^{-1} ( \| f \|_{2,2} + \| f \|_{2,2}^2 ) .$$

Finally, differentiating (33) with respect to $x_3$, we get (27)

q.e.d.

The existence and uniqueness of a solution path, i.e., of $u$

from $C([0,t_{cr}], [W^{3,2}(\Omega)]^m)$, can be easily proved by the implicit function theorem and (26), provided that $f_i^r \in C([0,T], W^{2,2}(\Omega))$, $u_r^0 \in C([0,T], W^{3,2}(\Omega))$, see [14] and [15]. $t_{cr} \leq T$ and is maximal, i.e., if $t_{cr} < T$, then $\max_{r,i} \left( \max_{x \in \bar{\Omega}} \frac{\partial u_r}{\partial x_i}(x)\right) = a$.

The papers [3], [9], [10], [11] and the papers by J. Nečas [18], J. Kadlec, J. Nečas [19] contain, in fact, estimates of the condition number, i.e. the estimate

$$\frac{c_1}{c_2} > h(n) \geq 0, \tag{36}$$

implying the regularity. Here, as before,

$$c_1 |\eta|^2 \leq a_{ij}^{rs}(\xi)\eta_i^r\eta_j^s \leq c_2|\eta|^2, \tag{37}$$

where, for simplicity, we suppose $a_{ij}^{rs} = a_{ji}^{sr}$. In all the mentioned papers, $h(n)$ is not evaluated, each time only $h(2) = 0$. A precise evaluation is done by A.I. Koševev, see [20], where for systems a generalization of H.O. Cordes's condition is given, see [21]. Koševev's condition implies that the weak solution belongs to $[C^{(0),\mu}(\bar{\Omega})]^m$ which follows also from the fact that the weak solution belongs to $\bigcap_{p>1}[W^{1,p}(\Omega)]^m$, provided that some asymptote type conditions are valid for the functions $a_i^r(\xi)$, see J. Nečas [22].

We shall sketch the proof of

Theorem 4. Let

$$\frac{c_1}{c_2} > \frac{\sqrt{1+\frac{(n-2)^2}{n-1}} - 1}{\sqrt{1+\frac{(n-2)^2}{n-1}} + 1}. \tag{38}$$

Then the weak solution $u$ to the equation

$$\int_\Omega a_i^r(\nabla u)\frac{\partial \varphi_r}{\partial x_i}\,dx = \int_\Omega f_i^r \frac{\partial \varphi_r}{\partial x_i}\,dx \tag{39}$$

lies in $[C^{(1),\mu}(\Omega)]^m$, provided that $f_i^r \in W^{1,p}(\Omega)$, $p > n$, and $\mu = 1 - \frac{n}{p}$. We have for $\bar{\Omega}' \subset \Omega$:

$$\|u\|_{[C^{1,\mu}(\bar{\Omega}')]^m} \leq c(\Omega')\left[\|f\|_{1,p} + \|u\|_{1,2}\right], \tag{40}$$

| $n$ | 2 | 3 | 4 | 5 |
|---|---|---|---|---|
| $\frac{c_1}{c_2}$ | 0 | 0,101 | 0,209 | 0,286 |

.

<u>Lemma 1.</u> Let $u \in W^{1,2}(K_\varepsilon) \cap C^{(0),\gamma}(K_\varepsilon)$ be the solution of

$$-\triangle u = -\frac{\partial g_i}{\partial x_i}\,. \tag{41}$$

Let $n-2 < \lambda < n-2+2\gamma$. Then

$$\int_{K_\varepsilon} \frac{\partial u}{\partial x_i}\frac{\partial u}{\partial x_i} r^{-\lambda}\,dx \le \alpha(\lambda)\left(1+\frac{(n-2)^2}{n-1}\right). \tag{42}$$

$$\left[\int_{K_\varepsilon} g_i g_i r^{-\lambda}\,dx + c_3\int_{K_\varepsilon} g_i g_i\,dx + \right.$$

$$\left. + c_3\int_{K_\varepsilon}\frac{\partial u}{\partial x_i}\frac{\partial u}{\partial x_i}\,dx + c_3|u(0)|^2\right],$$

where $\alpha(\lambda) \to 1$ as $\lambda \to (n-2)$.

<u>Proof.</u> Put $v(x) \equiv (u(x)-u(0))\,\psi(x)$, $\psi \in \mathcal{D}(K_\varepsilon)$, $\psi(x) = 1$ for $|x| \le \frac{\varepsilon}{2}$, $0 \le \psi \le 1$. We have

$$\int_{R_n} \frac{\partial v}{\partial x_i}\frac{\partial u}{\partial x_i}\,dx = \int_{R_n} h_i\frac{\partial u}{\partial x_i}\,dx, \tag{43}$$

with

$$\left[\int_{R_n} h_i h_i (1+r^{-\lambda})\,dx\right]^{1/2} \le \left[\int_{K_\varepsilon} g_i g_i r^{-\lambda}\,dx\right]^{1/2} + \tag{44}$$

$$+ c_4\left[\int_{K_a} g_i g_i\,dx\right]^{1/2}.$$

In polar coordinates (employing our notation) we get for $v$:

$$\int_0^\infty\int_S \partial_i v\,\partial_i\varphi\, r^{n-1}dr\,dS = \int_0^\infty\int_S m_i\,\partial_i\varphi\, r^{n-1}\,dr\,dS, \tag{45}$$

where $m_i = a_{ij}h_j$. Putting $p' = p-(n-3)$, $p = x'+iy$, $x = -\frac{1}{2}\,[\lambda-(n-2)]$, $y \in R_1$, $x < x' < 0$, $x \ge -\gamma$, $\varphi = w(\vartheta)r^{p'-1}$, $w \in \mathcal{E}(S)$ and denoting by

$$(46)\quad V(p,\vartheta) \equiv \int_0^\infty r^{p-1} v(r,\vartheta)\, dr, \quad M_i(p,\vartheta) \equiv \int_0^\infty r^p m_i(r,\vartheta)\, dr,$$

Mellin's transforms of $v$ and $m_i$, we obtain from (45)

$$(47)\quad \int_S [-p(p'-1)Vw + \bar{\partial}_{i'} V \bar{\partial}_{i'} w]\, dS = \int_S [M_1(p'-1)w + M_{i'} \bar{\partial}_{i'} w]\, dS.$$

Let us decompose $W^{1,2}(S) = W_1 + W_2$, $W_1 \equiv \{\mathrm{const}\}$, let $V = V_1 + V_2$, $M = M_1^1 + M_1^2$ be orthogonal decompositions, the latter in $L_2(S)$. Put $w = -\dfrac{\bar{p}}{p'-1}\bar{V}_1$ in (47). We obtain

$$(48)\quad \int_S |p|^2 |V_1|^2\, dS \leq \int_S |M_1|^2\, dS.$$

Put $w = \bar{V}_2$. Because we have

$$(49)\quad (n-1)\int_S |w|^2\, dS \leq \int_S \bar{\partial}_{i'} w \bar{\partial}_{i'} \bar{w}\, dS,$$

for $w$ from $W_2$ we get in virtue of (48):

$$(50)\quad \int_S [|p|^2 |V|^2 + \bar{\partial}_{i'} V \bar{\partial}_{i'} \bar{V}]\, dS \leq$$

$$\leq \left[1 + \frac{(n-2)^2}{n-1} - 2x' \frac{n-2}{n-1}\right] \int_S M_i \bar{M}_i\, dS.$$

Parseval's identities

$$(51)\quad \frac{1}{2\pi}\int_{-\infty}^{\infty} [|p|^2 |V|^2 + \bar{\partial}_{i'} V \bar{\partial}_{i'} \bar{V}]\, dy = \int_0^\infty (\partial_i v \partial_i v) r^{-\lambda' + n-1}\, dr,$$

$$\frac{1}{2\pi}\int_{-\infty}^{\infty} M_i \bar{M}_i\, dy = \int_0^\infty m_i m_i r^{-\lambda' + n-1}\, dr,$$

together with (50) imply the result, q.e.d.

<u>Lemma 2.</u> Let $v \in [W^{1,2}(K_a)]^m$ be a weak solution to the linear system

$$(52)\quad \int_{K_\varepsilon} a_{ij}^{rs} \frac{\partial v_r}{\partial x_i} \frac{\partial \varphi_s}{\partial x_j}\, dx = \int_{K_\varepsilon} h_i^r \frac{\partial \varphi_r}{\partial x_i}\, dx,$$

$a_{ij}^{rs} = a_{ji}^{sr} \in L_\infty(K_a)$,

(53) $$c_1|\xi|^2 \le a_{ij}^{rs}\xi_i^r\xi_j^s \le c_2|\xi|^2.$$

Let $\lambda$ be chosen such that (see Lemma 1)

(54) $$\frac{c_1}{c_2} > \frac{\alpha(\lambda)^{1/2}\sqrt{1+\frac{(n-2)^2}{n-1}}-1}{\alpha(\lambda)^{1/2}\sqrt{1+\frac{(n-2)^2}{n-1}}+1}.$$

Then

(55) $$\sup_{x_0\in K_{\varepsilon/2}}\int_{K_\varepsilon}\frac{\partial v_r}{\partial x_i}\frac{\partial v_r}{\partial x_i}|x-x_0|^{-\lambda}\,dx \le$$

$$\le c_3\left[\sup_{x_0\in K_\varepsilon}\int_{K_\varepsilon} h_i^r h_i^r|x-x_0|^{-\lambda}\,dx+\int_{K_\varepsilon}\frac{\partial v_r}{\partial x_i}\frac{\partial v_r}{\partial x_i}\,dx\right].$$

Proof. Smoothing $h_i^r$ and $a_{ij}^{rs}$ by a positive mollifier, we can suppose $v\in[C^{(1)}(K_\varepsilon)]^m$. Let us remark that (53) remains true after mollifying. Put $\gamma=\frac{2}{c_1+c_2}$. Using the equation (where $K(x_0,\delta)\equiv$ $\equiv|x-x_0|<\delta$)

(56) $$\int_{K(x_0,\frac{\varepsilon}{2})}\frac{\partial v_r}{\partial x_i}\frac{\partial\varphi_r}{\partial x_i}\,dx=\int_{K(x_0,\frac{\varepsilon}{2})}\left(\frac{\partial v_r}{\partial x_i}-\gamma a_{ij}^{rs}\frac{\partial v_s}{\partial x_j}\right)\cdot$$

$$\cdot\frac{\partial\varphi_r}{\partial x_i}\,dx+\gamma\int_{K(x_0,\frac{\varepsilon}{2})}g_i^r\frac{\partial\varphi_r}{\partial x_i}\,dx$$

and the relation

(57) $$\int_{K(x_0,\frac{\varepsilon}{2})}\left(\frac{\partial v_r}{\partial x_i}-\gamma a_{ij}^{rs}\frac{\partial v_s}{\partial x_j}\right)\left(\frac{\partial v_r}{\partial x_i}-\gamma a_{ij}^{rs}\frac{\partial v_s}{\partial x_j}\right)\cdot$$

$$\cdot\ x-x_0{}^{-\lambda}\,dx\le\left(\frac{c_2-c_1}{c_2+c_1}\right)^2\int_{K(x_0,\frac{\varepsilon}{2})}\frac{\partial v_r}{\partial x_i}\frac{\partial v_r}{\partial x_i}|x-x_0|^{-\lambda}\,dx,$$

we get the result (if we let the mollifier's parameter converge to zero), taking into account that , $\forall\,\delta>0$,

(58) $$v_r(x_0)v_r(x_0)\le\delta\sup_{x_0'\in K_{\frac{\varepsilon}{2}}}\int_{K(x_0',\frac{\varepsilon}{2})}\quad.$$

$$\frac{\partial v_r}{\partial x_i}\frac{\partial v_r}{\partial x_i}|x-x_0'|^{-\lambda}\,dx + \mu(\delta)\int_{K_\varepsilon}\left(\frac{\partial v_r}{\partial x_i}\frac{\partial v_r}{\partial x_i} + v_r v_r\right)dx ,$$

see for example J. Nečas [23] , q.e.d.

Theorem 4 can be proved by the standard method, starting from the equations (9) in variations, using Lemmas 1 and 2, and finally the standard results for the regularity of the weak solution to the linear systems with Hölder-continuous coefficients, see for example S. Agmon, A. Douglis, L. Nirenberg [24].

We conclude our explanation with a regularity theorem for systems of variational inequalities and Signorini's type unilateral conditions, see G. Fichera [25], J. Nečas [26], J. Frehse [28], [29].

Let $\Omega$ be a domain with the Lipschitz boundary $\partial\Omega$, let $\Gamma \subset \partial\Omega$ be a part of $\partial\Omega$ smooth enough. We suppose on $\Gamma$

$$(59)\qquad b_{rs}u_s \geq \psi_r, \quad r = 1,2,\dots k \leq m,$$

with $b_{rs}$, $\psi_r$ regular enough. Let $\partial\Omega = \bar\Gamma_1 \cup \bar\Gamma_2 \cup \bar\Gamma$, where $\Gamma_1, \Gamma_2, \Gamma$ are disjoint open sets in $\partial\Omega$. We suppose that there exists $u_1 \in [W^{1,2}(\Omega)]^m$ such that $b_{rs}u_s^1 = \psi_r$. Let $u^0 \in [W^{1,2}(\Omega)]^m$, $g \in [L_2(\Gamma_2 \cup \Gamma)]^m$, $f \in [L_2(\Omega)]^m$. Let the rank of $b_{rs}(x) = k$. Put

$$(60)\qquad K \equiv \{v \mid v = u^0 \text{ on } \Gamma_1,\ b_{rs}u_s \geq \psi_r,\ r = 1,2,\dots k \text{ on } \Gamma\}.$$

We suppose that $u_0 \in K$ and we look for $u \in K$ such that, for $\forall v \in K$, we have

$$(61)\qquad \int_\Omega a_i^r(\nabla u)\left(\frac{\partial v_r}{\partial x_i} - \frac{\partial u_r}{\partial x_i}\right) + \int_\Omega u_r(v_r - u_r)\,dx \geq$$

$$\geq \int_\Omega f_r(v_r-u_r)\,dx + \int_{\Gamma_2\cup\Gamma} g_r(v_r-u_r)\,dS.$$

We can answer the regularity question by the penalty method. We put

$$(62)\qquad (\beta(u),v) = -\int_\Gamma (b_{rs}u_s - \psi_r)^- b_{rt}v_t\,dS$$

and look for the solutions $u^\varepsilon$ of the equations with penalty.

By a standard difference method, see [26] in detail, we get

<u>Theorem 5.</u> Let $F = \bar F \subset U(F) \cap \bar\Omega \subset F^* \subset \bar F^* \subset \Omega \cup \Gamma$, where $U(F)$ is a neighbourhood of $F$. Under our assumptions, provided that

$u^1 \in [W^{3,2}(F^*)]^m$, $g \in [W^{1,2}(\Gamma)]^m$, we obtain

$$(63) \quad \|u^\varepsilon\|_{[W^{2,2}(F)]^m} \leq c(F)\Big[1 + \|u^\varepsilon\|_{[W^{1,2}(\Omega)]^m} + \|f\|_{[L_2(\Omega)]^m} + \|g\|_{[W^{1,2}(\Gamma)]^m} + \|u^1\|_{[W^{3,2}(F^*)]^m}\Big].$$

## Bibliography


[1] O.A. Ladyženskaja, N.N. Uralceva: Linějnyje i kvazilinějnyje uravněnija eliptičeskovo tipa, Moskva 1973 (the second edition),

[2] Ch.B. Morrey: Differentiability theorems for weak solutions of nonlinear elliptic differential equations, BAMS, Vol. 75, n.4 (1969), 684-705,

[3] Ch.B. Morrey: Existence and differentiability theorems for the solutions of variational problems for multiple integrals, BAMS 46 (1940), 439-458,

[4] E. Giorgi: Sulla differenziabilita e analiticita delle estremali degli integrali multipli regolari, Mem. Acad. Sci. Torino Cl. Sci. Fis. Mat. Nat. (3) 3 (1957), 25-43,

[5] J. Moser: A new proof of De Giorgi's theorem concerning the regularity problem for elliptic differential equations, Comm. Pure Appl. Math. 13 (1960), 457-468,

[6] G. Stampacchia: Problemi al contorno ellitici, con dali discontinui, dotati di soluzioni hölderiane, Ann. Mat. Pura ed Appl. 51 (1960), 1-38,

[7] J.L. Lions: Quelques méthodes de résolution des problèmes aux limites non linéaires, Paris 1969,

[8] S. Fučík, J. Nečas, J. Souček, V. Souček: Spectral analysis of nonlinear operators, lecture notes in Mathematics 346, Springer,

[9] J. Stará: Regularity Results for Nonlinear Elliptic Systems in two Dimensions, Annali Scuola Norm. Sup. Pisa Vol. XXV, Fase I (1971) 163-190,

[10] J. Nečas: Sur la régularité des solutions variationelles des équations elliptiques non-linéaires d'ordre 2k en deux dimensions, Annali Scuola. Norm. Sup. Pisa, fasc. III (1967) 427-457,

[11] J. Nečas: Sur la régularité des solutions des équations elliptiques non linéaires, Comm. Math. Univ. Cavol. 9,3 (1968), 365-413,

[12] E. Giusti, M. Miranda: Un esempio di soluzioni discontinue per un problema di minimo relatio ad un integrale regolare del calcolo delle variazioni, Boll. Un. Mat. Ital. (4) 2 (1968), 1-8,

[13] S.A. Arakčejev: O gladkosti obobščonnych rešenij některorovo klasa kvazilinějnych elliptičeskich uravněnij, Věstník Mosk. Univ. n. 1, 1975, 49-57,

[14] J. Nečas: Example of an irregular solution to a nonlinear elliptic system with analytic coefficients and conditions for regularity, to appear in Beiträge Zur Analysis, 1977,

[15] J. Nečas, J. Stará, R. Švarc: Classical solution to a second order nonlinear elliptic systems in $R_3$, to appear in Annali Scuola Norm. Sup. Pisa, 1978,

[16] C. Truesdell: A first course in rational continuum mechanics, 1972,

[17] E. Giusti: Un' aggiunta alla mia nota: Regollarità parziale delle soluzioni di sistemi ellittici quasi lineari di ordine arbitrario, Ann. Scuola Norm. Sup. Pisa Ser. 3, 27, n.2, 1973,

[18] J. Nečas: On the existence and regularity of solutions of non-linear elliptic equations, Acta fac. rer. nat. Univ. Com. Math. XVII, 1967, 101-119,

[19] J. Kadlec, J. Nečas: Sulla regularità delle soluzioni di equazioni ellitiche negli spazi $H^{k,\lambda}$, Annali Scuola Norm. Sup. Pisa, Vol. XXI, Fase. IV (1967), 527-545,

[20] A.I. Koševelev: O gladkosti rešenij kvazilinějnych elliptičeskich sistěm vtorovo porjadka, DAN 1976, T. 228, n.4, 787-789,

[21] H.O. Cordes: Über die erste Randwertaufgabe bei quasilinearen Differentialgleichungen zweiter Ordnung in mehr als zwei Variablen, Math. Annalen, Bd. 131, 278-312 (1956),

[22] J. Nečas: On the demiregularity of weak solutions of nonlinear elliptic equations, BAMS, Vol. 77, n. 1, (1971),

[23] J. Nečas: Les méthodes directes en théorie des équations elliptiques, 1967,

[24] S. Agmon, A. Dougles, L. Nirenberg: Estimates near boundary for solutions of elliptic partial diff. equat. satisfying general boundary conditions, I, II, Comm. Pure Appl. Math. 12 (1959), 623-727; 17(1964), 35-92,

[25] G. Fichera: Boundary value problems of elasticity with unilateral constraints, Handbuch der Physik Vol. VII a/2, 1972,

[26] J. Nečas: On regularity of solutions to nonlinear variational inequalities for second-order elliptic systems, Rendiconti di Matematica (2), Vol. 8, Ser. VI, 481-498,

[27] E. Shamir: Regularity of mixed second-order elliptic problems, Israel Math. J. 6 (1968),

[28] J. Frehse: On Variational Inequalities with Lower Dimensional Obstacles, Universität Bonn, Preprint n. 114,

[29] J. Frehse: On Signorini's Problem and Variational Problems with Thin Obstacles, Universität Bonn, Preprint n. 117.

Author's address: Jindřich Nečas, Matematicko-fyzikální fakulta UK, Malostranské nám. 25, Praha 1, Czechoslovakia

# THE SOLUTION OF PARABOLIC MODELS BY FINITE ELEMENT SPACE AND A-STABLE TIME DISCRETIZATION

J. Nedoma, Brno

In papers on solution of parabolic differential equations by the finite element method error bounds are given either in the case that the union of finite elements (straight or curved) matches exactly the given domain (e.g. in Zlámal's papers) or in the case of curved elements which do not cover, in general, the given domain (e.g. in Raviart's papers). In the former case the error bounds are given for fully (i.e. both in space and time) discretized approximate solutions. In the latter case the numerical integration is taken into account, however the error bounds are given only for semidiscrete (not discretized in time) approximate solutions. Error bounds introduced in this lecture are given for fully discretized approximate solutions and for arbitrary curved domains. Discretization in time is carried out by A-stable linear multistep methods. Isoparametric simplicial curved elements in n-dimensional space are applied. Degrees of accuracy of quadrature formulas are determined such that numerical integration does not worsen the optimal order of convergence in $L_2$ norm of the method.

## 1. The finite element space discretization of the problem

Let us first introduce the parabolic problem in the variational form. Let $x=(x_1,\dots x_n)\in R^n$. Let $\Omega$ be a bounded domain in $R^n$. Let the functions $g(x)$, $g_{ij}(x)$, $i,j=1,\dots n$ defined on $\Omega$ and the function $f(x,t)$ defined on $\Omega\times(0,T]$ be smooth enough. Let

(1) $g_{ij}(x) = g_{ji}(x)$, $g(x)\geq g_0(=\text{const})>0$, $\forall x\in\overline{\Omega}$

and let the differential operator

$$(2)\quad L=\sum_{i,j=1}^{n}\frac{\partial}{\partial x_j}\left(g_{ij}(x)\frac{\partial}{\partial x_i}\right)$$

be uniformly elliptic in $\Omega$. Let $a(u,v)$ be the bilinear form corresponding to operator L, i.e.

$$(3)\quad a(u,v)=\int_{\Omega}\sum_{i,j=1}^{n}g_{ij}\frac{\partial u}{\partial x_i}\frac{\partial v}{\partial x_j}dx.$$

We study the following problem:

Find a function $u(x,t)$ such that

$u \in L^{\infty}(H_0^1(\Omega))$, $\frac{\partial u}{\partial t} \in L^{\infty}(H^{-1}(\Omega))$,

(4) $(g\frac{\partial u}{\partial t}, v)_{0,\Omega} + a(u,v) = (f,v)_{0,\Omega}$, $\forall v \in H_0^1(\Omega)$ and $t \in (0,T]$

$u(x,0) = u_0(x) \in L^2(\Omega)$.

Here $H_0^1(\Omega)$ is the closure of the set $C_0^{\infty}(\Omega)$(i.e. of the set of infinitely differentiable functions with compact support in $\Omega$) in the Sobolev norm $\|\cdot\|_{1,\Omega}$.$H^{-1}(\Omega)$ is the space dual to $H_0^1(\Omega)$ (with dual norm).$L^{\infty}(H^m(\Omega))$ is the space of all functions $v(x,t), x=(x_1,..x_n)\in\Omega$, $t \in (0,T]$ such that $v(x,t) \in H^m(\Omega)$, $\forall t \in (0,T]$ and the function $\|v(x,t)\|_{m,\Omega}$ is bounded for almost all $t \in (0,T]$.

First we discretize the problem (4) by the finite element method with respect to x.For this we use a k-regular family of isoparametric simplicial curved elements in n-dimensional space which are constructed in Raviart's paper[1].Let $\mathcal{C}_h$ be a k-regular triangulation of the set $\Omega$ and let $V_h$ be the corresponding finite element space.The union of the elements e from $\mathcal{C}_h$ forms some set $\Omega_h$ which,in general,differs from $\Omega$.We extend the functions $g(x), g_{ij}(x), u_0(x)$ to a greater set $\tilde{\Omega} \supset \bar{\Omega}$ such that the conditions (1)and (2) are satisfied.In such a way we obtain the functions $\tilde{g}(x), \tilde{g}_{ij}(x)$ and $\tilde{u}_0(x)$.Obviously,for sufficiently small h,it is true

(5) $\Omega_h \subset \tilde{\Omega}$.

About the solution u of the problem (4) we suppose

(6) $u, \frac{\partial u}{\partial t} \in L^{\infty}(H^{k+3}(\Omega))$.

By the Calderon extension theorem,for every $t \in (0,T]$ there exist extensions $\tilde{u}(x,t)$, $\widetilde{\frac{\partial u}{\partial t}}$. Let us denote

(7) $\tilde{f}(x,t) = \tilde{g}(x)\widetilde{\frac{\partial u}{\partial t}} - \tilde{L}\tilde{u}$,

where

(8) $\tilde{L} = \sum_{i,j=1}^{n} \frac{\partial}{\partial x_j}(\tilde{g}_{ij}(x)\frac{\partial}{\partial x_i})$.

According to (4) we define now the following semidiscrete problem:

Find a function $u_s(x,t)$ such that

$$u_s, \frac{\partial u_s}{\partial t} \in L^\infty(V_h(\Omega_h)),$$

$$(9)\quad (\tilde{g}(x) \frac{\partial u_s}{\partial t}, v)_{0,\Omega_h} + \tilde{a}(u_s,v) = (\tilde{f},v)_{0,\Omega_h}, \ \forall v \in V_h, \ t \in (0,T],$$

$$u_s(x,0) = u_0 \in V_h,$$

where $u_0$ is an approximate of $u_0(x)$ and $\tilde{a}(u,v)$ is the bilinear form

$$(10)\quad \tilde{a}(u,v) = \int_{\Omega_h} \sum_{i,j=1}^{n} \tilde{g}_{ij}(x) \frac{\partial u}{\partial x_i} \frac{\partial v}{\partial x_j} \, dx.$$

## 2. The A-stable time discretization of the problem

We called the problem (9) semidiscrete because it is discretized with respect to x only. It is obvious that (9) is a system of ordinary differential equations with an unknown vector function of parameter t. From here we find the way how to discretize the problem with respect to t. We solve the system by $\nu$-step A-stable linear method (for $\nu$ =1,2) of order q. We divide the time interval $[0,T]$ into a finite number of intervals of the same length $\Delta t$. We introduce the following notation

$$(11)\quad \phi^m = \phi^m(x) = \phi(x,m\Delta t), \ m = 0,1,\dots$$

for any function $\phi(x,t)$.
If we apply to (9) a $\nu$-step ($\nu$ = 1,2) A-stable linear method we get the following discrete problem:

Find a function $u_d(x,t)$ such that

$u_d \in V_h$ for any $t = 0, \Delta t, 2\Delta t, \dots T$

$$(12)\quad (\tilde{g}(x) \sum_{j=0}^{\nu} \alpha_j u_d^{m+j}, v)_{0,\Omega_h} + \Delta t \tilde{a}( \sum_{j=0}^{\nu} \beta_j u_d^{m+j}, v)$$

$$= \Delta t( \sum_{j=0}^{\nu} \beta_j \tilde{f}^{m+j}, v)_{0,\Omega_h}, \ \forall v \in V_h, \ m = 0,1,\dots$$

$$u_d^0 = u_0 \in V_h;$$

here (see [11] and [12])
a) for one-step A-stable methods

$\gamma = 1$, $\alpha_1 = 1$, $\alpha_0 = -1$, $\beta_1 = 1-\theta$, $\beta_0 = \theta$, $\theta \leq \frac{1}{2}$ is any real number. If $\theta = 1/2$ then the method is of order $q = 2$, in all the other cases the method is of order $q = 1$.

b) for two-steps A-stable methods
$\gamma = 2$, $\alpha_2 = \theta$, $\alpha_1 = 1-2\theta$, $\alpha_0 = -1+\theta$, $\beta_2 = (1/2)\theta + \delta$, $\beta_1 = (1/2)-2\delta$, $\beta_0 = (1/2) - (1/2)\theta + \delta$, $\theta \geq (1/2)$, $\delta > 0$.

## 3. The numerical integration

Since it is either too costly or simply impossible to evaluate exactly the integrals $(.,.)_{0,\Omega_h}$, $\tilde{a}(.,.)$, we must now take into account the fact that approximate integration is used for their computation. For this purpose we use the isoparametric numerical integration (see [1]). We remember:

Every finite element $e \in \mathcal{C}_h$ is the image (i.e. $e = F_e(\hat{T})$) of the unit n-simplex $\hat{T}$ through the unique mapping $F_e: \hat{T} \to R^n$. Let us suppose that we have at our disposal a quadrature formula of degree d over the reference set $\hat{T}$. In other words,

$$(13) \quad \int_{\hat{T}} \hat{\phi}(\hat{x})d\hat{x} \approx \sum_r \hat{\omega}_r \hat{\phi}(\hat{b}_r)$$

for any function $\hat{\phi}(\hat{x})$ which is defined on $\hat{T}$ and for some specified points $\hat{b}_r \in \hat{T}$ and weights $\hat{\omega}_r$. Let $\phi(x)$ be any function defined on e. Then using the standard formula for the change of variables in multiple integrals, we find that

$$(14) \quad \int_e \phi(x)dx \approx \sum_r \hat{\omega}_r |J_e(\hat{b}_r)| \phi(F_e(\hat{b}_r))$$

where $J_e(\hat{x})$ is Jacobian of the transformation $e = F_e(\hat{T})$.
We see that the quadrature scheme (13) over the reference set $\hat{T}$ induces the quadrature scheme (14) over the element e, a circumstance which we call "isoparametric numerical integration".

In agreement with (14) we replace in (12)

$$(15) \quad (.,.)_{0,\Omega_h} \approx (.,.)_h, \quad \tilde{a}(.,.) \approx a_h(.,.)$$

According to (12) and (15) we define the following full discretized problem:

Find a function $u_h(x,t)$ such that

$u_h \in V_h$ for $t = 0, \Delta t, \ldots T$

(16) $$(g(x)\sum_{j=0}^{\nu} \alpha_j u_h^{m+j}, v)_h + \Delta t a_h(\sum_{j=0}^{\nu} \beta_j u_h^{m+j}, v)$$

$$= \Delta t(\sum_{j=0}^{\nu} \beta_j f^{m+j}, v)_h, \quad \forall v \in V_h, \; m = 0,1,\ldots$$

$$u_h^0 = u_0 \in V_h \,.$$

## 4. Error bounds

**Theorem.**

Let $u(x,t)$ be the solution of the problem (5) such that $u, \frac{\partial^r u}{\partial t^r} \in L^\infty(H^{k+3}(\Omega)), r = 1,\ldots q$. Let $\mathcal{C}_h$ be a k-regular triangulation of the set $\Omega_h$ where k is a positive integer such that $k > n/2 - 1$. Let the quadrature formulas on the reference set $\hat{T}$ for calculation of the forms $(.,.)_{0,\Omega_h}$ and $\tilde{a}(.,.)$ be of degree $d \geq 2k$ and $d \geq 2k - 1$, respectively. Let a given $\nu$ -step time discretization method be A-stable and of order q. Let $\nu = 1$ or 2.

Then the full discrete problem (16) has one and only one solution $u_h(x,t)$ and there exists a constant c independent of t and h such that

(17) $$\|u^s - u_h^s\|_{0,\Omega\cap\Omega_h} \leq c(\Delta t^q + h^{k+1} + |\varepsilon^0|_h + |\varepsilon^{\nu-1}|_h).$$

[Here $\varepsilon^0, \varepsilon^{\nu-1}$ are the errors on the first $\nu$ steps, $|\varepsilon|_h = \sqrt{(g\varepsilon, \varepsilon)_{h^\circ}}$]

Outline of the proof.

Let us denote

(18) $$\tilde{u}^j = \eta^j + \xi^j,$$

where $\eta^j = \eta(x, j\Delta t)$ is the Ritz approximation of the function $\tilde{u}^j = \tilde{u}(x, j\Delta t)$. We recall that $\tilde{u}$ and $\widetilde{\frac{\partial u}{\partial t}}$ are extensions of u and $\frac{\partial u}{\partial t}$ satisfying the inequalities

(19) $$\|\tilde{u}\|_{k+3,\tilde{\Omega}} \leq c\|u\|_{k+3,\Omega}, \quad \|\widetilde{\tfrac{\partial u}{\partial t}}\|_{k+3,\tilde{\Omega}} \leq c\|\tfrac{\partial u}{\partial t}\|_{k+3,\Omega}.$$

Next we recall that by the Ritz approximation of the function $\widetilde{u(x,t)}$

we mean the function $\eta(x,t) \in V_h$ $(V_h \subset H_0^1(\Omega_h))$, $\forall\, t \in (0,T]$ such that

(20) $(\tilde{g}(x)\, \frac{\partial \tilde{u}}{\partial t}\, ,v)_{0,\Omega_h} + \tilde{a}(\eta(x,t),v) = (\tilde{f}(x,t),v)_{0,\Omega_h}, \ \forall\, v \in V_h$.

It is easy to prove that $\eta(x,t)$ is an orthogonal projection onto $V_h$ of the function $\tilde{u}(x,t)$ in the energy norm given by the bilinear form $\tilde{a}(.,.)$,i.e.that it satisfies

(21) $\tilde{a}(\tilde{u} - \eta, v) = 0, \ \forall\, v \in V_h$.

For the Ritz approximation the following estimate can be derived

(22) $\|\tilde{u} - \eta\|_{i,\Omega_h} \leq ch^{k+1-i}\|u\|_{k+3,\Omega}, \ i = 0,1,$

where c is a constant independent of h and t.
From (22) and (18) we get

(23) $\|\xi^j\|_{0,\Omega_h} = \|\tilde{u}^j - \eta^j\|_{0,\Omega_h} \leq ch^k\|u\|_{k+3,\Omega}$.

From (18) it follows

(24) $\|u^j - u_h^j\|_{0,\Omega\cap\Omega_h} \leq \|\tilde{u}^j - u_h^j\|_{0,\Omega_h} \leq \|\xi^j\|_{0,\Omega_h} + \|\eta^j - u_h^j\|_{0,\Omega_h}$.

Hence,it is sufficient to give an estimation of error bounds for

(25) $\varepsilon^j = \eta^j - u_h^j$ .

By simple calculation we get from (14) and (16)

$$(g\sum_{j=0}^{\nu} \alpha_j \varepsilon^{m+j}\, ,v)_h + \Delta t a_h(\sum_{j=0}^{\nu} \beta_j \varepsilon^{m+j}\, ,v)$$

(26)
$$= (\Pi_\nu^m - \omega_\nu^m\, ,v)_{0,\Omega_h} + \Delta t E(v\sum_{j=0}^{\nu}\beta_j\tilde{f}^{m+j}) - E(\tilde{g}v\sum_{j=0}^{\nu}\alpha_j\eta^{m+j})$$
$$- \Delta t E(\sum_{i,j=1}^{n}\tilde{g}_{ij}\frac{\partial v}{\partial x_j}\sum_{r=0}^{\nu}\beta_1\frac{\partial \eta^{m+r}}{\partial x_i}), \ \forall\, v \in V_h ,$$

where

(27) $\Pi_\nu^m = \tilde{g}\sum_{j=0}^{\nu}(\alpha_j\tilde{u}^{m+j} - \Delta t\beta_j\frac{\partial \tilde{u}^{m+j}}{\partial t}), \ \omega_\nu^m = \tilde{g}\sum_{j=0}^{\nu}\alpha_j\xi^{m+j}, \ E(\phi) = \sum_{e\in\mathcal{C}_h} E_e(\phi)$

and where $E_e(\phi)$ is according to (14) the error given by the isoparametric integration,i.e.

(28) $E_e(\phi) = \int_e \phi(x)dx - \sum_r \hat{\omega}_r |J_e(\hat{b}_r)|\, \phi(F_e(\hat{b}_r))$.

We denote the expressions in identity (26) by $A_\nu^m(v)$, $B_\nu^m(v)$, $D_\nu^m(v)$, $F_\nu^m(v)$, $G_\nu^m(v)$, $H_\nu^m(v)$ respectively. Next we denote $Q_\nu^m = \Delta t F_\nu^m - G_\nu^m - \Delta t H_\nu^m$. The identity (26) is true for all $v \in V_h$, hence it is also true for

(29) $v = \gamma = \sum_{j=0}^{\nu} \beta_j \varepsilon^{m+j}$ .

From here we get the following basic identity

(30) $\sum_{m=0}^{s-\nu} A_\nu^m(\gamma) + \Delta t \sum_{m=0}^{s-\nu} B_\nu^m(\gamma) = \sum_{m=0}^{s-\nu} D_\nu^m(\gamma) + \sum_{m=0}^{s-\nu} Q_\nu^m(\gamma)$

valid for any s such that $s\Delta t \leq T$, $s \geq \nu$.
Using the similar technique as in [11] we prove

(31) $\sum_{m=0}^{s-\nu} A_\nu^m(\gamma) \geq c_2 \|\varepsilon^s\|_{0,\Omega_h}^2 - c_1 \left[ |\varepsilon^0|_h^2 + |\varepsilon^{\nu-1}|_h^2 \right]$, $\nu = 1,2.$

To this end we use the inequality

(32) $c_3 \|v\|_{0,\Omega_h} \leq |v|_h$ , $\forall\, v \in V_h$

valid under the assumption that the quadrature formula on the reference set $\hat{T}$ is of degree $d \geq 2k$. In the inequality the notation $|v|_h^2 = (g(x)v, v)_h$ is used.
It is easy to derive the following inequality

(33) $\sum_{m=0}^{s-\nu} B_\nu^m(\gamma) \geq c_4 \sum_{m=0}^{s-\nu} \left| \sum_{j=0}^{\nu} \beta_j \varepsilon^{m+j} \right|_{1,\Omega_h}^2$, $\nu = 1,2.$

For this purpose we use the inequality

(34) $c_5 |v|_{1,\Omega_h} \leq \|v\|_h$, $\forall\, v \in V_h$

valid under the assumption that the quadrature formula on the reference set $\hat{T}$ is of degree $d \geq 2k-2$. In this inequality the notation $\|v\|_h^2 = a_h(v,v)$ is used.
Next, we prove the inequality

(35) $\sum_{m=0}^{s-\nu} |D_\nu^m(\gamma)| \leq c_6 \Delta t (\Delta t^q + h^{k+1}) \sum_{m=0}^{s-\nu} \left\| \sum_{j=0}^{\nu} \beta_j \varepsilon^{m+j} \right\|_{0,\Omega_h}$, $\nu = j,2.$

For $Q_\nu^m(\gamma)$ the following estimate can be derived

$$(36)\quad \sum_{m=0}^{s-\nu}|Q_\nu^m(\gamma)| \leq c_7\Delta t h^{k+1}\sum_{m=0}^{s-\nu}\|\sum_{j=0}^{\nu}\beta_j\varepsilon^{m+j}\|_{1,\Omega_h},\quad \nu = 1,2.$$

To this end we use the inequalities

$$|E(wv)| \leq c_8 h^{k+1}\|w\|_{k+1,\Omega_h}\|v\|_{1,\Omega_h},$$

$$(37)\quad |E(b\eta v)| \leq c_9 h^{k+1}\|u\|_{k+3,\Omega}\|v\|_{1,\Omega_h},$$

$$|E(b\frac{\partial\eta}{\partial x_i}\frac{\partial v}{\partial x_j})| \leq c_{10}h^{k+1}\|u\|_{k+3,\Omega}\|v\|_{1,\Omega_h}$$

valid for all $w \in H^{k+1}(\Omega_h)$, $v \in V_h(\Omega_h)$, $u \in H^{k+3}(\Omega)$, $t \in (0,T]$ and $b(x) \in C^{k+1}(\Omega_h)$ under the assumption that the quadrature formula on the reference set $\hat{T}$ is of degree $d \geq 2k-1$.

From (31), (33), (35) and (36), using several times the inequality

$$(38)\quad |ab| \leq \frac{1}{2}\tau a^2 + \frac{1}{2\tau}b^2$$

we get

$$(39)\quad \|\varepsilon^s\|_{0,\Omega_h}^2 \leq c_{11}(\Delta t^{2q} + h^{2(k+1)} + |\varepsilon^0|_h^2 + |\varepsilon^{\nu-1}|_h^2) + \Delta t\sum_{m=0}^{s-1}\|\varepsilon^m\|_{0,\Omega_h}^2.$$

From (39) and from [9] (see Lemma 2.1, p.396) we get the estimate (17).

In the end let us add the following remarks:

Remark 1.

From (17) we see that the $L_2$-norm of the error is of a magnitude of the order $\Delta t^q$ $(q = 1,2)$ with respect to $\Delta t$ and of the order $h^{k+1}$ with respect to h.

Remark 2.

According to our result, for 1-regular triangulation (i.e. for linear isoparametric elements) the quadrature formula on the reference set $\hat{T}$ for calculation of the forms $(.,.)_{0,\Omega}$ and $a(.,.)$ must be, in general, of degree 2 and 1, respectively. It can be proved that using the quadrature formula

$$(40)\quad \int_{\hat{T}}\varphi(\hat{x})d\hat{x} \approx \frac{\text{mes}\,\hat{T}}{n}\left[\varphi(0,\ldots 0) + \varphi(0,1,\ldots 0) + \ldots + \varphi(0,0,\ldots 1)\right]$$

(which is of degree 1) for calculation of the form $(.,.)_{0,\Omega}$ we obtain the same estimate as in (17). In this case the mass matrix is diagonal.In the engineering literature this effect is called the mass lumping.

Remark 3.

For the three-dimensional space the simplicial curved elements have no practical use.For such case the theory using quadrilateral elements must be developed.We are working on this problem now.

References

[1] Ciarlet P.G.and Raviart P.A.:The combined effect of curved boundaries and numerical integration in isoparametric finite element methods.(Paper published in the book:A.K.Aziz: The mathematical foundations of the finite element method with applications to partial differential equations, Academic Press,New York and London,1972).

[2] Raviart P.A.:The use of numerical integration in finite element methods for solving parabolic equations.(Paper presented at the Conference on Numerical Analysis,Royal Irish Academic,Dublin,August 14-18,1972.)

[3] Nečas J.:Les Méthodes Directesen Théorie des Equations Elliptiques. (Mason,Paris,1967.)

[4] Smirnov V.J.:Kurs vysšej matěmatiki,tom V (Moskva 1960).

[5] Zlámal M.:Finite Element Multistep Discretizations of Parabolic Boundary Value Problems.(Mathematics of Computation, Vol.29,Number 130,April 1975,Pages 350-359)

[6] Zlámal M.:Curved Elements in the Finite Element Method I.(SIAM J.Numer.Anal.,Vol.11,No.1,March 1973.)

[7] Zlámal M.:Curved Elements in the Finite Element Method II.(SIAM J.Numer.Anal.,Vol.11,No.2,April 1974.)

[8] Zlámal M.:Finite Element Methods for Parabolic Equations.(Mathematics of Computation,Vol.28,No.126,April 1974.)

[9] Dupont T.,Fairweather G.,Johnson J.P.:Three-level Galerkin Methods for Parabolic Equations.(SIAM J.Numer.Anal.,Vol.11,No.2, April 1974.)

[10] Lees M.:A priori estimates for the solutions of difference approximations to parabolic differential equations.(Duke Math. J.,27,1960,pp.287-311.)

[11] Zlámal M.:Finite element method for nonlinear parabolic equations. (To appear in RAIRO.)

[12] Liniger W.:A criterion for A-stability of linear multistep integration formulae.(Computing 3,1968,pp.280-285.)

Author's address: Laboratoř počítacích strojů, Vysoké učení technické
Třída Obránců míru 21, 602 00 Brno,
Czechoslovakia

# GLOBAL PROPERTIES OF THE nTH ORDER LINEAR DIFFERENTIAL EQUATIONS

F. Neuman, Brno

In my lecture I should like to describe an approach to problems concerning global properties of linear homogeneous differential equations (LDEs) of the nth order, $n \geqq 2$, and some basic facts of the theory.

Let me start with a few historical remarks. Investigations concerning LDEs of the nth order began in the middle of the last century and were connected with the names of E. E. Kummer [5], E. Laguerre [7], F. Brioschi, G. H. Halphen, A. R. Forsyth, P. Stäckel [19], S. Lie, E. J. Wilczynski [20], and others. Between the main objects of their study were transformations, canonical forms and invariants of LDEs. Their investigations were of local character, which was already noticed by George D. Birkhoff [1] in 1910. He pointed out that not every 3rd order LDE can be reduced to its Laguerre-Forsyth canonical form on its whole interval of definition.

Of course, the local character of results is not suitable for global problems, like questions concerning boundedness of solutions, solutions of the classes $L^2$ and $L^p$, periodic solutions, solutions converging to zero, oscillatory behavior of solutions: conjugate points, disconjugate equations etc.

Except that G.Birkhoff [1] introduced a geometrical interpretation of solutions of the 3rd order LDEs using curves in the projective plane, and except for some isolated results of a global character, there was no theory describing global properties of LDEs, not even in the simplest cases $n=2$ and 3.

As a simple illustration that any question of a global character was difficult to solve let me mention the following one. There was a conviction that some properties of LDEs with variable coefficients might be modifications of properties of LDEs with constant coefficients. E.g., the 3rd order LDE with real constant coefficients has always at least one nonvanishing solution; one might expect that in the case of variable coefficients at least one solution of any LDE of the 3rd order would have only finite number of zeros. That this is not the case was discovered by G. Sansone [18] in 1948.

In the last twenty years O. Borůvka [2] developed the theory of global properties of LDEs of the 2nd order as you have heard in his plenar lecture at the conference.

For the nth order LDEs there are now results of N. V. Azbelev and Z. B. Caljuk, J. H. Barrett, T. A. Burton and W. T. Patula, W. A. Coppel, W. N. Everitt, M. Greguš, H. Guggenheimer, G. B. Gustafson, M. Hanan, M. K. Kwong, V. A. Kondrat'jev, A. C. Lazer, A. Ju. Levin, M. Ráb, G. Sansone, C. A. Swanson, M. Švec and others having global character but mainly devoted to oscillatory behavior of solutions, conjugate points and disconjugacy. However there was still no theory of global properties of LDEs of the nth order enabling us to foretell the possible behavior of solutions, to exclude the impossible cases, to enable us to see globally the whole situation.

Global structure of linear differential equations

All our considerations will be in the real domain. Consider a LDE of the nth order, $n \geqq 2$:

$$y^{(n)} + p_{n-1}(x)y^{(n-1)} + \dots + p_0(x)y = 0$$

on an open (bounded or unbounded) interval $I \subset \mathbb{R}$ that will be shortly denoted by $\underline{P}$ (together with its interval of definition, that is important when studying situation globally). The coefficients are supposed to be real and continuous. Let $\underline{Q}$ be another LDE of the same order, say

$$z^{(n)} + q_{n-1}(t)z^{(n-1)} + \dots + q_0(t)z = 0 \quad \text{on } J \subset \mathbb{R}.$$

Let $\underline{y} = (y_1, \dots, y_n)^T$ be a column vector of n linearly independent solutions of $\underline{P}$ considered again on the whole interval I; similarly $\underline{z}$ is defined for $\underline{Q}$.

We say that $\underline{P}$ is <u>globally transformable</u> into $\underline{Q}$ if there exist
1. a bijection h of J onto I of the class $C^n$ with $dh(t)/dt \neq 0$ on J,
2. a nonvanishing scalar function $f\colon J \to \mathbb{R}$ of the class $C^n$, and
3. an n by n regular constant matrix A such that

$$(\alpha) \qquad \underline{z}(t) = A.f(t).\underline{y}(h(t)) \quad \text{on } J$$

for some (then every) $\underline{y}$ and $\underline{z}$ of $\underline{P}$ and $\underline{Q}$, resp.
Due to Stäckel, $(\alpha)$ is the most general pointwise transformation that for $n \geqq 2$ keeps the kind of our differential equations (i.e., the order and the linearity) unchanged.

The h and f in $(\alpha)$ will be called <u>transformator</u> and <u>multiplicator</u> of the transformation $(\alpha)$, resp. We shall also simply write

$$\alpha \underline{P} = \underline{Q}$$

to express the fact that $\underline{P}$ is globally transformed into $\underline{Q}$ by $\alpha$.

The relation of global transformability is an equivalence and we often call the $\underline{P}$ and $\underline{Q}$ <u>globally equivalent</u> equations. We come to a decomposition of all LDEs of all orders $n \geqq 2$ into classes of globally equivalent equations.

Let D be one of the classes, $\underline{P} \in D$, $\underline{Q} \in D$, and $\alpha \underline{P} = \underline{Q}$. For $\underline{R} \in D$ and $\beta \underline{Q} = \underline{R}$ we may define $(\beta\alpha)\underline{P} := \beta(\alpha \underline{P}) = \underline{R}$. It is easy to check that we have introduced a structure of <u>Brandt groupoid</u> into each class of globally equivalent equations. From the theory of categories it is known that each Brandt groupoid essentially depends on the <u>stationary group</u> of its arbitrary element, e.g. on the group $B(\underline{P})$ of all morphisms (or transformations) of the equation $\underline{P}$ into itself. This stationary group $B(\underline{P})$ for $n = 2$ coincides with the group of dispersions of $\underline{P}$ introduced by O. Borůvka.

If we consider transformations that not only globally transform $\underline{P}$ into itself but, moreover, that transform each solution of the equation $\underline{P}$ into itself (i.e. A is the unit matrix in ($\alpha$)), then we get a subgroup $C(\underline{P})$ of $B(\underline{P})$.

When studying global properties of solutions then transformations with increasing transformators $h$, $h' > 0$, are extremely important. Let $B^{+}(\underline{P})$ and $C^{+}(\underline{P})$ be the subgroups of $B(\underline{P})$ and $C(\underline{P})$ with increasing transformators.

The fundamental results of that part of the theory are the following ones.

<u>Theorem 1</u>. $B^{+}(\underline{P})$ <u>is not trivial if and only if</u> D <u>contains an equation with periodic coefficients.</u>

<u>Theorem 2</u>. $C^{+}(\underline{P})$ <u>is not trivial if and only if there is an equation in</u> D <u>having only periodic solutions with the same period.</u>

<u>Theorem 3</u>. <u>For</u> $\alpha\underline{P} = \underline{Q}$ <u>it holds</u> $B(\underline{Q}) = \alpha B(\underline{P}) \alpha^{-1}$, <u>and similarly for</u> $B^{+}(\underline{Q})$, $C(\underline{Q})$ <u>and</u> $C^{+}(\underline{Q})$.

<u>Theorem 4</u>. <u>All transformations of</u> $\underline{P}$ <u>into</u> $\underline{Q}$ <u>form the set</u>

$$\alpha B(\underline{P}) = B(\underline{Q})\alpha = B(\underline{Q})\alpha B(\underline{P}).$$

<u>Proofs</u> of the theorems are essentially based on methods of the theory of categories and can be seen from the following picture.

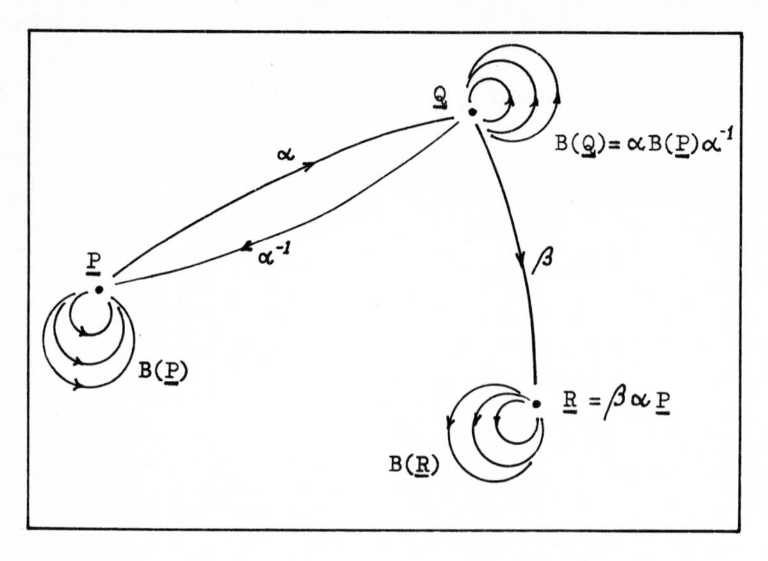


To each class D of globally equivalent LDEs we may assign a (<u>canonical</u>) equation $\underline{E}(D)$. Then for each equation $\underline{P} \in D$ there exists a transformation $\alpha$ (not necessarily unique, it depends on $B(\underline{P})$) that transforms $\underline{E}(D)$ into $\underline{P}$. The transformator h and the multiplicator f of the $\alpha$ are called <u>phase</u> and <u>amplitude</u> of $\underline{P}$ (with respect to the canonical $\underline{E}(D)$). Hence we have introduced "polar coordinates" in each class of globally equivalent LDEs.

The just mentioned categorial description of global structure of LDEs of arbitrary order n, $n \geqq 2$, has its geometrical aspects that enable us to understand the sense of analytic constructions in the theory of global transformations, to solve open problems, and, sometimes, to find occasional inaccuracies in the mathematical literature occurring in complicated and lengthy analytic processes without necessity of a tiresome calculation.

The essence of our geometrical approach is the following observation first introduced in [10] and [11].

<u>Theorem 5. Consider LDE P and its n linearly independent solutions $y_1,\dots,y_n$ forming the coordinates of the vector function $\underline{y}$, now considered as a curve in n-dimensional vector space. There is a 1-1 correspondence between all solutions of equation P and all hyperplanes passing through origin, in which parameters of intersections of the curve $\underline{y}$ with a particular hyperplane are zeros of the correspond-</u>

<u>ing solution and vice versa, counting multiplicities that occur as the order of contacts.</u>

This result is essentially used in recent literature, see e.g. [4] .

Moreover, see again [11], if $y$ is considered in n-dimensional euclidean space, the central projection of the curve $y$ onto the unit sphere $S_{n-1}$ has the same property. But now all intersections are on the unit sphere, and if instead of hyperplanes main circles are under consideration we have all the situation in a compact space, where strong tools of topology are to our disposal. Some open problems were already solved by the method ([13]).

Furthermore, having the central projection of the curve $y$ on the unit sphere in n-dimensional euclidean space we introduce a new parametrization as the length of the projection. We could see that, firstly, by the projection the multiplicator was eliminated, and secondly, by specifying the parametrization we unify the transformator. Hence we get a special curve $u$ on the sphere. LDEs which conversely correspond to these special curves are called <u>canonical</u>. The explicit forms of the canonical equations are obtained using Frenet formulae of the special curves.

I should like to stress that these special equations are canonical in the global sense, that means, each LDE can be transformed on its whole interval of definition into its canonical form without any restrictions on the smoothness of its coefficients.

E.g.

$$y'' + y = 0 \qquad \text{on } I$$

are all canonical differential equations for $n = 2$ (there are still several equivalent classes depending on the length of I);

$$y''' - \frac{a'}{a} y'' + (1 + a^2)y' - \frac{a'}{a} y = 0 \qquad \text{on } I,$$

$a \in C^1$, $a > 0$, are all canonical forms for $n = 3$ (they depend on a function a and an interval I), etc.

<u>Examples</u>

Let me demonstrate the above few facts from the groundwork of the theory of global properties of LDEs on special problems.

Let us see the following picture of "a prolonged cycloid" $y$ infinitely many times surrounding the equator of the unit sphere in

3-dimensional space:

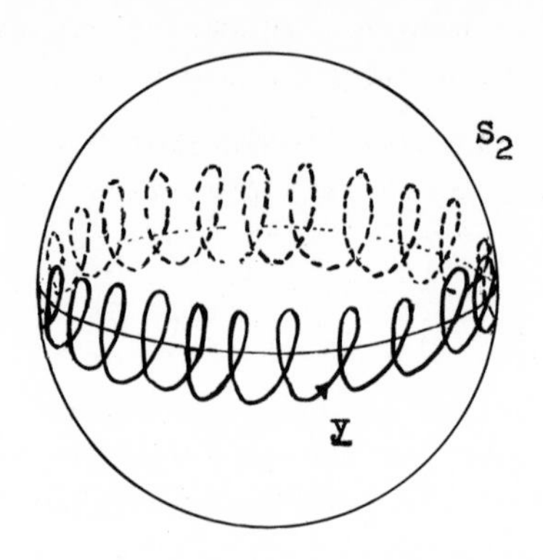


If a curve $\underline{y}$ is three times differentiable and without points of inflexion (that corresponds to nonvanishing Wronskian of its coordinates), then its coordinates may be considered as 3 linearly independent solutions of a LDE of the 3rd order. Since each plane going through the origin intersects $\underline{y}$ infinitely many times, each solution of the LDE has infinitely many zeros. We have Sansone's interesting result using our approach.

Considering again LDEs of the 3rd order with only constant coefficients we can observe that if one oscillatory solution occurs, then necessarily there must be two linearly independent oscillatory solutions. One may ask whether for general LDEs of the 3rd order (with variable coefficients) the same situation holds. Using our method, we want to know whether a curve of the class $C^3$ without points of inflexion on the unit sphere $S_2$ of 3-dimensional space exists such that it is intersected infinitely many times just by one plane passing through origin, whereas any other plane passing through origin has only finite number of intersections with our curve.

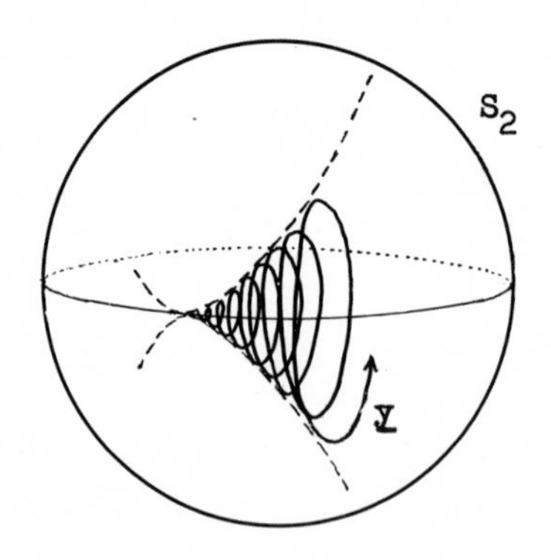

The above picture gives the answer: there exists such an equation; there is again no analogy to the case of constant coefficients.

One may ask, why is the situation for $n = 3$ different from the case $n = 2$, where there are so many analogies. The answer follows from our results: because for $n = 2$ each class of globally equivalent equations has a global representation (e.g. its canonical equation) with constant coefficients (i.e. $y'' + y = 0$), however this is not the case for $n > 2$.

Let me come to other type of applications of our approach. Many recent problems and results concern LDE of the 2nd order in the form

$$(1) \qquad u'' + q(t)u = 0 \quad \text{on } I$$

having all solutions square integrable. There was a problem whether in this case all solutions of (1) are also bounded, see [17] and [6].

Using our method we may proceed as follows.

$$(2) \qquad y'' + y = 0 \quad \text{on } J$$

is a canonical form of (1). The curve $\underline{y} = \binom{\sin x}{\cos x}$ corresponds to (2), hence the curve $\underline{u}(t) = \binom{f(t)\ \sin h(t)}{f(t)\ \cos h(t)}$, $f, h \in C^2$, $f.h' \neq 0$ on $I$, corresponds to LDE of the second order. Since the coefficient by $u'$ in (1) is zero, we have $f(t) = \text{const.}\cdot|h'(t)|^{-1/2}$ (cf. O. Borůvka's lecture). Hence $f \in C^2$ implies $h \in C^3$ and $|h'(t)|^{-1/2}\cdot\sin h(t)$, $|h'(t)|^{-1/2}\cdot\cos h(t)$ are two linearly independent solutions of (1).

It is easy to derive the following succession of implications:

Each solution of (1) is square integrable <u>iff</u>
Two linearly independent solutions of (1) are square integrable <u>iff</u>
$\int_I |h'(t)|^{-1}\cdot\sin^2 h(t)dt < \infty$ and $\int_I |h'(t)|^{-1}\cdot\cos^2 h(t)dt < \infty$ <u>iff</u>
$\int_I |h'(t)|^{-1}dt < \infty$ .

Analogously

Each solution of (1) is bounded <u>iff</u>
Two linearly independent solutions of (1) are bounded <u>iff</u>
Both $|h'(t)|^{-1}\cdot\sin^2 h(t)$ and $|h'(t)|^{-1}\cdot\cos^2 h(t)$ are bounded on $I$ <u>iff</u>
$|h'(t)|^{-1}$ is bounded on $I$,
where $h' \neq 0$ and $h \in C^3$.

And we ask whether (1) with all square integrable solutions has only bounded solutions. In our model it is equivalent to the question, whether

$$\int_I |h'(t)|^{-1} dt \quad \Rightarrow \quad |h'|^{-1} \text{ is bounded on } I$$

for $h \in C^3$, $h' \neq 0$; see [8] and [9].

Of course, the implication is not true. Taking suitable h' we can explicitly construct an example of such an equation if it is necessary. Similarly we may construct explicitly examples of LDEs with certain properties using coordinates of the corresponding curves and making some boring computation.

I should like to conclude my lecture by the following remark. The above sketched method and results are suitable for reviewing globally the whole situation, to see what can and what cannot happen, they are applicable in cases when problems concern behaviour of solutions, distribution of their zeros and other properties of this kind. On the other hand, within the reach of our approach there are only few results for both second and higher order equations which make use of conditions on coefficients.

## References

[1] G. D. Birkhoff, On the solutions of ordinary linear homogeneous differential equations of the third order. Annals of Math. 12 (1910/11), 103-127.

[2] O. Borůvka, Linear differential transformations of the second order. The English Univ. Press, London 1971.

[3] O. Borůvka, Teorija global'nych svojstv obyknovennych linejnych differencial'nych uravnenij vtorogo porjadka. Differencial'nyje uravnenija 12 (1976), 1347-1383.

[4] H. Guggenheimer, Distribution of zeros and limit behavior of solutions of differential equations. Proc. AMS 61 (1976), 275-279.

[5] E. E. Kummer, De generali quadam aequatione differentiali tertii ordinis. Progr. Evang. Royal & State Gymnasium Liegnitz 1834, reprinted in J. Reine Angew. Math. (Crelle Journal) 100 (1887), 1-10.

[6] M. K. Kwong, On boundedness of solutions of second order differential equations in the limit circle case. Proc. AMS 52 (1975), 242-246.

[7] M. Laguerre, Sur les équations différentielles linéaires du troisième ordre. Comptes rendus 88 (1879), 116-119.

[8] F. Neuman, Relation between the distribution of the zeros of the solutions of a 2nd order linear differential equation and the boundedness of these solutions. Acta Math. (Hung.) 19 (1968), 1-6.

[9] F. Neuman, $L^2$-solutions of $y'' = q(t)y$ and a functional equation. Aequationes Math. 6 (1971), 66-70.

[10] F. Neuman, Some results on geometrical approach to linear differential equations of the n-th order, Comm. Math. Univ. Carol. 12 (1971), 307-315.

[11] F. Neuman, Geometrical approach to linear differential equations of the n-th order. Rend. Mat. 5 (1972), 579-602.

[12] F. Neuman, On n-dimensional closed curves and periodic solutions of linear differential equations of the n-th order. Demonstratio Math. 6 (1973), 329-337.

[13] F. Neuman, On two problems about oscillation of linear differential equations of the third order, J. Diff. Equations 15 (1974), 589-596.

[14] F. Neuman, Global transformations of linear differential equations of the n-th order. Knižnice odb. a věd. spisů VUT Brno, B-56 (1975), 165-171.

[15] F. Neuman, On solutions of the vector functional equation $y(\xi(x)) = f(x).A.y(x)$. To appear in Aequationes Math. in 1977.

[16] F. Neuman, Categorial approach to global transformations of the n-th order linear differential equations, Časopis Pěst. Mat. 102 (1977), 350-355.

[17] W. T. Patula, J. S. W. Wong, An $L^p$-analogue of the Weyl alternative. Math. Ann. 197 (1972), 9-28.

[18] G. Sansone, Studi sulle equazioni differenziali lineari omogenee di terzo ordine nel campo reale. Revista Mat. Fis. Teor. Tucuman 6 (1948), 195-253.

[19] P. Stäckel, Über Transformationen von Differentialgleichungen. J. Reine Angew. Math. (Crelle Journal) 111 (1893), 290-302.

[20] E. J. Wilczynski, Projective differential geometry of curves and ruled surfaces. Teubner - Leipzig 1906.

Author's address: Mathematical Institute of the Czechoslovak Academy of Sciences, Branch Brno,

662 95 Brno, Janáčkovo nám. 2a,
Czechoslovakia

# A FORCED QUASILINEAR WAVE EQUATION WITH DISSIPATION

J. A. Nohel, Madison, Wisconsin*

1. Introduction. We study the global existence, uniqueness and continuous dependence on data of smooth solutions of the initial value problem

$$(1.1)\qquad y_{tt} + \alpha y_t - (\sigma(y_x))_x = g \qquad (0 < t < \infty,\ x \in \mathbb{R}) ,$$

$$(1.2)\qquad y(0,x) = y_0(x),\ y_t(0,x) = y_1(x) \qquad (x \in \mathbb{R}) ,$$

where the subscripts $t, x$ denote partial differentiation, $\alpha > 0$ is a fixed constant, $\sigma : \mathbb{R} \to \mathbb{R}$, $g : [0,\infty) \times \mathbb{R} \to \mathbb{R}$ and $y_0, y_1 : \mathbb{R} \to \mathbb{R}$ are given smooth functions. We shall assume throughout that

$$(\sigma)\qquad \sigma \in C^2(\mathbb{R}),\ \sigma(0) = 0,\ \sigma'(\xi) \geq \varepsilon > 0 \qquad (\xi \in \mathbb{R};\ \varepsilon > 0) ;$$

the case $\sigma''(\xi) \not\equiv 0$ is of primary interest.

If $\alpha = 0$, $g \equiv 0$ it is known [4], [7] that solutions of the Cauchy problem (1.1), (1.2) will in general develop singularities in the first derivatives even for smooth data, and smooth solutions will not exist for large $t$. If $\alpha > 0$, $g \equiv 0$ Nishida [10], has established the existence and uniqueness of global smooth solutions of (1.1) for smooth and sufficiently small data (1.2) by a remarkably simple method.

It is the purpose of this note to (i) extend Nishida's method to obtain the global existence and uniqueness of smooth solutions of (1.1), (1.2) with $g \not\equiv 0$, and (ii) study the continuous dependence of solutions of (1.1), (1.2) on the data $y_0$, $y_1$, $g$. The result (i) is implicit in a recent paper of MacCamy [5]; however, his proof of the analogue of the important Lemma 2.3 below is not entirely complete. The result (ii) is new.

We remark that our results (i) and (ii) can be used to obtain a local existence and uniqueness result for smooth solutions of the functional differential equation

$$(1.3)\qquad y_{tt} + \alpha y_t - (\sigma(y_x))_x = G(y) \qquad (0 \leq t \leq T,\ x \in \mathbb{R}) ,$$

subject to the initial condition (1.2), for some $T > 0$. In (1.3) $G$ is a given mapping defined on a suitable function space, and $G$ satisfies a Lipschitz type condition. While limitations of space do not allow us to present this problem in detail, we point out that if $F(g)$ denotes the solution of (1.1), (1.2) on $[0,T] \times \mathbb{R}$, then a solution of (1.3), (1.2) is a fixed point of the composition map $K$ defined by $K(y) = F(G(y))$. Such a fixed point can be found with the aid of our continuous dependence result for smooth solutions of (1.1), (1.2) for sufficiently small data in a manner similar to the method we used with Crandall

*Research sponsored by the United States Army under Grant No. DAAG29-77-G-0004 and Contract No. DAAG29-75-C-0024 and the National Science Foundation under Grant No. MSC75-21868.

in [1] to solve a functional differential equation in which, however, the underlying problem was an evolution equation of parabolic, rather than hyperbolic type. The details will be presented in a forthcoming joint paper with C. Dafermos.

The Cauchy problem (1.3), (1.2) has arisen in certain applications in heat flow and viscoelastic motion for "materials with memory" studied by MacCamy [5], [6]; the functional $G$ has the form

$$G(y)(t,x) = \Psi(t,x) + \beta y(t,x) + \int_0^t b(t-\tau)y(\tau,x)d\tau \,, \tag{1.4}$$

where $\Psi$ is a real smooth function on $[0,\infty) \times \mathbb{R}$ such that

$$\sup_{x\in\mathbb{R}} |\Psi(t,x)| \in L^1(0,\infty) \cap L^\infty(0,\infty), \ \sup_{x\in\mathbb{R}} |\Psi_x(t,x)| \in L^\infty(0,\infty) \,,$$

$\beta > 0$ is a constant, and $b \in L^1(0,\infty;\mathbb{R})$, the value of $G$ at $(t,x)$ depends on the restriction of $y(\cdot,x)$ to $[0,t]$. In [5], [6] the interest is in the existence of global smooth solutions of the Cauchy problem (1.3), (1.4), (1.2); this is carried out by combining Nishida's method with certain delicate a priori estimates obtained by energy methods. However, the proof in [5], [6] appears to us to be incomplete, because the local existence problem for (1.3), (1.4), (1.2), which can be handled by the method outlined above, is essentially ignored.

In Section 2 we obtain the desired results for a "diagonal" strictly hyperbolic system of first order equations equivalent to (1.1), (1.2); the results for (1.1), (1.2) follow as an easy corollary and these are stated in Section 3. We acknowledge useful discussions with M. G. Crandall, C. Dafermos, and R. J. DiPerna during the preparation of this paper.

Finally, we mention related work of Matsumara [8], [9] received after the completion of this paper; the author generalizes Nishida's results for (1.1), (1.2) with $g \equiv 0$ from one space dimension to quasilinear hyperbolic equations in several space dimensions, and he obtains global existence of weak solutions and results concerning their decay (Nishida's method does not apply in this case).

2. Equivalent Systems and Preliminary Results. We assume that $\sigma$ in (1.1) satisfies assumptions $(\sigma)$. In addition, assume that $g$, and the initial functions $y_0$, $y_1$ in (1.1), (1.2) satisfy:

$$\text{(g)} \qquad g, g_x \in C([0,\infty) \times \mathbb{R}), \ g(t) = \sup_{x\in\mathbb{R}} |g(t,x)| \in L^\infty(0,\infty) \cap L^1(0,\infty) \,,$$

$$g_1(t) = \sup_{x\in\mathbb{R}} |g_x(t,x)| \in L^\infty(0,\infty) \,,$$

$$\text{(I)} \qquad y_0 \in \beta^2(\mathbb{R}), \ y_1 \in \beta^1(\mathbb{R}) \,,$$

where $\beta^m$ denotes the set of real functions with continuous and bounded derivatives up to and including order $m$.

Following Nishida [10] we reduce the Cauchy problem (1.1), (1.2) to the equivalent system (2.3) below. Putting $y_x = v$ and $y_t = w$ in (1.1), (1.2) yields the equivalent Cauchy problem

$$(2.1)\qquad \begin{cases} v_t - w_x = 0,\ w_t - \sigma'(v)v_x + \alpha w = g & (0 < t < \infty,\ x \in \mathbb{R}) \\ v(0,x) = y_0'(x),\ w(0,x) = y_1(x) & (x \in \mathbb{R}) \ . \end{cases}$$

The eigenvalues of the matrix of (2.1)

$$\begin{pmatrix} 0 & 1 \\ -\sigma'(v) & 0 \end{pmatrix}$$

are $\lambda = -\sqrt{\sigma'(v)}$, $\mu = \sqrt{\sigma'(v)}$; by assumptions $(\sigma)$, $\lambda$ and $\mu$ are real and distinct so that (2.1) is a strictly hyperbolic problem in the region $\{(v,w) : v \in \mathbb{R}, w \in \mathbb{R}\}$. To diagonalize (2.1) introduce the Riemann invariants

$$(2.2)\qquad r = w + \phi(v),\ s = w - \phi(v),\ \phi(v) = \int_0^v \sqrt{\sigma'(\xi)}\, d\xi \ ;$$

by $(\sigma)$ the mapping $(v,w) \to (r,s)$ defined by (2.2) is one to one from $\mathbb{R} \times \mathbb{R}$ onto $\mathbb{R} \times \mathbb{R}$. A simple calculation shows that (2.1) is equivalent to the Cauchy problem for the diagonal, strictly hyperbolic system

$$(2.3)\qquad \begin{cases} r_t + \lambda r_x + \frac{\alpha}{2}(r+s) = g & \\ & (0 < t < \infty;\ x \in \mathbb{R}) \\ s_t + \mu s_x + \frac{\alpha}{2}(r+s) = g & \\ r(0,x) = r_0(x),\ s(0,x) = s_0(x) & (x \in \mathbb{R}) \ , \end{cases}$$

where by (2.2) $\lambda = \lambda(r-s)$, $\mu = \mu(r-s) \in C'(\mathbb{R})$, and where by (2.1)

$$(2.4)\qquad r_0(x) = y_1(x) + \phi(y_0'(x)),\ s_0(x) = y_1(x) - \phi(y_0'(x)) \qquad (x \in \mathbb{R}) \ ;$$

by assumptions $(\sigma)$ and I, the initial data $r_0, s_0 \in \beta^1(\mathbb{R})$. It is also seen that if $r, s$ is a smooth $(\beta^1)$ solution of the problem (2.3) for $(t,x) \in \Omega \subseteq ([0,\infty) \times \mathbb{R})$, then $y$, determined by the relations $y_t = w(r,s)$, $y_x = v(r,s)$ (where $v, w$ are uniquely determined by (2.2)), will be a smooth $(\beta^2)$ solution of the Cauchy problem (1.1), (1.2) and conversely; we shall therefore deduce our results for (1.1), (1.2) from (2.3).

The following local result for (2.3) is known [2; Sec. 8], [3, Theorem VI]:

Lemma 2.1. *Let $r_0, s_0 \in \beta^1(\mathbb{R})$, let assumptions $(\sigma)$ hold, and assume that $g, g_x \in \beta^0$ for $(t,x) \in [0,T] \times \mathbb{R}$, where $T > 0$. Then there exists a number $0 < T_1 \leq T$ such that the Cauchy problem (2.3) has a unique smooth solution $r, s \in \beta^1([0,T_1] \times \mathbb{R})$.*

The objective of the next two lemmas is to obtain apriori estimates on $r, s, r_x, s_x$ (and hence by (2.3) on $r_t, s_t$), independent of $T$, which enable us to continue the local $\beta^1$-solution in $t$ by a standard method.

Lemma 2.2. *Let the assumptions of Lemma 2.1 hold. In addition, assume that $g(t) = \sup_{x \in \mathbb{R}} |g(t,x)| \in L^1(0,\infty)$. Define the a priori constant $M_0 > 0$ by*

$$M_0 = r_0 + s_0 + 2\int_0^\infty g(\xi)d\xi,\ r_0 = \sup_{x \in \mathbb{R}} |r_0(x)|,\ s_0 = \sup_{x \in \mathbb{R}} |s_0(x)| \ .$$

*For as long as the* $\beta^1$*-solution* $r, s$ *of* (2.3) *exists one has*

$$\sup_{x\in \mathbb{R}} |r(t,x)| \le M_0, \quad \sup_{x\in \mathbb{R}} |s(t,x)| \le M_0 . \tag{2.5}$$

*Sketch of Proof*. The proof is similar to that of [10, Lemma 1], [5, Lemma 6.2]. Define the $\lambda$ and $\mu$ characteristics of (2.3) respectively by

$$x = x_1(t,\beta) = \beta + \int_0^t \lambda d\tau, \quad x = x_2(t,\gamma) = \gamma + \int_0^t \mu d\tau \quad (\beta,\gamma \in \mathbb{R}) , \tag{2.6}$$

where $\lambda = \lambda[r(t,x_1(t,\beta)) - s(t,x_1(t,\beta))]$, $\mu = \mu[r(t,x_2(t,\gamma)) - s(t,x_2(t,\gamma))]$ .

Let $' = \frac{\partial}{\partial t} + \lambda\frac{\partial}{\partial x}$ , $` = \frac{\partial}{\partial t} + \mu \frac{\partial}{\partial x}$ denote differentiation along the $\lambda$ and $\mu$ characteristics respectively, thus $r'(t,x_1) = \frac{d}{dt} r(t,x_1(t,\beta))$, $s`(t,x_2) = \frac{d}{dt} s(t,x_2(t,\gamma))$. Equations (2.3) become the ordinary differential equations

$$\begin{cases} \frac{d}{dt} r(t,x_1(t,\beta)) + \frac{\alpha}{2} (r(t,x_1(t,\beta)) + s(t,x_1(t,\beta))) = g(t,x_1(t,\beta)) \\ \frac{d}{dt} s(t,x_2(t,\gamma)) + \frac{\alpha}{2} (r(t,x_2(t,\gamma)) + s(t,x_2(t,\gamma))) = g(t,x_2(t,\gamma)) \end{cases} ; \tag{2.7}$$

note that solutions of (2.7) will exist for as long as the slopes $\lambda, \mu$ of the characteristics $x_1(t,\beta)$ and $x_2(t,\gamma)$ remain bounded. Put

$$R(t) = e^{\frac{\alpha}{2}t} \left[\sup_{x\in \mathbb{R}} |r(t,x)| + \sup_{x\in \mathbb{R}} |s(t,x)|\right]$$

$$r_0 = \sup_{x\in \mathbb{R}} |r_0(x)|, \quad s_0 = \sup_{x\in \mathbb{R}} |s_0(x)| .$$

Integrate each of the equations (2.9) using $r(0,x_1(0,\beta)) = r_0(\beta)$, $s(0,x_2(0,\gamma)) = s_0(\gamma)$ (see (2.3), (2.6)), add the resulting equations and take absolute values; a standard argument yields the inequality

$$R(t) \le r_0 + s_0 + \frac{\alpha}{2} \int_0^t R(\xi)d\xi + 2 \int_0^t e^{\frac{\alpha}{2}\xi} g(\xi)d\xi . \tag{2.8}$$

Gronwall's inequality applied to (2.8) gives

$$R(t) \le (r_0+s_0)e^{\frac{\alpha}{2}t} + 2e^{\frac{\alpha}{2}t} \int_0^t g(\xi)d\xi , \tag{2.9}$$

and thus finally

$$\sup_{x\in \mathbb{R}} |r(t,x)| + \sup_{x\in \mathbb{R}} |s(t,x)| \le M_0 \tag{2.10}$$

and the proof is complete.

*Lemma 2.3. Let the assumptions of Lemma 2.1 and* (g) *be satisfied. Define the constant* $D_1 > 0$ *by*

$$D_1 = r_0 + s_0 + \sup_{x\in\mathbb{R}} |r_0'(x)| + \sup_{x\in\mathbb{R}} |s_0'(x)| + \|g\|_{L^1(0,\infty)} + \|g\|_{L^\infty(0,\infty)} + \|g_1\|_{L^\infty(0,\infty)} .$$

*For as long as the $\beta^1$-solution $r, s$ of (2.3) exists and if $D_1$ is sufficiently small, there exists a constant $M_1 = M_1(D_1) > 0$ where $M_1(D_1) \to 0$ as $D_1 \to 0$, such that*

$$\sup_{x\in\mathbb{R}} |r_x(t,x)| \le M_1, \quad \sup_{x\in\mathbb{R}} |s_x(t,x)| \le M_1 . \tag{2.11}$$

*Sketch of Proof.* (Compare [10, Lemma 2], [5, Lemma 6.3].) Differentiate the first equation in (2.3) obtaining (recall $\lambda = \lambda(r-s)$)

$$r_{xt} + \lambda r_{xx} = -\lambda_r r_x^2 - \lambda_s r_x s_x - \frac{\alpha}{2}(r_x + s_x) + g_x . \tag{2.12}$$

We remark that although Lemma 2.1 does not assert the existence of $r_{xx}$ and $r_{xt}$, note that the left side of (2.12) is $r_x'$ and this does exist for as long as the $\beta^1$-solution $r, s$ of (2.3) exists. This observation also justifies the validity of equations (2.12)-(2.18) which follow. Since $\mu = -\lambda$ the second equation in (2.3) gives

$$s_x = \frac{s'}{2\lambda} + \frac{\alpha}{4\lambda}(r+s) - \frac{g}{2\lambda} \qquad \left(' = \frac{\partial}{\partial t} + \lambda \frac{\partial}{\partial x}\right) . \tag{2.13}$$

Define

$$h = \frac{1}{2}\log(-\lambda(r-s)) . \tag{2.14}$$

Differentiating $h$ along the $\lambda$-characteristic and using $\lambda_s = -\lambda_r$ gives

$$h' = \frac{\lambda_r}{2\lambda}\left(-\frac{\alpha}{2}(r+s) + g - s'\right) . \tag{2.15}$$

Substitution of (2.13), (2.15) into (2.12) yields

$$r_x' + \left(\frac{\alpha}{2} + \lambda_r r_x + h'\right) r_x = -\frac{\alpha}{4\lambda} s' - \frac{\alpha^2}{8\lambda}(r+s) + \frac{\alpha}{4\lambda} g + g_x ,$$

or equivalently

$$(e^h r_x)' + \left(\frac{\alpha}{2} + \lambda_r r_x\right) e^h r_x = \left(-\frac{\alpha}{4\lambda} s' - \frac{\alpha^2}{8\lambda}(r+s) + \frac{\alpha}{4\lambda} g + g_x\right) e^h . \tag{2.16}$$

Define the function $z$ by

$$z(r-s) = \int_0^{r-s} \frac{\alpha}{4\lambda(\xi)} e^{h(\xi)} d\xi ; \tag{2.17}$$

then $z' = -\frac{\alpha^2}{8\lambda} e^h (r+s) + \frac{\alpha}{4\lambda} e^h g - \frac{\alpha}{4\lambda} e^h s'$ and (2.16) becomes

$$(e^h r_x)' + \left(\frac{\alpha}{2} + \lambda_r r_x\right) e^h r_x = z' + e^h g_x . \tag{2.18}$$

To integrate (2.18) along the $\lambda$-characteristic put

$$\begin{cases} k(t) = \frac{\alpha}{2} + \lambda_r(r(t,x_1(t,\beta)) - s(t,x_1(t,\beta)))\, r_x(t,x_1(t,\beta)) \\ \rho(t) = r_x(t,x_1(t,\beta)) \exp[h(t,x_1(t,\beta))] \\ p(t) = z'(t,x_1(t,\beta)) + g_x(t,x_1(t,\beta)) \exp[h(t,x_1(t,\beta))] . \end{cases} \tag{2.19}$$

Then

$$\rho(t) = \rho(0)\exp[-\int_0^t k(\tau)d\tau] + \int_0^t p(\xi)\exp[-\int_\xi^t k(\tau)d\tau]d\xi . \tag{2.20}$$

Suppose we can show that for any solution $r, s$ of (2.3)

$$|\lambda_r r_x| \le \frac{\alpha}{4} . \tag{2.21}$$

Then $k(t) = \frac{\alpha}{2} + \lambda_r(\cdot)r_x(\cdot) \ge \frac{\alpha}{4}$ and by an easy calculation

$$|\rho(t)| \le |\rho(0)| + 3 \sup_{0\le\tau\le t} |z(\tau,x_1(\tau,\beta))| \tag{2.22}$$

$$+ \frac{4}{\alpha} \sup_{0\le\tau\le t} |g_x(\tau,x_1(t,\beta))\exp h(\tau,x_1(\tau,\beta)| .$$

We next show that (2.21) holds for any solution $r, s$ of (2.3), provided the constant $D_1 > 0$ is sufficiently small. Indeed, at $t = 0$ and for any $\beta \in \mathbb{R}$ $\lambda_r(r(0,\beta) - s(0,\beta))r_x(0,\beta)$ satisfies

$$|r_0'(\beta)\lambda_r(r_0(\beta) - s_0(\beta))| \le \sup_{x\in\mathbb{R}} |r_0'(x)\lambda_r(r_0(x) - s_0(x))| < \frac{\alpha}{4} \tag{2.23}$$

provided $D_1 > 0$ is sufficiently small. By (2.19) $\lambda_r r_x = \lambda_r e^{-h}\rho$ and by $(\sigma)$, (2.14) and Lemma 2.2 $|\lambda_r e^{-h}|$ is uniformly bounded in $(t,x)$ by a constant $K_1(D_1) > 0$, where $K_1(D_1) \to 0$ as $D_1 \to 0$, for any solution $r, s$ of (2.3). By (2.14), (2.17), (2.22) and Lemma 2.2 the quantity $\rho$ is uniformly bounded for any solution $r, s$ of (2.3) as follows:

$$|\rho(t)| \le |\rho(0)| + 3 \sup_{\substack{0\le\tau\le t\\ x\in\mathbb{R}}} |z[r(\tau,x) - s(\tau,x)]| \tag{2.24}$$

$$+ \frac{4}{\alpha} \sup_{\substack{0\le\tau\le t\\ x\in\mathbb{R}}} \sqrt{-\lambda[r(\tau,x) - s(\tau,x)]}\, |g_x(\tau,x)|$$

$$\le K_2(D_1), \text{ where } K_2(D_1) \to 0 \text{ as } D_1 \to 0 .$$

Therefore $|\lambda_r r_x| \le K_1(D_1)K_2(D_1)$ uniformly in $(t,x)$ for any solution $r, s$ of (2.3). The assertion (2.21) holds at $t = 0$, for $D_1$ sufficiently small (by (2.23)). Choosing $D_1$ smaller if necessary so that $K_1(D_1)K_2(D_1) \le \frac{\alpha}{4}$, we conclude that (2.21) continues to hold for as long as the solution $r, s$ exists. Returning to (2.22), (2.24) and using $r_x = \rho e^{-h}$ establishes the estimate for $r_x$ in (2.11); a similar argument yields the estimate for $s_x$ and completes the proof of Lemma 2.3.

The a priori estimates of Lemmas 2.2 and 2.3, together with $||g||_{L^\infty(0,\infty)} < \infty$ yield uniform a priori estimates for $r_t$, $s_t$, for any solution $r, s$ of (2.3), provided $D_1 > 0$ is sufficiently small. Then Lemmas 2.1, 2.2, 2.3 and a standard continuation argument give the first part of the following global result for the Cauchy problem (2.3).

Theorem 2.1. Let the assumption $(\sigma)$, $(g)$ be satisfied, and let the initial data $r_0, s_0 \in \beta^1(\mathbb{R})$. If the constant $D_1$ (see Lemma 2.3) is sufficiently small, then the Cauchy problem (2.3) has a unique $\beta^1$-solution $r, s$ for $0 \le t < \infty$, $x \in \mathbb{R}$ and the a priori estimates (2.5), (2.11) are satisfied for $0 \le t < \infty$.

Let the above assumptions be satisfied by initial data $r_0, s_0$ and $\bar{r}_0, \bar{s}_0$ and forcing functions $g, \bar{g}$; denote by $r, s$ and $\bar{r}, \bar{s}$ the corresponding $\beta^1$-solutions of (2.3) on $[0,\infty) \times \mathbb{R}$. Define

$$\zeta(t) = \sup_{x\in\mathbb{R}} |r(t,x) - \bar{r}(t,x)| + \sup_{x\in\mathbb{R}} |s(t,x) - \bar{s}(t,x)| .$$

Then there exists a constant $M_2 = M_2(\sigma, M_0) > 0$ such that

$$(2.25) \quad \zeta(t) \le e^{2M_2M_1t}\Big(\zeta(0) + \int_0^t e^{-2M_2M_1\tau} \sup_{x\in\mathbb{R}} |g(\tau,x) - \bar{g}(\tau,x)| d\tau\Big) \quad (0 \le t < \infty) ,$$

where $M_0, M_1$ are the bounds in (2.5), (2.11) respectively.

Remark. The continuous dependence result (2.25) also holds for local solutions $r, s$ and $\bar{r}, \bar{s}$ on $[0,T_1] \times [0, \ )$ of the Cauchy problem (2.3) (see Lemma 2.1), but only for $0 \le t \le T_1$.

Proof of Theorem 2.1. It remains only to prove (2.25). If $r, s$ and $\bar{r}, \bar{s}$ are $\beta^1$-solutions of (2.3) for the situation in the theorem, one has

$$(2.26) \quad \begin{cases} (r-\bar{r})_t + \lambda r_x - \bar{\lambda}\bar{r}_x = -\frac{\alpha}{2}[(r-\bar{r}) + (s-\bar{s})] + g - \bar{g} \\ (s-\bar{s})_t + \mu s_x - \bar{\mu}\bar{s}_x = -\frac{\alpha}{2}[(r-\bar{r}) + (s-\bar{s})] + g - \bar{g} \end{cases} \quad \begin{matrix} 0 < t < \infty \\ x \in \mathbb{R} \end{matrix}$$

subject to the initial conditions

$$(2.27) \quad r(0,x) - \bar{r}(0,x) = r_0(x) - \bar{r}_0(x),\ s(0,x) - \bar{s}(0,x) = s_0(x) - \bar{s}_0(x) \qquad (x \in \mathbb{R})$$

where $\lambda = \lambda(r-s)$, $\mu = \mu(r-s)$, $\bar{\lambda} = \lambda(\bar{r}-\bar{s})$, $\bar{\mu} = \mu(\bar{r}-\bar{s})$. But

$$\lambda r_x - \bar{\lambda}\bar{r}_x = \lambda(r-\bar{r})_x + (\lambda-\bar{\lambda})\bar{r}_x$$

$$\mu s_x - \bar{\mu}\bar{s}_x = \mu(s-\bar{s})_x + (\mu-\bar{\mu})\bar{s}_x ;$$

Therefore (2.26) can be written as

$$(2.28) \quad \begin{cases} (r-\bar{r})_t + \lambda(r-\bar{r})_x = -\frac{\alpha}{2}[(r-\bar{r}) + (s-\bar{s})] + g - \bar{g} - (\lambda-\bar{\lambda})\bar{r}_x \\ (s-\bar{s})_t + \mu(s-\bar{s})_x = -\frac{\alpha}{2}[(r-\bar{r}) + (s-\bar{s})] + g - \bar{g} - (\mu-\bar{\mu})\bar{s}_x . \end{cases}$$

Recalling the definitions of $\lambda, \mu$ and using $(\sigma)$ and the mean value theorem one has

$$\lambda - \bar{\lambda} = \lambda(r-s) - \lambda(\bar{r}-\bar{s}) = \frac{d\lambda}{d\xi}(\bar{r} - \bar{s} + \theta_1[r - s - (\bar{r}-\bar{s})])(r - s - (\bar{r}-\bar{s}))$$

$$\mu - \bar{\mu} = \mu(r-s) - \mu(\bar{r}-\bar{s}) = \frac{d\mu}{d\xi}(\bar{r} - \bar{s} + \theta_2[r - s - (\bar{r}-\bar{s})])(r - s - (\bar{r}-\bar{s})) ,$$

for some $0 < \theta_1, \theta_2 < 1$. Therefore, (2.28) becomes

$$(2.29)\quad \begin{cases} (r-\bar{r})' + \frac{\alpha}{2}(r-\bar{r}) = -\frac{\alpha}{2}(s-\bar{s}) - \frac{d\lambda}{d\xi}(\cdot)(r-\bar{r}-(s-\bar{s}))\bar{r}_x + g - \bar{g} \\ (s-\bar{s})^{\grave{}} + \frac{\alpha}{2}(s-\bar{s}) = -\frac{\alpha}{2}(r-\bar{r}) - \frac{d\mu}{d\xi}(\cdot)(r-\bar{r}-(s-\bar{s}))\bar{s}_x + g - \bar{g}, \end{cases}$$

where $' = \frac{\partial}{\partial t} + \lambda\frac{\partial}{\partial x}$, $\grave{} = \frac{\partial}{\partial t} + \mu\frac{\partial}{\partial x}$. We next note that assumption $(\sigma)$ and Lemmas 2.2 and 2.3 imply the existence of a constant $M_2 = M_2(\sigma, M_0)$ such that

$$(2.30)\qquad \left|\frac{d\lambda}{d\xi}(\cdot)\bar{r}_x\right| \le M_2M_1, \quad \left|\frac{d\mu}{d\xi}(\cdot)\bar{s}_x\right| \le M_2M_1,$$

uniformly in $(t,x) \in [0,\infty) \times \mathbb{R}$ where $M_1$ is the bound in Lemma 2.3. Integrating the first equation in (2.29) along any $\lambda$-characteristic and the second along any $\mu$-characteristic and making simple estimates one obtains the pair of inequalities

$$e^{\frac{\alpha}{2}t}\sup_{x\in\mathbb{R}}|r(t,x)-\bar{r}(t,x)| \le \sup_{x\in\mathbb{R}}|r_0(x)-\bar{r}_0(x)| + \int_0^t e^{\frac{\alpha}{2}\tau}\sup_{x\in\mathbb{R}}|g(\tau,x)-\bar{g}(\tau,x)|d\tau$$
$$+ \int_0^t e^{\frac{\alpha}{2}\tau}\left(\frac{\alpha}{2}+M_2M_1\right)\sup_{x\in\mathbb{R}}|s(\tau,x)-\bar{s}(\tau,x)|d\tau + M_1M_2\int_0^t e^{\frac{\alpha}{2}\tau}\sup_{x\in\mathbb{R}}|r(\tau,x)-\bar{r}(\tau,x)|d\tau,$$

$$e^{\frac{\alpha}{2}t}\sup_{x\in\mathbb{R}}|s(t,x)-\bar{s}(t,x)| \le \sup_{x\in\mathbb{R}}|s_0(x)-\bar{s}_0(x)| + \int_0^t e^{\frac{\alpha}{2}\tau}\sup_{x\in\mathbb{R}}|g(\tau,x)-\bar{g}(\tau,x)|d\tau$$
$$+ \int_0^t e^{\frac{\alpha}{2}\tau}\left(\frac{\alpha}{2}+M_2M_1\right)\sup_{x\in\mathbb{R}}|r(\tau,x)-\bar{r}(\tau,x)|d\tau + M_2M_1\int_0^t e^{\frac{\alpha}{2}\tau}\sup_{x\in\mathbb{R}}|s(\tau,x)-\bar{s}(\tau,x)|d\tau.$$

Adding these inequalities one obtains on using the definition of $\zeta$

$$(2.31)\qquad \zeta(t)e^{\frac{\alpha}{2}t} \le \zeta(0) + 2\int_0^t e^{\frac{\alpha}{2}\tau}\sup_{x\in\mathbb{R}}|g(\tau,x)-\bar{g}(\tau,x)|d\tau$$
$$+ \int_0^t \left(\frac{\alpha}{2}+2M_1M_2\right)e^{\frac{\alpha}{2}\tau}\zeta(\tau)d\tau \qquad (0 \le t < \infty).$$

Finally, applying Gronwall's inequality to (2.31) yields the result (2.25) completing the proof.

## 3. Global Existence, Uniqueness, and Continuous Dependence for the Cauchy Problem (1.1), (1.2).

As an immediate consequence of Theorem 2.1 and of the equivalence of the Cauchy problems (1.1), (1.2) and (2.3), (2.4) we obtain the main result of this paper.

Theorem 3.1. *Let the assumptions* $(\sigma)$, $(g)$, $(I)$ *be satisfied. Define the constant*

$$(3.1)\quad D = \sup_{x\in\mathbb{R}}|y_0'(x)| + \sup_{x\in\mathbb{R}}|y_1(x)| + \sup_{x\in\mathbb{R}}|y_0''(x)| + \|g\|_{L^1(0,\infty)} + \|g\|_{L^\infty(0,\infty)} + \|g_1\|_{L^\infty(0,\infty)}.$$

*If* $D$ *is sufficiently small, then the Cauchy problem* (1.1), (1.2) *has a unique* $\beta^2$*-solution* $y$ *on* $[0,\infty) \times \mathbb{R}$, *and the solution* $y$ *satisfies the a priori estimates:*

$$(3.2)\quad \sup_{x\in\mathbb{R}} |y_x(t,x)|,\ \sup_{x\in\mathbb{R}} |y_t(t,x| \le \sup_{x\in\mathbb{R}} |y_0'(x)| + \sup_{x\in\mathbb{R}} |y_1(x)| + 2\int_0^\infty g(\xi)d\xi = M_0$$

$$(0 \le t < \infty),$$

and there exists a constant $M_1 = M_1(D) > 0$ (which $\to 0$ as $D \to 0$) such that

$$(3.3)\quad \sup_{x\in\mathbb{R}} |y_{xx}(t,x)|,\ \sup_{x\in\mathbb{R}} |y_{xt}(t,x)|,\ \sup_{x\in\mathbb{R}} |y_{tt}(t,x)| \le M_1 \qquad (0 \le t < \infty).$$

Let, in addition, the assumptions (g) and (I) be satisfied also by functions $\bar{g}$ and $\bar{y}_0$, $\bar{y}_1$ and let $\bar{y}$ denote the corresponding $\beta^1$-solution on $[0,\infty) \times \mathbb{R}$. Define

$$\zeta(t) = \sup_{x\in\mathbb{R}} |y_x(t,x) - \bar{y}_x(t,x)| + \sup_{x\in\mathbb{R}} |y_t(t,x) - \bar{y}_t(t,x)|;$$

then there exists a constant $M_2 = (\sigma, M_0) > 0$ such that

$$(3.4)\quad \zeta(t) \le e^{2M_2M_1t}\left(\zeta(0) + \int_0^t e^{-2M_2M_1\tau} \sup_{x\in\mathbb{R}} |g(\tau,x) - \bar{g}(\tau,x)|d\tau\right) \qquad (0 \le t < \infty),$$

where $M_0$, $M_1$ are bounds in (3.2), (3.3) respectively; moreover (using $y(t,x) = y_0(x) + \int_0^t y_t(\tau,x)d\tau$)

$$\sup_{x\in\mathbb{R}} |y(t,x) - \bar{y}(t,x)| \le \sup_{x\in\mathbb{R}} |y_0(x) - \bar{y}_0(x)|$$
$$+ \frac{e^{2M_1M_2t}}{M_1M_2}\Big\{\sup_{x\in\mathbb{R}} |y_0'(x) - \bar{y}_0'(x)| + \sup_{x\in\mathbb{R}} |y_1(x) - \bar{y}_1(x)|$$
$$+ \int_0^t e^{-2M_1M_2\tau} \sup_{x\in\mathbb{R}} |g(\tau,x) - \bar{g}(\tau,x)|d\tau\Big\} \qquad (0 \le t < \infty).$$

## References


[1] M. G. Crandall and J. A. Nohel, An abstract functional differential equation and a related nonlinear Volterra equation. Israel J. Math. 29 (1978), 313-328; also Math. Res. Center, Univ. of Wisconsin, Madison, Tech. Summary Report No. 1765, July 1977).

[2] A. Douglis, Some existence theorems for hyperbolic systems of partial differential equations in two independent variables. Comm. Pure Appl. Math. 5 (1952), 119-154.

[3] P. Hartmann and A. Wintner, On hyperbolic partial differential equations. American J. Math. 74 (1952), 834-864.

[4] P. D. Lax, Development of singularities of solutions of nonlinear hyperbolic partial differential equations. J. Math. Phys. 5 (1964), 611-613.

[5] R. C. MacCamy, An integro-differential equation with applications in heat flow. Q. Appl. Math. 35 (1977), 1-19.

[6] R. C. MacCamy, A model for one-dimensional, nonlinear viscoelasticity. Q. Appl. Math. 35 (1977), 21-33.

[7] R. C. MacCamy and V. Mizel, Existence and nonexistence in the large of solutions of quasilinear wave equations. Arch. Rational Mech. and Anal. 25 (1967), 299-320.

[8] A. Matsumura, Global existence and asymptotics of the solutions of the second-order quasilinear hyperbolic equations with first order dissipation, (to appear).

[9] A. Matsumura, Energy decay of solutions of dissipative wave equations (to appear).

[10] T. Nishida, Global smooth solutions for the second-order quasilinear wave equations with the first-order dissipation (unpublished).

Author's address: Mathematics Research Center, University of Wisconsin-Madison,
610 Walnut Street, Madison, Wisconsin 53706 U.S.A.
and Department of Mathematics, University of Wisconsin-Madison,
Van Vleck Hall, Madison, Wisconsin 53706 U.S.A.

# ENERGETIC ESTIMATES ANALOGOUS TO THE SAINT-VENANT PRINCIPLE AND THEIR APPLICATIONS

O.A. Oleinik, Moscow

In 1855 Saint-Venant [1] formulated a principle which is of an exceptional importance in the theory of elasticity as well as in its applications in the construction mechanics. During the last hundred years numerous studies have been devoted to the Saint-Venant principle and to the clarification of conditions of its applicability. A strict mathematical formulation of the Saint-Venant principle together with its justification for cylindrical bodies was given by Toupin [2] in 1965 and for arbitrary twodimensional bodies by Knowles [3]. A survey of investigations concerning this problem is found in Gurtin's paper [4].

The Saint-Venant principle can be expressed in the form of an a priori energetic estimate of the solution of the system of equations of the elasticity theory. It was found that estimates of this type can be established for wide classes of partial differential equations and systems. Theorems of the Phragmen-Lindelöf type, existence and uniqueness theorems for solutions of boundary value problems in both bounded and unbounded domains in the class of functions with unbounded energy integrals, theorem on the behavior of solutions in the neighborhood of non-regular points of the boundary (in the neighborhood of angles, ribs etc.) and in the neighborhood of infinity can be obtained as consequences of the energetic estimates which express the Saint-Venant principle. A number of such results was obtained in [5] - [11].

As the simplest example let us consider the Saint-Venant principle for the Laplace equation in a domain $\Omega$ of a special shape.

<u>Theorem 1.</u> Let a bounded domain $\Omega$ from the class $C^1$ coincide for $|x_n| < T$ with a cylinder $\{x: x' \in \Omega', -T < x_n < T\}$ where $x = (x_1,\dots,x_n)$, $x' = (x_1,\dots,x_{n-1})$, $T = \text{const.}$, $\Omega'$ is a domain in the space $R^{n-1}_{x'}$. Assume that $u \in C^2(\Omega) \cap C^1(\bar{\Omega})$,

$$(1) \quad \Delta u = f \quad \text{in} \quad \Omega, \quad \frac{\partial u}{\partial \nu}\Big|_{\partial\Omega} = \psi, \quad \Delta u \equiv \sum_{j=1}^{n} u_{x_j x_j},$$

where $f \equiv 0$ in $\Omega_T$, $\psi \equiv 0$ on $\partial\Omega \cap \partial\Omega_T$ and, moreover,

$$(2) \quad \int_{\Omega^+} f dx - \int_{\partial\Omega \cap \partial\Omega^+} \psi ds = 0, \quad \int_{\Omega^-} f dx - \int_{\partial\Omega \cap \partial\Omega^-} \psi ds = 0,$$

where $\Omega_\tau = \Omega \cap \{x: |x_n| < \tau\}$, $\tau = \text{const} > 0$, $\tau \leqq T$, $\Omega^+ = \Omega \cap \{x: x_n > T\}$, $\Omega^- = \{x: x_n < -T\}$, $\partial\Omega$ is the boundary of $\Omega$, $\nu$ is the direction of the unit outer normal to $\partial\Omega$. Then

$$(3) \qquad \int_{\Omega_{\tau_0}} |\nabla u|^2\, dx \leqq \exp\left\{-2\lambda^{\frac{1}{2}}(\tau_1 - \tau_0)\right\} \int_{\Omega_{\tau_1}} |\nabla u|^2\, dx ,$$

where $|\nabla u|^2 \equiv \sum_{j=1}^{n} u_{x_j}^2$, $\tau_0$, $\tau_1 = \text{const} > 0$, $\tau_0 < \tau_1 \leqq T$,

$$(4) \qquad \lambda = \inf_{v \in M} \left\{ \int_{\Omega'} \sum_{j=1}^{n-1} v_{x_j}^2\, dx' \left[\int_{\Omega'} v^2 dx'\right]^{-1} \right\},$$

$M$ is the family of all functions $v(x')$ continuously differentiable on $\bar{\Omega}'$ which satisfy the condition

$$(5) \qquad \int_{\Omega'} v(x')dx' = 0 .$$

Proof. With regard to (1) we obtain for $-T < a < T$

$$\int_{\Omega^+} f dx = \int_{\Omega \cap \{x: x_n < a\}} \Delta u dx = \int_{\partial\Omega \cap \partial\Omega^+} \psi ds + \int_{x_n = a} u_{x_n} dx'.$$

This together with the conditions (2) implies

$$(6) \qquad \int_{x_n = a} u_{x_n} dx' = 0 , \quad -T < a < T .$$

Let $S_\tau^+ = \{x: x_n = \tau\}$, $S_\tau^- = \{x: x_n = -\tau\}$, $S_\tau = S_\tau^+ \cup S_\tau^-$. According to the Green formula, we have for arbitrary positive $\tau \leqq T$

$$0 = \int_{\Omega_\tau} u \Delta u dx = - \int_{\Omega_\tau} |\nabla u|^2 dx + \int_{S_\tau^+} u u_{x_n} dx' - \int_{S_\tau^-} u u_{x_n} dx' .$$

Taking into account the relation (6), we conclude that for any constants $C_\tau^+$ and $C_\tau^-$

$$\int_{\Omega_\tau} |\nabla u|^2 dx = \int_{S_\tau^+} (u + C_\tau^+) u_{x_n} dx' - \int_{S_\tau^-} (u + C_\tau^-) u_{x_n} dx' .$$

Consequently

$$(7)\qquad \int\limits_{\Omega_\tau} |\nabla u|^2 dx \leqq \Big[\int\limits_{S_\tau^+} (u+C_\tau^+)^2 dx'\Big]^{\frac{1}{2}} \Big[\int\limits_{S_\tau^+} u_{x_n}^2\, dx'\Big]^{\frac{1}{2}} + \Big[\int\limits_{S_\tau^-} (u+C_\tau^-)^2 dx'\Big]^{\frac{1}{2}} \Big[\int\limits_{S_\tau^-} (u_{x_n})^2 dx'\Big]^{\frac{1}{2}} .$$

Let us choose the constants $C_\tau^+$ and $C_\tau^-$ so that

$$\int\limits_{S_\tau^+} (u + C_\tau^+)\, dx' = 0 , \qquad \int\limits_{S_\tau^-} (u + C_\tau^-)\, dx' = 0 .$$

Then the inequality (7) and the relation (8) imply that

$$(8)\qquad \int\limits_{\Omega_\tau} |\nabla u|^2 dx \leqq \lambda^{-\frac{1}{2}} \Big[\int\limits_{S_\tau^+} \sum_{j=1}^{n-1} u_{x_j}^2\, dx'\Big]^{\frac{1}{2}} \Big[\int\limits_{S_\tau^+} u_{x_n}^2\, dx'\Big]^{\frac{1}{2}} + \lambda^{-\frac{1}{2}} \Big[\int\limits_{S_\tau^-} \sum_{j=1}^{n-1} u_{x_j}^2\, dx'\Big]^{\frac{1}{2}} \Big[\int\limits_{S_\tau^-} u_{x_n}^2\, dx'\Big]^{\frac{1}{2}} \leqq$$

$$\leqq \frac{1}{2} \lambda^{-\frac{1}{2}} \int\limits_{S_\tau} \Big(\sum_{j=1}^{n-1} u_{x_j}^2 + u_{x_n}^2\Big) dx' = \frac{1}{2} \lambda^{-\frac{1}{2}} \int\limits_{S_\tau} |\nabla u|^2 dx' .$$

Set $F(\tau) = \int\limits_{\Omega_\tau} |\nabla u|^2 dx$ , $0 \leqq \tau \leqq T$ . We obtain from (8) that

$$F(\tau) \leqq \frac{1}{2} \lambda^{-\frac{1}{2}} \frac{dF}{d\tau} .$$

Multiplying this inequality by $\exp\{-2\lambda^{\frac{1}{2}} \tau\} \cdot 2\lambda^{\frac{1}{2}}$ and integrating from $\tau_0$ to $\tau_1$ we obtain the inequality (3). The theorem is proved.

The conditions (2) for a membrane correspond in the Saint-Venant principle for an elastic body to the condition that the forces acting at the ends are statically equivalent to zero. The number $\lambda$ defined by the conditions (4), (5) equals to the first non-zero eigenvalue of the Neumann boundary value problem

$$(9) \qquad \sum_{j=1}^{n-1} v_{x_j x_j} + \lambda v = 0 \quad \text{in} \quad \Omega', \quad \frac{\partial v}{\partial \nu}\Big|_{\partial\Omega'} = 0 .$$

It is easily seen that $\lambda = \pi^2/l^2$ for $n = 2$, where $l$ is the length of the interval $S_\tau^+$. The following theorem of the Phragmen--Lindelöf type (a uniqueness theorem) for the solution of the Neumann problem in an infinite cylinder $\Omega$ is a consequence of the estimate (3).

Theorem 2. Let $\Omega = \{x: x' \in \Omega', -\infty < x_n < +\infty\}$, $\Delta u = 0$ in $\Omega$, $\frac{\partial u}{\partial \nu}\Big|_{\partial\Omega} = 0$, $u \in C^2(\Omega) \cap C^1(\bar{\Omega})$ and let for a constant b

$$\int_{x_n=b} u_{x_n}\, dx' = 0 .$$

Then $u \equiv \text{const}$ in $\Omega$ provided there is a sequence $R_j \to \infty$ with

$$(10) \qquad \int_{\Omega_{R_j}} |\nabla u|^2 dx \leqq \mathcal{E}(R_j) \exp\{2\lambda^{\frac{1}{2}} R_j\} ,$$

where $\mathcal{E}(R_j) \to 0$ for $R_j \to \infty$.

This theorem is an immediate consequence of Theorem 1.

The constant $2\lambda^{\frac{1}{2}}$ which appears in the exponential function in the inequalities (3) and (10) is the best possible, i.e. it cannot be replaced by a greater constant. This is demonstrated by the following example. Let $v(x')$ be a non-trivial solution of the problem (9) corresponding to the first non-zero eigenvalue $\lambda$. Then

$$\int_{\Omega'} v(x')\, dx' = 0 .$$

Put $u(x) = v(x') \exp\{\lambda^{\frac{1}{2}} x_n\}$. The function $u(x)$ satisfies all assumptions of Theorem 2 except the condition (10). The inequality (10) holds for $u(x)$ provided $\mathcal{E}(R_j) = \text{const} > 0$ and hence $u \equiv$ $\equiv$ const. Indeed, for any $R > 0$ we have

$$\int_{\Omega_R} |\nabla u|^2 dx = C_1 \int_{-R}^{R} \exp\{2\lambda^{\frac{1}{2}} x_n\} dx_n \leqq C_2 \exp\{2\lambda^{\frac{1}{2}} R\}, \quad C_1, C_2 = \text{const}.$$

Analogously to the proof of Theorem 1 we can prove an estimate of the type (3) and a Phragmen-Lindelöf theorem for solutions of the Dirichlet problem for the Laplace equation. The following assertion holds.

Theorem 3. Let $\Omega$ be the domain defined in Theorem 1, $\Delta u = f$ in $\Omega$, $u|_{\partial\Omega} = \psi$ with $f \equiv 0$ in $\Omega_T$, $\psi \equiv 0$ on $\partial\Omega \cap \partial\Omega_T$, $u \in C^2(\Omega) \cap C^1(\bar{\Omega})$. Then $u(x)$ satisfies the inequality (3) with

$$(11) \qquad \lambda = \inf_{v \in M_1} \left\{ \int_{\Omega'} \sum_{j=1}^{n-1} v_{x_j}^2 \, dx' \left[ \int_{\Omega'} v^2 dx' \right]^{-1} \right\},$$

$M_1$ being the family of all functions $v(x')$ continuously differentiable in $\bar{\Omega}'$ and such that $v|_{\partial\Omega'} = 0$.

Let us notice that no conditions are put on $f$ in $\Omega \setminus \Omega_T$ and on $\psi$ in $\partial\Omega \setminus \partial\Omega_T$ in Theorem 3 in contradistinction to Theorem 1.

Theorem 4. Let $\Omega = \{x: x' \in \Omega', \ -\infty < x_n < +\infty\}$, $\Delta u = 0$ in $\Omega$, $u|_{\partial\Omega} = 0$, $u \in C^2(\Omega) \cap C^1(\bar{\Omega})$, $\partial\Omega \in C^1$. Then $u \equiv 0$ in $\Omega$ provided there is a sequence $R_j \to \infty$ satisfying the inequality (10) with $\lambda$ defined by the relation (11) and $\mathcal{E}(R_j) \to 0$ for $R_j \to \infty$.

Similarly as in the case of the Neumann problem the constant $2\lambda^{\frac{1}{2}}$ in the inequality (10) is the best possible which is demonstrated by the example of the solution $u(x) = w(x') \exp\{\lambda^{\frac{1}{2}} x_n\}$ of the equation $\Delta u = 0$ where $w(x')$ is the eigenfunction corresponding to the first eigenvalue of the Dirichlet problem

$$\sum_{j=1}^{n-1} w_{x_j x_j} + \lambda w = 0 \ \text{ in } \ \Omega', \quad w|_{\partial\Omega'} = 0 .$$

Theorem 5. Let $\Omega$ be a subset of the half-space $x_n > 0$ and let $S_\tau = \Omega \cap \{x: x_n = \tau\}$ be a domain in the plane $x_n = \tau$. Let $\Delta u = f$ in $\Omega$, $u|_{\partial\Omega} = \psi$ with $f \equiv 0$ and $\psi \equiv 0$ for $x_n \leqq T$, $u \in C^2(\Omega) \cap C^1(\bar{\Omega})$. Then

$$(12)\qquad \int_{\Omega_{\tau_0}} |\nabla u|^2 dx \leqq \exp\Big\{-\int_{\tau_0}^{\tau_1} 2\lambda^{\frac{1}{2}}(\tau)d\tau\Big\}\int_{\Omega_{\tau_1}} |\nabla u|^2 dx ,$$

where

$$\Omega_\tau = \Omega \cap \{x:\ x_n < \tau\},\qquad \tau_0 < \tau_1 \leqq T ,$$

$$\lambda(\tau) = \inf_{v\in N_\tau}\Big\{\int_{S_\tau}\sum_{j=1}^{n-1} v_{x_j}^2\, dx'\Big[\int_{S_\tau} v^2 dx'\Big]^{-1}\Big\},$$

$N_\tau$ is the family of all functions $v(x')$ continuously differentiable in $\bar{S}_\tau$ satisfying $v = 0$ on $\partial\Omega \cap \bar{S}_\tau$ .

In this way the exponential factor occuring in the inequality (12) analogous to the Saint-Venant principle, can have an arbitrary character of decrease depending on the metric properties of the domain. Generalizations of Theorems 1 to 5 to the case of elliptic and parabolic equations of the second order in domains $\Omega$ of general shapes are given in [5] - [7] .

Let us now consider the Saint-Venant principle for the biharmonic equation which results from plane problems of the linear elasticity theory. Let $\Omega$ be a bounded domain in the plane $(x_1, x_2)$ from the class $C^1$ such that $\Omega \subset \{x:\ x_2 > 0\}$ and the intersection of the domain $\Omega$ with the straight line $x_2 = \tau$ is a set $S_\tau$ which consists of a finite number of intervals. Let $l(\tau)$ equal the length of the largest interval from $S_\tau$ , $\tau = \mathrm{const} > 0$ . In the domain $\Omega$ let us consider the equation

$$(13)\qquad \Delta\Delta u = f\ ,\qquad \Delta\Delta u = u_{x_1x_1x_1x_1} + 2u_{x_1x_1x_2x_2} + u_{x_2x_2x_2x_2}\ ,$$

with boundary conditions

$$(14)\qquad u\big|_{\partial\Omega} = \psi_1\ ,\qquad \frac{\partial u}{\partial\nu}\Big|_{\partial\Omega} = \psi_2\ ,$$

assuming that $f \equiv 0$ in $\Omega_T$ , $\psi_1 \equiv 0$ , $\psi_2 \equiv 0$ on $\partial\Omega \cap \cap\, \partial\Omega_T$ where $\Omega_\tau = \Omega \cap \{x:\ x_2 < \tau\}$ , $\nu$ is the direction of the outer normal to $\partial\Omega$ , $T = \mathrm{const} > 0$ . In the domain $\Omega$ we obtain an estimate for $u(x)$ which expresses the Saint-Venant principle for a two-dimensional elastic body. Special cases of this estimate are established in [3], [12] in a different way.

<u>Theorem 6.</u> Let $u(x)$ be a solution of the problem (13), (14)

in a domain $\Omega$, $f \equiv 0$ in $\Omega_T$, $\psi_1 \equiv \psi_2 \equiv 0$ on $\partial\Omega \cap \partial\Omega_T$, $u \in C^4(\Omega) \cap C^3(\bar{\Omega})$. Then an estimate

$$(15) \qquad \int_{\Omega_{\tau_0}} E(x)dx \leqq \left[\phi(\tau_0, \tau_1)\right]^{-1} \int_{\Omega_{\tau_1}} E(x)dx$$

holds, where $E(x) = u^2_{x_1x_1} + 2u^2_{x_1x_2} + u^2_{x_2x_2}$, $\tau_0 < \tau_1 \leqq T$, the function $\phi(x_2, \tau_1)$ satisfies the identity

$$(16) \qquad \phi_{x_2x_2} - \mu(x_2)\phi = 0$$

for $\tau_0 \leqq x_2 \leqq \tau_1$ and the initial conditions

$$(17) \qquad \phi(\tau_1, \tau_1) = 1, \quad \phi_{x_2}(\tau_1, \tau_1) = 0,$$

where $\mu(\tau)$ is an arbitrary continuous function satisfying

$$(18) \quad 0 < \mu(\tau) \leqq \lambda(\tau) = \inf_{v \in N} \left\{ \int_{S_\tau} E dx_1 \left[ \left| \int_{S_\tau} (v^2_{x_2} - vv_{x_2x_2} + v^2_{x_1}) dx_1 \right| \right]^{-1} \right\},$$

$N$ is the family of functions $v(x_1,x_2)$ twice continuously differentiable in a neighborhood of $\bar{S}_\tau$ and such that $v = 0$, $v_{x_1} = 0$, $v_{x_2} = 0$ at the endpoints of the intervals from $S_\tau$.

Proof. Integrating by parts we obtain

$$0 = \int_{\Omega_{\tau_1}} \phi u \Delta\Delta u dx = \int_{\Omega_{\tau_1}} E\phi dx - \int_{\Omega_{\tau_1}} (u^2_{x_2} - uu_{x_2x_2} + u^2_{x_1}) \phi_{x_2x_2} dx +$$

$$+ \int_{S_{\tau_1}} (u_{x_2x_2x_2}u - u_{x_2x_2}u_{x_2} - u_{x_1x_2}u_{x_1})\phi dx_1 +$$

$$+ \int_{S_{\tau_1}} (u^2_{x_2} - uu_{x_2x_2} + u^2_{x_1})\phi_{x_2} dx_1 .$$

This implies

$$(19) \qquad \int_{\Omega_{\tau_1}} E\phi dx = \int_{\Omega_{\tau_1}} (u^2_{x_2} - uu_{x_2x_2} + u^2_{x_1})\phi_{x_2x_2} dx -$$

$$- \int_{S_{\tau_1}} (u_{x_2x_2x_2}u - u_{x_2x_2}u_{x_2} - u_{x_1x_2}u_{x_1})\phi \, dx_1 -$$

$$- \int_{S_{\tau_1}} (u_{x_2}^2 - uu_{x_2x_2} + u_{x_1}^2)\phi_{x_2} dx_1 .$$

Taking here $\phi \equiv 1$ we conclude

$$(20) \quad \int_{\Omega_{\tau_1}} E(x)dx = - \int_{S_{\tau_1}} (u_{x_2x_2x_2}u - u_{x_2x_2}u_{x_2} - u_{x_1x_2}u_{x_1})dx_1 .$$

Let us introduce a function $\phi = \phi(x_2, \tau_1)$ defined for $\tau_0 \leqq x_2 \leqq \tau_1$ by the equation (16) and initial conditions (17) and continued linearly for $0 \leqq x_2 \leqq \tau_0$ so that for $x_2 = \tau_0$ the function $\phi$ is continuous and has a continuous derivative $\phi_{x_2}$. Taking into account (18) we obtain from (19), (20)

$$\int_{\Omega_{\tau_1}} E\phi \, dx \leqq \int_{\Omega_{\tau_1} \setminus \Omega_{\tau_0}} E\mu^{-1}(x_2)\,\phi_{x_2x_2}dx + \int_{\Omega_{\tau_1}} Edx .$$

Hence with regard to (16)

$$(21) \quad \int_{\Omega_{\tau_0}} E(x)\,\phi(x_2, \tau_1)dx \leqq \int_{\Omega_{\tau_1}} E(x)dx .$$

Let us now study the function $\phi(x_2, \tau_1)$. We shall show that $\phi > 0$, $\phi_{x_2} < 0$ for $0 \leqq x_2 < \tau_1$. Integrating the equation

$$\phi_{x_2x_2} - \mu(x_2)\phi = 0$$

from $x_2$ to $\tau_1$ we obtain

$$(22) \quad \phi_{x_2}(x_2, \tau_1) = - \int_{x_2}^{\tau_1} \mu(x_2)\,\phi(x_2, \tau_1)dx_2 , \quad \tau_0 \leqq x_2 \leqq \tau_1 .$$

If the inequality $\phi(x_2, \tau_1) > 0$ is not valid for $\tau_0 \leqq x_2 \leqq \tau_1$ then there exists a point $x_2 = \alpha$ such that $\phi(\alpha, \tau_1) = 0$, $\phi(x_2, \tau_1) > 0$ for $x_2 > \alpha$. Obviously $\phi_{x_2}(\alpha, \tau_1) \geqq 0$. On the other hand, the relation (22) implies that $\phi_{x_2}(\alpha, \tau_1) < 0$.

The contradiction just obtained proves that $\Phi(x_2, \tau_1) > 0$, $x_2 \leqq \tau_1$. The identity (22) implies that $\Phi_{x_2}(x_2, \tau_1) < 0$ for $\tau_0 \leqq x_2 < \tau_1$. Consequently, the inequality (21) implies the estimate (15). The theorem is proved.

Let us now assume that $\lambda(x_2) \geqq \mu = \text{const} > 0$. Then for $\tau_0 \leqq x_2 \leqq \tau_1$

$$\Phi(x_2, \tau_1) = \frac{1}{2}\left[\exp\{\mu^{\frac{1}{2}}(\tau_1 - x_2)\} + \exp\{-\mu^{\frac{1}{2}}(\tau_1 - x_2)\}\right].$$

In this case the estimate (15) implies the inequality

$$(23) \qquad \int_{\Omega_{\tau_0}} E(x)dx \leqq 2\exp\left\{-\mu^{\frac{1}{2}}(\tau_1 - \tau_0)\right\} \int_{\Omega_{\tau_1}} E(x)dx .$$

Let us now estimate $\mu$. It is known that if $v(x_1,x_2)$ is such that $v = 0$, $v_{x_1} = 0$, $v_{x_2} = 0$ at the endpoints of the intervals from $S_\tau$ then

$$\int_{S_\tau} v_{x_2}^2 dx_1 \leqq \lambda_1^{-1}(\tau) \int_{S_\tau} v_{x_2x_1}^2 dx_1 , \qquad \lambda_1 = \frac{\pi^2}{l^2(\tau)} ,$$

$$\int_{S_\tau} v^2 dx_1 \leqq \lambda_2^{-1}(\tau) \int_{S_\tau} v_{x_1x_1}^2 dx_1 , \qquad \lambda_2 = \frac{(4,73)^4}{l^4(\tau)} ,$$

$$\int_{S_\tau} v_{x_1}^2 dx_1 \leqq \lambda_3^{-1}(\tau) \int_{S_\tau} v_{x_1x_1}^2 dx_1 , \qquad \lambda_3 = \frac{4\pi^2}{l^2(\tau)} .$$

Here $l(\tau)$ is the length of the largest interval from $S_\tau$. Let $l = \sup_{0 \leqq \tau \leqq T} l(\tau)$. With regard to the above inequalities we obtain

$$\left|\int_{S_\tau} (v_{x_2}^2 - vv_{x_2x_2} + v_{x_1}^2)dx_1\right| \leqq \frac{1}{2}\lambda_1^{-1} \int_{S_\tau} 2v_{x_1x_2}^2 dx_1 +$$

$$+ \lambda_3^{-1}\int_{S_\tau} v_{x_1x_1}^2 dx_1 + \frac{\theta}{2}\int_{S_\tau} v_{x_2x_2}^2 dx_1 + \frac{\lambda_2^{-1}}{2\theta}\int_{S_\tau} v_{x_1x_1}^2 dx_1 ,$$

where $\theta = \text{const} > 0$. Let us choose $\theta$ so that $\frac{1}{2}\theta = \lambda_3^{-1} + \frac{1}{2}(\theta\lambda_2)^{-1}$. By an easy computation we have $\frac{1}{2}\lambda_1^{-1} > \lambda_3^{-1} + \frac{1}{2}(\theta\lambda_2)^{-1} = \frac{\theta}{2}$. Therefore $\lambda(\tau) \geqq 2\lambda_1(\tau) \geqq 2\frac{\pi^2}{l^2} = \mu$. The estimate (23) implies the inequality

$$\int_{\Omega_{\tau_0}} E(x)dx \leqq 2 \exp\left\{ -2 \frac{\pi}{l\sqrt{2}}(\tau_1 - \tau_0)\right\} \int_{\Omega_{\tau_1}} E(x)dx .$$

This estimate is better than the corresponding ones obtained in [2], [12]. The following theorem is analogous to the Phragmen-Lindelöf theorem for the biharmonic equation.

Theorem 7. Let $\Omega \subset \{x: x_2 > 0\}$, let the set $S_\tau = \Omega \cap \cap \{x: x_2 = \tau\}$ be nonempty for all $\tau > 0$, $f \equiv 0$ in $\Omega$, $\psi_1 \equiv 0$, $\psi_2 \equiv 0$ on $\partial\Omega$. Let $u(x)$ be a solution of the problem (13), (14) and $u \in C^4(\Omega) \cap C^3(\bar{\Omega})$. Then $u \equiv 0$ in $\Omega$ provided there is a sequence of numbers $R_j \to \infty$ for $j \to \infty$ and a constant $d > 0$ such that

$$\int_{\Omega_{R_j}} E(x)dx \leqq \varepsilon(R_j) \left[\phi(d,R_j)\right]^{-1} , \tag{24}$$

where $\varepsilon(R_j) \to \infty$ for $R_j \to \infty$.

Proof. By virtue of Theorem 6 and the condition (24) we have

$$\int_{\Omega_d} E(x)dx \leqq \left[\phi(d,R_j)\right]^{-1} \int_{\Omega_{R_j}} E(x)dx \leqq \varepsilon(R_j)$$

for any $R_j$. Hence

$$\int_{\Omega_d} E(x)dx = 0$$

and consequently, $u \equiv 0$ in $\Omega_d$ since $u = 0$, $u_{x_1} = 0$, $u_{x_2} = 0$ on $\partial\Omega$. It is known that a solution of the equation $\Delta\Delta u = 0$ is an analytic function in $\Omega$. Hence $u \equiv 0$ in $\Omega$.

Given an unbounded domain $\Omega$ such that $\lambda(\tau) \geqq \mu = \text{const} > > 0$, the condition (24) can be written in the form

$$\int_{\Omega_{R_j}} E(x)dx \leqq \varepsilon(R_j) \exp\left\{\mu^{\frac{1}{2}} R_j\right\}. \tag{25}$$

The problem whether the constant $\mu^{\frac{1}{2}}$ in the condition (25) is the best possible remains open. Theorems analogous to Theorem 6 and 7 can be established in the same way also for more complicated domains $\Omega$, in particular, for the case of a domain $\Omega$ which has several branches which stretch to infinity along various directions. Such domains are studied for elliptic equations of the second order in [5], [6]. The method used here for investigating the problems (13), (14) was former used in [10] to study the behavior of solutions of the system of equations of the elasticity theory at non-regular points of the boundary. Analogous results may be established also for solutions of the problem (13), (14). In particular, the following theorem holds.

Theorem 8. Let a bounded domain $\Omega$ belong to the halfplane $\{x: x_2 > 0\}$, $\sigma = \bar{\Omega} \cap \{x: x_2 = 0\}$ being nonempty. Let $u(x)$ be a solution of the problem (13), (14), $u \in H_2(\Omega) \cap C^4(\Omega) \cap \cap C^3(\bar{\Omega} \setminus \sigma)$ and let the curve $\partial\Omega \setminus \sigma$ belong to the class $C^1$, $f \equiv 0$, $\psi_1 \equiv 0$, $\psi_2 \equiv 0$ in a certain neighborhood of the set $\sigma$. Then

$$\int_{\Omega} E(x)\, \phi(x_2)dx < \infty ,$$

where $\phi(x_2)$ satisfies the equation $\phi_{x_2x_2} - \frac{1}{2}\mu(x_2)\phi = 0$ and the initial conditions $\phi(\alpha) = 1$, $\phi_{x_2}(\alpha) = 0$, $0 < x_2 \leqq \alpha$ where $\alpha$ is a constant, the function $\mu(x_2)$ is defined by the relation (18) and by the assumption $\mu(x_2) \to \infty$ for $x_2 \to 0$.

It is possible to establish estimates for the function $\phi(x_2)$ which characterize the growth of $\phi(x_2)$ for $x_2 \to 0$ in dependence on the geometric properties of the domain $\Omega$ in a neighborhood of the set $\sigma$.

Let us remark that estimates analogous to the Saint-Venant principle for solutions of the Dirichlet problem for the system of equations of the elasticity theory are established in [10] while for the mixed problem they are given in [11]. Inequalities analogous to the Saint-Venant principle as well as theorems of Phragmen-Lindelöf type which are their consequences, hold under certain conditions for solutions of general boundary value problems for both elliptic and parabolic equations. These estimates are given in [13] - [15]. In these papers an approach is used which is connected with a study of analytic continuation of solutions in a domain of variation of one of the independent variables of some specially constructed auxiliary systems.

## References


[1] de Saint-Venant A.J.C.B.: De la torsion des prismes, Mem. présentés par divers savants a l'Acad. des Sci. XIV(1855), Paris

[2] Toupin R.: Saint-Venant's principle. Arch.Rat.Mech.Anal. 18 (1965), 83-96

[3] Knowles J.K.: On Saint-Venant's principle in the two-dimensional linear theory of elasticity. Arch.Rat.Mech.Anal. 21 (1966), 1-22

[4] Gurtin M.E.: The linear theory of elasticity. Handbuch der Physik Vol. VIa/2, Springer 1972

[5] Oleinik O.A., Yosifian G.A.: Energetic estimates of generalized solutions of boundary value problems for elliptic equations of the second order and their applications. Dokl.Akad. Nauk SSSR 232(1977), No.6, 1257-1260 (Russian)

[6] Oleinik O.A., Yosifian G.A.: Boundary value problems for second order elliptic equations in unbounded domains and Saint-Venant's principle. Annali Sc.Norm.Super.Pisa, Classe di Sci., Ser.IV, IV(1977), No.2, 269-290

[7] Oleinik O.A., Yosifian G.A.: Analogue of Saint-Venant's principle and uniqueness of solutions of boundary value problems in unbounded domains for parabolic equations. Uspechi Mat.Nauk 31(1976), No.6, 142-166 (Russian)

[8] Oleinik O.A., Yosifian G.A.: On removable singularities on boundary and uniqueness of solutions of boundary value problems for elliptic and parabolic equations of the second order. Funkcion.Anal. i Prilož. 2(1977), No.3, 54-67 (Russian)

[9] Oleinik O.A., Yosifian G.A.: On some properties of solutions of equations of hydrodynamics in domains with moving boundary. Vestnik Moskov.Univ., Mat. i Mech. No.5, (1977) (Russian)

[10] Oleinik O.A., Yosifian G.A.: A priori estimates of solutions of the first boundary value problem for the system of equations of the elasticity theory and their applications. Uspechi Mat. Nauk 32(1977), No.5, 197-198 (Russian)

[11] Oleinik O.A., Yosifian G.A.: Saint-Venant's principle for the mixed problem of the elasticity theory and its applications. Doklady Akad.Nauk SSSR 233(1977), No.5, 824-827 (Russian)

[12] Flavin J.N.: On Knowles' version of Saint-Venant's principle in two-dimensional elastostatics. Arch.Rat.Mech.Anal. 53(1974), No.4, 366-375

[13] Oleinik O.A.: On the behaviour of solutions of the Cauchy problem and the boundary value problem for parabolic systems of partial differential equations in unbounded domains. Rendiconti di Mat., Ser.VI, 8(1975), fasc.2, 545-561

[14] Oleinik O.A., Radkevič E.V.: Analyticity and theorems of Liouville and Phragmen-Lindelöf types for general parabolic systems of differential equations. Funkcion.Anal.Prilož. 8(1974), No.4, 59-70 (Russian)

[15] Oleinik O.A., Radkevič E.V.: Analyticity and theorems of Liouville and Phragmen-Lindelöf types for general elliptic systems of differential equations. Matem.Sbornik 95(137:1)(1974), No.9, 130-145 (Russian)

Author's address: Moskovskiĭ universitet, Mechaniko-matematičeskiĭ fakultet, Moskva B-234, USSR

A priori bounds for a semilinear wave equation

Paul H. Rabinowitz

The purpose of this note is to describe some recent results on the existence and regularity of solutions of semilinear wave equations of the form

$$
(1) \qquad \begin{cases} u_{tt} - u_{xx} + f(u) = 0\,, & 0 < x < \pi\,, \quad t \in \mathbb{R} \\ u(0,t) = 0 = u(\pi,t) \end{cases}
$$

where $f(0) = 0$. Of interest is the existence of time periodic solutions of (1). Note that $u \equiv 0$ is a trivial such solution; nontrivial solutions are often called free vibrations for (1). The same methods we shall describe below can be used to treat the forced vibration case where $f$ also depends explicitly on $t$ in a time periodic fashion.

There is a substantial literature on forced and free vibration problems for (1), mainly for the former case with $f$ being a perturbation term, i.e. $f = \varepsilon g(x,t,u)$ where $\varepsilon$ is small. See e.g. [1-8] and the references cited there. The work to be discussed here can be found in detail in [9].

Our main result for (1) is:

Theorem 2: Suppose $f \in C^k$, $k \geq 2$, and satisfies

(f1) $f$ is strictly monotone increasing with $f(0) = 0$,

(f2) $f$ is superlinear at $0$ and $\infty$, i.e.

(i) $f(z) = o(|z|)$ at $z = 0$

(ii) $F(z) = \int_0^z f(s)ds \leq \theta z f(z)$ for $|z| \geq \overline{z}$ where $\theta \in [0, \frac{1}{2})$.

Then for any period $T$ which is a rational multiple of $\pi$, (1) possesses a nontrivial $T$-periodic solution $u \in C^k$.

$$\|u\|_E^2 = \iint_D (u_t^2 + u_x^2)\,dx\,dt$$

where $D = [0,\pi] \times [0,2\pi]$ . Ignoring questions of where it is defined, formally critical points of

$$(3) \qquad \iint_D [\tfrac{1}{2}(u_t^2 - u_x^2) - F(u)]\,dx\,dt$$

in $E$ are weak solutions of (1) . Unfortunately we know of no direct way of determining nontrivial critical points of (3) . However if a finite dimensional approximation argument is used, i.e. (3) is restricted to $E_m = \text{span}\{\sin jx \ \sin kt\ ,\ \sin jx \ \cos kt \,|\, 0 \le j,k \le m\}$ , the form of (3) and hypotheses on $f$ imply the existence of a nontrivial critical point. The problem then becomes that of finding bounds for this critical point

Before sketching the proof, a few remarks are in order. We do not know if the restriction on $T$ is essential. The reason for this assumption here is that for such $t$ , the spectrum of

$\square = \dfrac{\partial^2}{\partial t^2} - \dfrac{\partial^2}{\partial x^2}$ in the class of $T$ periodic functions in $t$ satisfying the boundary conditions in $x$ is discrete with $0$ being an isolated point in the spectrum while if $T$ is irrational, the spectrum is dense and $0$ is an accumulation point of the spectrum. Condition (f2)(i) can be eliminated provided that (fl) is retained. We do not know if the monotonicity of $f$ can be relaxed in any essential way. See however [7] . Lastly condition (f2)(ii) implies $|f(z)| \ge a_1 |z|^{\frac{1}{\theta}-1} - a_2$ ; hence the terminology superlinearity at $\infty$ .

Turning now to the proof of Theorem 2 , the basic idea is to try to find a solution of (1) as a critical point of a corresponding functional. For convenience we take $T = 2\pi$ . Consider the set of smooth functions $2\pi$ periodic in $t$ and having compact support (in $(0,\pi)$) in $x$ . Let $E$ denote the closure of this set with respect to which enable us to pass to a limit and get a nontrivial solution of (1). Two technical problems impede this program. To describe them, consider the linear problem

$$(4) \qquad \begin{cases} \Box w = g(x,t), \\ w(0,t) = 0 = w(\pi,t); \; w(x,t+2\pi) = w(x,t). \end{cases}$$

Under these boundary and periodicity conditions, $\Box$ has a null space whose closure in $L^2(T) \equiv L^2$ is $N = \{p(x+t) - p(-x+t) \mid p \in L^2(S^1)\}$. If $N^\perp$ denotes the orthogonal complement of $N$ in $L^2$, then for $g \in N^\perp$, (4) is uniquely invertible in $N^\perp$ with a gain of one derivative in either the $L^2$ or sup norms [4,5]. Thus we have some compactness for the projection of (1) in $N^\perp$; however there is none in $N$.

The two technical difficulties mentioned above are: (i) the unrestricted growth of $f$ at $\infty$ does not permit us to obtain the necessary estimates for $w$, the component of $u$ in $N^\perp$; (ii) the lack of compactness of the projection of (1) in $N$.

To get around these difficulties, we modify (1) and (3). For $u \in E$, we can write $u = v + w$ where $v \in N$ and $w \in N^\perp$. Let $\beta > 0$. Consider

$$(5) \qquad \begin{cases} \Box u - \beta v_{tt} + f_K(u) = 0, \quad 0 < x < \pi, \quad t \in \mathbb{R} \\ u(0,t) = 0 = u(\pi,t); \; u(x,t+2\pi) = u(x,t) \end{cases}$$

where $f_K$ satisfies (f1)-(f2), $f_K(z) = f(z)$ for $|z| \le K$, and $f_K$ grows at a prescribed rate, e.g. cubically, at $\infty$. The $\beta$ term essentially compactifies the projection of (1) on $N$. Corresponding to (5) we have the functional:

$$(6) \qquad I(u) = \iint_D \left[\tfrac{1}{2}(u_t^2 - u_x^2 - \beta v_t^2) - F_K(u)\right] dx\, dt$$

where $F_K$ is the primitive of $f_K$. The idea now is to: $1^o$ find an appropriate critical point of $I|_{E_m}$; $2^o$ get suitable estimates for this critical point; $3^o$ pass to a limit and solve (5); $4^o$ get $\beta$ and K independent estimates for solutions of (5); and $5^o$ let $\beta \to 0$ and $K \to \infty$ to solve (1). This is too lengthy a process for us to carry out now so we will content ourselves with just trying to give the flavor of a few of the estimates that are involved. To do this we return to (1) and argue *a priori*. This is much simpler than the actual procedure carried out in the existence argument.

Thus suppose we have a smooth solution u, of (1). We will obtain bounds for u in terms of c, the critical value of I corresponding to u. Thus suppose $I(u) = c$. The first estimate gives a bound for $\|f(u)\|_{L^1}$. Since $I'(u) = 0$ (where $I'(u)$ denotes the Frechet derivative of I at u),

(7) $$c = I(u) - \frac{1}{2} I'(u)u = \iint_D [\frac{1}{2} f(u)u - F(u)]\, dx\, dt$$

Invoking (f2)(ii) then gives

(8) $$\|f(u)u\|_{L^1} \leq M_1$$

for some constant $M_1$ depending on c. By (f1), $|f(z)| \leq f(1) - f(-1) + f(z)z$. Hence (8) implies a bound for $\|f(u)\|_{L^1}$.

Next writing $u = v + w$, $v \in N$, $w \in N^\perp$, we have

(9) $$\Box w = -f(w) .$$

There is a representation theorem [5] for solutions of (4) which implies:

(10) $$\|w\|_{L^\infty} \leq a_3 \|g\|_{L^1} .$$

Consequently we conclude

(11) $$\|w\|_{L^\infty} \le a_3 \|f(u)\|_{L^1} \le a_3 M_1 \equiv M_2 .$$

The next step is to get an estimate for $\|v\|_{L^\infty}$. This is more subtle. We assume $v \not\equiv 0$ or there is nothing to prove. From (9) we conclude that

(12) $$\iint_D f(u)\varphi \, dx \, dt = 0$$

for all $\varphi \in N$. By choosing $\varphi$ to be an appropriate nonlinear function of $v$, we will obtain the desired estimate for $\|v\|_{L^\infty}$. By rewriting (12) we get

(13) $$\iint_D (f(v+w) - f(w))\varphi \, dx \, dt = - \iint_D f(w)\varphi \, dx \, dt$$
$$\le \|f(w)\|_{L^\infty} \iint_D |\varphi| \, dx \, dt$$

From the definition of $N$ we have $\varphi = p(x+t) - p(-x+t)$ where $p \in L^2(S^1)$. Clearly $p$ is only determined up to an additive constant. We make $p$ unique by requiring that

$$[p] \equiv \int_0^{2\pi} p(s)ds = 0 .$$

Define a function $q(s)$ by $q(s) = 0$ if $|s| \le M$, $q(s) = s - M$ if $s > M$; $q(s) = s + M$ if $s < -M$. With the above normalization on $p$, we write $v(x,t) = p(x+t) - p(-x+t) \equiv v^+ - v^-$ and chose $\varphi = q(v^+) - q(v^-) \equiv q^+ - q^- \in N$. Therefore for any $\delta > 0$, by (f1),

(14) $$\iint_{D_\delta} (f(v+w) - f(w))(q^+ - q^-) \, dx \, dt \le \|f(w)\|_{L^\infty} (\|q^+\|_{L^1} + \|q^-\|_{L^1})$$

where $D_\delta = \{(x,t) \in D \,\big|\, |v| \geq \delta\}$. Let $D^+ = \{(x,t) \in D_\delta | v \geq \delta\}$ and $D^- = D_\delta \backslash D^+$. Define

$$\psi(z) = \begin{cases} \min\limits_{|\zeta| \leq M_2} f(z+\zeta) - f(\zeta) & z \geq 0 \\ \max\limits_{|\zeta| \leq M_2} f(z+\zeta) - f(\zeta) & z < 0 \end{cases}$$

Then by (fl), $\psi$ is strictly monotone increasing and $|\psi(z)| \to \infty$ as $|z| \to \infty$. From the definition of $\psi$ we get

$$\iint_{D^+} (f(v+w) - f(w))(q^+ - q^-)\,dx\,dt \geq \tag{15}$$

$$\geq \iint_{D^+} \frac{\psi(v)}{v} v(q^+ - q^-)\,dx\,dt \geq \frac{\psi(\delta)}{\|v\|_{L^\infty}} \iint_{D^+} v(q^+ - q^-)\,dx\,dt$$

since $v(q^+ - q^-) \geq 0$. Similar estimates for the $T^-$ integral lead to

$$\iint_{D_\delta} (f(v+w) - f(w))(q^+ - q^-)\,dx\,dt \geq \frac{\mu(\delta)}{\|v\|_{L^\infty}} \iint_{D_\delta} v(q^+ - q^-)\,dx\,dt \tag{16}$$

where for $z \geq 0$, $\mu(z) = \min(\psi(z), -\psi(-z))$. Note that $\mu$ is strictly monotone increasing and $\mu(z) \to \infty$ as $z \to \infty$. Now

$$\iint_{D_\delta} v(q^+ - q^-)\,dx\,dt \geq \iint_D v(q^+ - q^-)\,dx\,dt - \delta(\|q^+\|_{L^1} + \|q^-\|_{L^1}) . \tag{17}$$

Since $[v^\pm] = 0$, it is easy to verify that

$$\iint_D v^+ q^-\,dx\,dt = 0 = \iint_D v^- q^+\,dx\,dt .$$

Therefore

$$\iint_T v(q^+ - q^-)\,dx\,dt = \iint_T (v^+ q^+ + v^- q^-)\,dx\,dt . \tag{18}$$

By the definition of $q$, we have $sq(s) \geq M|q(s)|$. Hence

(19) $$\iint_D (v^+q^+ + v^-q^-)dx\,dt \geq M(\|q^+\|_{L^1} + \|q^-\|_{L^1}) .$$

Combining (14), (16)-(19) yields

(20) $$\frac{(M-\delta)}{\|v\|_{L^\infty}} \mu(\delta)(\|q^+\|_{L^1} + \|q^-\|_{L^1}) \leq \|f(w)\|_{L^1}(\|q^+\|_{L^1} + \|q^-\|_{L^1}) .$$

Choosing any $M < \|v^+\|_{L^\infty} = \|v^-\|_{L^\infty}$, the $L^1$ terms are positive so they can be cancelled and

(21) $$\frac{(M-\delta)}{\|v\|_{L^\infty}} \mu(\delta) \leq \|f(w)\|_{L^\infty} .$$

Since this is true for all $M < \|v^\pm\|_{L^\infty}$, we can take $M = \|v^\pm\|_{L^\infty}$. Further noting that $\|v\|_{L^\infty} \leq 2\|v^\pm\|_{L^\infty}$ and taking $\delta = \frac{1}{2}\|v\|^{\pm}_{L^\infty}$ yields

(22) $$\|v^\pm\|_{L^\infty} \leq 2\mu^{-1}(4\|f(w)\|_{L^\infty}) .$$

Thus (22) and our estimate for $\|w\|_{L^\infty}$ give the desired bound for $\|v\|_{L^\infty}$. Therefore we have a bound for $\|u\|_{L^\infty}$.

To get further estimates, from (9) and the properties of $\square^{-1}$ we have

(23) $$\|w\|_{C^1} \leq a_4\|f(w)\|_{L^\infty} \leq M_3 .$$

Next the arguments used to obtain the bound for $\|v\|_{L^\infty}$ can be modified to estimate the modulus of continuity of $v$. In the framework of (5), these bounds enable us to pass to a limit to get a continuous weak solution of (1). A separate argument shows $u \not\equiv 0$. To verify that $u$ is indeed a smooth solution of (1) requires further arguments which we will not carry out here.

## References

[1] Vejvoda, O., Periodic solutions of a linear and a weakly nonlinear wave equation in one dimension, I, Czech. Math. J. 14, (1964), 341-382.

[2] Vejvoda, O., Periodic solutions of nonlinear partial differential equations of evolution, Proc. Sym. on Diff. Eq. and Applic. Brataslava—1966, (1969), 293-300.

[3] Kurzweil, J., Van der Pol perturbation of the equation for a vibrating string, Czech. Math. J., 17, (1967), 558-608.

[4] Rabinowitz, P. H., Periodic solutions of nonlinear hyperbolic partial differential equations, Comm. Pure Appl. Math., 20, (1967), 145-205.

[5] Lovicarová, H., Periodic solutions of a weakly nonlinear wave equation in one dimension, Czech. Math. J., 19, (1969), 324-342.

[6] Rabinowitz, P. H., Time periodic solutions of a nonlinear wave equation, Manus. Math., 5, (1971), 165-194.

[7] Štědrý, M. and O. Vejvoda, Periodic solutions to weakly nonlinear autonomous wave equations, Czech. Math. J., 25, (1975), 536-555.

[8] Brezis, H. and L. Nirenberg, Forced vibrations for a nonlinear wave equation, to appear.

[9] Rabinowitz, P. H., Free vibrations for a semilinear wave equation, to appear Comm. Pure Appl. Math.

This research was sponsored in part by the Office of Naval Research under Contract No. N00014-76-C-0300 and in part by the U.S. Army under Contract No. DAAG29-75-C-0024. Any reproduction in part or in full is permitted for the purposes of the U.S. Government.

Author's address: Department of Mathematics University of Wisconsin-Madison, Madison, Wis.53706, USA

# THE METHOD OF LEAST SQUARES ON THE BOUNDARY AND VERY WEAK SOLUTIONS OF THE FIRST BIHARMONIC PROBLEM

K. Rektorys, Praha

In this paper, the so-called Method of Least Squares on the Boundary is presented and its application to an approximate solution of the first biharmonic problem is shown. This method is applicable even if the boundary conditions are so general that the existence of a weak solution is not ensured, so that current variational methods (the Ritz method, the finite element method, etc.) cannot be applied. Moreover, it enables to solve the first problem of plane elasticity by reducing it into the first biharmonic problem also in the case of multiply connected regions, where other current methods meet with well-known difficulties even in the case of smooth boundary conditions.

Because the origin of this method lies in solving problems of the theory of plane elasticity, let us recall, in brief, basic concepts and results of this theory.

Throughout this paper, G is a bounded region in $E_2$ with a Lipschitzian boundary $\Gamma$.

Under the first problem of plane elasticity we understand a problem to find three functions

$$\sigma_x, \ \sigma_y, \ \tau_{xy}, \tag{1}$$

the so-called components of the stress-tensor, sufficiently smooth in G (to be made more precise later), fulfilling in G the equations of equilibrium

$$\frac{\partial \sigma_x}{\partial x} + \frac{\partial \tau_{xy}}{\partial y} = 0, \tag{2}$$

$$\frac{\partial \tau_{xy}}{\partial x} + \frac{\partial \sigma_y}{\partial y} = 0 \tag{3}$$

and the equation of compatibility

$$\Delta ( \sigma_x + \sigma_y ) = 0 \tag{4}$$

(where $\Delta$ is the Laplace operator), and on $\Gamma$ the boundary conditions

$$\sigma_x \nu_x + \tau_{xy} \nu_y = X(s), \tag{5}$$

$$\tau_{xy} \nu_x + \sigma_y \nu_y = Y(s). \tag{6}$$

Here $\nu_x$, $\nu_y$ are components of the unit outward normal to $\Gamma$ (existing almost everywhere on $\Gamma$, because $\Gamma$ is Lipschitzian), X and Y are components of the outward loading which acts on the boundary, s is the length of arc on $\Gamma$. If G is multiply connected, it is required, moreover, that the vector of displacement corresponding to the stress-tensor (1) is a single-valued function in G.
In what follows, we assume that the loading is in the static equilibrium (both in forces and moments).

## I. Simply connected regions

In this case, the first problem of plane elasticity can be easily transformed into the first biharmonic problem

$$\Delta^2 u = 0 \text{ in } G, \tag{7}$$

$$u = g_0(s), \quad \frac{\partial u}{\partial \nu} = g_1(s) \text{ on } \Gamma. \tag{8}$$

The functions $g_0$, $g_1$ are derived, in a simple way, from the functions X,Y (for details see [5]). In this paper, we assume

$$g_0 \in W_2^{(1)}(\Gamma), \quad g_1 \in L_2(\Gamma) \tag{9}$$

only. This assumption is sufficiently general from the point of view of the theory of elasticity and sufficiently interesting from the mathematical point of view. Indeed, (9) does not ensure existence of a weak solution of (7), (8). But (see the Nečas monography [3] ) it ensures existence of the so-called wery weak solution: In fact, traces (in the sense of (9)) of functions from the space $W_2^{(2)}(G)$ are dense in $W_2^{(1)}(\Gamma) \times L_2(\Gamma)$. Consequently, a sequence of functions $v_n \in W_2^{(2)}(G)$ exists such that

$$\left(v_n, \frac{\partial v_n}{\partial \nu}\right) \to (g_0, g_1) \text{ in } W_2^{(1)}(\Gamma) \times L_2(\Gamma). \tag{10}$$

Then (see [3] again) the sequence of weak solutions $\tilde{v}_n$ of the problem (7), (8) with $g_0, g_1$ replaced by $v_n$, $\partial v_n/\partial \nu$ converges, in $L_2(G)$, to a function $u \in L_2(G)$. This function is uniquely determined by the functions $g_0, g_1$ and is called the <u>very weak solution</u> of the problem (7), (8). The function u can be shown to be a classical solution of (7) inside of G. The components of the desired stress-tensor are then given by the relations

$$\sigma_x = \frac{\partial^2 u}{\partial y^2}, \quad \sigma_y = \frac{\partial^2 u}{\partial x^2}, \quad \tau_{xy} = -\frac{\partial^2 u}{\partial x \partial y}. \tag{11}$$

Because a very weak solution of (7), (8) need not be a weak solution and, consequently, need not belong to $W_2^{(2)}(G)$, usual variational methods are not applicable, in general, to get an approximate solution of the problem (7), (8). Therefore, in [1] the above mentioned method of least squares on the boundary has been developed by K. Rektorys and V. Zahradník:

Let

(12) $$z_1(x,y),\ z_2(x,y),\ \ldots,\ z_n(x,y)$$

be the system of basic biharmonic polynomials. (For details see [1] ; note that for $n \geqq 2$ there are precisely $4n - 2$ basic biharmonic polynomials of degree $\leqq n$.) Let $n \geqq 2$ be fixed. Denote by M the set of all functions of the form

(13) $$v(x,y) = \sum_{i=1}^{4n-2} b_{ni}\, z_i(x,y)$$

with $b_{ni}$ arbitrary (real) and let

(14) $$Fv = \int_\Gamma (v-g_0)^2\, ds + \int_\Gamma \left(\frac{\partial v}{\partial s} - \frac{dg_0}{ds}\right)^2 ds + \int_\Gamma \left(\frac{\partial v}{\partial \nu} - g_1\right)^2 ds$$

be a functional on M. (Because of (10) and of the Lipschitzian boundary, all integrals in (14) have a sense.) Let us look for an approximate solution in the form

(15) $$u_n = \sum_{i=1}^{4n-2} a_{ni} z_i(x,y),$$

where the coefficients $a_{ni}$ are determined from the condition

(16) $$Fu_n = \text{min. on } M.$$

The functional F being quadratic, the condition (16) leads to the solution of a system of $4n - 2$ linear equations for $4n - 2$ coefficients $a_{ni}$.

Theorem 1. The above mentioned system is uniquely solvable.

The proof is relatively simple: On the set M of all functions (13) one defines the scalar product $(u,v)_\Gamma$ by

$$(u,v)_\Gamma = \int_\Gamma uv\, ds + \int_\Gamma \frac{\partial u}{\partial s}\frac{\partial v}{\partial s}\, ds + \int_\Gamma \frac{\partial u}{\partial \nu}\frac{\partial v}{\partial \nu}\, ds.$$

It turns out that the determinant of the above mentioned system is the Gram determinant of the linearly independent functions (13), and, consequently, it is different from zero.

Theorem 2. For $n \to \infty$ we have

$$u_n \to u \text{ in } L_2(G),$$

where $u(x,y)$ is the very weak solution of (7), (8). Moreover, on every subregion $G' \subset G$ the convergence is uniform. The same holds for the convergence of the derivatives $D^i u_n$ to $D^i u$ in $G'$.

The proof is not simple (see [1] , pp. 119-130). It is based on the following two lemmas:

Lemma 1. Let $u_0$ be a weak solution of a first biharmonic problem in G. Then to every $\varepsilon > 0$ there exist a region $\tilde{G} \supset G$ and a function $\tilde{u}$ biharmonic in $\tilde{G}$ such that for its restriction on G we have

$$\| \tilde{u} - u_0 \|_{W_2^{(2)}(G)} < \varepsilon .$$

(Because u is biharmonic in $\tilde{G}$, it has continuous derivatives of all orders in $\bar{G}$; thus, Lemma 1 says that every weak biharmonic function in G can be approximated, in $W_2^{(2)}(G)$, with an arbitrary accuracy, by a very smooth biharmonic function in $\bar{G}$.)

For the proof of this lemma, one constructs a sequence of bounded regions $G_j$,

$$\bar{G} \subset G_j, \quad G_{j+1} \subset G_j \text{ for every } j = 1,2, \dots ,$$
$$\lim_{j \to \infty} \operatorname{mes} (G_j - \bar{G}) = 0$$

(thus $G_j$ converge for $j \to \infty$ to G "from outside"), extends the function $u_0$ to $G_1$ so that this extension - let us denote it by $U_0$ - belongs to $W_2^{(2)}(G_1)$ (and, consequently, to every $W_2^{(2)}(G_j)$, as the restriction on $G_j$, $j = 2,3, \dots$ ). This is possible (cf. [3] , p. 80). On every $G_j$ one constructs the (uniquely determined) weak solution $u_j$ of the first biharmonic problem with boundary conditions given by the function $U_0$ and proves for the restrictions $\tilde{u}_j$ of $u_j$ on G that

$$\lim_{j \to \infty} \| u_0 - \tilde{u}_j \|_{W_2^{(2)}(G)} = 0.$$

For the function $\tilde{u}$ it is then sufficient to take the restriction $\tilde{u}_j$ of a function $u_j$ with a sufficiently high index j. (For details see [1] , pp. 122 - 128.)

Lemma 2 (on density). The traces of biharmonic polynomials are dense in $W_2^{(1)}(\Gamma) \times L_2(\Gamma)$. In detail : To every pair of functions $g_0 \in W_2^{(1)}(\Gamma)$, $g_1 \in L_2(\Gamma)$ and to every $\varepsilon > 0$ there exists a biharmonic polynomial p satisfying

$$\| p - g_0 \|_{W_2^{(1)}(\Gamma)} < \varepsilon , \quad \| \frac{\partial p}{\partial \nu} - g_1 \|_{L_2(\Gamma)} < \varepsilon .$$

The proof is relatively simple and is based on Lemma 1, on the well-known representation of biharmonic functions by holomorphic functions (see [5] ) and application of the Walsh theorem on approximation of holomorphic functions by polynomials. For details see [1] .

Having proved Lemmas 1 and 2, the proof of the first assertion of Theorem 2 is only a technical matter. (One applies a procedure similar to that described in the text following (10) and some almost obvious properties of the method of least squares.) For details see [1] pp. 129-130.
The second assertion of this theorem is an easy consequence of Theorem 4.1.3 from [3] , p. 200 (on the behaviour, in the interior of G, of solutions of equations with sufficiently smooth coefficients).

Remark 1. In [1] also a numerical example can be found. Note that the second integral on the right-hand side of (14) plays an essential role in the proof of convergence as well as in the numerical process (as a "stabilizator").

II. Multiply connected regions

Let G be a bounded (k+1)-tuply connected region in $E_2$ with a Lipschitzian boundary

(17) $$\Gamma = \Gamma_0 \cup \Gamma_1 \cup \ldots \cup \Gamma_k,$$

$\Gamma_1, \ldots, \Gamma_k$ being inner boundary curves. Let a loading be acting on each of the boundary curves. As well as in the case of the simply connected region, the functions $g_{i0}$, $g_{i1}$ (i = 0,1, ... ,k) can be constructed and the problem

(18) $$\Delta^2 u = 0 \text{ in } G,$$

(19) $$u = g_{i0}, \quad \frac{\partial u}{\partial \nu} = g_{i1} \text{ on } \Gamma_i, \quad i = 0,1, \ldots, k$$

can be solved. Assuming that

(20) $$g_{i0} \in W_2^{(1)}(\Gamma_i), \quad g_{i1} \in L_2(\Gamma_i), \quad i = 0,1, \ldots, k,$$

it can be shown, in a quite similar way as in the case of the simply connected region, that a (unique) very weak solution of (18), (19) exists. It is a classical solution inside of G again. But in contrast to the case of a simply connected region, the functions (11) need not be components of a stress-tensor, because the corresponding vector of displacement need not be a single-valued function in G. (For details and for an example see [2] , Part I.)
Definition 1. A(very weak) biharmonic function to which there corresponds - through the functions (11) - a single-valued displacement is called an Airy function. In the opposite case we speak of a singular biharmonic function.
In a simply connected region, every biharmonic function is an Airy function. In a multiply connected region it need not be the case. From the point of view of the theory of elasticity, we are interested

in Airy functions only.
From the construction of the functions $g_{i0}$, $g_{i1}$ it follows (see [2], Part I) that the functions

(21) $$\overline{g}_{i0} = g_{i0} + \ell_i(x,y), \quad \overline{g}_{i1} = g_{i1} + \frac{\partial \ell_i}{\partial \nu},$$

where

(22) $\ell_i(x,y) = a_i x + b_i y + c_i$ ($a_i$, $b_i$, $c_i$ real constants), correspond to the same loading on $\Gamma_i$. A question arises if it is possible to find, on $\Gamma_i$ $(i = 1,\dots,k)$, the constants $a_i$, $b_i$, $c_i$ in such a way that the very weak solution of the problem

(23) $$\Delta^2 u = 0,$$

(24) $$u = g_{i0} + \ell_i, \quad \frac{\partial u}{\partial \nu} = g_{i1} + \frac{\partial \ell_i}{\partial \nu}, \quad i = 0,1,\dots,k,$$

be an Airy function. Here

$$\ell_i = 0 \text{ for } i = 0, \quad \ell_i = a_i x + b_i y + c_i \text{ for } i = 1,\dots,k.$$

Definition 2. An Airy function which is the (very weak) solution of (23), (24), is called an Airy function corresponding to the given loading (given by the functions $g_{i0}$, $g_{i1}$).

Formulation of the problem: The functions g being given, to find an Airy function corresponding to the given loading. In detail: To find the constants $a_i$, $b_i$, $c_i$ $(i = 1,\dots,k)$ in such a way that the solution of (23), (24) be an Airy function, and to find this function.

Theorem 3. Let the functions $g_{i0}$, $g_{i1}$ satisfy (20). Then there exists precisely one (very weak) Airy function corresponding to the given loading.

The idea of the proof is the following: Let $u_0$ be the very weak solution of (18), (19). This solution need not be an Airy function. Now, if it is not, the so-called basic singular biharmonic functions $r_{ij}$ $(i = 1,\dots,k\ j = 1,2,3)$ are constructed which are weak solutions of the first biharmonic problem with functions of the form (22) as boundary conditions. It is shown that there exists a linear combination of these functions which "removes" the singularity from the solution $u_0$, i.e., if added to this solution, an Airy function is obtained. In this way, we get the required Airy function corresponding to the given loading. Uniqueness: It is shown that the difference U(x,y) of two Airy functions corresponding to the given loading is a linear combination of basic singular biharmonic functions. At the same time, U(x,y) - as a difference of two Airy functions - should be an Airy function. But this is possible, as shown in the work, only if all the coefficients of the above-mentioned linear combination are equal to zero. - For a detailed proof of Theorem 3 see [2], Part I.

Also in the case of multiply connected regions, the method of least squares on the boundary can be applied and is shown to be very convenient as a numerical method. The approximate solution cannot be assumed in the simple form (15) only, but in the form

$$u_{mn}(x,y) = U_{mn}(x,y) + V_{mn}(x,y), \tag{25}$$

where

$$U_{mn}(x,y) = \sum_{i=1}^{4n-2} a_{mni}\, z_i(x,y) + \sum_{i=1}^{k} \sum_{q=1}^{4m} b_{mniq}\, v_{iq}(x,y) + \sum_{i=1}^{k} c_{mni} \ln\left[(x-x_i)^2 + (y-y_i)^2\right] \tag{26}$$

and

$$V_{mn}(x,y) = \sum_{i=1}^{k} \sum_{j=1}^{3} \alpha_{mnij}\, r_{ij}(x,y). \tag{27}$$

Here, $z_i(x,y)$ are basic biharmonic polynomials,

$$v_{i,4\ell+1}(x,y) = \mathrm{Re}\left[\frac{\bar{z}}{(z-z_i)^{\ell+1}}\right], \quad v_{i,4\ell+2}(x,y) = \mathrm{Im}\left[\frac{\bar{z}}{(z-z_i)^{\ell+1}}\right],$$

$$v_{i,4\ell+3}(x,y) = \mathrm{Re}\left[\frac{1}{(z-z_i)^{\ell+1}}\right], \quad v_{i,4\ell+4}(x,y) = \mathrm{Im}\left[\frac{1}{(z-z_i)^{\ell+1}}\right],$$

$z_j = x_j + y_j$ is an (arbitrary) point lying inside of the inner boundary curve $\Gamma_j$ $(j = 1,\ldots,k)$ and $r_{ij}(x,y)$ are the above mentioned basic singular biharmonic functions. (These functions cause no difficulties in the numerical process, because in this process there appear only their values on the boundary curves $\Gamma_i$ $(i = 1,\ldots,k)$, and these are extremely simple.) (26) represents the "Airy part" and (27) the "singular part" of the approximation, respectively.

The coefficients $a_{mni}$, $b_{mniq}$, $c_{mni}$, $\alpha_{mnij}$ are determined from the condition (16) again, M being the set of functions of the form (25) with arbitrary (real) coefficients. The condition (16) leads to the solution of a system of linear equations for the unknowns $a_{mni}$, $b_{mniq}$, $c_{mni}$, $\alpha_{mnij}$ (for details see [2], Part I).

Theorem 4. The above mentioned system is uniquely solvable.

The proof is simple and is an analogue of the proof of Theorem 1.

Theorem 5. For $m,n \longrightarrow \infty$ we have

$$u_{mn}(x,y) \longrightarrow u(x,y) \text{ in } L_2(G),$$

where $u(x,y)$ is the very weak solution of the problem (18), (19). At the same time, the "Airy part" $U_{mn}(x,y)$ converges, in $L_2(G)$, to the Airy function corresponding to the given loading, and thus to the desired solution of the first problem of plane elasticity. The convergence is uniform on every subregion $G'$ of $G$ such that $\overline{G'} \subset G$. The same holds for the convergence of the derivatives $D^i u_{mn}$ or $D^i U_{mn}$.

The proof follows the same idea as the proof of Theorem 2. Only the technique is more pretentious because of the multiple connectivity of the region. Especially, the Walsh theorem on approximation of a holomorphic function by a polynomial should be replaced by a more general theorem on approximation by a rational function (this is also the cause why functions $v_{iq}$ appear in (26)), etc.

Remark 2. Because, solving the problem of plane elasticity, we are interested only in the Airy function corresponding to the given loading, it is not at all necessary to construct, actually, the "singular" functions $r_{ij}(x,y)$.

Remark 3. The method of least squares on the boundary suggested by the authors proved to be a very effective approximate method when solving problems of the theory of elasticity and of related fields. Especially, it has been applied with success to the solution of some rather difficult problems of wall-beams in soil mechanics. In the case of the biharmonic problem, it takes advantage of the form of the biharmonic equation. However, it can be applied as well, when properly modified, to the solution of other problems.


[1] Rektorys, K. - Zahradník, V.: Solution of the First Biharmonic Problem by the Method of Least Squares on the Boundary. Aplikace matematiky 19 (1974), No 2, 101-131.

[2] Rektorys, K, - Danešová, J. - Matyska, J. - Vitner, Č.: Solution of the First Problem of Plane Elasticity for Multiply Connected Regions by the Method of Least Squares on the Boundary. Aplikace matematiky 22 (1977); Part I, No 5, 349-394; Part II, No 6, 425-454.

[3] Nečas, J.: Les méthodes directes en théorie des équations elliptiques. Praha, Academia 1967.

[4] Rektorys, K.: Variational Methods in Mathematics, Science and Engineering. Dortrecht (Holland), Boston (USA), Reidel 1977.

[5] Babuška, I. - Rektorys, K. - Vyčichlo, F.: Mathematische Elastizitätstheorie der ebenen Probleme. Berlin, Akademieverlag 1960.


Author's address: Technical University Prague, Faculty of Civil Engineering, Chair of Mathematics, 121 34 Praha 2, Trojanova 13, Czechoslovakia

# APPLICATION OF BOUNDED OPERATORS AND LYAPUNOV'S MAJORIZING EQUATIONS TO THE ANALYSIS OF DIFFERENTIAL EQUATIONS WITH A SMALL PARAMETER

Yu. Ryabov, Moscow

## 1. Introduction

Given a system of ordinary differential equations with a small parameter $\varepsilon$ $(\varepsilon \geq 0)$

$$(1) \qquad \dot{z} = F(z,t,\varepsilon)\ , \quad (\dot{\ } = d/dt)\ ,$$

let us consider the problems of existence, of estimating the domain of existence and of the construction of solutions of a certain class, for example, periodic or satisfying some initial conditions. Following the usual methods of small parameter we assume that the solution $z^{0}(t)$ of the system (1) for $\varepsilon = 0$ is known and that the solution $z(t,\varepsilon)$ is continuous at $\varepsilon = 0$. Moreover we assume that the function $F(z,t,\varepsilon)$ is continuous in $t,\varepsilon$ and differentiable with respect to $z$ in a neighborhood of $z^{0}(t)$.

A well known method of investigating the problem of existence and uniqueness of a solution consists in proving the possibility of transforming the system (1) into an operator system of the form $x = Sx$ where $S$ is the corresponding operator and $x$ is a new variable, and further in an application of the contractive mapping principle. Our approach which develops further the Lyapunov methods [1], [2] consists in associating the system $x = Sx$ with finite (as a rule, algebraic) equations which will be called Lyapunov's majorizing equations. Constructing these equations, we write the given operator system equivalent to the system (1) on the corresponding set of functions $\mathcal{H}$ in the form

$$(2) \qquad x = LW(x,t,\varepsilon)$$

where $L$ is a linear bounded matrix operator in $\mathcal{H}$ while $W(x,t,\varepsilon)$ is a function continuous in $t,\varepsilon$ and differentiable in $x$ in a domain $D(\|x\| \leq R,\ 0 \leq t \leq T,\ 0 \leq \varepsilon \leq \varepsilon_{*})$. The variable $x$ is such that

$$(3) \qquad W(0,t,0) = 0\ , \quad \partial W(0,t,0)/\partial x = 0\ .$$

Then in the general case, the system of majorizing equations is

$$(4) \qquad u = \Lambda\phi(u,\varepsilon) = 0$$

where $\Lambda$ is a constant matrix such that the following vector condition of boundedness of the operator L is satisfied:

$$(5) \qquad (\|L\varphi(t)\|) \leq \Lambda(\|\varphi(t)\|), \qquad \varphi(t) \in \mathcal{H}$$

while $\phi(u, \varepsilon)$ is the so called Lyapunov's majorant with respect to $W(x,t,\varepsilon)$; all components $W_i$, $\phi_i$ satisfy the inequalities

$$(6) \qquad \|W_i(x,t,\varepsilon)\| \leq \phi_i(u,\varepsilon), \quad \|\partial W_i(x,t,\varepsilon)/\partial x_j\| \leq \leq \partial\phi_i(u,\varepsilon)/\partial u_j,$$

provided $x,t,\varepsilon \in D$, $\|u\| \leq R$, $|x_j| \leq u_j$, $j=1,2,\dots,n$.

By means of the majorizing system (4) it is not only possible to establish the convergence of the iterative process

$$(7) \qquad x_k = LW(x_{k-1}, t, \varepsilon), \; k=1,2,\dots, \quad x_o \equiv 0$$

in a certain domain of variation of $\varepsilon$ to a unique solution but also to obtain estimates of this domain as well as of the error of the approximative solutions constructed on the basis of (7).

Various modifications of the system (2) and of the majorizing equations (4) may be studied. A number of results based on this approach are given in [3] - [10].

Notation used throughout the paper: (i) The symbol $\|\varphi(t)\|$ means the usual norm $\sup_t |\varphi(t)|$ of a scalar function in the space $C^o$; (ii) The symbol $\|\varphi(t)\|_*$ stays for the so called trigonometric norm in the space of functions which are expressed by absolutely convergent Fourier series:

$$\|\varphi(t)\|_* = \sum_{|k| \geq 0} |a_k|$$

where $a_k$ are the coefficients of the complex Fourier series (or polynomial) for $\varphi(t)$; (iii) By $(\|x(t)\|)$ we denote the vector whose components are $\|x_1(t)\|,\dots,\|x_n(t)\|$; (iv) A vector inequality $x \leq y$ is considered equivalent to the same inequalities for all the components of the vectors $x,y$. Vectors which satisfy the inequality $x > 0$ or $x \geq 0$ are called positive or nonnegative, respectively. The other notation is standard.

## 2. Fundamental theorems for the systems (2) and (4)

Generally, the function $\Lambda\phi(u,\varepsilon)$ is continuous in $\varepsilon$, continuously differentiable in $u$ and belongs to the class of nonlinear vector functions which are positive for $\varepsilon > 0$, $u > 0$, none of the elements of the matrix $\Lambda\partial\phi/\partial u$ is negative and there is at least one element which is an increasing function of at least one component of the vector $u$. Moreover,

$$\phi(0,0) = 0 , \quad \partial\phi(0,0)/\partial u = 0 .$$

Further we assume that the system (4) is non singular, i.e., it neither splits into separate subsystems nor has a solution for $\varepsilon > 0$ with some components of the vector $u$ equal to zero and the others positive. Then we have the following

Theorem 1.

I. The system (4) has a positive solution $u = u(\varepsilon)$ in and only in the domain $[0, \varepsilon_*]$ whose upper limit $\varepsilon_*$ and the corresponding vector $u_* = u(\varepsilon_*)$ satisfy simultaneously the relation (4) and the relation

$$\det\left[E - \Lambda\partial\phi(u, \varepsilon)/\partial u\right] = 0 \tag{8}$$

where $E$ is the unit matrix.

II. For $\varepsilon \in [0, \varepsilon_*]$ the system (4) has a unique solution $u = \bar{u}(\varepsilon) \in C^0[0, \varepsilon_*]$ such that $\bar{u}(\varepsilon) > 0$ for $\varepsilon > 0$ and $\bar{u}(0) = 0$; for $\varepsilon \in (0, \varepsilon_*)$ and the corresponding $\bar{u}(\varepsilon)$ the determinant (8) as well as all its principal minors are positive.

III. For $\varepsilon \in [0, \varepsilon_*]$ the iterations

$$u_k = \Lambda\phi(u_{k-1}, \varepsilon) , \; k=1,2,\ldots, \quad u_0=0 \tag{9}$$

form a nondecreasing sequence and converge to $\bar{u}(\varepsilon)$.

Theorem 2.

For a given system (2) and the corresponding system of inequalities

$$v \leq \Lambda\phi(v, \varepsilon) \tag{10}$$

let $u = \bar{u}(\varepsilon)$ be the solution of the system (2) from Theorem 1, and let $v = v(\varepsilon) \in C^0[0, \varepsilon_*]$ satisfy (10), $v(\varepsilon) > 0$ for $\varepsilon > 0$ and $v(0) = 0$.

Then $v(\varepsilon) \leq \bar{u}(\varepsilon)$, $\varepsilon \in [0, \varepsilon_*]$.

The proof of the above assertions is first carried out for the case when (2) is a scalar equation $(u = f(u, \varepsilon))$. Simple geometric arguments are used (graphs of the curves $y = f(u, \varepsilon)$ for $\varepsilon < \varepsilon_*$, $\varepsilon = \varepsilon_*$, $\varepsilon > \varepsilon_*$ on the surface $(u,y)$ and the graph of the straight line $y=u$ are considered) together with the monotonicity of $f(u, \varepsilon)$, $f'_u(u, \varepsilon)$ for increasing $u, \varepsilon$ and with the boundedness of $u, \varepsilon$ from above which is a consequence of an equation analogous to (8). The method of induction allows us to extend the results to systems of arbitrary orders (see [6], [10]). At the same time it is proved that (2), (8) together form a system of equations with respect to $u, \varepsilon$ possessing a unique positive solution $u=u_*$, $\varepsilon = \varepsilon_*$.

Hence to find $\varepsilon_*$ is an algebraic problem and it is known that its solution exists and is unique.

The fundamental result concerning the system (2) is the following

Theorem 3.

Let us consider the system (2) in the domain $D(\|x\| \leq R,\ 0 \leq t \leq T,\ 0 \leq \varepsilon \leq \varepsilon^0)$ and let (4) be the corresponding majorizing system in the domain. Let $u = \bar{u}(\varepsilon)$ be the solution of the system (4) and $[0, \varepsilon_*]$ the domain $(\varepsilon_* \leq \varepsilon^0)$ from Theorem 1, $\|\bar{u}(\varepsilon)\| \leq R$ for $\varepsilon \in [0, \varepsilon_*]$.

Then for $\varepsilon \in [0, \varepsilon_*]$ (i) the sequence $\{u_k(\varepsilon)\}$ defined according to (9) is majorizing with respect to the sequence $\{x_k(t, \varepsilon)\}$ defined according to (7); (ii) the sequence $x_k(t, \varepsilon)$ converges on the segment $0 \leq t \leq T$ to the solution $x(t, \varepsilon)$ of the system (2) and this solution is unique in the class of functions belonging to $C^0[0, \varepsilon_*]$ and equal to zero for $\varepsilon = 0$.

The proof is based on comparing the expressions (7), (9) for $x_k(t, \varepsilon)$, $u_k(\varepsilon)$ with regard to the inequalities (5), (6). We obtain

$$(11) \qquad (\|x_k(t, \varepsilon)\|) \leq u_k(\varepsilon), \quad (\|x_{k+1}(t, \varepsilon) - x_k(t, \varepsilon)\|) \leq \leq u_{k+1}(\varepsilon) - u_k(\varepsilon)$$

and this implies in virtue of Theorem 1 the uniform convergence of the sequence $\{x_k(t, \varepsilon)\}$ to the solution $x(t, \varepsilon)$ of the system (2) on the segment $[0, T]$ for $\varepsilon \in [0, \varepsilon_*]$ and moreover, $x(t, \varepsilon) \in \in C^0[0, \varepsilon_*]$, $x(t, 0) = 0$. The uniqueness of solution is established via Theorem 2. Assuming the existence of a solution $\tilde{x}(t, \varepsilon) \in C^0[0, \varepsilon_*]$ $(\tilde{x}(t, 0) \equiv 0)$ of the system (2) we obtain that the vector $v(\varepsilon) = =(\|\tilde{x}(t, \varepsilon)\|)$ satisfies the system of inequalities (10). In virtue of Theorem 2 we have $v(\varepsilon) \leq \bar{u}(\varepsilon)$. Further we establish the inequalities

$$(\|\tilde{x}(t, \varepsilon) - x_k(t, \varepsilon)\|) \leq \bar{u}(\varepsilon) - u_k(\varepsilon), \quad k = 1, 2, \ldots .$$

Hence we conclude that for $\varepsilon \in [0, \varepsilon_*]$ the system (2) has no solution different from the limit of the sequence $\{x_k(t, \varepsilon)\}$ which belongs to the class $C^0[0, \varepsilon_*]$ and equal to zero for $\varepsilon = 0$.

Remark 1. Theorem 3 offers directions how to construct the number $\varepsilon_*$ which gives a lower estimate of the values $\varepsilon$ for which the desired solution $x(t, \varepsilon)$ of the system (2) exists and is unique. Furthermore, we obtain an estimate for the solution itself and for the error of the k-th approximation $x_k(t, \varepsilon)$ since

$$(12) \qquad (\|x(t, \varepsilon)\|) \leq \bar{u}(\varepsilon), \quad (\|x(t, \varepsilon) - x_k(t, \varepsilon)\|) \leq \bar{u}(\varepsilon) - u_k(\varepsilon).$$

Remark 2. The majorizing equations (4) may be simplified by replacing them in the simplest case by a single equation. At the same time these equations and consequently, also the established estimates may be improved by immediate estimation of a certain number of approximations $x_k(t, \varepsilon)$. In the case of periodic solutions and analytic in $x$ right hand sides of (2) this improvement can be achieved by taking into account the structure of the desired solution and using the norm $\|\cdot\|_*$ in the conditions (5), as well as owing to the possibility of distinguishing the main harmonics in the solution (cf. [10]).

## 3. Analytic case

If the function $W(x,t,\varepsilon)$ in (2) is an analytic function of $x, \varepsilon$ in the domain $D$ and if an analytic with respect to $W(x,t,\varepsilon)$ majorant of Lyapunov is used in the construction of the system (4), then the number $\varepsilon_*$ from Theorem 3 gives an estimate of the domain of convergence of the power series in $\varepsilon$, in the form of which the solution of the system (2) can be sought. This series always converges for $|\varepsilon| \leq \varepsilon_*$. This follows from the fact that in this case all $x_k(t,\varepsilon)$ are expressed by series (polynomials) in $\varepsilon$ and they are majorized by the corresponding series (polynomials) for $u_k(\varepsilon)$ with nonnegative coefficients; the sequence $\{u_k(\varepsilon)\}$ converges for $0 \leq \varepsilon \leq \varepsilon_*$ to a function which can be expressed by a series of the same character.

## 4. Connection with the condition of contractivity of the mapping

The system (2) being written in the form $x = Sx$, Theorem 3 may be viewed as a theorem on existence and uniqueness of a fixed point of the operator (mapping) $S$ on the segment $0 \leq t \leq T$ for $\varepsilon \in \in [0, \varepsilon_*]$ in the class of functions $C^0[0,\varepsilon_*]$ equal to zero for $\varepsilon = 0$. Nonetheless the proof of Theorem 3 proceeds without using the notion of contractivity of the operator.

Dealing with (2), let us choose arbitrary consistent norms of vectors and matrices, formulate the conditions (5) of boundedness of the operator $L$ and construct the system (4). Let the latter be written in the form

$$(13) \qquad u = Qu\,, \qquad (Qu \equiv \Lambda \phi(u, \varepsilon))$$

where $Q$ is a finite functional operator. This operator majorizes the operator $S$ so that the contractivity of $Q$ is a sufficient condition of contractivity of $S$. However, it may happen that the operator $Q$ does not satisfy the condition of contractivity on the whole segment $[0, \varepsilon_*]$ from Theorem 1, if the above chosen norm is considered. It is not difficult to find examples of such systems of the form (4) (see [10]). In these cases it cannot be guaranteed that the operator S fulfils the condition of contractivity in the given norm for all $0 \le \varepsilon \le \varepsilon_*$. In other words, the requirement of contractivity of the operator with respect to the given norm may prove more restrictive than that which follows from the analysis of the majorizing system.

At the same time there exists such a vector norm that the condition of contractivity of the operator Q is fulfilled for all $0 < \varepsilon < \varepsilon_*$. Indeed, let us consider the matrix $Q'(u) = \Lambda \partial \phi(u, \varepsilon)/ / \partial u$ and its spectral radius $\rho(\varepsilon)$ for a fixed $u = \tilde{u} > 0$. The matrix $Q'(\tilde{u})$ is nonnegative and the assumptions of Theorem 1 imply that for $0 < \varepsilon < \varepsilon_*$ and $u = \bar{u}(\varepsilon)$, $\rho$ remains less than one, while $\rho = 1$ for $\varepsilon = \varepsilon_*$. According to [11] there is a vector norm such that the corresponding norm $\|Q'(\bar{u})\|$ is greater than $\rho(\varepsilon)$ by an arbitrarily small number. Hence for every $\varepsilon$ from the interior of the segment $[0, \varepsilon_*]$ there is a vector norm such that $\|Q'(\bar{u})\| < \delta < 1$. This is sufficient for the validity of the assertion on contractivity of the operator Q, and hence also of the majorizing operator S. However, for $\varepsilon = \varepsilon_*$ none of the norms $\|Q'(\bar{u})\|$ can be less than one so that the operator Q fails to be contractive. Consequently, for $\varepsilon = \varepsilon_*$ the operator S can also be non contractive. However, this does not influence our proof of Theorem 3.

## 5. Examples

1. Let us consider the periodic solution of the equation

$$\ddot{z} + \omega^2 z = \varepsilon z^3 + \sin t , \qquad \omega^2 = 5 . \tag{14}$$

We have $z^o = q \sin t$, $q = \frac{1}{4}$ and putting $z = z^o + x$ we obtain the equation

$$\begin{aligned} \ddot{x} + \omega^2 x &= \varepsilon f(x,t) = \\ &= \varepsilon \left[ q^3 \sin^3 t + 3q^2 \sin^2 tx + 3q \sin tx^2 + x^3 \right] . \end{aligned} \tag{15}$$

The equivalent operator equation is $x = \varepsilon Lf(x,t)$, where L is considered in the class of functions representable by trigonometric polynomials and

$$Le^{imt} = e^{imt}/(\omega^2 - m^2) . \tag{16}$$

All approximations $x_k(t, \varepsilon)$ are Fourier polynomials which include only odd harmonics. An estimate for L on the family of such functions with $\omega^2 = 5$ with respect to the norm $\|.\|_*$ is

$$(17) \qquad \|L\varphi(t)\|_* \le \|\varphi(t)\|_*/(\omega^2-1) = \tfrac{1}{4}\|\varphi(t)\|_*$$

while the majorizing Lyapunov's equation (being scalar) has the form

$$(18) \qquad u = \tfrac{1}{4}(q^3 + 3q^2u + 3qu^2 + u^3) , \quad q = \tfrac{1}{4} .$$

This equation yields $\varepsilon_* = 9.48$ , $u_* = 0.12$ (with an error not greater than 0.01) so that the convergence of the sequence $\{x_k(t,\varepsilon)\}$ to the unique solution is guaranteed for rather great $\varepsilon$. The solution $\bar{u}(\varepsilon)$ of the equation (18) yields an estimate with respect to the norm $\|\cdot\|_*$ of the deviation of the periodic solution $z(t,\varepsilon)$ of the equation (14) from the generating solution $z^0(t)$. The quantity $\delta_1 = \bar{u}(\varepsilon) - u_1(\varepsilon) = \bar{u}(\varepsilon) - \frac{1}{4}\varepsilon q^3$ represents an estimate of the first approximation $z_1(t,\varepsilon) = z^0(t) + x_1(t,\varepsilon)$. For example, for $\varepsilon = 8$ we obtain (with an error not greater than 0.001): $\bar{u}(\varepsilon) = 0.059$ , $u_1 = 0.031$ , $\delta_1 = 0.028$.

2. Let us consider the Mathieu equation

$$(19) \qquad \ddot{z} + (a + \varepsilon \cos 2t)z = 0 .$$

The Mathieu functions as well as the corresponding eigenvalues $a_{cn}$ , $a_{sn}$ can be found in the form of power series in $\varepsilon$. The literature offers only few facts concerning the radii of convergence of these series.

Let us discuss e.g. the case $ce_1(t)$ , setting $a = 1 + \varepsilon h$ , $z = ce_1(t) = \cos t + x$. The equation obtained for $x$ is

$$(20) \qquad \ddot{x} + x = \varepsilon f(x,t,h) = \\ = \varepsilon\left[-(h + \tfrac{1}{2})\cos t - \tfrac{1}{2}\cos 3t - (h + \cos 2t)x\right] .$$

We seek for the approximations $x_k(t,\varepsilon)$ , $h_k(\varepsilon)$ , $k=1,2,\ldots$ from the equations

$$(21) \qquad \ddot{x}_k + x_k = \varepsilon f(x_{k-1}, t, h_k) , \quad x_0 \equiv 0$$

having determined $h_k$ from the periodicity conditions for $x_k(t,\varepsilon)$. The operator equations corresponding to the given iteration process have the form

$$(22) \qquad x = \varepsilon L_1\left[-\tfrac{1}{2}\cos 3t - (h + \cos 2t)x\right] , \\ h = -\tfrac{1}{2} - L_2(x \cos 2t) .$$

Since all $x_k(t,\varepsilon)$ are expressed in the form of Fourier polynomials in cosines $jt$ , $j = 3,5,7,\ldots,$ $L_1$ and $L_2$ are considered to be in the same class of functions and hence

$$(23) \qquad \|L_1\varphi(t)\|_* \le \tfrac{1}{8}\|\varphi(t)\|_* , \ \|L_2\varphi(t)\|_* = |L_2\varphi(t)| \le \tfrac{1}{2}\|\varphi(t)\|_* .$$

Using these estimates we write the majorizing equation

(24) $$u = \frac{1}{8}\,\varepsilon\left[\frac{1}{2} + (v+1)u\right] , \qquad v = \frac{1}{2} + \frac{1}{2}\,u$$

where $u$ and $v$ majorize $x(t,\varepsilon)$ and $h(\varepsilon)$, respectively. For these equations we obtain $u_* = 1$ , $v_* = 1$ , $\varepsilon_* = 0.2$ . Since the analytic case is considered, the convergence of the sequence $\{x_k(t,\varepsilon)\}$ to a function $x(t,\varepsilon)$ analytic in $\varepsilon$ is guaranteed for $|\varepsilon| \le 0.2$ .

Let us note that the majorizing equations (24) can be improved so that an estimate $|\varepsilon| \le 0.35$ is obtained [10] . Further, let us mention that the given algorithm for the construction of $ce_1(t)$ was used for machine computation. Practical convergence occurs for $0 < \varepsilon < 0.50$ . For $\varepsilon = 0.55$ the process proved to be divergent.

3. Let us consider a system with an additional small parameter $\mu > 0$ at the derivative

$$\mu\dot{x} = Ax + \varepsilon F(x,t)$$

and its solution which vanishes for $\varepsilon = \mu = 0$ . The matrix $A$ is assumed to be constant and to satisfy the so called stability condition: the real parts of all its eigenvalues are negative. The function $F(x,t)$ is assumed to be continuous in $t$ and differentiable in $x$ in a domain. The equivalent operator system reads

(25) $$x = \varepsilon L_\mu F(x,t) , \quad L_\mu \varphi(t) = \frac{1}{\mu}\int_0^t \exp\left[\frac{A}{\mu}(t-s)\right]\varphi(s)ds .$$

In virtue of the stability condition we have

$$\|\exp(At)\| \le c\exp(-\alpha t) , \qquad c > 0 , \qquad \alpha > 0$$

so that $L_\mu$ is a bounded operator on the whole half-axis $t \ge 0$ for any $\mu > 0$ and the estimate $\sup_t \|L\varphi(t)\| \le \rho \sup_t \|\varphi(t)\|$ , $\rho = \frac{c}{\alpha}$ is independent of $\mu$. Hence the system (25) has the same character as the system (2). Its solution may be constructed by an iterative process of the form (7) whose convergence is guaranteed on a segment $[0, \varepsilon_*]$ and for all $\mu > 0$ . For example, for the scalar equation

$$\mu\dot{x} = -2x + \varepsilon(\cos t + 2x + x^2)$$

we have an estimate of the operator $\|L\varphi(t)\| \le \frac{1}{2}\|\varphi(t)\|$ and the majorizing equation $u = \frac{1}{2}\,\varepsilon(1 + 2u + u^2)$ for which $u_* = 1$ , $\varepsilon_* = \frac{1}{2}$ .

## References

(All references in Russian)


[1] Lyapunov A.M.: The general problem of the stability of motion. Gostehizdat, Moscow 1950

[2] Lyapunov A.M.: On the series proposed by Hill to represent the motion of the Moon. Collected works vol. 1. Akad.Nauk SSSR 1954

[3] Ryabov Yu.A.: An application of the small parameter method to the theory of nonlinear oscillations in the case of discontinuous characteristics of the nonlinearity. Trudy Vses. Zaočn.Energ.Inst. 16, 1960, 68-80

[4] Ryabov Yu.A.: An application of the small parameter method to the construction of solutions of differential equations with a delayed argument. Dokl.Akad.Nauk SSSR vol. 133, No.2, 1960, 288-291

[5] Ryabov Yu.A.: On a method of estimating the domain of applicability of the small parameter method in the theory of nonlinear oscillations. Inženernyj žurn.Akad.Nauk SSSR vol. 1, No.1, 1961, 16-28

[6] Ryabov Yu.A.: Some problems of application of the small parameter method and an estimate of the domain of its applicability in the theory of nonlinear oscillations and systems with delay. Doctor thesis, Moscow 1962

[7] Ryabov Yu.A., Lika D.K.: Convergence of the small parameter method for the construction of periodic solutions of ordinary differential equations of neutral type with a small delay. Diff.urav. 8 (1972), No.11, 1977-1987

[8] Ryabov Yu.A., Magomadov A.R.: On periodic solutions of a class of differential equations with "maxima". Izv.Akad.Nauk Azerb. SSR 1973, 1-19

[9] Ryabov Yu.A.: On periodic solutions and on the small parameter method for singularly perturbed differential equations. Nonlinear Vibration Problems vol. 15, Warsaw 1974, 67-73

[10] Ryabov Yu.A., Lika D.K.: Iteration methods and majorizing equations of Lyapunov in the theory of nonlinear oscillations. Štiinca, Kišinev 1974

[11] Krasnoselskij M.A., Vajnikko G.M. et al.: Approximate solution of operator equations. Nauka, Moscow 1969

Author's address: Moscow 125319, Leningrads.prosp. 64, MADI. USSR

ON LINEAR PROBLEMS IN THE SPACE BV

Š.Schwabik, M.Tvrdý, Praha

BV denotes the Banach space of column n-vector valued functions $x : [0,1] \to R_n$ endowed with the norm $x \in BV \to \|x\|_{BV} = |x(0)| + \text{var}_0^1 x$. A great variety of linear equations in BV is connected with linear operators of the form

(1) $$K : x \in BV \to Kx = \int_0^1 d_s[K(t,s)]\, x(s)$$

where $K : J = [0,1] \times [0,1] \to L(R_n)$ is an $n \times n$-matrix valued function defined on the square $J$ and the integral is taken in the Perron-Stieltjes sense. The main assumption is

(2) $$v_J(K) + \text{var}_0^1 K(0,.) < \infty$$

with $v_J(K)$ the twodimensional Vitali variation of $K$ on $J$ and $\text{var}_0^1 K(0,.)$ the usual variation with respect to the second variable of $K : J \to L(R_n)$ on $[0,1]$. Without any loss of generality we may assume in addition to (2)

(3) $$K(t,1) = 0,\ K(t,s+) = K(t,s) \quad \text{for any} \quad t \in [0,1]\ ,\ s \in [0,1)$$

because any kernel satisfying (2) can be replaced by a new one which satisfies also (3) and for which the operator (1) remains unchanged. Set

$$NBV = \{y : [0,1] \to R_n^*;\ y \in BV,\ y(s+) = y(s) \ \text{ for } \ s \in (0,1),\ y(1) = 0\}$$

($R_n^*$ is the space of row n-vectors, the star indicates the transposition of a matrix) and let

(4) $$K^* : y \in NBV \to \int_0^1 d[y^*(t)]\, K(t,s)\ .$$

The following facts are known (see [2]): If (2) holds then the linear operator $K : BV \to BV$ given by (1) is compact. If (2), (3) are satisfied then $K : NBV \to NBV$ given by (4) is also compact. The spaces BV and NBV form a dual pair with respect to the bilinear form

$$(x,y^*)\in BV\times NBV \longrightarrow \langle x,y^*\rangle = \int_0^1 d[y^*(t)]\,x(t)$$

and we have

(5) $\langle Kx,y^*\rangle = \langle x,K^*y^*\rangle$ for all $x\in BV$, $y\in NBV$ .

For the linear equation

(6) $x - Kx = f$, $f\in BV$

the Fredholm theory works in our case provided the operator $K^*$ is used instead of the usual adjoint to $K$ (see [1]). In this situation this is useful because neither the analytic description of the elements of the dual space to BV nor the analytic form of the adjoint operator to $K$ are available. Here we are interested in the resolvent formula for the equation (6).

Assume in addition to (2) that we have $N(I-K) = \{0\}$ for the kernel of the operator $I-K$, i.e. for its range we have $R(I-K) = BV$ ( $I$ is the identity operator on BV). Then we obtain easily that

(7) $$x(t) = f(t) + \int_0^1 d_s[R(t,s)]\,f(s)$$

(or shortly $x = f + Rf$) is for any $f\in BV$ a solution to (6) if and only if $x = x - Kx + R(x - Kx)$ for every $x\in BV$, i.e.

$$\int_0^1 d_s\Big[R(t,s) - K(t,s) - \int_0^1 d_r[R(t,r)]\,K(r,s)\Big]x(s)$$

for all $x\in BV$, $t\in[0,1]$. Hence we get

<u>Proposition 1.</u> Let $K : J\longrightarrow L(R_n)$ fulfil (2) and let $N(I-K) = \{0\}$. Then (7) yields a solution to (6) for any $f\in BV$ if and only if

$$R(t,s) - K(t,s) - \int_0^1 d_r[R(t,r)]\,K(r,s)\in S$$

for all $t\in[0,1]$ where $S$ is the linear space of $n\times n$-matrix valued functions $A : [0,1]\longrightarrow L(R_n)$ such that $A\in BV$, $A(0) = A(0+) = A(t-) = A(t+) = A(1-) = A(1)$ for all $t\in(0,1)$.

<u>Theorem 1.</u> If $K : J\longrightarrow L(R_n)$ satisfies (2), (3) and $N(I-K) = \{0\}$, then there exists a uniquely determined $n\times n$-matrix valued function

$R : J \longrightarrow L(R_n)$ such that

$$(8) \qquad R(t,s) = K(t,s) + \int_0^1 d_r[K(t,r)]\,R(r,s), \quad t,s \in [0,1]$$

$$(9) \qquad R(t,s) = K(t,s) + \int_0^1 d_r[R(t,r)]\,K(r,s), \quad t,s \in [0,1]$$

and $\mathrm{var}_0^1 R(.,s) < \infty$ for each $s \in [0,1]$. Furthermore $v_J(R) +$ $+ \mathrm{var}_0^1 R(0,.) < +\infty$, $R(t,1) = 0$ for all $t \in [0,1]$ and, given $f \in BV$, (7) gives the unique solution to $x - Ky = f$ in BV.

P r o o f. According to the Bounded Inverse Theorem there is a unique $R : J \longrightarrow L(R_n)$ fulfilling (8) and such that $R(.,s) \in BV$ for all $s \in [0,1]$. Moreover,

$$\|R(.,s)\|_{BV} \le c\|K(.,s)\|_{BV} \le m < \infty \quad (c < \infty)$$

for all $s \in [0,1]$ and it may be shown that $v_J(R) < \infty$. Let us put

$$\tilde{R}(t,s) = \begin{cases} R(t,s) & \text{if } s = 0 \text{ or } 1,\ t \in [0,1] \\ R(t,s+) & \text{if } s \in (0,1),\ t \in [0,1] \end{cases}$$

and $\hat{R}(t,s) = R(t,s) - \tilde{R}(t,s)$ on J. Then

$$\hat{R}(t,s) - \int_0^1 d_r[K(t,r)]\,\hat{R}(r,s) = 0$$

for all $t,s \in [0,1]$ and thus $\hat{R}(t,s) \equiv 0$ on J. This means that the rows of $R(t,.)$ belong for every $t \in [0,1]$ to the class NBV. Moreover, it is easy to verify that $x = f + Rf$ is for any $f \in BV$ a solution to (6). According to Proposition 1 we have $Q(t,.) \in S$ for all $t \in [0,1]$, where

$$Q(t,s) = R(t,s) - K(t,s) - \int_0^1 d_r[R(t,r)]\,K(r,s) \quad \text{on } J.$$

Since $R(t,.) \in NBV$ and $K(t,.) \in NBV$ for all $t \in [0,1]$, $Q(t,.) \in NBV$ for all $t \in [0,1]$, i.e. $Q(t,s) \equiv 0$ on J. This yields (9).

Remark. If $K : J \longrightarrow L(R_n)$ satisfies (2), (3) but $\dim N(I-K) = k > 0$, then a "pseudoresolvent technique" can be used for showing that there exists $R : J \longrightarrow L(R_n)$ with $v_J(R) < \infty$, $\mathrm{var}_0^1 R(0,.) < \infty$, $R(t,.) \in NBV$ for all $t \in [0,1]$ such that if for a given $f \in BV$ the equation (6) possesses a solution, then $x = f + Rf$ is also its solution. All the solutions to (6) may be then written in the form $x(t) = x(t) + \sum_{i=1}^{k} d_i x_i(t)$, where $x_1, x_2, \ldots, x_k$ is a basis in $N(I-K)$ and $d_i$, $i=1,\ldots,k$ are real numbers.

Let an $n \times n$-matrix valued function $A : [0,1] \longrightarrow L(R_n)$ be of bounded variation on $[0,1]$ and $f \in BV$. We consider the integral equation

$$(10) \qquad x(t) - x(0) - \int_0^t d[A(s)]x(s) = f(t) - f(0), \quad t \in [0,1]$$

called the generalized linear ordinary differential equation. It is known (see [3]) that if $x : [0,1] \longrightarrow R_n$ satisfies (10) then $x \in BV$. Furthermore if

$$(11) \qquad \det[I - \Delta^{-}A(t)] \neq 0 \quad \text{on} \quad (0,1]$$

($I \in L(R_n)$ is the identity matrix, $\Delta^{-}A(t) = A(t) - A(t-)$) then for any $f \in BV$ and $c \in R_n$ the equation (10) possesses a unique solution $x(t)$ on $[0,1]$ such that $x(0) = c$. If we assume, in addition,

$$(12) \qquad \det[I + \Delta^{+}A(t)] \neq 0 \quad \text{on} \quad [0,1] \quad (\Delta^{+}A(t) = A(t+) - A(t))$$

then there exists an $n \times n$-regular matrix valued function $X : [0,1] \longrightarrow L(R_n)$ of bounded variation on $[0,1]$ such that for any $t,s \in [0,1]$

$$(13) \qquad X(t) = X(s) + \int_s^t d[A(r)]X(r) .$$

$X^{-1}(t) : [0,1] \longrightarrow L(R_n)$ is also of bounded variation and satisfies

$$(14) \qquad X^{-1}(t) = X^{-1}(s) - X^{-1}(t)A(t) + X^{-1}(s)A(s) + \int_s^t d[X^{-1}(r)]A(r)$$

for all $t,s \in [0,1]$. For given $f \in BV$ and $c \in R_n$ the corresponding solution $x(t)$ of (10) with $x(0) = c$ is given by the variation-of-constants formula

$$(15) \qquad x(t) = X(t)X^{-1}(0)c+f(t)-f(0)-X(t)\int_0^t d[X^{-1}(s)](f(s)-f(0)) .$$

Let us consider the boundary value problem (P) of determining a solution $x : [0,1] \longrightarrow R_n$ of (10) which fulfils the side condition

(16) $\quad Mx(0) + Nx(1) = r$

where $M,N$ are $m\times n$-matrices and $r\in R_m$. Inserting the variation-of-constants formula (15) into (16) we get that our problem (P) is solvable if and only if

$$d^*Nf(1) - d^*NX(1)X^{-1}(0)f(0) - d^*NX(1)\int_0^1 d[X^{-1}(s)]f(s) = d^*r$$

for any $d\in R_m$ such that

(17) $\quad d^*[MX(0) + NX(1)] = 0$ .

It follows from the properties of the matrix function $X^{-1}(s)$ (see (14)) that, given $d\in R_m$ fulfilling (17), the couple $(y^*(s),d^*), y^*(s) = d^*NX(1)X^{-1}(s)$, satisfies the system

(18) $$y^*(s) - y^*(1) + \int_s^1 d[y^*(t)]A(t) - y^*(1)A(1) + y^*(s)A(s) = 0,$$

$$y^*(0) + d^*M = 0, \quad y^*(1) - d^*N = 0 .$$

Hence if

(19) $$y^*(1)f(1) - y^*(0)f(0) - \int_0^1 d[y^*(t)]f(t) = d^*r$$

for any solution $(y(s),d)\in BV\times R_m$ of (18), then our problem (P) has a solution. On the other hand, if (P) has a solution $x$ , then

$$y^*(1)f(1) - y^*(0)f(0) - \int_0^1 d[y^*(t)]f(t) =$$

$$= (y^*(1) - d^*N)x(1) - (y^*(0) + d^*M)x(0) +$$

$$+ \int_0^1 d[y^*(s) - \int_0^s d[y^*(t)]A(t) + y^*(s)A(s)]x(s) = 0 .$$

Theorem 2. Under our assumptions the problem (P) has a solution if and only if (19) holds for every couple $(y(s),d)\in BV\times R_m$ satisfying (18).

The system (18) is called "the conjugate problem to (P)".

Remarks. 1. If $A(t-) = A(t)$ on $(0,1]$, $A(0+) = A(0)$ and $B(0) = A(0)$, $B(t) = A(t+)$ on $(0,1)$, $B(1-) = A(1)$, then the first equation from (18) reduces to

$$y^*(s) = y^*(1) - \int_s^1 y^*(t)dB(t) .$$

2. Under our assumptions, given $d \in R_n$ , the function $y^*(s) = d^*NX(1)X^{-1}(s)$ is a unique solution of the first equation from (18) on $[0,1]$ such that $y^*(1) = d^*$ .

3. It may be shown that if the homogeneous problem corresponding to (P)($f(t) \equiv 0$, $r = 0$) has exactly $k$ linearly independent solutions in BV, then its conjugate problem possesses exactly $k^* = k+m-n$ linearly independent solutions in $BV \times R_m$ , i.e. the index of the problem is $n - m$ .

4. The homogeneous problem corresponding to (P) possesses only the trivial solution if and only if

$$\text{rank}\,[MX(0) + NX(1)] = n .$$

Similarly as in the classical case we may show

Theorem 3. Let in addition $m=n$ and $\det D = \det [MX(0) + NX(1)] \neq 0$. Let us put

$$G(t,s) = \begin{cases} -X(t)D^{-1}MX(0)X^{-1}(s) & \text{for } s<t , \\ X(t)D^{-1}NX(1)X^{-1}(s) & \text{for } s>t \end{cases}$$

($G(0,0) = X(0)D^{-1}NX(1)X^{-1}(0)$, $G(1,1) = -X(1)D^{-1}MX(0)X^{-1}(1)$, the values $G(t,t)$, $t \in (0,1)$ need not be defined at this moment),

$$H(t) = X(t)D^{-1} \quad \text{on } [0,1] .$$

Then for any $f \in BV$ and $r \in R_m$ the function

$$x(t) = H(t)r + G(t,1)f(1) - G(t,0)f(0) - \int_0^1 d_s[G(t,s)]f(s)$$

is a unique solution of the problem (P).

In virtue of the properties of $X^{-1}$ we have

Theorem 4. Let $P(t,s) = G(t,s)$ if $t,s \in [0,1]$, $t \neq s$, $P(1,1) = G(1,1)$, $P(t,t) = X(t)D^{-1}NX(1)X^{-1}(t)$ if $0 \leq t < 1$ . Then under the assumptions of Theorem 3 the functions $H(t)$, $P(t,s)$ are such that $v_J(P) + \text{var}_0^1 P(0,.) + \text{var}_0^1 P(.,0) < \infty$, $\text{var}_0^1 H < \infty$ and moreover

$$P(t,s) - P(t,1) + \int_s^1 d_r[P(t,r)]A(r) + P(t,s)A(s) - P(t,1)A(1) =$$

$= \Delta(t,s)$ for $t \in (0,1)$, $s \in [0,1]$, $P(t,0) = -H(t)M$, $P(t,1) = H(t)N$ for $t \in (0,1)$ where $\Delta(t,s) = 0$ if $t \leq s$ and $\Delta(t,s) = -I$ if $t > s$.

($P(t,.)$, $H(t)$ is for any $t \in [0,1]$ a solution of the conjugate problem with $\Delta(t,s)$ on the right hand side.)

Remark. Theorem 4 describes the behaviour of the functions occuring in the solution formula for the problem (P) given in Theorem 3. The connection of the matrix $G(t,s)$ with the original problem (P) involves the first variable $t$. The proofs of these relations are straightforward but tedious, they will be given in a separate paper of the authors.

Boundary value problems of the form (P) are of interest since by generalized linear differential equations (10) some special interface problems may be described.

The more general boundary value problem with the side condition

$$Mx(0) + Nx(1) + \int_0^1 d[K(s)]x(s) = r$$

can be also handled in the same manner or it can be transferred to a boundary value problem with a side condition of the type (16) using the Jones transform for the boundary value problem similarly as was done in the paper [4].

## References

[1] Heuser H., Funktionalanalysis, B.G.Teubner, Stuttgart 1975.

[2] Schwabik Š., On an integral operator in the space of functions with bounded variation, II, Časopis pěst.mat. 102, 1977, 189-202.

[3] Schwabik Š., Tvrdý M., Vejvoda O.: Differential and integral equations. Boundary value problems and adjoints, Academia, Praha 1978, to appear.

[4] Tvrdý M., Vejvoda O., Existence of solutions to a linear integro-boundary-differential equation with additional conditions, Ann.Mat. Pura Appl., 89, 1971, 169-216.

Authors' address: Matematický ústav ČSAV, Žitná 25, 115 67 Praha 1,

ČSSR

# A PARTIALLY ORDERED SPACE CONNECTED WITH THE DE LA VALLÉE POUSSIN PROBLEM

V. Šeda, Bratislava

In the paper the m-point BVP

$$(1) \qquad (L(x) \equiv ) \sum_{j=0}^{n} p_j(t)x^{(n-j)} = f(t,x,x',\dots,x^{(n-1)}),$$

$$(2) \qquad x^{(i-1)}(t_k) = A_{i,k} (i=1,\dots,r_k,\ k=1,2,\dots,m)$$

is considered where $n \geqq 2$, $2 \leqq m \leqq n$, $1 \leqq r_k$ $(k=1,2,\dots,m)$ are natural numbers such that $r_1 + r_2 + \dots + r_m = n$, and $-\infty < a < t_1 < \dots < t_m < b < \infty$, $A_{i,k}(i=1,\dots,r_k,\ k=1,2,\dots,m)$ are real numbers. First the properties of the Green function corresponding to the homogeneous BVP

$$(1') \qquad L(x) = 0,$$

$$(2') \qquad x^{(i-1)}(t_k) = 0 \quad (i=1,\dots,r_k,\ k=1,2,\dots,m)$$

and their consequences are studied. The regularity properties determine continuity and compactness of the integrodifferential operator associated to (1), (2). The sign of the Green function decides the monotonicity properties of this operator in a partially ordered space which is described here. Using the fixed point theory in partially ordered spaces we obtain new results for the mentioned BVP connected with the notion of the lower (upper) solution of (1).

Throughout the paper the following two assumptions are satisfied:
$(A_1)$ $p_j\colon (a,b) \to R$, $p_j \in L_{loc}(a,b)$ $(j=0,1,\dots,n)$, $p_0(t) = 1$ in $(a,b)$ a.e. and $f\colon (a,b) \times R^n \to R$ satisfies Carathéodory conditions, i.e. $f(\cdot,x_0,x_1,\dots,x_{n-1})$ is measurable in $(a,b)$ for each fixed $(x_0,x_1,\dots,x_{n-1}) \in R^n$, for almost all $t_0 \in (a,b)$ $f(t_0,\cdot,\dots,\cdot)$ is continuous in $R^n$ and $\sup\{|f(t,x_0,x_1,\dots,x_{n-1})| : (x_0^2 + x_1^2 + \dots + x_{n-1}^2)^{1/2} \leqq r\} \in L_{loc}(a,b)$ for each $r>0$.
$(A_2)$ The differential equation (1') is disconjugate in $(a,b)$.

With respect to Assumption $(A_1)$, by a solution (1) on an interval $j \subset (a,b)$ each function $x$ with $x^{(n)} \in L_{loc}(j)$ which satisfies (1) a.e. in $j$ is understood.

Under the assumptions $(A_1)$, $(A_2)$ for any $m$, $2 \leqq m \leqq n$, any $r_k$ $(k = 1,2,\dots,m)$ with $1 \leqq r_k$, $r_1 + r_2 + \dots + r_m = n$, and any $a < t_1 < t_2 < \dots < t_m < b$ the problem (1'), (2') has only the trivial solution and there exists a unique function $G = G(t,s)$, the so-called Green function of (1'), (2') defined on $[t_1,t_m] \times \bigcup_{\ell=1}^{m-1} [t_\ell ,t_{\ell+1}]$

and such that if $r \in L(t_1,t_m)$ and $w$ is the unique solution of the problem (1'), (2), then the unique solution $y$ of the problem $L(y) = r(t)$, (2) is given by $y(t) = w(t) + \int_{t_1}^{t} G(t,s)r(s)\,ds$ ([4], p. 137). Hence the BVP (1), (2) is equivalent to the integrodifferential equation

$$(3)\quad x(t)=w(t)+\int_{t_1}^{t_m} G(t,s)f\left[s,x(s),x'(s),\dots,x^{(n-1)}(s)\right] ds \quad (t\in[t_1,t_m])$$

in the sense that each solution $x$ of (3) with $x^{(n-1)} \in L(t_1,t_m)$ possesses $x^{(n)} \in L(t_1,t_m)$ and is a solution of (1), (2) and conversely.

Lemma 1. The function $G$ has the following properties:

1. $\dfrac{\partial^i G}{\partial t^i}$, $i = 0,\dots,n-2$, is continuous in $[t_1,t_m] \times [t_1,t_m]$.
2. $\dfrac{\partial^{n-1} G}{\partial t^{n-1}}$ is continuous in the domains $t_1 \leqq t \leqq s \leqq t_m$ and $t_1 \leqq s \leqq t \leqq t_m$. Further,

$$(4)\qquad \lim_{t\to s+} \frac{\partial^{n-1}G(t,s)}{\partial t^{n-1}} - \lim_{t\to s-} \frac{\partial^{n-1}G(t,s)}{\partial t^{n-1}} = 1 \qquad (t_1 < s < t_m).$$

3. $G(\cdot,s)$ satisfies (1') in $[t_1,t_m]$ a.e. and the boundary conditions (2').
4. The sign of $G$ is determined by the inequality

$$G(t,s)(t-t_1)^{r_1}(t-t_2)^{r_2}\dots(t-t_m)^{r_m} \geqq 0,\quad t_1 \leqq t \leqq t_m,\ t_1 \leqq s \leqq t_m$$

and

$$G(t,s) \neq 0 \ \text{ for } \ t_k < t < t_{k+1},\ \ t_1 < s < t_m,\ \ k = 1,\dots,m-1.$$

5. $$\lim_{s\to t_1+} \frac{\partial^i G(t,s)}{\partial t^i} = \lim_{s\to t_m-} \frac{\partial^i G(t,s)}{\partial t^i} = 0 \ \text{ for } \ t_1 < t < t_m,\quad 0 \leqq i \leqq n-1$$

and

$$\lim_{s\to t_1+} \frac{\partial^i G(t,s)}{\partial t^i} = \begin{cases} 0, & 0 \leqq i \leqq n-2 \\ -1, & i = n-1 \end{cases},$$

$$\lim_{s\to t_m-} \frac{\partial^i G(t,s)}{\partial t^i} = \begin{cases} 0, & 0 \leqq i \leqq n-2 \\ 1, & i = n-1 \end{cases}.$$

Proof. Let $u = u(t,s)$ be the Cauchy function for (1'), i.e. $u(\cdot,s)$ is the solution of (1') determined by the conditions $x^{(i-1)}(s) = 0$ $(i = 1,\ldots,n-1)$, $x^{(n-1)}(s) = 1$. Let $v_{j\ell}$, $\ell = 1,2,\ldots,m$, $j = 1,\ldots,r_\ell$, be the solution of (1') satisfying the conditions

$v_{j\ell}^{(i-1)}(t_k) = 0$ $(k \neq 1,\ i = 1,\ldots,r_k,\ k = 1,2,\ldots,m)$

$v_{j\ell}^{(i-1)}(t_\ell) = \delta_{ji}$ $(i = 1,\ldots,r_\ell)$. $\delta_{ji}$ is the delta Kronecker symbol.

Then, by [4], p. 137,

$$(5)\quad G(t,s) = \begin{cases} \displaystyle\sum_{\ell=1}^{k}\sum_{j=1}^{r_\ell} v_{j\ell}(t)\,\frac{\partial^{j-1}u(t_\ell,s)}{\partial t^{j-1}}, & t \geqq s \\ \displaystyle -\sum_{\ell=k+1}^{m}\sum_{j=1}^{r_\ell} v_{j\ell}(t)\,\frac{\partial^{j-1}u(t_\ell,s)}{\partial t^{j-1}}, & t \leqq s, \end{cases}$$

$$t_k \leqq s \leqq t_{k+1} \quad (k = 1,\ldots,m-1)$$

and the equality

$$(6)\quad \sum_{\ell=1}^{m}\sum_{j=1}^{r_\ell} v_{j\ell}(t)\,\frac{\partial^{j-1}u(t_\ell,s)}{\partial t^{j-1}} = u(t,s) \quad \text{for each} \quad (t,s) \in [t_1,t_m] \times [t_1,t_m]$$

is true ([4], p. 136). By (5), $\frac{\partial^i G}{\partial t^i}$, $i = 0,1,\ldots,n-1$, is piecewise continuous in $[t_1,t_m] \times [t_1,t_m]$ and its only discontinuities may lie on the lines $s = t$ and $s = t_k$, $k = 2,\ldots,m-1$ (if $m > 2$). Since $\frac{\partial^{j-1}u}{\partial t^{j-1}}$ is continuous in $s$ and $\frac{\partial^{j-1}u(t_k,t_k)}{\partial t^{j-1}} = 0$ for $j = 1,\ldots,r_k$, by (5) we get that $\lim\limits_{s\to t_k+} \frac{\partial^i G(t,s)}{\partial t^i} = \lim\limits_{s\to t_k-} \frac{\partial^i G(t,s)}{\partial t^i}$ $(t \in [t_1,t_m],\ t \neq t_k,\ k = 2,\ldots,m-1,\ 0 \leqq i \leqq n-1)$. Further,

$$\lim_{s\to t-} \frac{\partial^i G(t,s)}{\partial t^i} - \lim_{s\to t+} \frac{\partial^i G(t,s)}{\partial t^i} = \begin{cases} 0, & 0 \leqq i \leqq n-2 \\ 1, & i = n-1 \end{cases} \qquad (t_1 < t < t_m)$$

follows from (5), (6). This implies the statements 1 and 2 as well as (4). 3 is a direct consequence of (5). The sign of $G$ has been determined in [5], pp. 80-81. (5) and (6) also imply the statement 5.

Denote now the set of all functions $x$ with $x^{(n-1)} \in L(t_1,t_m)$ (with $x^{(n)} \in L(t_1,t_m)$) by $L^{(n-1)}(t_1,t_m)$ $(L^{(n)}(t_1,t_m))$. Further, let $C_{n-1} = C_{n-1}([t_1,t_m])$ be provided with the norm $\|x\| =$

$= \max_{k=0,1,\dots,n-1} \{\max_{t\in[t_1,t_m]} |x^{(k)}(t)|\}$. Similarly, the norm $\|x\|_0 = \max_{t\in[t_1,t_m]} |x(t)|$ will be used in $C = C([t_1,t_m])$.

The right-hand side of (3) defines in $L^{(n-1)}(t_1,t_m)$ an operator $T$ which is given by

$$(7) \quad T(x)(t) = w(t) + \int_{t_1}^{t_m} G(t,s)f[s,x(s),x'(s),\dots,x^{(n-1)}(s)]\,ds \quad (t\in[t_1,t_m]).$$

By $(A_1)$ and Lemma 1, $T: L^{(n-1)}(t_1,t_m) \to L^{(n)}(t_1,t_m)$. Since $L^{(n)}(t_1,t_m) \subset C_{n-1} \subset L^{(n-1)}(t_1,t_m)$, $T$ maps $C_{n-1}$ into itself.

<u>Lemma 2.</u> $T$ as a mapping from $C_{n-1}$ into itself is continuous and compact.

Proof. Suppose that $S \subset C_{n-1}$ is bounded. Then, by $(A_1)$, there exists a function $m \in L(t_1,t_m)$ such that for any $x \in S$, $|f(t,x(t),\dots,x^{(n-1)}(t)| \leqq m(t)$ $(t\in[t_1,t_m])$. On the basis of (7), this implies that there exist constants $A_j$, $j = 0,1,\dots,n-1$, such that $T(x) = y$ satisfies the inequalities $|y^{(j)}(t)| \leqq A_j$ $(j = 0,1,\dots,n-1$, $t\in[t_1,t_m])$ and, since $y$ is the solution of $L(y) = f[t,x(t),x'(t),\dots,x^{(n-1)}(t)]$, (2), $|y^{(n)}(t)| \leqq \sum_{j=1}^{n} |p_j(t)|A_{n-j} + m(t)$. Thus the sets $\{y^{(j)}\}$, $j = 0,1,\dots,n-1$, are equicontinuous and uniformly bounded on $[t_1,t_m]$. From this the relative compactness of $T(S)$ in $C_{n-1}$ follows. Thus $T$ is compact.

If $x_n \to x$ in $C_{n-1}$, then $[T(x_n)]^{(j)}(t) \to [T(x)]^{(j)}(t)$ $(j = 0,1,\dots,n-1)$ pointwise in $[t_1,t_m]$ and the sequences $\{[T(x_n)]^{(j)}\}$ are equicontinuous on that interval. Hence $[T(x_n)]^{(j)}$ converge uniformly to $[T(x)]^{(j)}$ $(j = 0,1,\dots,n-1)$ on $[t_1,t_m]$ and $T$ is continuous.

Many consequences follow from the sign of the Green function. In order to show them, similarly as in [7], a partial ordering in $C_{n-1}$, and more generally, in $C$, will be introduced by the rule: If $x, y \in C_{n-1}$ or in $C$, then $x \leqq y$ iff $(-1)^{r_{k+1}+\dots+r_m}[y(t) - x(t)] \geqq 0$ for all $t\in[t_k,t_{k+1}]$ and all $k = 1,2,\dots,m-1$. By Lemma 1, $x \leqq y$ means that $x(t) \leqq y(t)$ $(x(t) \geqq y(t))$ on each interval $[t_k,t_{k+1}]$ where $G(t,s) \geqq 0$ $(G(t,s) \leqq 0)$. Denote $H_1 = \{k\in\{1,\dots,m-1\} : r_{k+1}+\dots+r_m \text{ is even}\}$, $H_2 = \{k\in\{1,\dots,m-1\} : r_{k+1}+\dots+r_m \text{ is odd}\}$.

Let $P = \{x \in C_{n-1}: x \geqq 0\}$ $(P_0 = \{x \in C: x \geqq 0\})$. Clearly $P$, $P_0$ are cones in the corresponding spaces and $(C_{n-1},P),(C,P_0)$, respectively, ordered Banach spaces with positive cones $P$ and $P_0$, respectively. While $P$ is not normal, $P_0$ already is. If $x_0 \leqq y_0$ in $C_{n-1}$ and in $C$, respectively, then by the interval $[x_0,y_0]$ the set $\{z \in C_{n-1}: x_0 \leqq z \leqq y_0\}$ and $\{z \in C: x_0 \leqq z \leqq y_0\}$, respectively, will be understood.

Further, $x \in L^{(n)}(t_1,t_m)$ will be called a <u>lower solution</u> ($y \in L^{(n)}(t_1,t_m)$ an <u>upper solution</u>) <u>of the differential equation (1)</u> if it satisfies the differential inequality $L(x) \leqq f(t,x,x',\ldots,x^{(n-1)})$ $(L(y) \geqq f(t,y,y',\ldots,y^{(n-1)}))$ a.e. in $[t_1,t_m]$.

Finally for each $x \in L^{(n)}(t_1,t_m)$ the solution $v_x$ of the problem

$$L(v) = 0,$$

$$v^{(i-1)}(t_k) = x^{(i-1)}(t_k) \quad (i = 1,\ldots,r_k, \quad k = 1,2,\ldots,m)$$

will be said to be the <u>solution of (1') associated with the function $x$.</u>

With help of these notions the following condition will be expressed which plays an important role in subsequent considerations.

$(A_3)$ There exist a lower and an upper solution $x_0$, $y_0$, respectively, of the equation (1) such that

(8) $$x_0 \leqq y_0$$

and the solutions $v_{x_0}$, $v_{y_0}$ of (1') associated with the function $x_0$, $y_0$, respectively, satisfy

(9) $$v_{x_0} \leqq w \leqq v_{y_0}$$

where $w$ is the solution of the problem (1'), (2).

The inequalities (8), (9) are considered in the partially ordered space $C_{n-1}$.

<u>Theorem 1.</u> Let the differential equation (1) satisfy the conditions $(A_1$ - $(A_3)$ and let

(10) $$L(x_0)(t)-f(t,x,x_1,\ldots,x_{n-1}) \leqq 0 \leqq L(y_0(t)-f(t,x,x_1,\ldots,x_{n-1})$$

be true for all $(t,x,x_1,\ldots,x_{n-1}) \in W$ where $W$ is given by

$$W = \{(t,x,x_1,\ldots,x_{n-1}): x_0(t) \leqq x \leqq y_0(t), -\infty < x_i < \infty, \ i = 1,\ldots,n-1\} \cup \{(t,x,x_1,\ldots,x_{n-1}): y_0(t) \leqq x \leqq x_0(t), -\infty < x_i < \infty, \ i = 1,\ldots,n-1\}.$$

Then the BVP (1), (2) has a solution $x$ which lies in the interval $[x_0,y_0]$.

Proof. The interval $[x_0,y_0]$ is a closed and convex subset of $C_{n-1}$. If $x \in [x_0,y_0]$, then $(t,x(t),x'(t),\dots,x^{(n-1)}(t)) \in W$ for all $t \in [t_1,t_m]$ and by (9), (10), for all $t \in [t_k,t_{k+1}]$, $k \in H_1$,

$$T(x)(t) = w(t) + \int_{t_1}^{t_m} G(t,s)\, f[s,x(s),x'(s),\dots,x^{(n-1)}(s)]\, ds \leqq v_{y_0}(t) + \int_{t_1}^{t_m} G(t,s)\, L(y_0)(s)\, ds = y_0(t).$$

For $k \in H_2$ we get $T(x)(t) \geqq y_0(t)$, $t \in [t_k,t_{k+1}]$. In a similar way, $T(x)(t) \geqq x_0(t)$ $(T(x)(t) \leqq x_0(t))$ for $t \in [t_k,t_{k+1}]$, $k \in H_1 (k \in H_2)$. Thus we see that (8) - (10) imply that $T([x_0,y_0]) \subset [x_0,y_0]$. By Lemma 2, $T$ is continuous.

Now we prove that $T([x_0,y_0])$ is relatively compact in $C_{n-1}$. Let $1 \leqq j \leqq n-1$ be a natural number. By the continuity properties of $G$ there exists a constant $G_j > 0$ such that $\left| \frac{\partial^j G(t,s)}{\partial t^j} \right| \leqq G_j$ in $[t_1,t_m] \times [t_1,t_m]$, and, on the basis of (10), the inequality

$$(11) \qquad \left| [T(x)]^{(j)}(t) \right| \leqq \left| w^{(j)}(t) \right| + G_j \int_{t_1}^{t_m} (|L(x_0)(s)| + |L(y_0)(s)|)\, ds = H_j$$

is true for $x_0 \leqq x \leqq y_0$. Let $H_0 = \max( \max_{t \in [t_1,t_m]} |x_0(t)|, \max_{t \in [t_1,t_m]} |y_0(t)|)$. Since $y = T(x)$ satisfies the equation $L(y) = f[t,x(t),x'(t),\dots,x^{(n-1)}(t)]$, (11) with $(A_1)$ give that

$$\left| [T(x)]^{(n)}(t) \right| \leqq \sum_{j=1}^{n} |p_j(t)| H_{n-j} + m(t)$$

where $m \in L(t_1,t_m)$. This proves that the functions $\{[T(x)]^{(j)}\}$, $j = 0,1,\dots,n-1$, are uniformly bounded and equicontinuous in $[t_1,t_m]$ which means that $T([x_0,y_0])$ is relatively compact. The Schauder fixed point theorem ensures a fixed point of $T$ in $[x_0,y_0]$.

In [7] two sufficient conditions are given for $x_0$ and $y_0$ to satisfy (9). Here they will be stated as

Lemma 3. Any of the following conditions is sufficient for the lower and the upper solutions $x_0, y_0$, respectively, of (1) to satisfy the inequalities (9).

1. $x_0$ and $y_0$ satisfy the boundary conditions

$$x_0^{(i-1)}(t_k) = A_{i,k} = y_0^{(i-1)}(t_k), \quad i = 1,\dots,r_k, \quad k = 2,\dots,m-1$$
(if such points exist),

$x_0^{(i-1)}(t_j) = A_{i,j} = y_0^{(i-1)}(t_j)$, $i = 1,\dots,r_j-1$, $j = 1,m$ (if $r_1, r_m \geqq 2$), $(-1)^{n+r_1-1}[x_0^{(r_1-1)}(t_1) - A_{r_1,1}] \geqq 0 \geqq (-1)^{n+r_1-1}[y_0^{(r_1-1)}(t_1) - A_{r_1,1}]$,

$$x_0^{(r_m-1)}(t_m) \geqq A_{r_m,m} \geqq y_0^{(r_m-1)}(t_m).$$

2. $r_1 = n-1$, $r_2 = 1$, $m = 2$, $L(x) = x^{(n)}$ and the lower and the upper solutions $x_0$, $y_0$, respectively, of

$$x^{(n)} = 0 \tag{12'}$$

satisfy the boundary conditions

$$y_0^{(i-1)}(t_1) \leqq A_{i,1} \leqq x_0^{(i-1)}(t_1), \quad i = 1,\dots,n-1,$$

$$y_0(t_2) \leqq A_{1,2} \leqq x_0(t_2).$$

On the basis of Lemma 3, Theorem 1 generalizes Theorem 3.1 from [6], pp. 672-673.

In some cases the uniqueness of the solution of the BVP (1), (2) can be ensured by the Kellog uniqueness theorem [3]. Since its proof can be based on the theory of the Leray-Schauder degree in locally convex spaces where the degree is defined also for unbounded sets ([2], p. 101), a slightly modified version of that theorem is given here as

Lemma 4. Let $X$ be a real Banach space with a convex open (not necessarily bounded) subset $D$ and let $F: \bar{D} \to \bar{D}$ be a continuous mapping which is continuously Fréchet differentiable on $D$ and for which $\overline{F(\bar{D})}$ is compact. Suppose that

(a) for each $x \in D$, 1 is not an eigenvalue of $F'(x)$, and

(b) for each $x \in \partial D$, $x \neq F(x)$.

Then $F$ has a unique fixed point.

Theorem 2. Suppose that

1. there exist two functions $x_0, y_0 \in C_n([t_1,t_2])$, $a < t_1 < t_2 < b$, such that $y_0(t) < x_0(t)$ $(t \in [t_1,t_2])$ and $y_0^{(i-1)}(t_1) < x_0^{(i-1)}(t_1)$, $i = 1,\dots,n-1$,

2. $f$, together with $\frac{\partial f}{\partial x_i}$, $i = 0,1,\dots,n-1$, is continuous on the set $W_1 = \{(t,x,x_1,\dots,x_{n-1}): y_0(t) \leqq x \leqq x_0(t), \ -\infty < x_i < \infty,$

$i = 1,\dots,n-1, \quad t_1 \leqq t \leqq t_2\}$,

3.

(13) $\quad x_0^{(n)}(t) - f(t,x,x_1,\dots,x_{n-1}) < 0 < y_0^{(n)}(t) - f(t,x,x_1,\dots,x_{n-1})$

on $W_1$,

4. the BVP

$$y^{(n)} = \sum_{i=0}^{n-1} \frac{\partial f[t,x(t),x'(t),\dots,x^{(n-1)}(t)]}{\partial x_i} y^{(i)},$$

(14') $\quad y^{(i-1)}(t_1) = 0, \quad i = 1,\dots,n-1, \quad y(t_2) = 0$

has only the trivial solution for all $x \in C_{n-1}([t_1,t_2])$ such that $y_0(t) < x(t) < x_0(t), \quad t \in [t_1,t_2]$.

Then for all numbers $A_{i,1}$, $i = 1,\dots,n-1$, $A_{1,2}$ such that $y_0^{(i-1)}(t_1) < A_{i,1} < x_0^{(i-1)}(t_1)$, $i = 1,\dots,n-1$, $y_0(t_2) < A_{1,2} < x_0(t_2)$ there exists a unique solution of the problem

(12) $\quad x^{(n)} = f(t,x,x',\dots,x^{(n-1)})$

(14) $\quad x^{(i-1)}(t_1) = A_{i,1}, \quad i = 1,\dots,n-1, \quad x(t_2) = A_{1,2}$.

Proof. Consider the ordered Banach space $D_{n-1} = C_{n-1}([t_1,t_2])$ with the norm and ordering similar to those in $C_{n-1}$. Let $D = \{x \in D_{n-1}: y_0(t) < x(t) < x_0(t), \quad t_1 \leqq t \leqq t_2\}$. $D$ is open and convex in $D_{n-1}$. Then $\bar{D} = [x_0,y_0]$ (we remind that $x \leqq y$ in $D_{n-1}$ iff $x(t) \leqq y(t)$ in $[t_1,t_2]$). Lemma 3 gives that the assumptions 1 and 3 of the theorem imply $(A_3)$ and (10). $(A_1)$ is ensured by the assumption 2. $(A_2)$ is clearly satisfied. Hence, from the proof of Theorem 1 it follows that the mapping $T_1$ which corresponds to the problem (12), (14) maps $\bar{D}$ into itself, is continuous and $\overline{T_1(\bar{D})}$ is compact. The Fréchet derivative of $T_1$ at $x \in D$ is

$$T'(x)(h(t)) = \int_{t_1}^{t_2} G_1(t,s) \left[\sum_{i=0}^{n-1} \frac{\partial f[s,x(s),x'(s),\dots,x^{(n-1)}(s)]}{\partial x_i} \cdot h^{(i)}(s)\right] ds, \qquad (t \in [t_1,t_2])$$

where $h \in D_{n-1}$ and $G_1$ is the Green function of (12'), (14'). $T_1'(x)$ continuously depends on $x$ in $D$. Condition (a) in Lemma 4 means that the linear operator $T_1'(x)$ has no nontrivial fixed point which is equivalent to the assumption 4. The condition (b) follows from $T_1(\bar{D}) \subset D$ which is guaranteed by (13). Thus all assumptions of Lemma 4 being satisfied, by this lemma Theorem 2 is true.

In the sequel a special case of (1) will be considered, namely

$$(15) \qquad L(x) = f(t,x),$$

where $f$ shows the following monotonicity property.

$(A_4)$ The function $f = f(t,x)$ is nondecreasing in $x$ for each $t \in [t_k, t_{k+1}]$, $k \in H_1$, and nonincreasing in $x$ for each $t \in [t_k, t_{k+1}]$, $k \in H_2$.

With respect to (15), the operator $T$ defined by (7) which now has the form $T(x)(t) = w(t) + \int_{t_1}^{t_m} G(t,s)f[s,x(s)]\,ds \quad (t \in [t_1,t_m])$ can be considered either as a mapping from $C_{n-1}$ into itself or as a mapping from $C$ into $C$. In both cases it is continuous and compact. The advantage of using $C$ lies in the fact that in this space an interval is a bounded set. The condition $(A_4)$ implies that $T$ is isotone which means that if $x \leqq y$ are two comparable elements of $C$ or $C_{n-1}$, then $T(x) \leqq T(y)$. When $x$ and $y$ are a lower and an upper solution of (15), $(A_4)$ gives (10), and by Theorem 1 the following lemma is true.

<u>Lemma 5.</u> Let the differential equation (15) satisfy the conditions $(A_1) - (A_4)$. Then the BVP (15), (2) has a solution $x \in [x_0, y_0]$.

Remarks. 1. By Theorem 2, [7], there exist a least $\bar{x}$ and a greatest solution $\bar{y}$ of (15), (2) in the interval $[x_0, y_0]$. They can be obtained by an iterative procedure starting with $x_0$ and $y_0$, respectively.

2. In Theorem 1 and Lemma 5 instead of $(A_2)$ it suffices to assume that (1') is disconjugate in $[t_1, t_m]$. On the basis of Lemma 7, [1], p. 93, the latter assumption implies that (1') is disconjugate in a greater interval $(a,b) \supset [t_1, t_m]$ and thus, the use of $(A_2)$ does not weaken the results.

<u>Theorem 3.</u> Assume that (15) satisfies the conditions $(A_1) - (A_3)$ and

$$(16) \qquad \begin{aligned} f(t,x) - f(t,\bar{x}) &\geqq -p(t)(x-\bar{x}) \quad (x \geqq \bar{x},\ t \in [t_k, t_{k+1}],\ k \in H_1) \\ f(t,x) - f(t,\bar{x}) &\leqq -p(t)(x-\bar{x}) \quad (x \geqq \bar{x},\ t \in [t_k, t_{k+1}],\ k \in H_2) \end{aligned}$$

where $p \in L(t_1, t_m)$ and it is such that the differential equation

$$(17) \qquad L(x) + p(t)x = 0$$

is disconjugate in $[t_1, t_m]$.

Then the BVP (15), (2) has the least $\bar{x}$ and the greatest solution $\bar{y}$ in $[x_0, y_0]$.

Proof. Defining the operator $L_1(x) = L(x) + p(t)x \quad (x \in$

$\in L^{(n)}(t_1,t_m))$ and $f_1(t,x) = f(t,x) + p(t)x$ $(t \in [t_1,t_m],$ $x \in \in (-\infty,\infty))$ we can write (15) in the form

$$(15_1) \qquad L_1(x) = f_1(t,x).$$

By (17), (16) and Remark 2, $(15_1)$ satisfies $(A_1)$ - $(A_4)$. Hence Lemma 5, together with Remark 1, brings the result.

Remark. By Proposition 9, [1], p. 95, there exists a $\delta > 0$ depending on (1') as well as on $[t_1,t_m]$ such that for $|p(t)| < \delta$ on $[t_1,t_m]$ the equation (17) is disconjugate.

References


[1] W.A. Coppel, Disconjugacy, Springer-Verlag, Berlin-Heidelberg-New York 1971.

[2] K. Deimling, Nichtlineare Gleichungen und Abbildungsgrade, Springer-Verlag, Berlin-Heidelberg-New York 1974.

[3] R.B. Kellog, Uniqueness in the Schauder fixed point theorem, Proc. Amer. Math. Soc. 60 (1976), 207-210.

[4] I.T. Kiguradze, Some singular boundary value problems for ordinary differential equations. Izdat. Tbilisskogo Univ., Tbilisi 1975 (Russian).

[5] A.Yu. Levin, Non-oscillation of solutions of the equation $x^{(n)}+p_1(t)x^{(n-1)}+\ldots+p(t)x=0$, Uspehi mat. nauk XXIV, 2(146) (1969), 44-96 (Russian)

[6] K. Schmitt, Boundary value problems and comparison theorems for ordinary differential equations, Siam J. Appl. Math., 26 (1974), 670-678.

[7] V. Šeda, Two remarks on boundary value problems for ordinary differential equations, J. Differential Equations (in print).



Author's address: Department of Mathematical Analysis, Comenius University, 816 31 Bratislava, Czechoslovakia.

# ABSTRACT CAUCHY PROBLEM

M. Sova, Praha

The aim of this lecture is to present some results from the theory of the Cauchy problem for linear differential equations in Banach spaces with unbounded coefficients.

## I. PRELIMINARIES

We denote by $E$ an arbitrary Banach space over the complex number field $C$. By an operator we always mean a linear operator acting on a linear subspace of $E$ into $E$.

For a function $f:(0,\infty)\to E$ we introduce the notion of the r-th integral of $f$ by the following definition: if $f$ is integrable on bounded subintervals of $(0,\infty)$, then we put for $t>0$ and $r\in\{1,2,\dots\}$,

$$\underline{|r}\int_0^t f(\tau)d\tau = \int_0^t\int_0^{\tau_1}\dots\int_0^{\tau_{r-1}} f(\tau_r)d\tau_1 d\tau_2\dots d\tau_r .$$

Moreover, for $t>0$ we shall write $\underline{|0}\int_0^t f(\tau)d\tau = f(t)$.

## II. SETTING OF THE CAUCHY PROBLEM

Indispensable definitions and properties connected with the basic notion of correctness of the abstract Cauchy problem are shortly summarized in this section.

Let $A_1,A_2,\dots,A_n$ $(n\in\{1,2,\dots\})$ be an arbitrary n-tuple of linear operators in $E$ which will be fixed throughout the whole lecture.

A function $u:(0,\infty)\to E$ will be called a solution of the Cauchy problem for $A_1,A_2,\dots,A_n$ if

($S_1$) $u$ is n-times differentiable on $(0,\infty)$,

($S_2$) $u^{(n-i)}(t)\in D(A_i)$ for $t>0$ and $i\in\{1,2,\dots,n\}$,

($S_3$) the functions $A_iu^{(n-i)}$ are continuous on $(0,\infty)$ and bounded on $(0,1)$ for every $i\in\{1,2,\dots,n\}$,

($S_4$) $u^{(n)}(t)+A_1u^{(n-1)}(t)+\dots+A_nu(t)=0$ for $t>0$,

($S_5$) $u(0_+)=u'(0_+)=\dots=u^{(n-2)}(0_+)=0$, $u^{(n-1)}(0_+)$ exists.

The notion of a solution in our sense is at the first sight restrictive because we consider only special initial values. However, it turned out that it is sufficiently general for the study of the

Cauchy problem since the solutions with arbitrary initial values can be calculated by means of these special solutions.

We shall say that the Cauchy problem for $A_1, A_2, \dots, A_n$ is determined if every solution $u$ satisfying $u^{(n-1)}(0_+) = 0$ is identically zero on $(0, \infty)$.

Further, we shall say that the Cauchy problem for $A_1, A_2, \dots, A_n$ is extensive if there exists a dense subset $D \subseteq E$ such that for every $x \in D$ there is a solution $u$ with $u^{(n-1)}(0_+) = x$.

Now let $m$ be a nonnegative integer. Then the Cauchy problem for $A_1, A_2, \dots, A_n$ will be called correct of class $m$ if

(I) it is extensive,

(II) there exist nonnegative constants $M, \omega$ so that for every solution $u$, for every $t > 0$ and $i \in \{1, 2, \dots, n\}$,

$$\left\| \lfloor m+1 \int_0^t A_i u^{(n-i)}(\tau)\, d\tau \right\| \leqq M\, e^{\omega t} \| u^{(n-1)}(0_+) \| .$$

Finally, the Cauchy problem for $A_1, A_2, \dots, A_n$ will be called correct if it is correct of class $m$ for some nonnegative integer $m$.

The operators $A_1, A_2, \dots, A_n$ will be sometimes called the generating operators (of the Cauchy problem associated with them).

It follows from the condition (II) that

(II′) $\left\| \lfloor m \int_0^t u^{(n-1)}(\tau)\, d\tau \right\| \leqq (1 + nM) e^{\omega t} \| u^{(n-1)}(0_+) \|$ for every $t > 0$.

If $A_1 = A_2 = \dots = A_{n-1} = 0$, then the conditions (II) and (II′) are equivalent, but in the general case, it is necessary to introduce the condition (II) in a more complicated form to get a satisfactory theory.

Theorem 1. Every correct Cauchy problem is determined as well.

Proof. Immediately from the property (II′). :::

Our definition of correctness was motivated by the classical Hadamard's considerations on the correctness of the initial value problem and by certain well-known special cases which have been examined lately. The first and second order Cauchy problems were studied in a different setting as generation problems for operator semigroups, distribution operator semigroups and cosine and sine operator functions (see [1] - [6]). Further, the Timoshenko equation of transverse vibrations of a beam or a plate led to the study of the fourth order Cauchy problem and its correctness (cf. [7]).

Let $u, h:(0,\infty) \to E$. The function $u$ will be called a <u>response to the excitation</u> $h$ <u>for the Cauchy problem for</u> $A_1, A_2, \ldots, A_n$ if

($R_1$) $u$ is n-times differentiable on $(0,\infty)$,
($R_2$) $u^{(n-i)}(t) \in D(A_i)$ for every $t > 0$ and $i \in \{1,2,\ldots,n\}$,
($R_3$) the functions $A_i u^{(n-i)}$ are continuous on $(0,\infty)$ and bounded on $(0,1)$ for every $i \in \{1,2,\ldots,n\}$,
($R_4$) $u^{(n)}(t)+A_i u^{(n-1)}(t)+\ldots+A_n u(t) = h(t)$ for every $t > 0$,
($R_5$) $u(0_+) = u'(0_+) = \ldots = n^{(n-1)}(0_+) = 0$.

Further, we define the space $L_{loc}((0,\infty),E)$ as the space of all functions $f:(0,\infty) \to E$ integrable over bounded subsets of $(0,\infty)$ and topologized by the natural Fréchet topology determined by the seminorms $\int_0^T \|f(\tau)\| d\tau$, $T > 0$.

<u>Theorem</u> 2. <u>Let</u> $m$ <u>be a nonnegative integer</u>. <u>If the operators</u> $A_1, A_2, \ldots, A_n$ <u>are closed</u>, <u>then the following two statements are equivalent</u>:

(A) <u>the Cauchy problem for</u> $A_1, A_2, \ldots, A_n$ <u>is correct of class</u> $m$,

(B) (I) <u>there is a dense subset</u> $J$ <u>of</u> $L_{loc}((0,\infty),E)$ <u>so that for every</u> $h \in J$ <u>there exists a response</u> $u$ <u>to the excitation</u> $h$,

(II) <u>there exist nonnegative constants</u> $M, \omega$ <u>so that for every</u> $u, h$ <u>such that</u> $u$ <u>is a response to the excitation</u> $h$ <u>and</u> $h \in L_{loc}((0,\infty),E)$, <u>for every</u> $t > 0$ <u>and</u> $i \in \{1,2,\ldots,n\}$,

$$\| \underline{|m+1} \int_0^t A_i u^{(n-i)}(\tau) d\tau \| \leq M e^{\omega t} \int_0^t \|h(\tau)\| d\tau .$$

Proof. See [8], p. 124 and 128. :::

## III. SPECTRAL PROPERTIES OF THE CAUCHY PROBLEM

Our aim in this section will be to clarify the relations between the correctness and the spectral properties of the generating operators.

Under spectral properties of operators $A_1, A_2, \ldots, A_n$ we understand various properties of the operator polynomial $z^n I + z^{n-1} A_1 + \ldots + A_n$, $z \in C$, which will be called the <u>spectral polynomial of the operators</u> $A_1, A_2, \ldots, A_n$ and denoted by $P(.;A_1, A_2, \ldots, A_n)$.

We introduce some notions and notation related to the spectral polynomial of the operators $A_1,A_2,\dots,A_n$.

We denote by $\rho(A_1,A_2,\dots,A_n)$ the set of all $z \in C$ for which the operator $P(z;A_1,A_2,\dots,A_n)$ is one-to-one and its inverse is everywhere defined and bounded. This inverse $P(z;A_1,A_2,\dots,A_n)^{-1}$ will be denoted by $R(z;A_1,A_2,\dots,A_n)$. Finally, we put $\sigma(A_1,A_2,\dots,A_n) = C \setminus \rho(A_1,A_2,\dots,A_n)$.

The set $\rho(A_1,A_2,\dots,A_n)$ is called the <u>resolvent set</u> and $\sigma(A_1,A_2,\dots,A_n)$ the <u>spectrum of the operators</u> $A_1,A_2,\dots,A_n$. The function $R(.;A_1,A_2,\dots,A_n)$, defined on $\rho(A_1,A_2,\dots,A_n)$, will be called <u>the resolvent function of these operators</u>.

In this notation, the specification of operators $A_1,A_2,\dots,A_n$ is sometimes omitted in proofs.

It should be still noted that the above introduced notions and notation do not coincide with those usually defined for one operator. They are slightly modified.

<u>Theorem</u> 3. <u>If the operators</u> $A_1,A_2,\dots,A_n$ <u>are closed</u>, <u>then</u>
(a) <u>the set</u> $\rho(A_1,A_2,\dots,A_n)$ <u>is open</u>,
(b) <u>the function</u> $R(.;A_1,A_2,\dots,A_n)$ <u>is analytic on the set</u> $\rho(A_1,A_2,\dots,A_n)$,
(c) <u>the operators</u> $A_iR(z;A_1,A_2,\dots,A_n)$ <u>are everywhere defined and bounded for every</u> $z \in \rho(A_1,A_2,\dots,A_n)$ <u>and</u> $i \in \{1,2,\dots,n\}$,
(d) <u>for every</u> $i \in \{1,2,\dots,n\}$, <u>the function</u> $A_iR(.;A_1,A_2,\dots,A_n)$ <u>is analytic on the set</u> $\rho(A_1,A_2,\dots,A_n)$.

Proof. See [9], p. 233-236. :::

<u>Theorem</u> 4 (<u>uniqueness for the Cauchy problem</u>). <u>We assume that there exist nonnegative constants</u> $\omega$, d, D <u>such that</u>
($\alpha$) <u>every</u> $\lambda > \omega$ <u>belongs to</u> $\rho(A_1,A_2,\dots,A_n)$,
($\beta$) $\|R(\lambda;A_1,A_2,\dots,A_n)\| \leq De^{d\lambda}$ <u>for every</u> $\lambda > \omega$.

<u>Under these assumptions</u>, <u>the Cauchy problem for</u> $A_1,A_2,\dots,A_n$ <u>is determined</u>.

Proof. See [10], p. 38. :::

In what follows we need still the notion of <u>higher domains of operators</u> $A_1,A_2,\dots,A_n$ which will be defined by induction in the following way:

$D_0(A_1,A_2,\dots,A_n) = E$,
$D_{k+1}(A_1,A_2,\dots,A_n) = \{x : x \in D(A_1) \cap D(A_2) \cap \dots \cap D(A_n)$ and $A_ix \in D_k(A_1,A_2,\dots,A_n)$ for every $i \in \{1,2,\dots,n\}\}$, $k \in \{0,1,\dots\}$.

Finally, we define $D_\infty(A_1,A_2,\dots,A_n)$ as the intersection $\bigcap_{k=0}^{\infty} D_k(A_1,A_2,\dots,A_n)$.

Theorem 5 (existence for the Cauchy problem). Let $m$ be a nonnegative integer. We assume that

($\alpha$) the operators $A_1,A_2,\dots,A_n$ are closed,

($\beta$) the set $D_{m+1}(A_1,A_2,\dots,A_n)$ is dense in $E$,

($\gamma$) there exist nonnegative constants $M, \omega$ so that

(I) $(\omega,\infty) \subseteq \rho(A_1,A_2,\dots,A_n)$,

(II) $$\left\| \frac{d^p}{d\lambda^p} \lambda^{n-m-i-1} A_i R(\lambda;A_1,A_2,\dots,A_n) \right\| \leq \frac{Mp!}{(\lambda-\omega)^{p+1}}$$

for every $\lambda > \omega$, $i \in \{1,2,\dots,n\}$ and $p \in \{0,1,\dots\}$.

Under these assumptions, the Cauchy problem for $A_1,A_2,\dots,A_n$ is correct of class $m$.

Proof. See [10], p. 46. :::

The preceding existence Theorem 5 for the Cauchy problem can be essentially converted in the following way:

Theorem 6 (converse). Let $m$ be a nonnegative integer. We assume that

($\alpha$) the operators $A_1,A_2,\dots,A_n$ are closed,

($\beta$) the Cauchy problem for $A_1,A_2,\dots,A_n$ is correct of class $m$.

Under these assumptions,

(a) the set $D_1(A_1,A_2,\dots,A_n)$ is dense in $E$,

(b) the condition ($\gamma$) of Theorem 4 holds.

Proof. See [10], p. 55. :::

The preceding Theorems 5 and 6 cover the well-known Hille-Yosida theorem on generation of strongly continuous semigroups, the related theorem on generation of cosine and sine operator functions and the Lions theorem on generation of exponential distribution semigroups (cf. [1] - [6]).

The assumption ($\gamma$) of Theorem 5 can be replaced by another condition which does not involve the derivatives of the resolvent function on a real halfaxis but makes use of the behavior of this function on a complex halfplane.

Theorem 7. If the operators $A_1,A_2,\dots,A_n$ are closed, then the condition ($\gamma$) of Theorem 5 is equivalent with

($\gamma'$) there exist nonnegative constants $M, \omega$ so that

(I') $\{z:\mathrm{Re}\, z > \omega\} \subseteq \rho(A_1,A_2,\dots,A_n)$,

(II′) $\|z^{n-m-i-1}A_iR(z;A_1,A_2,\dots,A_n)\| \leq \frac{M}{\mathrm{Re}\,z-\omega}$ <u>for every</u> Rez <u>and</u> $i \in \{1,2,\dots,n\}$,

(III′) $\left\| \int_{\alpha-i\infty}^{\alpha+i\infty} \frac{z^{n-m-i-1}A_iR(z;A_1,A_2,\dots,A_n)}{(1-s(z-\alpha))^r}\,dz \right\| \leq M$ <u>for every</u> $\alpha > \omega$, $s > 0$, $i \in \{1,2,\dots,n\}$ <u>and</u> $r \in \{2,3,\dots\}$.

Proof. We shall need the following

L e m m a . Let $\alpha \in R$, $J \in \{z:\mathrm{Re}\,z \geq \alpha\} \to E$, $k \in \{0,1,\dots\}$. If the function $J$ is continuous for $\mathrm{Re}\,z \geq \alpha$, analytic for $\mathrm{Re}\,z > \alpha$ and $\|J(z)\| \leq K(1+|z|)^k$ for $\mathrm{Re}\,z \geq \alpha$ with $K \geq 0$, then for every $\lambda > \alpha$ and $p \in \{k+1,k+2,\dots\}$,

$$[*] \quad J^{(p)}(\lambda) = (-1)^p \frac{p!}{2\pi i} \int_{\alpha-i\infty}^{\alpha+i\infty} \frac{J(z)}{(\lambda-z)^p}\,dz .$$

The verification of this Lemma is easy by means of the Cauchy integral formula.

Now to the proof itself. We fix $i \in \{1,2,\dots,n\}$ and write $H(z) = z^{n-m-i-1}A_iR(z)$ for $z \in \varrho$.

$(\gamma') \Rightarrow (\gamma)$ According to Theorem 3, our Lemma is applicable with arbitrary $\alpha > \omega$ to corresponding restriction of the function $H$ and then the identity [*] gives immediately the estimate (II) of $(\gamma)$ for $p \in \{1,2,\dots\}$ if we finally let $\alpha \to \omega_+$. For $p = 0$, we deduce the desired estimate from the case $p = 1$ because $H(\lambda) \to 0$ $(\lambda \to \infty)$ according to (II′).

$(\gamma) \Rightarrow (\gamma')$ Fix $z$ in the halfplane $\mathrm{Re}\,z > \omega$.

It is easy to see from the inequality (II) in $(\gamma)$ that, for sufficiently large $\lambda > \omega$, the series $\sum_{k=0}^{\infty} \frac{(z-\lambda)^k}{k!} R^{(p)}(\lambda)$ and $\sum_{k=0}^{\infty} \frac{(z-\lambda)^k}{k!} A_iR^{(p)}(\lambda)$ converge and define $R(z)$ and $A_iR(z)$.

This shows in particular that (I) is true.

On the other hand, after a little calculation, we find that there exists a $\lambda_0 \geq \omega$ such that $\lambda - \omega > |z-\lambda|$ for $\lambda > \lambda_0$. By means of this inequality and of the inequality (II) we can now estimate the above series for $A_iR(z)$ and we easily obtain the desired inequality (II′) letting $\lambda \to \infty$.

The facts proved above together with Theorem 3 enable us to apply again our Lemma with arbitrary $\alpha > \omega$ to corresponding restriction of the function $H$ and now it is a matter of routine to get (III′) from (II) by means of the identity [*]. :::

The following two theorems concern the general notion of correctness.

Theorem 8 (existence for the Cauchy problem). We assume that

($\alpha$) the operators $A_1, A_2, \ldots, A_n$ are closed,

($\beta$) the set $D_\infty(A_1, A_2, \ldots, A_n)$ is dense in $E$,

($\gamma$) there exist nonnegative constants $L$, $l$, $\varkappa$ so that

(I) $\{z : \text{Re}\, z \geq \varkappa\} \subseteq \rho(A_1, A_2, \ldots, A_n)$,

(II) $\|A_i R(z; A_1, A_2, \ldots, A_n)\| \leq L(1+|z|)^l$ for every $\text{Re}\, z > \varkappa$ and $i \in \{1, 2, \ldots, n\}$.

Under these assumptions, the Cauchy problem for $A_1, A_2, \ldots, A_n$ is correct.

**Proof.** Choosing $m$ sufficiently large, for example $m = l+2$, we verify easily that the condition ($\gamma'$) in Theorem 7 is satisfied. Consequently, by Theorem 7, the assumptions of Theorem 5 hold and this implies the assertion of the present theorem. :::

Theorem 9 (converse). We assume that

($\alpha$) the operators $A_1, A_2, \ldots, A_n$ are closed,

($\beta$) the Cauchy problem for $A_1, A_2, \ldots, A_n$ is correct.

Under these assumptions,

(a) the set $D_1(A_1, A_2, \ldots, A_n)$ is dense in $E$,

(b) the condition ($\gamma$) of Theorem 8 holds.

Proof. Immediate consequence of Theorems 6 and 7 ( $l$ may be taken equal to $m$ chosen for Theorem 6). :::

In the following theorem we shall make some a priori restrictions concerning the basic space $E$ and the generating operators $A_1, A_2, \ldots, A_n$. Then the Cauchy problem for these operators is not only always determined and extensive but above all, its correctness is fully specified merely by the location of the spectrum of generating operators. All the new notions used (normal operator, abelian system, spectral measure, spectral integral) can be found in [11] (see in particular Chap. VII, VIII and X).

Theorem 10. If the operators $A_1, A_2, \ldots, A_n$ are normal and form an abelian system in a Hilbert space $E$, then the Cauchy problem for $A_1, A_2, \ldots, A_n$ is always determined and extensive.

Moreover, it is correct if and only if there exists a constant $\omega$ such that $\sigma(A_1, A_2, \ldots, A_n) \subseteq \{z : \text{Re}\, z \leq \omega\}$.

Proof. We denote by $\mathcal{B}$ the family of all Borel subsets of $C$.

According to [11] (in particular Chap. X), we can find a spectral measure $\mathcal{E}$ on $\mathcal{B}$ and Borel measurable functions $a_1, a_2, \ldots, a_n : C \to C$ so that $A_i x = \int_C a_i(\sigma)\, \mathcal{E}(d\sigma)x$ for $x \in D(A_i)$ in the sense of spectral integration described in [11], Chap. VII. In the rest of this proof we shall frequently use this integration without special reference to [11].

We begin with proving that our Cauchy problem is determined and extensive. Let $\mathcal{B}_0$ be the family of bounded sets from $\mathcal{B}$. It is easy to prove that the set $Q = \{x : \mathcal{E}(X)x = x$ for some $X \in \mathcal{B}_0\}$ is dense in $E$. Moreover, $Q \subseteq D_\infty(A_1, A_2, \ldots, A_n)$. On the other hand, the operators $A_i\,\mathcal{E}(X)$ are bounded and $A_i\,\mathcal{E}(X) \supseteq \mathcal{E}(X)A_i$ for $X \in \mathcal{B}_0$. These facts enable us to construct easily a solution for every $x \in Q$ and so to prove that the problem is extensive. To prove that it is determined we use Theorem 4 for bounded operators $A_i\,\mathcal{E}(X)$, $X \in \mathcal{B}_0$, (since in this case it is clearly valid) and the fact that there is a sequence $X_k \in \mathcal{B}_0$ such that $\mathcal{E}(X_k)x \to x$ for any $x \in E$.

Now to the proof of the last assertion of our theorem.

We shall write $p(z,s) = z^n + a_1(s)z^{n-1} + \ldots + a_n(s)$ for $z, s \in C$. Further for $X \in \mathcal{B}$ we put $K(X) = \{z : p(z,s) = 0$ for some $s \in C \setminus X\}$.

We first need to prove that

[*] there exists $N \in \mathcal{B}$ such that $\mathcal{E}(X) = 0$ and $K(N) \subseteq \subseteq \sigma(A_1, A_2, \ldots, A_n)$.

To this aim, let us denote $N_z = \{s : p(z,s) < \|R(z)\|^{-1}\}$ for $z \in \rho = \rho(A_1, A_2, \ldots, A_n)$.

It is clear that $N_z \in \mathcal{B}$ for any $z \in \rho$. Now we shall show that $\mathcal{E}(N_z) = 0$ for $z \in \rho$. Proceeding indirectly, we fix $z \in \rho$ such that $\mathcal{E}(N_z) \neq 0$. We have shown above that $D_\infty(A_1, A_2, \ldots, A_n)$ is dense in $E$ and hence we can find an $x \neq 0$, $x \in D_\infty(A_1, A_2, \ldots, A_n)$ and $\mathcal{E}(N_z)x = x$. Since then $C \setminus N_z$ is a null set for the measure $\|\mathcal{E}(.)x\|^2$ we obtain $\|P(z)x\| =$

$$= \left\|\int_C p(z,\sigma)\, \mathcal{E}(d\sigma)x\right\| = \sqrt{\int_C |p(z,\sigma)|^2\, \|\mathcal{E}(d\sigma)x\|^2} =$$

$$= \sqrt{\int_{N_z} |p(z,\sigma)|^2\, \|\mathcal{E}(d\sigma)x\|^2} < \sqrt{\int_{N_z} \|R(z)\|^{-2}\, \|\mathcal{E}(d\sigma)x\|^2} =$$

$= \|R(z)\|^{-1}\, \|\mathcal{E}(N_z)x\| = \|R(z)\|^{-1}\|x\|$. But this inequality is contradictory since it implies that $\|R(z)\|^{-1}\|x\| = \|R(z)\|^{-1}\|R(z)P(z)x\| \leq$

$\|R(z)\|^{-1}\|R(z)\|\,\|P(z)x\| = \|P(z)x\| < \|R(z)\|^{-1}\|x\|$.

Now we put $N = \bigcup N_z$ where $z$ runs through $z \in \rho$ with rational real and imaginary parts. It is immediate from the preceding result that $N \in \mathcal{B}$ and $\mathcal{E}(N) = 0$. With regard to the continuity of $p(.,s)$ we obtain further that $|p(z,s)| \geqq \|R(z)\|^{-1}$, i.e. in particular $p(z,s) \neq 0$, for $z \in \rho$ and $s \in C \setminus N$. But this implies that $K(N) \cap \rho = \emptyset$, i.e. $K(N) \subseteq \sigma$ which proves [*].

Let our Cauchy problem be correct. Then by Theorem 9, there is an $\omega$ such that $\rho(A_1,A_2,\ldots,A_n) \subseteq \{z:\mathrm{Re}z \leqq \omega\}$.

Conversely, let $\omega$ be such that $\rho(A_1,A_2,\ldots,A_n) \subseteq \{z:\mathrm{Re}z \leqq \leqq \omega\}$. According to [*] we can find an $N \in \mathcal{B}$ so that $\mathcal{E}(N) = 0$ and $K(N) \subseteq \{z:\mathrm{Re}z \leqq \omega\}$. On the other hand, we can write $p(z,s) = = (z-z_1(s))(z-z_2(s))\ldots(z-z_n(s))$ for $z, s \in C$. For $s \in C \setminus N$ we obtain that $z_i(s) \in K(N)$, i.e. $\mathrm{Re}z_i \leqq \omega$. Consequently $\left|\frac{1}{z-z_i(s)}\right| \leqq \leqq \frac{1}{\mathrm{Re}z-\omega}$ for every $s \in C \setminus N$ and $\mathrm{Re}z > \omega$. This yields $\left|\frac{z_i(s)}{z-z_i(s)}\right| \leqq 1 + \frac{|z|}{\mathrm{Re}z-\omega}$ for $s \in C \setminus N$ and $\mathrm{Re}z > \omega$. Using Vièta's formulas expressing $a_i(s)$ in terms of $z_i(s)$ we obtain that there are constants $L \geqq 0$ and $l \in \{0,1,\ldots\}$ such that $\left|\frac{a_i(s)}{p(z,s)}\right| \leqq \leqq L(1+|z|)^l$ for every $s \in C \setminus N$ and $\mathrm{Re}z > \omega+1$. From this estimate we deduce easily that $\|A_iR(z)x\| = \left\| \int_{C \setminus N} \frac{a_i(\sigma)}{p(z,\sigma)} \mathcal{E}(d\sigma)x \right\| \leqq$

$\leqq L(1+|z|)^l\|x\|$ for $x \in E$ and $\mathrm{Re}z > \omega+1$. This fact together with the above proved density of $D_\infty(A_1,A_2,\ldots,A_n)$ shows that Theorem 8 is applicable and hence our Cauchy problem is correct.

Finally, let us remark that a more accurate result can be proved, namely that our problem is in fact correct of class n-1, but this requires a little lengthy estimates. :::

The a priori restrictions in the preceding theorem as to the basic space $E$ and the generating operators $A_1,A_2,\ldots,A_n$ can be weakened on the ground of the theory of <u>scalar-type operators</u> in Banach spaces, extensively studied in [12].

<u>Theorem</u> 11. <u>The preceding Theorem</u> 10 <u>is valid also in an arbitrary Banach space</u> $E$ <u>if we replace normal operators by scalar-type ones</u>.

Proof. Similar argument as in Theorem 10 can be used. :::

## IV. ASYMPTOTIC PROPERTIES OF THE CAUCHY PROBLEM

In this section, we shall deal with examining the growth of solutions of the Cauchy problem on the whole time halfaxis in dependence upon their boundedness on finite time intervals.

It is known that in certain special cases, the condition of exponential growth of solutions, required in our definition of correctness of the Cauchy problem, can be replaced by a formally weaker condition of local boundedness. This concerns the cases $n=1$, $m=0$ and $n=2$, $m=0$ with $A_1=0$ where it is possible to use the functional equations either for operator semigroups or for cosine operator functions to obtain exponential estimates of growth from local boundedness. However, in general such an approach is not available because we do not know any functional equation for solutions of the general Cauchy problem. Despite of this, we have succeeded recently to prove a general theorem of this type for the class of correctness $m = 0$.

Theorem 12. We assume that

($\alpha$) the operators $A_1, A_2, \dots, A_n$ are closed,

($\beta$) the Cauchy problem for $A_1, A_2, \dots, A_n$ is extensive,

($\gamma$) for every $T > 0$, there is a nonnegative constant $K$ so that for every solution $u$, every $0 < t \leqq T$ and $i \in \{1,2,\dots,n\}$,

$$\left\| \int_0^t A_i u^{(n-i)}(\tau)\,d\tau \right\| \leqq K \| u^{(n-1)}(0_+) \| .$$

Under these assumptions, there exist nonnegative constants $M$, $\omega$ so that for every solution $u$, every $t > 0$ and $i \in \{1,2,\dots,n\}$,

$$\left\| \int_0^t A_i u^{(n-i)}(\tau)\,d\tau \right\| \leqq M e^{\omega t} \| u^{(n-1)}(0_+) \| .$$

In other words, we can say that the Cauchy problem for $A_1, A_2, \dots, A_n$ is correct of class $m = 0$.

Proof. We denote by $A_0$ the identity operator.

It is easy to prove from our assumptions that there exists a function $W: (0,\infty) \times E \to E$ such that

(1) $W$ is continuous in both variables, linear in the second,

(2) $\displaystyle\sum_{j=0}^{n} A_j \lfloor j \int_0^t W(\tau,x)\,d\tau = x$ for $x \in E$ and $t > 0$,

(3) $W(0_+,x) = x$ for every $x \in E$,

(4) $\| A_j \lfloor j \int_0^t W(\tau,x)\,d\tau \| \leqq K(T) \|x\|$ for $x \in E$, $0 < t \leqq T$ and $j \in \{0,1,\dots,n\}$ with some constant $K(T)$.

Further we choose a function $\vartheta : (0, \infty) \to R$ such that

(5) $\vartheta(t) = 1$ for $0 < t \leqq 1$, $0 \leqq \vartheta(t) \leqq 1$ for $1 < t \leqq 2$ and $\vartheta(t) = 0$ for $t > 2$,

(6) $\vartheta$ is n-times continuously differentiable on $(0, \infty)$.

Let us multiply (2) by $\vartheta$ and consider the Laplace transform to the identity so obtained. Simple but careful calculations based on current properties of the Laplace transform (see [10], p. 16) and on interchange of closed operators and integrals (see [10], p. 10) enable us to infer, with regard to the properties ($\alpha$), (1), (3), (5) and (6), that for every $x \in E$ and $\lambda > 0$,

$$(7) \quad \sum_{j=0}^{n} \lambda^{n-j} A_j \Big( \int_0^\infty e^{-\lambda\tau} \vartheta(\tau) \Big( \underline{|n} \int_0^\tau W(\sigma, x) d\sigma \Big) d\tau \Big) = x -$$

$$- \Big[ - \int_0^\infty e^{-\lambda\tau} \vartheta'(\tau) d\tau\, x + \sum_{j=0}^{n-1} (-1)^{n-j} A_j \cdot$$

$$\cdot \Big( \sum_{k=0}^{n-j-1} \binom{n-j}{k} (-\lambda)^{k+1} \int_0^\infty e^{-\lambda\tau} \vartheta^{(n-j-k)}(\tau) \Big( \underline{|n} \int_0^\tau W(\sigma, x) d\sigma \Big) d\tau \Big) \Big].$$

Let us denote by $G(\lambda)x$ the left hand side of (7) and by $H(\lambda)x$ the expression in square brachets on the right hand side of (7). Then

$$(8) \quad \sum_{j=0}^{n} \lambda^{n-j} A_j G(\lambda) = I - H(\lambda) \quad \text{for} \quad \lambda > 0 .$$

The inequalities (4) together with some assumed or already proved properties enable us to deduce analogous estimates for higher iterated integrals of solutions. Using these estimates and the properties (5) and (6) we obtain, after a little lenghty calculations, that there is a $K \geqq 0$ such that

$$(9) \quad \Big\| \frac{d^p}{d\lambda^p} \lambda^{n-i-1} G(\lambda) \Big\| \leqq \frac{Kp!}{\lambda^{p+1}}, \quad \Big\| \frac{d^p}{d\lambda^p} H(\lambda) \Big\| \leqq \frac{Kp!}{\lambda^{p+1}} \quad \text{for every}$$

$\lambda > 0$, $i \in \{1, 2, \dots, n\}$ and $p \in \{0, 1, \dots\}$.

From the second inequality in (8) we see that there is a constant $\omega_0 \geqq 0$ such that $\|H(\lambda)\| < 1$ for $\lambda > \omega_0$ which implies together with (8) that

$$(10) \quad (\omega_0, \infty) \subseteq \rho(A_1, A_2, \dots, A_n),$$

$$(11) \quad R(\lambda; A_1, A_2, \dots, A_n) = \Big( \sum_{j=0}^{n} \lambda^{n-j} A_j \Big)^{-1} = G(\lambda)(I - H(\lambda))^{-1} \quad \text{for}$$

$\lambda > \omega_0$.

Using the lemma from [13], p. 49, we obtain from (9) and (11) that there are two constants $M \geqq 0$ and $\omega \geqq \omega_0$ such that

(12) $\left\| \frac{d^p}{d\lambda^p} R(\lambda; A_1, A_2, \ldots, A_n) \right\| \leq \frac{Mp!}{(\lambda - \omega)^{p+1}}$ for $\lambda > \omega$ and $p \in$ $\in \{0,1,\ldots\}$.

Further, we deduce easily from (3) that

(13) $D_1(A_1, A_2, \ldots, A_n)$ is dense in $E$.

Now the properties (10), (12) and (13) permit to apply Theorem 5 with $m = 0$ and the assertion of this theorem completes the proof. :::

It would be desirable to have a theorem analogous to Theorem 12 for higher classes of correctness. However, the following Theorem 13 suggests that such a result will hardly hold because the estimates of growth, obtained in this theorem, do not guarantee the correctness in our sense and moreover, as can be easily shown by examples, cannot be generally improved.

Theorem 13. We assume that

($\alpha$), ($\beta$) as in Theorem 12,

($\gamma$) for every $T > 0$, there exist a nonnegative constant $K$ and a nonnegative integer $r$ so that for every solution $u$, every $0 < t \leqq T$ and $i \in \{1,2,\ldots,n\}$,

$$\left\| \underline{\lfloor r} \int_0^t A_i u^{(n-i)}(\tau) d\tau \right\| \leqq K \| u^{(n-1)}(0_+) \| .$$

Under these assumptions, there exist nonnegative constants $M$, $\omega$ and nonnegative integers $\varkappa$, $m$ so that for every solution $u$, every $\nu \in \{1,2,\ldots\}$, $0 < t \leqq \nu$ and $i \in \{1,2,\ldots,n\}$,

$$\left\| \underline{\lfloor \varkappa\nu + m} \int_0^t A_i u^{(n-i)}(\tau) d\tau \right\| \leqq M e^{\omega t} \| u^{(n-1)}(0_+) \| .$$

Roughly speaking, our Theorem guarantees an exponential growth only of certain linearly increasing (with respect to time) iterated integrals of solutions.

Proof. We begin similarly as in the first part of the proof of Theorem 12. We express the resolvent function $R$ again in the form (11) but for $\lambda$ from a logarithmic domain $\{\lambda : \mathrm{Re}\,\lambda \geqq \alpha \log(1 + |\mathrm{Im}\,\lambda|) + \varkappa\}$ where the constants $\alpha \geqq 0$ and $\varkappa$ depend on $A_1, A_2, \ldots, A_n$ only. In this domain, we obtain easily $\|R(\lambda)\| \leqq K(1 + |\lambda|)^l$, $K \geqq 0$, $l \geqq 0$ (derivatives of $R$ need not be estimated). On the other hand, the iterated integrals of solutions which we have to estimate can be expressed in terms of the resolvent function $R$ by means of a modified Laplace complex inverse integral whose integral path is the boundary of our logarithmic domain. This formula together with the above proved estimate of the resolvent function R yields the required estimates of solutions almost immediately. :::

## References


[1] Hille, E.: Functional analysis and semigroups, 1948.

[2] Yosida, K.: On the differentiability and the representation of one-parameter semi-groups, J. Math. Soc. Japan, 1(1948), 15-21.

[3] Lions, J. L.: Les semi-groupes distributions, Portugal. Math., 19(1960), 141-164.

[4] Sova, M.: Cosine operator functions, Rozprawy Matematyczne, 49(1966).

[5] Sova, M.: Problèmes de Cauchy paraboliques abstraits de classes supérieures et les semi-groupes distributions, Ricerche di Mat., 18(1969), 215-238.

[6] Sova, M.: Encore sur les équations hyperboliques avec petit paramètre dans les espaces de Banach généraux (appendice), Colloquium Math., 25(1972), 155-161.

[7] Sova, M.: On the Timoshenko type equations, Čas. pěst. mat., 100(1975), 217-254.

[8] Sova, M.: Inhomogeneous linear differential equations in Banach spaces, Čas. pěst. mat., 103(1978), 112-135.

[9] Obrecht, E.: Sul problema di Cauchy per le equazioni paraboliche astratte di ordine n, Rend. Sem. Mat. Univ. Padova, 53(1975), 231-256.

[10] Sova, M.: Linear differential equations in Banach spaces, Rozpravy Československé akademie věd, Řada mat. a přír. věd, 85(1975), No 6.

[11] v. Sz.-Nagy, B.: Spektraldarstellung linearer Transformationen des Hilbertschen Raumes, 1942.

[12] Dunford, N., Schwartz, J. T.: Linear operators III, 1971.

[13] Sova, M.: Equations différentielles opérationelles linéaires du second ordre à coefficients constants, Rozpravy Československé akademie věd, Řada mat. a přír. věd, 80(1970), No 7.



Author's address: Matematický ústav ČSAV, Žitná 25, 115 67 Praha 1, Czechoslovakia.

# SOLUTION OF SYMMETRIC POSITIVE SYSTEMS OF DIFFERENTIAL EQUATIONS

U.M.Sultangazin, Alma-Ata

S.K.Godunov and the author showed in 1969 that the hyperbolic system obtained from the one-velocity kinetic equation in $P_{2n+1}$ - approximations of the method of spherical harmonics under boundary conditions of Vladimirov's type is symmetric positive. Writing the system of equations of the method of spherical harmonics in the form of a symmetric system in the sense of Friedrichs together with a proof of dissipativity of the boundary conditions have made it possible to discover new qualitative laws of the theory of spherical harmonics. Under general assumptions concerning the dissipation indicatrix, the author proved weak convergence. A little later V.Skoblikov and A.Akišev studied the problem of strong convergence of the method of spherical harmonics. It is also important that the symmetry of the system and the positivity of the boundary conditions allowed to construct effective computing algorithms for the solution of the three-dimensional system of equations of spherical harmonics. The present paper offers a survey of results of the study of symmetric positive systems which appear in the method of spherical harmonics.

## 1. Formulating the problem

Let $G$ be a convex domain in the three-dimensional Euclidean space $R_3$ whose boundary is a smooth surface $\Gamma$. Let us assume that the surface $\Gamma$ belongs to the class $C^1$ and has a bounded radius of curvature at any point. In the cylindric domain $S_T = [0,T]\times G\times\Omega$ with the base $Q = G\times\Omega$ we consider the following initial-boundary value problem for non-stationary one-velocity kinetic transport equation:

$$(1)\qquad Lu \equiv \frac{\partial u}{\partial t} + \xi\frac{\partial u}{\partial x} + \eta\frac{\partial u}{\partial y} + \zeta\frac{\partial u}{\partial z} + \sigma u - \frac{\sigma_s}{4\pi}\int_\Omega g(\bar\omega,\bar\omega')u\,d\bar\omega' = f ,$$

$$(2)\qquad u(0,\bar r,\bar\omega) = \Phi(\bar r,\bar\omega) ,$$

$$(3)\qquad u(t,r,\omega) = 0 \quad \text{for} \quad (\bar\omega,\bar n)\le 0, \quad \bar r\in\Gamma$$

where $\bar r = (x,y,z)$ are space coordinates, $\omega = (\xi,\eta,\zeta)$ are angular variables which may be written in terms of spherical coordinates as $\xi = \cos\varphi \sin\theta$ , $\eta = \sin\varphi \sin\theta$ , $\zeta = \cos\theta$ , $\Omega = \{\varphi, \theta : 0 \le \varphi \le 2\pi , 0 \le \theta \le \pi\}$ , $u(t,\bar{r},\bar{\omega})$ is the function of distribution of particles, $f(t,\bar{r},\bar{\omega})$ is the source function, $\sigma(\bar{r})$, $\sigma_s(\bar{r})$ are the macroscopic sections characterizing the properties of the medium, $\frac{1}{4\pi} g(\bar{\omega},\bar{\omega}')$ is the dissipation indicatrix, $\bar{\omega} \in \Omega$, $\bar{\omega}' \in \Omega'$ .

Definition 1. A function $u(t,\bar{r},\bar{\omega}) \in C([0,T]; L_2(G\times\Omega))$ is called a generalized solution of the problem (1)-(2) if it satisfies the integral identity

$$(4) \qquad \int_{S_T} L^* vu\,dt\,d\bar{r}\,d\bar{\omega} + \int_{t=0} \Phi(\bar{r},\omega) v(0,\bar{r},\omega)\,d\bar{r}\,d\bar{\omega} = \int_{S_T} vf\,dt\,d\bar{r}\,d\bar{\omega}$$

for all continuously differentiable $v(t,\bar{r},\bar{\omega})$ with

$$(5) \qquad v(T,\bar{r},\bar{\omega}) = 0 ,$$

$$(6) \qquad v(t,\bar{r},\bar{\omega}) = 0 \quad \text{for} \quad (\bar{n},\bar{\omega}) \ge 0, \quad \bar{r} \in \Gamma .$$

The formula (4) which is the basis of the above definition requires that the coefficients of the equation (1) and the initial data fulfil the following minimal conditions:

(a) $\sigma$ , $\sigma_s$ are measurable in $G$ and $\sigma, \sigma_s \in L_2(G)$ ,

(b) $f \in C([0,T] ; L_2(G\times\Omega))$ ,

(c) $\Phi \in L_2(G\times\Omega)$ .

It is easy to see that the integrals in (4) exist provided the conditions (a)-(c) are fulfilled.

Theorem 1. Let us assume that

(i) the coefficients $\sigma(\bar{r})$, $\sigma_s(\bar{r})$ are bounded and fulfil the Lipschitz condition with respect to $x,y,z$ with constants $\sigma_x$ , $\sigma_y$ , $\sigma_z$ , $\sigma_{sx}$ , $\sigma_{sy}$ , $\sigma_{sz}$ , respectively (e.g. the Lipschitz condition in $x$:

$$|\sigma(x',y,z) - \sigma(x'',y,z)| \le \sigma_x |x' - x''|);$$

(ii) the source function $f(t,\bar{r},\bar{\omega})$ is bounded, i.e. $|f(t,\bar{r},\bar{\omega})| \le f_0$ and it fulfils the Lipschitz condition with respect to $t$ and $x,y,z$ with constants $f_t$ ,$f_x$ ,$f_y$ ,$f_z$ , respectively;

(iii) the initial function $\Phi(\bar{r},\bar{\omega})$ is bounded, i.e. $|\Phi(\bar{r},\bar{\omega})| \le \Phi_0$ and it fulfils the Lipschitz condition with respect to $x,y,z$ with constants $\Phi_x$ , $\Phi_y$ , $\Phi_z$ , respectively.

Then there exists a bounded generalized solution $u(t,\bar{r},\bar{\omega})$ which is absolutely continuous with respect to $t,x,y,z$ and, moreover, Lipschitz continuous in the domain $S_T$ with respect to $t,x,$ $y,z$ .

A proof requires only to establish a priori estimates for the differences $u(t',\bar{r},\omega) - u(t'',\bar{r},\omega)$, $u(t,\bar{r}',\omega) - u(t,\bar{r}'',\omega)$ by virtue of the integral equation

$$(7) \qquad u(t,\bar{r},\omega) = e^{-\int_{t^*}^{t} \sigma(\bar{r}-\omega(t-s))ds} u(t^*,\bar{r}^*,\omega) + \int_{t^*}^{t} e^{-\int_{s}^{t} \sigma(\bar{r}-\omega(t-s'))ds'} Bu(s,\bar{r}-\omega(t-s),\omega)ds.$$

Here $t^*$, $\bar{r}^*$ correspond to the point of intersection of the characteristic $\bar{r} - \omega t = \text{const}$ with the side boundary of the domain $S_T$ and with the plane $t=0$, respectively,

$$u(t^*,\bar{r}^*,\omega) = \begin{cases} 0 & \text{if } t^k > 0 \\ \Phi(\bar{r},\omega) & \text{if } t^k = 0 \end{cases},$$

$$Bu = \frac{\sigma_s}{4\pi} \int_{\Omega} g(\bar{\omega},\bar{\omega}')u(t,\bar{r},\bar{\omega}')d\bar{\omega}' + f .$$

We have

$$|u(t',\bar{r},\bar{\omega}) - u(t'',\bar{r},\bar{\omega})| \leq M_1 |t'-t''| ,$$
$$d(\bar{r}',\bar{r}'',\bar{\omega})|u(t,\bar{r}',\bar{\omega}) - u(t,\bar{r}'',\bar{\omega})| \leq M_2 |\bar{r}'-\bar{r}''|$$

where $M_i$ depend on $T$ and the constants which appear in the assumptions of Theorem 1, $d(\bar{r}',\bar{r}'',\bar{\omega}) = \min\{d(\bar{r}',\bar{\omega}), d(\bar{r}'',\bar{\omega})\}$, $d(\bar{r},\bar{\omega})$ is the distance from the point whose coordinates are $\bar{r} = (x,y,z)$ to the boundary $\Gamma$ along the direction $\bar{\omega}$. A detailed proof of these estimates is to be found in [6], [7].

## 2. Method of spherical harmonics

Let us introduce a projection operator

$$\mathcal{P}_n u = \frac{1}{2\pi} \sum_{k=0}^{n} \sum_{m=0}^{k} [C_k^{(m)}(u,C_k^{(m)}) + S_k^{(m)}(u,S_k^{(m)})] ,$$

where

$$C_k^{(m)} = (2k+1) \frac{(k-m)!}{(k+m)!} P_k^{(m)}(\zeta) \cos m\varphi ,$$

$$S_k^{(m)} = (2k+1) \frac{(k-m)!}{(k+m)!} P_k^{(m)}(\zeta) \sin m\varphi ,$$

$$(u,v) = \int_0^{\pi}\int_0^{2\pi} uv \sin\theta d\theta d\varphi .$$

Then using the method of spherical harmonics, we determine the approximate solution

$$v_n = \frac{1}{2\pi} \sum_{k=0}^{n} \sum_{m=0}^{k} \left[ C_k^{(m)} \varphi_k^{(m)} + S_k^{(m)} \psi_k^{(m)} \right]$$

from the equation

(8) $$\mathcal{P}_n L v_n = \mathcal{P}_n f \ .$$

This system of equations together with the corresponding initial and boundary conditions can be written in the form

(9) $$\frac{\partial v_n}{\partial t} + \frac{\partial [\xi] v_n}{\partial x} + \frac{\partial [\eta] v_n}{\partial y} + \frac{\partial [\zeta] v_n}{\partial z} + \sigma v_n - \frac{\sigma_s}{4\pi} \int_\Omega g_n v_n d\omega' = f_n \ ,$$

(10) $$v_n|_{t=0} = \Phi_n(\bar{r}, \bar{\omega}) \ ,$$

(11) $$v_n \in \mathcal{N}_+^n(s) \ , \quad s \in \Gamma$$

where

$$\mathcal{N}_+^n(s) = \{ v_n : (v_n \ , (n_x \xi + n_y \eta + n_z \zeta) v_n) \geq 0 \} \ ,$$

$n_x, n_y, n_z$ are the components of the outer normal $n$ ; $\xi, \eta, \zeta$ are operators which map a harmonic polynomial $v_n$ onto another such polynomial without increasing its degree [2] .

It is known that the system of equations of the method of spherical harmonics may be written in another form which enables us to express the system in the form of a symmetric hyperbolic system in the sense of Friedrichs, namely

(12) $$B \frac{\partial v_n}{\partial t} + A_1 \frac{\partial v_n}{\partial t} + A_2 \frac{\partial v_n}{\partial y} + A_3 \frac{\partial v_n}{\partial z} + D v_n = F_n \ ,$$

(13) $$v_n(0, \bar{r}) = \Phi_n \ ,$$

(14) $$M v_n(t, \bar{r}) = 0 \qquad \text{for} \quad r \in \Gamma$$

where $v_n = \{ \varphi_n^{(m)}, \psi_n^{(m)} \}$ , $B$ is a positive definite matrix, $A_i$ are symmetric matrices [2] , $M$ is a rectangular matrix satisfying boundary conditions of the type of Vladimirov-Maršak. The boundary matrix $A = n_x A_1 + n_y A_2 + n_z A_3$ has a constant rank on the boundary $\Gamma$ . In virtue of dissipativity of the boundary conditions (11) it is possible to establish a priori estimates

$$\max_t \int_G \int_\Omega v_n^2 d\bar{r} d\omega \leq M_3 \ ,$$

$$\max_t \iint_{\Gamma\,\Omega} (\omega, n) v_n^2 ds d\bar{\omega} \le M_4$$

where $M_3$, $M_4$ are independent of the number of harmonics $n$. On the basis of the a priori estimates we can prove also the absolute continuity of $v_n(t,\bar{r},\bar{\omega})$ with respect to $t,\bar{r}$ under the conditions (i)-(iii).

Let $u(t,\bar{r},\bar{\omega})$ be a generalized solution of the problem (1) - (3) which is absolutely continuous with respect to $t,\bar{r}$ and let $v_n(t,\bar{r},\bar{\omega})$ be the corresponding generalized solution of the problem (9) - (11). Then the following theorem [4] is valid.

Theorem 2. The sequence $\{v_n(t,\bar{r},\bar{\omega})\}$ approximates $u(t,\bar{r},\bar{\omega})$ with respect to the norm of the space $C([0,T]; L_2(G\times\Omega))$.

In order to prove Theorem 2 we introduce $W_n = u_n - v_n$ where $u_n = \mathcal{P}_n u$. Then the Green formula

$$\int_G\int_\Omega W_n^2(t,\bar{r},\bar{\omega}) d\bar{r} d\omega = \int_G\int_\Omega W_n^2(0,\bar{r},\bar{\omega}) d\bar{r} d\bar{\omega} - \tag{15}$$
$$- 2\int_0^t\int_G\int_\Omega |\sigma W_n - \frac{\sigma_s}{4\pi}\int_\Omega g_n W_n d\omega| W_n dt d\bar{r} d\omega +$$
$$+ 2\int_0^t\int_G\int_\Omega (f - f_n + R_n) W_n dt d\bar{r} d\bar{\omega} - \int_0^t\int_\Gamma\int_\Omega (\bar{\omega},\bar{n}) W_n^2 dt ds d\omega ,$$

holds for the difference $W_n$ with $f_n = \mathcal{P}_n f$, $g_n = \mathcal{P}_n g$ and

$$R_n = \frac{1}{2\pi}\Big\{\sum_{m=0}^{n}\Big[\frac{1}{2}\Big(-\frac{(n-m+2)!}{(n+m)!}C_{n+1}^{m-1} + \frac{(n-m)!}{(n+m)!}C_{n+1}^{m+1}\Big)\frac{\partial\tilde{\varphi}_n^{(m)}}{\partial x} +$$
$$+ \frac{1}{2}\Big(\frac{(n-m+2)!}{(n+m)!}S_{n+1}^{m-1} + \frac{(n-m)!}{(n+m)!}S_{n+1}^{m+1}\Big)\frac{\partial\tilde{\varphi}_n^{(m)}}{\partial y} +$$
$$+ \frac{(n-m+1)!}{(n+m)!}C_{n+1}^{m}\frac{\partial\tilde{\varphi}_n^{(m)}}{\partial z}\Big] + \sum_{m=1}^{n}\Big[\frac{1}{2}\Big(-\frac{(n-m+2)!}{(n+m)!}S_{n+1}^{m-1} +$$
$$+ \frac{(n-m)!}{(n+m)!}S_{n+1}^{m+1}\Big)\frac{\partial\tilde{\psi}_n^{(m)}}{\partial x} - \frac{1}{2}\Big(\frac{(n-m+2)!}{(n+m)!}C_{n+1}^{m} +$$
$$+ \frac{(n-m)!}{(n+m)!}C_{n+1}^{m+1}\Big)\frac{\partial\tilde{\psi}_n^{(m)}}{\partial y} + \frac{(n-m+1)!}{(n+m)!}\frac{\partial\tilde{\psi}_n^{(m)}}{\partial z}\Big]\Big\}$$

where $\tilde{\varphi}_n^{(m)}$, $\tilde{\psi}_n^{(m)}$ are the coefficients of the representation of $u$ at the last harmonics.

We estimate (15) by

$$\|W_n\|_t^2 \le \|W_n\|_0^2 + k\int_0^t \|W_n\|_\tau^2 d\tau + \|f-f_n+R_n\|^2 t - \int_0^t\int_T\int_\Omega (\bar{n}, \bar{\omega}) W_n^2 dt ds d\omega$$

with

$$\|W_n\|_t^2 = \int_G\int_\Omega W_n^2 d\bar{r} d\omega \ .$$

In virtue of the properties of harmonic polynomials we obtain the inequality

$$\int_0^t\int_T\int_\Omega (\bar{n}, \bar{\omega}) W_n^2 dt d\bar{r} d\bar{\omega} \le \|u-u_n\|^2_{C([0,T];L_2)} t \ .$$

Using this inequality, we conclude finally

$$\|u_n - v_n\|^2_{C([0,T];L_2)} \le C_1(T) \left\{ \| \operatorname{grad} |u-u_n| \|^2_{C([0,T];L_2)} + \right.$$
$$\left. + \|u-u_n\|^2_{C([0,T];L_2)} \right\}$$

which proves the theorem.

3. Estimate of error

Let us consider the stationary problem for the kinetic equation in the case of the plane geometry. Here the boundary value problem for the equations of the method of spherical harmonics in $P_{2n+1}$ - approximations assumes the form

$$[\mu] \frac{\partial v_n}{\partial z} + \sigma v_n = \frac{\sigma_s}{2}\int_{-1}^{1} v_n d\mu' + f_n \ ,$$

$$\int_0^1 P_{2s+1}(\mu) v_n(0,\mu) d\mu = 0 \ ,$$

$$\int_{-1}^0 P_{2s+1}(\mu) v_n(H,\mu) d\mu = 0 \ , \qquad s=0,1,\ldots,n \ ;$$

where

$$[\mu] v_n = \frac{1}{2}\sum_{k=0}^{n-1} (2k+1)\mu P_k(\mu)\varphi_k + \frac{1}{2} n P_{n-1}(\mu)\varphi_n \ .$$

If $u$ is the exact solution of the original problem, then $W_n = v_n - u$ satisfies the equation

$$\mu\frac{\partial W_n}{\partial z} + \sigma W_n = \frac{\sigma_s}{2}\int_{-1}^{1} W_n d\mu' + \frac{P_{n+1}(\mu)}{P_{n+1}(0)} \left[ \sigma v_n(z,0) - \right. \tag{16}$$
$$\left. - \frac{\sigma_s}{2}\int_{-1}^{1} v_n d\mu' - f_n(z,0) \right] + f - f_n \ .$$

The function $W_n$ introduced above satisfies the boundary conditions

(17) $W_n(0,\mu) = v_n(0,\mu)$ if $\mu > 0$ ,

(18) $W_n(H,\mu) = v_n(H,\mu)$ if $\mu < 0$ .

Thus we have formulated a boundary value problem (16) - (18) with respect to $W_n(z,\mu)$ . By virtue of the maximum principle [5] this yields an estimate for $W_n$

$$|W_n| \le \max \left\{ \max_{z,\mu} \left| \frac{P_{n+1}(\mu)}{P_{n+1}(0)} (Lv_n - f)_{\mu=0} + R_n(f) \right| , \max_{\mu} |v_n(0,\mu)| , \max_{\mu} |v_n(H,\mu)| \right\}$$

where

$$R_n(f) = f - f_n \; ; \quad Lv_n = \mu \frac{\partial v_n}{\partial z} - \sigma v_n - \frac{\sigma_s}{2} \int_{-1}^{1} v_n d\mu' \; .$$

## 4. The splitting method for equations of the method of spherical harmonics

Let us now proceed to the problems of construction of the difference schemes for the solution of the problem (12) - (14). To this aim we divide the interval $[0,T]$ into $m$ equal parts with the step $\tau = T/m$ and replace the problem on the interval $[m\tau,(m+1)\tau]$ by a sequence of one-dimensional problems. The convergence of this method was studied in our former papers [8] . Here we show some new approaches to the realisation of symmetric systems on each intermediate step. We consider the one-dimensional problem

(19) $B \frac{\partial u}{\partial t} + A \frac{\partial u}{\partial x} + Du = f$,

(20) $u(m\tau, x) = \Phi$ ,

(21) $M_1 u = 0$ for $x=0$ , $M_2 u = 0$ for $x=H$ .

We introduce the spaces

$$\mathcal{N}_+ = \{u : M_1 u = 0\} , \quad \mathcal{N}_- = \{v : M_2 v = 0\} , \quad \mathcal{N}_0 = \{w : Aw = 0\} .$$

In virtue of the dissipativity of the boundary conditions $(Au,u) = 0$, $u \in \mathcal{N}_+$ , $(Av,v) \le 0$, $v \in \mathcal{N}_-$ . This implies that the spaces $\mathcal{N}_+$ , $\mathcal{N}_-$ are orthogonal with respect to the metric $[u,v] = (Au,v)$ . Let $\{u\}$, $\{v\}$, $\{w\}$ form orthogonal systems of bases of the spaces $\mathcal{N}_+$ , $\mathcal{N}_-$ , $\mathcal{N}_0$ , respectively.

Denote the matrix formed by these basis vectors by $L = [\bar{u},\bar{v},\bar{w}]$. Then using the mapping $u=LV$ we can rewrite the system (19) in the form

$$\widetilde{B}\frac{\partial V}{\partial t} + \Lambda\frac{\partial V}{\partial x} + \widetilde{D}V = g$$

where $\widetilde{B} = L^*BL$ , $\Lambda = L^*AL$ is a diagonal matrix. The vector $V$ may be written in the form $V = (V^+, V^-, V^0)$ with $V^+, V^-, V^0$ corresponding to the blocks $\bar{u}, \bar{v}, \bar{w}$.

The boundary conditions (21) are

$V^+ = 0$ for $x=0$ ,

$V^- = 0$ for $x=H$ .

After transforming the system into the canonical form it is not difficult to find stable difference schemes.

## References

(All references except [7] in Russian)


[1] Marčuk G.I., Lebedev V.I.: Numerical methods in the theory of transport of neutrons, Atomizdat 1971

[2] Godunov S.K., Sultangazin U.M.: On dissipativity of boundary conditions of Vladimirov for the symmetric system of the method of spherical harmonics. Žur.vyčisl.matem. i matem.fiz. 11 (1971), No.3

[3] Sultangazin U.M.: To the problem of convergence of the method of spherical harmonics for the nonstationary equation of transport, Žur.vyčisl.matem. i matem.fiz. 14 (1974), No.1

[4] Skoblikov V.S.: Strong convergence of the method of spherical harmonics, Preprint, Novosibirsk 1972

[5] Germogenova T.A.: Maximum principle for the transport equation, Žur.vyčisl.matem. i matem.fiz. 2 (1962), No.1

[6] Sultangazin U.M.: Differential properties of solutions of the mixed Cauchy problem for the nonstationary kinetic equation, Preprint, Novosibirsk 1971

[7] Douglis A.: The solution of multidimensional generalized transport equations and their calculation by difference methods in Numerical Solutions of Part.Diff.Eq.,Acad.Press,New York 1966

[8] Sultangazin U.M.: To the problem of justification of the method of weak approximation for the equations of spherical harmonics, Preprint, Novosibirsk 1971


Author's address: Kazachskij gosudarstv.univ.im. C.M.Kirova
Alma-Ata 480091, Kirova 136, USSR

# SOME PROBLEMS CONCERNING THE FUNCTIONAL DIFFERENTIAL EQUATIONS

M.Švec, Bratislava

We consider the equation

$$(1) \qquad \dot{x}(t) = f(t, x_t) ,$$

where $\dot{x}(t)$ means the derivative of the vector function $x(t)$ at the point $t$ . If $x(t)$ is a function defined on the interval $[t_o-h,T)$, where $h>0$, $t_o$ ,$T$ are real numbers, then $x_t = x(t+\theta)$, $\theta\in[-h,0]$ for $t\in[t_o ,T)$ as usual. Let us explain the meaning of the notation used in this paper. Let $n$ be a natural number, $R^n$ the n-dimensional vector space of the points $X = (x_1,x_2,\dots,x_n)$ with a suitable norm $|.|$ , $C = C([-h,0],R^n)$ the Banach space of all continuous functions $\psi$ with the norm $\|\psi\| = \max \{|\psi(t)|, t\in[-h,0]\}$ , $C_o = \{\Phi\in C : \Phi(0) = 0\}$ a subspace of $C$ . Furthermore, let $B = B([t_o ,T),R^n)$ be the Banach space of all functions continuous and bounded on $[t_o ,T)$ with the uniform norm $\|u\|_u = \sup\{|u(t)|, t\in[t_o ,T)\}$ and $B_o = \{u(t)\in B : u(t_o) = 0\}$ a subspace of $B$ . Let $\Phi\in C_o$ be fixed. Then $B_{\Phi} = \{z(t) : [t_o-h,T)\to R^n , z(t) = \Phi(t-t_o)$ for $t\in[t_o-h,t_o]$ , $z(t) = u(t)$ for $t\in[t_o,T)$; $u(t)\in B_o\}$ is a complete metric space with the metric $\varrho(z_1,z_2) = \|u_1-u_2\|_u$ , where $z_1(t) = u_1(t)$, $z_2(t) = u_2(t)$ for $t\in[t_o ,T)$, $u_1(t),u_2(t)\in B_o$ .

As usual, the initial problem for (1) is formulated as follows: For given $t_o\in R$, $\psi\in C$ find a function $x\in C([t_o-h,A),R^n)$ such that $x(t) = \psi(t-t_o)$ for $t\in[t_o-h,t_o]$ and $\dot{x}(t) = f(t,x_t)$ for $t\in[t_o ,A)$. We shall denote this solution by $x(t,t_o ,\psi)$ and say that it is given by $(t_o ,\psi)$. Because every $\psi\in C$ can be written as $\psi(t) = X_o+\Phi(t)$, where $X_o= \psi(0)$, $\Phi\in C_o$ , we shall write $x(t,t_o ,X_o+\Phi)$ to express that the solution $x$ passes through the point $X_o$ at $t=t_o$ .

Now, the main problem we will discuss is the following:

(P) Let be given $T\in R$, $t_o< T$, $X_o,X_1\in R^n$ . Find $\Phi\in C_o$ such that the solution $x(t,t_o,X_o+\Phi)$ exists on $[t_o ,T)$ and $\lim x(t,t_o,X_o+\Phi) = X_1$ as $t\to T-$ .

The function $f$ is subjected to the following hypotheses:

$(H_1)$ $f(t,\psi)$ is continuous on $[t_o ,T)\times C$ and $\int_{t_o}^{T} |f(t,0)|\,dt = K<\infty$ .

($H_2$) There is a function $\beta(t)$ continuous on $[t_o, T)$ such that $|f(t,\psi_1) - f(t,\psi_2)| \le \beta(t)\|\psi_1 - \psi_2\|$ for every

$$\psi_1, \psi_2 \in C \quad \text{and} \quad t \in [t_o, T) \quad \text{and} \quad \int_{t_o}^{T} \beta(t)dt = k < 1 .$$

The hypotheses $H_1$ and $H_2$ being satisfied, we can conclude

Theorem 1 [1] . Let $H_1$ and $H_2$ be satisfied. Then the solution $x(t,t_o,\psi)$, $\psi \in C$ exists on $[t_o, T)$, is unique and $\lim x(t,t_o,\psi) = X_1 \in R^n$ as $t \to T-$ .

Theorem 2 [1] . Let $H_1$ and $H_2$ with $k < 1/2$ be satisfied. Let be given $X_1 \in R^n$ , $\Phi \in C_o$ . Then there exists a unique $X_o \in R^n$ such that $\lim x(t,t_o,X_o+\Phi) = X_1$ as $t \to T-$ .

Now we define a map $F(X_o,\Phi) : R^n \times C_o \to R^n$ by the relation $F(X_o,\Phi) = \lim x(t,t_o,X_o+\Phi) = X_1$ as $t \to T-$ . The following theorem mentions some properties of this map.

Theorem 3 [1] . Let $H_1$ and $H_2$ with $k < 1/2$ be valid. Then

a) the map $F(X_o,\Phi)$, by fixed $\Phi$, is a one-to-one map of $R^n$ onto $R^n$ ;

b) $F(X_o,\Phi)$ fulfils the Lipschitz condition:
$$|F(X_{o1},\Phi_1) - F(X_{o2},\Phi_2)| \le e^k |X_{o1} - X_{o2}| + (e^k - 1)\|\Phi_1 - \Phi_2\| .$$

Our problem (P) was partially solved in the papers [1], [2], [3] . In [1] we obtained some results of negative character, e.g. if $H_1$ and $H_2$ are valid and if $X_o, X_2 \in R^n$, $|X_o| + K \neq 0$ and $|X_1| > [|X_o| + K] \frac{a}{1-k}$ , $0 < a < 1$, then there is no solution of the problem (P) for $\Phi \in C_o$ , $\|\Phi\| < K \frac{a}{1-k}$ .

For further purposes we need an estimation of $|x(t,t_o,X_o+\Phi)|$ and $\|x_t(t_o,X_o+\Phi)\|$ . Let us use the notation $x(t) = x(t,t_o,X_o+\Phi)$. Then, $x_t(\theta)$, $\theta \in [t-h,t]$ being continuous, there exists $v \in [t-h,t]$ such that $|x(v)| = \|x_t\|$ . Suppose that $t \ge t_o$ . Then either $v \ge t_o$ or $v \in [t_o-h, t_o]$ .

Let $v \ge t_o$ . Then we get

$$\|x_t\| = |x(v)| = \left|X_o + \int_{t_o}^{v} f(s,x_s)ds\right| \le |X_o| + \int_{t_o}^{v} |f(s,0)|\, ds +$$
$$+ \int_{t_o}^{v} |f(s,x_s) - f(s,0)|\, ds \le |X_o| + K + \|\Phi\| + \int_{t_o}^{t} \beta(s)\|x_s\| ds .$$

If $v \in [t_o-h, t_o]$ , we have
$$\|x_t\| = |x(v)| = |X_o + \Phi(v-t_o)| \le |X_o| + \|\Phi\| + K + \int_{t_o}^{t} \beta(s)\|x_s\| ds .$$

Thus, for $t \in [t_o ,T)$ we have

$$\|x_t\| \le |X_o| + K + \|\Phi\| + \int_{t_o}^{t} \beta(s)\|x_s\| ds \quad .$$

The application of Gronwall-Bellman lemma gives

$$(2) \qquad \|x_t\| \; [|(X_o| + K + \|\Phi\|] \; \exp \; ( \int_{t_o}^{t} \beta(s)ds), \quad t \in [t_o ,T) \; ,$$

which implies

$$(3) \qquad |x(t)| \le \|x_t\| \le [|X_o| + K + \|\Phi\|] \; \exp \; ( \int_{t_o}^{t} \beta(s)ds, \; t \in [t_o ,T).$$

Let us turn our attention to the dependence of $F(X_o, \Phi)$ on $\Phi$ by fixed $X_o$ . It follows from Theorem 3 that, if $\|\Phi_1 - \Phi_2\| = 0$, we have $F(X_o, \Phi_1) = F(X_o, \Phi_2)$. If $\|\Phi_1 - \Phi_2\| \ne 0$, it may happen that $F(X_o, \Phi_1) \ne F(X_o, \Phi_2)$, but also $F(X_o, \Phi_1) = F(X_o, \Phi_2)$. If the former case occurs, it influences both solutions $x(t,t_o,X_o+\Phi_1)$ and $x(t,t_o,X_o+\Phi_2)$. The following theorem holds.

Theorem 4. Let $\|\Phi_1 - \Phi_2\| \ne 0$ and $F(X_o, \Phi_1) = F(X_o, \Phi_2)$. Then either

a) $0 < \|x(t,t_o,X_o+\Phi_1) - x(t,t_o,X_o+\Phi_2)\|_u < \|\Phi_1 - \Phi_2\|$

or

b) $\|x(t,t_o,X_o+\Phi_1) - x(t,t_o,X_o+\Phi_2)\|_u = 0$ .

Proof. The function $|H(t)| = |x(t,t_o,X_o+\Phi_1) - x(t,t_o,X_o+\Phi_2)|$; $t \ge t_o$ is nonnegative and $|H(t_o)| = 0 = |H(T)|$ . Thus there exists $t_1 \in [t_o ,T)$ such that $H(t_1) = \max\{|H(t)|, \; t \in [t_o ,T)\}$ . If $t_1 = t_o$ , the second case (b) occurs. If $t_1 \in [t_o+h,T)$ and $H(t_1) \ne 0$ , the hypothesis $H_2$ yields

$$(4) \qquad |\dot{H}(t)| \le \beta(t)\|x_t(t_o,X_o+\Phi_1) - x_t(t_o,X_o+\Phi_2)\| \le \beta(t)|H(t_1)| \; ,$$
$$t \in [t_o+h,T) \; .$$

Hence we get

$$(5) \qquad |H(t_1)| = |\int_{t_1}^{T} \dot{H}(t)dt| \le \int_{t_1}^{T} |\dot{H}(t)|dt \le \int_{t_1}^{T} \beta(t)dt|H(t_1)| \le$$
$$\le k|H(t_1)|.$$

Therefore $t_1 \bar{\in} [t_o + h,T)$. Suppose that $t_1 \in (t_o, t_o+h)$ and that $|H(t_1)| \ge \|\Phi_1 - \Phi_2\|$. In this case the inequalities (4) and (5) hold as well and the same reasoning as above gives a contradiction which completes the proof.

In the following let $X_o$ be fixed. We are going to examine the properties of the set of images of the set $G = \{\Phi \in C_o : \|\Phi\| \leq r\}$, $r > 0$, by the map $F$. We shall denote this set of images by $F(X_o, G)$.

<u>Theorem 5.</u> Let $H_1$, $H_2$ and $H_3$ be valid with

($H_3$) There exists a constant $d$, $0 < d \leq 1$, such that for any $X_o \in R^n$ and any $y_i \in B_o$, $i=1,2$ and any $\Phi_i \in C_o$, $i=1,2$ the inequality

$$d\|\Phi_1 - \Phi_2\| \leq \left|\int_{t_o}^{t_o+h} [f(s, X_o + z_{1s}) - f(s, X_o + z_{2s})] ds\right|$$

holds where $z_i(t) = \Phi_i(t-t_o)$ for $t \in [t_o-h, t_o]$, $z_i(t) = y_i(t)$ for $t \in [t_o, t_o+h]$, $i=1,2$.

Then the set $F(X_o, G)$ is bounded, closed and connected.

Proof. The boundedness of $F(X_o, G)$ follows immediately from (3) or from Theorem 3, (b). Consider now the set of solutions $S = \{x(t, t_o, X_o + \Phi), \Phi \in G\}$ on $[t_o, T)$. From (3) we have that these solutions are uniformly bounded on $[t_o, T)$ by $[|X_o| + K + r]e^k$. The same holds also for the set $\{x_s(t_o, X_o + \Phi), s \in [t_o, T)\}$ as follows from (2). Further, for $t, t' \in [t_o, T)$, $t < t'$ we have

$$|x(t', t_o, X_o + \Phi) - x(t, t_o, X_o + \Phi)| \leq \int_t^{t'} |f(s, 0)| ds + \int_t^{t'} \beta(s)\|x_s\| ds .$$

Now, from the existence of $\int_{t_o}^{T} |f(s,0)| ds$ and $\int_{t_o}^{T} \beta(s) ds$ and from the uniform boundedness of $\|x_s\|$ we get the equicontinuity of the elements of $S$ on $[t_o, T)$. Thus we may apply on $S$ the theorem of Ascoli-d'Arzelà on every compact set from $[t_o, T)$. Suppose that $X_i \in F(X_o, G)$, $i=1,2,\ldots$ and that $\lim X_i = Y$ as $i \to \infty$. We are going to show that $Y \in F(X_o, G)$. Let $\{x(t, t_o, X_o + \Phi_i), \Phi_i \in G\}$ be the sequence of solutions of (1) such that $\lim x(t, t_o, X_o + \Phi_i) = X_i$ as $t \to T-$, $i=1,2,\ldots$. Applying the Ascoli-d'Arzela theorem we get that we can choose a subsequence $\{x(t, t_o, X_o + \Phi_{i_k}), \Phi_{i_k} \in G\}$ from $\{x(t, t_o, X_o + \Phi_i)\}$ which converges to a continuous function $u(t)$ uniformly on every closed subinterval of $[t_o, T)$. Let $\lim x(t, t_o, X_o + \Phi_{i_k}) = X_{i_k}$ as $t \to T-$. Evidently $\lim X_{i_k} = Y$ as $k \to \infty$. The solutions $x(t, t_o, X_o + \Phi_{i_k})$ satisfy the equations

$$x(t,t_o,X_o+\Phi_{i_k}) = X_{i_k} - \int_t^T f(s,x_s(t_o,X_o+\Phi_{i_k}))ds \ , \ k=1,2,\ldots \ .$$

The application of Lebesgue's dominated convergence theorem gives

$$u(t) = Y - \int_t^T f(s,u_s)ds \qquad \text{for} \quad t\in[t_o+h,T) \ .$$

Thus, we have got that $u(t)$ satisfies (1) on $[t_o+h,T)$ and $\lim u(t) = Y$ as $t\to T-$ . The problem which appears here is: How to ensure that $u(t)$ satisfies (1) on $[t_o\ ,T)$; if this is possible, to which function $\Phi\in C_o$ this solution will correspond ? The validity of $H_3$ represents one of the possibilities. In fact, we know that the sequence $\{x(t_o+h,t_o,X_o+\Phi_{i_k})\}$ converges to $u(t_o+h)$. Therefore it is a Cauchy sequence. Using the hypothesis $H_3$, we get

$$|x(t_o+h,t_o,X_o+\Phi_{i_m}) - x(t_o+h,t_o,X_o+\Phi_{i_n})| =$$

$$= \left|\int_{t_o}^{t_o+h}[f(s,x_s(t_o,X_o+\Phi_{i_m}))-f(s,x_s(t_o,X_o+\Phi_{i_n}))]ds\right| \geq d\|\Phi_{i_m}-\Phi_{i_n}\| \ .$$

Hence we get that $\{\Phi_{i_k}\}$ is a Cauchy sequence and therefore it converges to a function $\Phi$ in the complete space $C_o$ . This convergence is uniform on $[-h,0]$ .

Now take the function $v_k(t)$ defined on $[t_o-h,T)$ as follows: $v_k(t) = X_o+\Phi_{i_k}(t-t_o)$, $t\in[t_o-h,t_o]$ , $v_k(t) = x(t,t_o,X_o+\Phi_{i_k}) = X_o +$

$$+ \int_{t_o}^{t} f(s,x_s(t_o,X_o+\Phi_{i_k}))ds = X_{i_k} - \int_t^T f(s,x_s(t_o,X_o+\Phi_{i_k}))ds \ , \ t\in[t_o,T),$$

$k=1,2,\ldots$ . We get that $v_k(t)$ converges to $v(t)$ : $v(t) = X_o+ + \Phi(t-t_o)$ for $t\in[t_o-h,t_o]$ , $v(t) = u(t)$ for $t\in[t_o\ ,T)$ uniformly on every closed subinterval of $[t_o-h,T)$. We get also that

$$v(t) = X_o + \int_{t_o}^{t} f(s,v_s)ds = Y - \int_t^T f(s,v_s)ds \ , \ t\in[t_o\ ,T) \ .$$

Thus $v(t) = x(t,t_o,X_o+\Phi)$ and $\lim v(t) = Y$ as $t\to T-$ . This proves that $Y\in F(X_o,G)$ and therefore $F(X_o,G)$ is closed.

Finally, we have to prove that $F(X_o,G)$ is connected. Suppose the contrary is true. Then $F(X_o,G)$ can be represented as $F(X_o,G)= =F_1\cup F_2$ , where $F_i$ , $i=1,2$ , are bounded, closed and disjoint sets. Let $G_i = \{\Phi\in G : F(X_o,\Phi)\in F_i\}$ , $i=1,2$ . Evidently $G = G_1\cup G_2$ and $G_1\cap G_2 = \emptyset$ and $G_1$ and also $G_2$ are nonvoid. Furthermore, the continuous dependence of solutions on the initial functions, Theorem 3 and the closedness of $F_i$ , i=1,2 imply the closedness of $G_i$ , i=1,2 . But then we have that the closed ball G is the union of two sets which are nonvoid, closed and disjoint which is in contradiction with the fact that G is connected.

Remark 1. The constant d in $H_3$ has to satisfy also the condition $d \leq \int_{t_o}^{t_o+h} \beta(s)ds$ for $H_2$, $H_3$ not to contradict each other. In fact, we have

$$d\|\Phi_1-\Phi_2\| \leq \left| \int_{t_o}^{t_o+h} [f(s,z_{1s})-f(s,z_{2s})]ds \right| \leq \int_{t_o}^{t_o+h} \beta(s)\|z_{1s}-z_{2s}\|ds .$$

If we take $z_i(t) = \Phi_i(t-t_o)$, $t\in[t_o-h,t_o]$ , $z_i(t) = y(t)$, $t\geq t_o$ , i=1,2 , we have that $\|z_{1s}-z_{2s}\| \leq \|\Phi_1-\Phi_2\|$ and from the preceding inequality we get that $d \leq \int_{t_o}^{t_o+h} \beta(s)ds$ .

Theorem 6. Let be valid $H_1,H_2,H_3$ with $\frac{k}{1+k} \leq d \leq \int_{t_o}^{t_o+h} \beta(s)ds$ and $H_4$:

($H_4$) For every two points $X_o,X\in R^n$ and every $y(t)\in B_o$ there is $\Phi\in C_o$ such that for $z(t) = \Phi(t-t_o)$, $t\in[t_o-h,t_o]$ , $z(t) = y(t)$, $t\in[t_o ,T)$ the equation

$$X = X_o + \int_{t_o}^{t_o+h} f(t,X_o+z_t)dt$$

holds.

Then the problem (P) has a solution.

Proof. Let $X_1,X_o\in R^n$ be given. Choose $y_1(t)\in B_o$ such that $\lim y_1(t) = X = X_1-X_o$ as $t\to T-$ . Then denote

$$(6) \qquad Y_1 = X_1 - \int_{t_o+h}^{T} f(t,X_o+y_{1t})dt .$$

With regard to $H_4$ applied to $X_o,Y_1\in R^n$ and $y_1(t)\in B_o$ there exists $\Phi_1\in C_o$ such that

$$(7) \qquad Y_1 = X_o + \int_{t_o}^{t_o+h} f(t,X_o+z_{1t})dt ,$$

$z_1(t) = \Phi_1(t-t_o)$ for $t\in[t_o-h,t_o]$ and $z_1(t) = y_1(t)$ for $t\in[t_o, T)$. From (6) and (7) we get

$$(8) \qquad X_1 = X_o + \int_{t_o}^{T} f(s, X_o + z_{1s})ds .$$

Denote

$$y_2(t) = \int_{t_o}^{t} f(s, X_o + z_{1s})ds , \qquad t \in [t_o, T) .$$

Evidently $y_2(t) \in B_o$ and $\lim y_2(t) = X_1 - X_o = X$ as $t \to T-$. Now we construct

$$Y_2 = X_1 - \int_{t_o+h}^{T} f(t, X_o + y_{2t})dt .$$

Then with regard to $H_4$ applied to $X_o, Y_2$ and $y_2(t)$ there exists $\Phi_2 \in C_o$ such that

$$Y_2 = X_o + \int_{t_o}^{t_o+h} f(t, X_o + z_{2t})dt ,$$

where $z_2(t) = \Phi_2(t - t_o)$ for $t \in [t_o - h, t_o]$, $z_2(t) = y_2(t)$ for $t \in [t_o, T)$. Once again we get

$$X_1 = X_o + \int_{t_o}^{T} f(t, X_o + z_{2t})dt .$$

Put

$$y_3(t) = \int_{t_o}^{t} f(s, X_o + z_{2s})ds , \qquad t \in [t_o, T) .$$

We have that $y_3(t) \in B_o$, $\lim y_3(t) = X_1 - X_o = X$ as $t \to T-$. Proceeding in this way we get the sequences, $n = 2, 3, \ldots$.

$$(9) \qquad y_n(t) = \int_{t_o}^{t} f(s, X_o + (z_{n-1})_s)ds , \qquad t \in [t_o, T) ,$$

$$(10) \qquad Y_n = X_1 - \int_{t_o+h}^{T} f(t, X_o + y_{nt})dt ,$$

$$(11) \qquad Y_n = X_o + \int_{t_o}^{t_o+h} f(t, X_o + (z_n)_t)dt ,$$

$z_n(t) = \Phi_n(t - t_o)$ for $t \in [t_o - h, t_o]$, $z_n(t) = y_n(t)$, $t \in [t_o, T)$

and

(12) $$X_1 = X_o + \int_{t_o}^{T} f(t, X_o + (z_n)_t)dt ,$$

(13) $$\lim y_n(t) = X_1 - X_o = X \quad \text{as} \quad t \to T- .$$

Now from (10), (11) applying $H_3$ and $H_2$ we have

(14) $$\|\Phi_n - \Phi_{n-1}\| \le \left| \int_{t_o}^{t_o+h} [f(t, X_o + z_{nt}) - f(t, X_o + (z_{n-1})_t)]dt \right| \frac{1}{d} =$$

$$= \left| \int_{t_o+h}^{T} [f(t, X_o + y_{nt}) - f(t, X_o + (y_{n-1})_t]dt \right| \frac{1}{d} \le$$

$$\le \frac{1}{d} \int_{t_o+h}^{T} \beta(t) \| [y_n - y_{n-1}]_t \| dt \le \frac{1}{d} \int_{t_o+h}^{T} \beta(s)ds \|y_n - y_{n-1}\|_u \le k \|y_n - y_{n-1}\|_u$$

From (9) using $H_2$ and (14) we get

(15) $$\|y_{n+1} - y_n\|_u \le \int_{t_o}^{T} \beta(t) \| [z_n - z_{n-1}]_t \| dt \le k \|y_n - y_{n-1}\|_u \quad .$$

Because $k < 1$, (15) means that the sequence $\{y_n(t)\}$ converges uniformly on $[t_o, T)$ to a function $y(t)$. But (14) implies the uniform convergence of the sequence $\{\Phi_n(t)\}$ to a function $\Phi \in C_o$. From all this we conclude that the sequence $\{z_n(t)\}$ converges uniformly on $[t_o - h, T)$ to the function $z(t)$: $z(t) = \Phi(t - t_o)$ for $t \in [t_o - h, t_o]$, $z(t) = y(t)$ for $t \in [t_o, T)$. Then from (9) we get

$$y(t) = \int_{t_o}^{t} f(s, X_o + z_s)ds , \qquad t \in [t_o, T) .$$

Therefore

(16) $$X_o + y(t) = X_o + \int_{t_o}^{t} f(s, X_o + z_s)ds .$$

Denoting $u(t) = X_o + z(t)$ for $t \in [t_o - h, T)$ we have

(17) $$u(t) = X_o + \Phi(t - t_o) \quad \text{for} \quad t \in [t_o - h, t_o] ,$$

$$u(t) = X_o + \int_{t_o}^{t} f(s, u_s)ds \quad \text{for} \quad t \in [t_o, T) .$$

Thus, $u(t)$ is the solution of (1) corresponding to the initial values $(t_o, X_o + \Phi)$ . From (12) and (16) we get that $\lim u(t) = X_1$ as $t \to T-$ and $u(t_o) = X_o$ . Thus, $u(t)$ is a solution of our problem (P).

Theorem 7. Let $H_1$, $H_2$, $H_3$ be valid and let

$$d > \left[\exp\left(\int_{t_o+h}^{T} \beta(s)ds\right) - 1\right] \exp\left(\int_{t_o}^{t_o+h} \beta(s)ds\right) . \tag{18}$$

Then the map $F(X_o, \Phi)$, by fixed $X_o$ , is a one-to-one map of $C_o$ into $R^n$ . This means that in this case the problem (P) has at most one solution.

Proof. Let $\Phi_1, \Phi_2 \in C_o$ , $\|\Phi_1 - \Phi_2\| \neq 0$ . Then

$$|F(X_o, \Phi_1) - F(X_o, \Phi_2)| = \left|\int_{t_o}^{T} \left[f(t, x_t(t_o, X_o + \Phi_1)) - f(t, x_t(t_o, X_o + \Phi_2))\right]\right.$$

$$\left.\cdot dt\right| \geq \left|\int_{t_o}^{t_o+h} \left[f(t, x_t(t_o, X_o + \Phi_1)) - f(t, x_t(t_o, X_o + \Phi_2))\right] dt\right| -$$

$$- \left|\int_{t_o+h}^{T} \left[f(t, x_t(t_o, X_o + \Phi_1)) - f(t, x_t(t_o, X_o + \Phi_2))\right] dt\right| \geq d\|\Phi_1 - \Phi_2\| -$$

$$- \int_{t_o+h}^{T} \beta(s)\|x_s(t_o, X_o + \Phi_1) - x_s(t_o, X_o + \Phi_2)\| ds .$$

Using Lemma 3 from [1] which asserts that, if $H_1$ and $H_2$ are valid, the inequality

$$\|x_t(t_o, X_o + \Phi_1) - x_t(t_o, X_o + \Phi_2)\| \leq \|\Phi_1 - \Phi_2\| \exp\left(\int_{t_o}^{t} \beta(s)ds\right)$$

holds, we get

$$|F(X_o, \Phi_1) - F(X_o, \Phi_2)| \geq \|\Phi_1 - \Phi_2\| \left\{d + \left[1 - \exp\int_{t_o+h}^{T} \beta(s)ds\right] \cdot \exp\int_{t_o}^{t_o+h} \beta(s)ds\right\} \tag{19}$$

which proves our theorem.

Remark 2. If we consider the scalar equation $\dot{x}(t) = a(t)x(t-h)$ where $a(t) \not\equiv 0$ for $t \in [t_o, t_o+h]$ , then $H_4$ will be valid if there is $\Phi \in C_o$ such that $\int_{t_o}^{t_o+h} a(t)\Phi(t-t_o-h)dt \neq 0$ . In fact, we have

$$X = X_o + \int_{t_o}^{t_o+h} a(t)(X_o + \lambda\Phi(t-t_o-h))dt = X_o + X_o \int_{t_o}^{t_o+h} a(t)dt + \lambda \int_{t_o}^{t_o+h} a(t)\Phi(t-t_o-h)dt .$$

From this we can calculate $\lambda$ and then $\lambda\Phi$ will be the sought function.

Remark 3. It can happen that for some given $X_o, X \in R^n$ and $y(t) \in B_o$ there are more than one function $\Phi \in C_o$ satisfying $H_4$. But if we suppose also the validity of $H_3$, there can be only one $\Phi \in C_o$ satisfying $H_4$. In fact, let $\Phi_1, \Phi_2 \in C_o$ be two functions satisfying $H_4$ for given $X_o, X \in R^n$ and $y(t) \in B_o$. Then we have

$$X = X_o + \int_{t_o}^{t_o+h} f(s, X_o + z_{1s})ds = X_o + \int_{t_o}^{t_o+h} f(s, X_o + z_{2s})ds ,$$

where $z_i(t) = \Phi_i(t-t_o)$ for $t \in [t_o-h, t_o]$, $z_i(t) = y(t)$ for $t \in [t_o, t_o+h]$, $i=1,2$. Applying $H_3$ we get

$$0 = \left| \int_{t_o}^{t_o+h} [f(s, X_o+z_{1s}) - f(s, X_o+z_{2s})]ds \right| \geq d\|\Phi_1 - \Phi_2\| .$$

Thus $\Phi_1 = \Phi_2$.

It would be desirable to clear up the relation between $H_3$ and $H_4$. It seems to us that both hypotheses $H_3$ and $H_4$ can be substituted by another one from which both $H_3$ and $H_4$ follow. This problem will be discussed in another paper.

## References

[1] Švec M.: Some Properties of Functional Differential Equations, Bolletino U.M.I. (4) 11, Suppl.fasc. 3 (1975), 467-477

[2] Seidov Z.B.: Boundary value problems with a parameter for differential equations in Banach space, Sibirskii mat.žur. IX, 223-228 (Russian)

[3] Mosyagin V.V.: Boundary value problem for differential equation with retarded argument in Banach space, Leningrad.Gos. Ped.Inst., Učennye Zapiski 387 (1968), 198-206 (Russian)

Author's address: Prírodovedecká fakulta Univerzity Komenského,
Matematický pavilón, Mlynská dolina,
816 31 Bratislava, Czechoslovakia

# A-STABILITY AND NUMERICAL SOLUTION OF ABSTRACT DIFFERENTIAL EQUATIONS

J. Taufer and E. Vitásek, Praha

## 1. Introduction

In this paper we try to give a survey of some results which have been achieved during a few last years in the Mathematical Institute of the Czechoslovak Academy of Sciences and which are related on the one hand to the problems connected with the numerical solution of stiff differential systems, on the other hand, to the problem of construction of methods for the numerical solution of partial differential equations of parabolic type which are of arbitrarily high order of accuracy with respect to the time integration step. It is well known that these two problems are very closely connected and thus, let us first of all mention this connection. Let us begin with recalling the concept of a stiff differential system. For the simplicity we speak only about the linear system with constant coefficients of the form

$$u' = Au. \tag{1.1}$$

This system is said to be stiff if some eigenvalues of the matrix $A$ have negative real parts the magnitudes of which are great in comparison with the magnitudes of the real parts of the other eigenvalues. The solution of the system (1.1) then contains rapidly decreasing components which are negligible in comparison with the other components and very often, we are not interested in them. But when solving the system (1.1) numerically we are generally in such a situation that the magnitude of the integration step is controlled exactly by these decreasing components. The practical consequences are that the integration step must be chosen unpractically small. Thus, it would be desirable to have at our disposal for solving stiff systems such methods in which the components of the approximate solution that correspond to the rapidly damped components of the exact solution would be negligible in comparison with other components even for relatively large values of the integration step. The A-stability introduced by Dahlquist is a property which guarantees such behaviour. Since it is now clear that this property will play an important role in our investigation let us also recall its definition. To define the A-stability, apply first the given method to one differential equation of the type (1.1) where $A$ is a complex constant with a negative real part. Then we say that the method is

A-stable if any approximate solution obtained with help of any integration step converges at infinity to zero. Now it is clear that the magnitude of the integration step is controlled in the case of an A-stable method only by the accuracy with which we want to approximate those components of the vector of the solution which we are interested in.

Let us now demonstrate on the trivial example of the heat conduction equation

$$\frac{\partial^2 u}{\partial x^2} = \frac{\partial u}{\partial t}, \quad 0 < x < 1, \ 0 < t < T \tag{1.2}$$

with the initial condition

$$u(x,0) = \eta(x) \tag{1.3}$$

and with the boundary conditions

$$u(0,t) = u(1,t) = 0 \tag{1.4}$$

the connection of the problem of solving a stiff differential system with the problem of constructing methods for solving parabolic differential equations.

Let us solve the problem (1.2) to (1.4) in such a way that we first discretize only the space variable, for the simplicity, by the finite-difference method. Putting $h = 1/m$ we obtain

$$\underline{u}_h' = A_h \underline{u}_h \tag{1.5}$$

where $\underline{u}_h = \underline{u}_h(t)$ is $(m-1)$-dimensional vector the components of which approximate the exact solution at the points $x_i = ih$, $i = 1, \dots, m-1$ and $A_h$ is the matrix given by

$$A_h = \frac{1}{h^2} \begin{bmatrix} -2 & 1 & 0 & \cdots & 0 \\ 1 & \ddots & \ddots & \ddots & \vdots \\ 0 & \ddots & \ddots & \ddots & 0 \\ \vdots & \ddots & \ddots & \ddots & 1 \\ 0 & \cdots & 0 & 1 & -2 \end{bmatrix} \tag{1.6}$$

It is commonly known that this method is convergent and that its error is of order $h^2$. If we want to replace the original problem by the finite dimensional one completely we must, moreover, solve the system (1.5). But the eigenvalues $\lambda_\nu$ of the matrix $A_h$ are given by the formula

$$(1.7) \qquad \lambda_\nu = -\frac{4}{h^2}\sin^2\frac{\nu\pi}{2m}, \qquad \nu = 1,\dots,m-1$$

and, consequently, the system (1.5) is stiff since $\lambda_{m-1}$ behaves for small $h$ as $-4/h^2$. Thus, if we apply to it a general method (not satisfying further special assumptions) the best result which we can expect is a relatively stable method, i.e., a method in which the magnitude of the time integration step is restricted by the magnitude of the space integration step. It may also happen that we get a completely unstable method. On the other hand, the application of an A-stable method will lead to an absolutely stable method and the convergence proof will cause no substantial difficulties.

Thus, we see that the problem of constructing absolutely stable methods the orders of which are arbitrarily high is very closely connected with the problem of constructing A-stable methods of arbitrarily high orders. Since it is known that the order of a classical method which is A-stable is at most 2 we must begin with introducing a class of methods which contain A-stable methods of arbitrarily high orders. There are various possibilities; we introduce the so called block onestep methods, especially for that reason that they are very simple and easily applicable to parabolic equations.

## 2. BO methods and A-stability

The method will be formulated for one differential equation

$$(2.1) \qquad u' = f(t,u), \qquad t \in \langle 0,T\rangle$$

with the initial condition

$$(2.2) \qquad u(0) = \eta .$$

The right-hand term of this differential equation is assumed to be defined, continuous and satisfying the Lipschitz condition with respect to $u$ in the strip $0 < t < T$, $-\infty < u < \infty$ so that the solution of the problem (2.1),(2.2) exists and is unique in the whole interval $\langle 0,T\rangle$.

Let an integer $k \geqq 1$, a matrix $C$ of order $k$ and a k-dimensional vector $\underline{d}$ be given. Further, put $t_i = ih$, $i = 0,1,\dots$, where $h > 0$ is the integration step and denote by $u_j$ the approximate solution at the point $t_j$. Then the block onestep method (BO method) is given by the formula

$$(2.3)\qquad \begin{bmatrix} u_{n+1} \\ \vdots \\ u_{n+k} \end{bmatrix} = \begin{bmatrix} u_n \\ \vdots \\ u_n \end{bmatrix} + hC \begin{bmatrix} f_{n+1} \\ \vdots \\ f_{n+k} \end{bmatrix} + hf_n\underline{d}, \quad n = 0,k,2k,\ldots$$

($f_j = f(t_j,u_j)$). One step of the BO method consists therefore in computing $k$ values of the approximate solution simultaneously from the generally nonlinear system of equations and the following step is started with the last one of these $k$ values. The Lipschitz property of $f$ guarantees that the method is practicable at least for sufficiently small h.

Defining now in the more or less usual way the local truncation error of the method and with its help the order it can be proved without substantial difficulties that the method of order at least 1 is convergent and that the method of order $p$ leads to the accuracy of order $h^p$ (supposing that the exact solution is sufficiently smooth).

If we now want to study the A-stability of a BO method we must apply it to the equation (1.1). If we eliminate unnecessary values of the approximate solution we get

$$(2.4)\qquad u_{(r+1)k} = \frac{P(z)}{Q(z)}u_{rk}, \quad r = 0,1,\ldots.$$

where

$$(2.5)\qquad z = hA,$$

$$(2.6)\qquad Q(z) = \det(I - zC)$$

and $P(z)$ is the determinant of the matrix which is obtained from the matrix $I - zC$ by replacing its last column by the vector $\underline{e} + z\underline{d}$ where $\underline{e} = (1,\ldots,1)^T$. Thus, the BO method leads in this special situation to the rational approximation of the exponential function $\exp(kz)$ and the fulfilment of the inequality

$$(2.7)\qquad \left|\frac{P(z)}{Q(z)}\right| < 1$$

for any $z$ with a negative real part forms obviously the necessary and sufficient condition for the A-stability of the BO method.

Further, the class of BO methods has such a property that to any rational approximation of the exponential there exists a BO method such that the ratio in (2.4) is exactly this approximation. This fact is very important and it implies among other that in the class of BO methods there exist A-stable methods of arbitrarily high orders.

## 3. Approximate solution of abstract differential equations

Let us pass now to the numerical solution of parabolic differential equations. As we have mentioned above the problem we are mostly interested in is the problem of the order of accuracy with respect to the time mesh-size. In order to emphasize this fact we will not deal in what follows with the partial differential equations of parabolic type but we will be interested in the abstract ordinary differential equation

$$\frac{du(t)}{dt} = Au(t) + f(t), \quad t \in (0,T) \tag{3.1}$$

with the initial condition

$$u(0) = \eta \tag{3.2}$$

where the unknown function $u(t)$ is a function of the real variable $t$ with values in a Banach space $B$, the given function $f(t)$ has also its values in $B$ and is assumed to be continuous while $A$ is generally an unbounded operator in $B$. We will suppose about it that its domain $\mathcal{D}(A)$ is dense in $B$, that $A$ is closed and that it is the generator of a strongly continuous semigroup of operators, i.e., that there exist (real) constants $M$ and $\omega$ such that

$$\|(\lambda I - A)^{-n}\| \leqq \frac{M}{(\mathrm{Re}\,\lambda - \omega)^n} \tag{3.3}$$

for any positive integer $n$ and for any (complex) $\lambda$ such that $\mathrm{Re}\,\lambda \geqq \omega$. In this situation, it is possible to speak also about the generalized solution of (3.1),(3.2) which is defined by the formula

$$u(t) = U(t)\eta + \int_0^t U(t-\tau)f(\tau)\,d\tau \tag{3.4}$$

where $U(t)$ is the semigroup generated by $A$. Consequently, this generalized solution exists for any $\eta \in B$.

Let us apply the BO method to the problem (3.1),(3.2). We get

$$\begin{bmatrix} u_{n+1} \\ \\ u_{n+k} \end{bmatrix} = \begin{bmatrix} u_n \\ \\ u_n \end{bmatrix} + hC \otimes A \begin{bmatrix} u_{n+1} \\ \\ u_{n+k} \end{bmatrix} + hD \otimes A \begin{bmatrix} u_n \\ \\ u_n \end{bmatrix} + hC \begin{bmatrix} f_{n+1} \\ \\ f_{n+k} \end{bmatrix} + hD \begin{bmatrix} f_n \\ \\ f_n \end{bmatrix} \tag{3.5}$$

where $D$ is the diagonal matrix with the components of the vector $\underline{d}$ on the main diagonal and the operator $C \otimes A$ mapping $\mathcal{D}(A) \times \dots \times \mathcal{D}(A)$ into $B \times \dots \times B$ is defined by

$$(3.6) \qquad C \otimes A = \begin{bmatrix} c_{11}A & \cdots & c_{1k}A \\ \vdots & & \vdots \\ c_{k1}A & \cdots & c_{kk}A \end{bmatrix}$$

and an analogous definition holds for the operator $D \otimes A$.

Here we cannot conclude as simply as above that (3.5) has a solution since here the operator $(I - hC \otimes A)$ is generally unbounded. Thus, the first question which must be answered is the question of the feasibility of our method. About this problem the following theorem can be easily proved.

Theorem 3.1 Let $A$ be the generator of a strongly continuous semigroup of operators and let $C$ have its eigenvalues in the right-hand half plane. Then there exists $h_o$ such that the operator $I - hC \otimes A$ has a bounded inverse for all $h \leqq h_o$ and it holds

$$(3.7) \qquad (I - hC \otimes A)^{-1} = \begin{bmatrix} M_{11} & \cdots & M_{1k} \\ \vdots & & \vdots \\ M_{k1} & \cdots & M_{kk} \end{bmatrix}$$

where $M_{ij}$ are rational functions of $hA$.

Strictly speaking, this theorem guarantees the feasibility of our method only in the case of the classical problem, i.e., in the case $\eta \in \mathcal{D}(A)$. But the operators $M_{ij}$ from Theorem 3.1 allow to rewrite (3.5) in the form which has sense also in the general case $\eta \in B$. The details will be omitted.

The practicability of the method does not guarantee the convergence. The convergence is controlled, as it can be expected, by the behaviour of the operator $R(hA) = P(hA)Q^{-1}(hA)$ where $P(z)$ and $Q(z)$ are the polynomials defined by (2.6).

Theorem 3.2 Let a BO method of order $p \geqq 1$ with a regular matrix $C$ be given and let $A$ be the generator of a strongly continuous semigroup of operators. Then the approximate solution obtained by this method converges at the point $t$ to the generalized solution of the problem (3.1).(3.2) if and only if

$$(3.8) \qquad \left\| R^n\left(\frac{t}{kn} A\right)\right\| \leqq M(t)$$

for $n = 0,1,\ldots$ Moreover, supposing that the generalized solution is sufficiently smooth the order of the error is $h^p$.

The proof is a simple consequence of expressing $R^n((t/kn)A)$ by the Dunford integral.

From this theorem it follows immediately that, e.g., in the case of a Hilbert space and a selfadjoint operator $A$ the A-stability is sufficient for the convergence. In general case, the results are not yet final. Nevertheless, the following theorem solves our problem in a special case.

Theorem 3.3 Let an A-stable BO method of order $p \geqq 1$ be given. Further, let $A$ be an operator with the domain which is dense in $B$ and let its resolvent $(\lambda I - A)^{-1}$ satisfy

$$(3.9) \qquad \|(\lambda I - A)^{-1}\| \leqq M(1 + |\lambda|)^q, \qquad q \geqq 0,$$

for $\operatorname{Re} \lambda > \omega$. Then it is possible to apply the method to the homogeneous problem (3.1),(3.2) with this operator and the sequence of elements obtained in this way forms for $\eta \in \mathcal{D}(A^{\ell})$ where $\ell > q + 1$ a convergent sequence.

Proof. Let $\ell > q + 1$ and let us fix $t$ and $\eta \in \mathcal{D}(A^{\ell})$. According to the preceding we have to prove that the sequence

$$(3.10) \qquad u_n = R^n\left(\frac{t}{kn} A\right)\eta$$

converges in $B$. To prove this fact, let us put first of all $U_n = \{\lambda, \operatorname{Re} \lambda \geq \omega_1, |\lambda| \leq Kn\}$ where $\omega_1 > \omega$ and $K$ is such a constant that the function $R((t/k)\lambda)$ is holomorphic outside the circle $|\lambda| \leqq K$. In virtue of this fact it follows immediately that $R((t/kn)\lambda)$ is holomorphic outside $U_n$ for sufficiently large $n$. Further, since $\eta \in \mathcal{D}(A^{\ell})$, there exists $z_o \in B$ such that

$$(3.11) \qquad \eta = (\lambda_o I - A)^{-\ell} z_o$$

and $\lambda_o$ is an arbitrary element from $U_1$. Thus, if we denote by $\Gamma_n$ the boundary of $U_n$ we can write, for any sufficiently large $n$ and for any $m \geqq n$,

$$(3.12) \qquad u_n = R^n\left(\frac{t}{kn} A\right)(\lambda_o I - A)^{-\ell} z_o =$$

$$\frac{1}{2\pi i}\int_{\Gamma_m} R^n(\tfrac{t}{kn}\lambda)(\lambda_0-\lambda)^{-\ell}(\lambda I - A)^{-1}z_0\,d\lambda .$$

The assumption (3.9) allows us to pass in (3.12) to the limit for $m \longrightarrow \infty$. We get

$$(3.13)\qquad u_n = \frac{1}{2\pi i}\int_{\omega_1-i\infty}^{\omega_1+i\infty} R^n(\tfrac{t}{kn}\lambda)(\lambda_0-\lambda)^{-\ell}(\lambda I - A)^{-1}z_0\,d\lambda .$$

The property $|R(\lambda)| < 1$ for $\mathrm{Re}\,\lambda < 0$ (following immediately from the A-stability of the given method) implies the existence of a constant $L$ (independent of $n$) such that

$$(3.14)\qquad \left|R^n(\tfrac{t}{kn}\lambda)\right| \leqq \exp(tL)$$

for $n = 1,2,\ldots$ and for $\mathrm{Re}\,\lambda = \omega_1$. Thus, the function

$$M\exp(tL)(\lambda_0-\lambda)^{-\ell}(1+|\lambda|^q)$$

forms an integrable majorant for the integrand in (3.13) and we can pass in (3.13) to the limit under the integral sign. We obtain

$$(3.15)\qquad \lim_{n\to\infty} u_n = \frac{1}{2\pi i}\int_{\omega_1-i\infty}^{\omega_1+i\infty} \exp(\lambda t)(\lambda_0-\lambda)^{-\ell}(\lambda I - A)^{-1}z_0\,d\lambda$$

and since the last integral converges absolutely the assertion of the theorem follows immediately.

Since in the case that $A$ is the generator of a strongly continuous semigroup of operators it is possible to choose for $q$ in Theorem 3.3 the value 0, it follows that in our situation the A-stable method is convergent for problems with sufficiently smooth initial data.

## References


[1] N. Dunford and J.T. Schwartz: Linear Operators, Interscience Publishers, New-York, London, 1958.

[2] M. Práger, J. Taufer and E. Vitásek: Overimplicit Multistep Methods, Aplikace matematiky 6(1973), 399 - 421.

[3] E. Vitásek and J. Taufer: Numerical Solution of Evolution Problems in Banach Spaces, Topics in Numerical Analysis II, Proceedings of the Royal Irish Academy Conference on Numerical

Analysis, 1974, edited by John J.H. Miller, 243 - 251.

Authors' address: Matematický ústav ČSAV, Opletalova 45,
110 00 Praha 2, Czechoslovakia

# MAPPING PROPERTIES OF REGULAR AND STRONGLY DEGENERATE ELLIPTIC DIFFERENTIAL OPERATORS IN THE BESOV SPACES $B^s_{p,p}(\Omega)$. THE CASE $0<p<\infty$

H. Triebel, Jena

## 1. Main Results

Let $\Omega$ be a bounded $C^\infty$ -domain in the Euclidean n-space $R_n$. Let A,

$$(Af)(x) = \sum_{|\alpha| \leq 2m} a_\alpha(x)\, D^\alpha f(x), \qquad a_\alpha(x) \in C^\infty(\bar{\Omega}),$$

be a properly elliptic differential operator of order 2m. Here m = 1,2,3,... Let $B_j$,

$$(B_j f)(x) = \sum_{|\alpha| \leq m_j} b_{j,\alpha}(x)\, D^\alpha f(x), \qquad b_{j,\alpha}(x) \in C^\infty(\partial\Omega),$$

j = 1,..., m, be m differential operators defined on the boundary $\partial\Omega$ of $\Omega$ . All functions in this paper, in particular the coefficients of the above differential operators, are complex-valued. As usual, $\{A, B_1, \ldots, B_m\}$ is said to be a regular elliptic problem if $0 \leq m_1 < m_2 < \ldots < m_m \leq 2m-1$ and if $\{B_j\}_{j=1}^m$ is a normal system satisfying the complementing condition with respect to A. For details concerning these well-known definitions we refer to [1] (cf. also [4] , pp. 361 - 363). It is convenient for our purpose to assume that the following additional assumption is satisfied.

Hypothesis. If $f(x) \in C^\infty(\bar{\Omega})$ such that $(Af)(x) = 0$ for $x \in \bar{\Omega}$ and $(B_j f)(x) = 0$ for $x \in \partial\Omega$ and $j = 1,\ldots,m$, then $f(x) \equiv 0$ in $\bar{\Omega}$.

Remark 1. In other words, it is assumed that the origin belongs to the resolvent set if $\{A, B_1, \ldots, B_m\}$ is considered as a mapping between appropriate function spaces.

Definition 1. (i) If

$$(1) \quad \begin{cases} \text{either } 1 < p < \infty & \text{and } 0 < s < \frac{1}{p} \\ \text{or } \quad 0 < p \leq 1 & \text{and } n(\frac{1}{p} - 1) < s < 1 \end{cases}$$

then $B^s_{p,p}(\Omega)$ is the completion of $C^\infty(\bar{\Omega})$ in the quasi-norm (norm if $p \geq 1$)

$$(2) \quad \|f\|_{B^s_{p,p}(\Omega)} = \left( \int_\Omega |f(x)|^p dx \right)^{\frac{1}{p}} + \left( \int_{\Omega\times\Omega} \frac{|f(x) - f(y)|^p}{|x-y|^{n+sp}} dx\, dy \right)^{\frac{1}{p}}$$

(ii) If p and s satisfy (1) and if $m = 1,2,3,\ldots$, then

$$(3) \quad B^{s+2m}_{p,p}(\Omega) = \{ f \mid D^{\alpha} f \in B^{s}_{p,p}(\Omega) \text{ for all } \alpha \text{ with } |\alpha| \leq 2m \}.$$

Remark 2. $B^{s+2m}_{p,p}(\Omega)$ is equipped in the usual way with a quasi-norm.

Remark 3. These are the underlying Besov spaces. The theory described below can be extended to an essentially larger class of Besov spaces $B^{s}_{p,q}(\Omega)$ and probably also to spaces of Hardy - Sobolev type defined on domains, cf. [8] . However the definitions are more complicated, cf. also Section 2. Furthermore, one can also include the case $p = \infty$ , which yields as a special case the famous Agmon-Douglis-Nirenberg theory in the Hölder - Zygmund spaces $\mathcal{C}^{s}(\Omega) = B^{s}_{\infty,\infty}(\Omega)$, where $s > 0$, cf. [8] .

Definition 2. Let $\nu$ be the (outer) normal with respect to $\partial\Omega$ . If p, s, and m have the meaning of Definition 1(ii) and if $k = 0,\ldots,2m-1$, then $B^{s+2m-k-\frac{1}{p}}_{p,p}(\partial\Omega)$ is the set of all distributions f on the compact $C^{\infty}$ - manifold $\partial\Omega$ for which there exists a function $g \in B^{s+2m}_{p,p}(\Omega)$ with $\frac{\partial^k g}{\partial \nu^k}\Big|_{\partial\Omega} = f$.

Remark 4. The spaces $B^{s}_{p,p}(R_n)$ (and more general $B^{s}_{p,q}(R_n)$) can be defined for all values of s, p (and q) with $-\infty < s < \infty$ , $0 < p \leq \infty$ (and $0 < q \leq \infty$ ). Using the standard method of local coordinates one can give a direct definition of the corresponding spaces on $\partial\Omega$, cf. [8] . In particular, the spaces in Definition 2 depend only on the difference 2m-k and not on the special choice of m and k. If $1 < p < \infty$ , then one has a well-known assertion, cf. e. g. [4] , p. 330, (cf. also Step 4 in Section 2, where further comments, also concerning the correctness of Definition 2, are given).

Theorem 1. Let $\{A, B_1, \ldots, B_m\}$ be regular elliptic and let the Hypothesis be satisfied. If p and s are given by (1) then $\{A, B_1, \ldots, B_m\}$ yields an isomorphic mapping from

$$(4) \quad B^{s+2m}_{p,p}(\Omega) \quad \text{onto} \quad B^{s}_{p,p}(\Omega) \times \prod_{j=1}^{m} B^{s+2m-m_j-\frac{1}{p}}_{p,p}(\partial\Omega).$$

Remark 5. The proof of this theorem is long and complicated. However in Section 2 we shall try to describe some of the main ideas and key-assertions of the proof. A more detailed version, including also more general spaces, will be published elsewhere, cf. [8].

In [4] , Chapter 6, we considered a rather general class of strongly degenerate elliptic differential operators in the framework of an $L_p$-theory, where $1 < p < \infty$ . On the one hand, we want to extend this theory to the spaces $B^s_{p,p}$ in the sense of Definition 1(i), on the other hand, in order to avoid technical difficulties, we restrict ourselves to a model case. Again, $\Omega$ is a bounded $C^\infty$-domain in $R_n$. The distance of $x \in \Omega$ from $\partial\Omega$ is denoted by $d(x)$.

Definition 3. If (1) is satisfied, $m = 1,2,3,\ldots$ and $\nu > 2m$, then $B^{s+2m}_{p,p}(\Omega, d^{-p\nu}(x))$ is the completion of $C_0^\infty(\Omega)$ in the quasi-norm (norm if $p \geq 1$)

$$(5) \qquad \|f\|_{B^{s+2m}_{p,p}(\Omega)} + \|d^{-\nu} f\|_{B^s_{p,p}(\Omega)} .$$

Theorem 2. If all the parameters have the same meaning as in Definition 3 and if $\lambda$ is a complex number with sufficiently large real part, then the operator $A + \lambda E$,

$$(6) \qquad (Af)(x) = (-\Delta)^m f + d^{-\nu}(x) f(x), \qquad E \text{ identity},$$

yields an isomorphic mapping from $B^{s+2m}_{p,p}(\Omega, d^{-p\nu}(x))$ onto $B^s_{p,p}(\Omega)$.

Remark 6. In Section 3 we sketch some main ideas of the proof. Theorem 2 can be extended essentially to more general operators and also to a wider class of underlying spaces. Detailed proofs and a precise description of the mentioned extensions will be published elsewhere, cf. [7] .

## 2. Outline of the Proof of Theorem 1

Step 1. ( Extension). If p and s satisfy (1) and if $\Omega$ in (2) and (3) is replaced by $R_n$, then one obtains corresponding spaces $B^{s+2m}_{p,p}(R_n)$. First of all we need properties of the spaces $B^{s+2m}_{p,p}(\Omega)$ and $B^{s+2m}_{p,p}(R_n)$. It can be shown that $B^{s+2m}_{p,p}(\Omega)$ is the restriction of $B^{s+2m}_{p,p}(R_n)$ to $\Omega$ (factor space) and that there exists a linear and bounded extension operator from $B^{s+2m}_{p,p}(\Omega)$ into $B^{s+2m}_{p,p}(R_n)$.

Step 2. (Fourier decomposition and Fourier multiplier). By Step 1 it is clear that properties of the spaces $B^\sigma_{p,q}(R_n)$ (in our case $\sigma = s + 2m$ and $p = q$ with the above restrictions) are of interest. Peetre's definition of the Besov spaces $B^\sigma_{p,q}(R_n)$ with $-\infty < \sigma < \infty$, $0 < p \leq \infty$ , and $0 < q \leq \infty$ is the following. Let $S(R_n)$ be the Schwartz space and let $S'(R_n)$ be the space of tempered distributions. Let $\varphi = \{\varphi_j(x)\}_{j=0}^\infty \subset S(R_n)$ be a smooth dyadic resolution of unity in $R_n$, i. e. $0 \leq \varphi_j(x) \leq 1$, $\sum_{j=0}^\infty \varphi_j(x) = 1$ for $x \in R_n$,

$$\operatorname{supp} \varphi_0 \subset \{y \mid |y| \leq 2\}, \quad \operatorname{supp} \varphi_j \subset \{y \mid 2^{j-1} \leq |y| \leq 2^{j+1}\}$$

if $j = 1,2,\ldots$; for any multi-index $\gamma$ there exists a constant $c_\gamma$ such that

$$|D^{\gamma}\varphi_j(x)| \leq c_\gamma 2^{-j|\gamma|}, \quad x \in R_n, \quad j = 0,1,2,\ldots$$

If $-\infty < \sigma < \infty$, $0 < p \leq \infty$, and $0 < q \leq \infty$, then

$$B^{\sigma}_{p,q}(R_n) = \{f \mid f \in S'(R_n), \ \|f\|^{\varphi}_{B^{\sigma}_{p,q}(R_n)} =$$

$$\Big[\sum_{j=0}^{\infty} 2^{j\sigma q}\Big(\int_{R_n} |F^{-1}[\varphi_j Ff](x)|^p dx\Big)^{\frac{q}{p}}\Big]^{\frac{1}{q}} < \infty$$

for all systems $\varphi\}$,

(usual modification if p or q equals $\infty$ ). Here F and $F^{-1}$ are the Fourier transform and its inverse on $R_n$, respectively. It can be shown that $B^{\sigma}_{p,q}(R_n)$ is a quasi-Banach space, where all the quasi-norms $\|f\|^{\varphi}_{B^{\sigma}_{p,q}(R_n)}$ for different choices of $\varphi$ are mutually equivalent. Furthermore, $B^{\sigma}_{p,q}(R_n)$ coincides with the above spaces $B^{s+2m}_{p,p}(R_n)$ if $\sigma = s+2m$ and $p = q$ (under the above restrictions of the parameters s, p and m). All the spaces $B^{\sigma}_{p,q}(R_n)$ satisfy the following weak Michlin - Hörmander Fourier multiplier property. There exists a natural number M and a positive number c (depending on $\sigma$, p and q) such that for all infinitely differentiable functions m(x) on $R_n$ and all $f \in B^{\sigma}_{p,q}(R_n)$

$$\|F^{-1}[m(\cdot)Ff]\|_{B^{\sigma}_{p,q}(R_n)} \leq c\big(\sup_{\substack{|\alpha| \leq M \\ x \in R_n}} (1+|x|^2)^{\frac{|\alpha|}{2}} |D^{\alpha}m(x)|\big) \|f\|_{B^{\sigma}_{p,q}(R_n)}.$$

(We omit the index $\varphi$ in $\|\cdot\|^{\varphi}_{B^{\sigma}_{p,q}(R_n)}$ because all these quasi-norms are mutually equivalent). Proofs of the assertions in this step may be found in [3] and [5].

<u>Step 3.</u> (Properties of the spaces $B^{\sigma}_{p,q}(R_n)$). The goal is to extend Arkeryd's proof (cf. [2] or [4], Chapter 5) for boundary value problems of $\{A, B_1, \ldots, B_m\}$ in the framework of an $L_p$-theory with $1 < p < \infty$ to the spaces $B^s_{p,p}$ in the sense of Definition 1(i). For this purpose, beside the extension property and the Fourier multiplier property described in the preceding steps, some other properties of the corresponding spaces on $R_n$ are indispensable. (i) (Diffeomorphic mappings, cf. [8] ). If $y = \psi(x)$ is an infinitely differentiable one-to-one mapping from $R_n$ onto itself such that $\psi(x) = x$ for large values of $|x|$, then $f(x) \longrightarrow f(\psi(x))$ yields an isomorphic mapping from $B^{\sigma}_{p,q}(R_n)$ onto itself. Here $-\infty < \sigma < \infty$, $0 < p \leq \infty$, and $0 < q \leq \infty$. (ii) (Multiplication property, cf. [5] ). If $g(x) \in C^{\infty}_0(R_n)$ then $f(x) \longrightarrow g(x) f(x)$ yields a linear and bounded mapping from $B^{\sigma}_{p,q}(R_n)$ into itself. Again $-\infty < \sigma < \infty$, $0 < p \leq \infty$ and $0 < q \leq \infty$.

Step 4. (Spaces on domains and manifolds). The two properties described in Step 3 (diffeomorphic mappings and multiplication property) are the basis for the well-known method of local coordinates. This gives the possibility to define the spaces $B^{\sigma}_{p,q}(\partial\Omega)$, where $-\infty < \sigma < \infty$ , $0 < p \leqq \infty$ and $0 < q \leqq \infty$ by standard arguments, cf. [4] , pp. 280/81 for the usual Besov spaces. The next step shows that these spaces coincide with the corresponding spaces in Definition 2 (under the restrictions of the parameters in the sense of Definition 2). Finally, by restriction of $B^{\sigma}_{p,q}(R_n)$ to $\Omega$ one can define spaces $B^{\sigma}_{p,q}(\Omega)$ for all values $-\infty < \sigma < \infty$ , $0 < p \leqq \infty$ and $0 < q \leqq \infty$ . All these spaces have the extension property described in Step 1, cf. [8] .

Step 5. (Traces). Let $\nu$ be the (outer) normal on $\partial\Omega$ and let $r = 0,1,2,\ldots$ By the above properties and the assertion in [5] , 2.4.2, it follows that R,

$$Rf = \left\{ f\big|_{\partial\Omega} , \frac{\partial f}{\partial\nu}\Big|_{\partial\Omega} , \ldots , \frac{\partial^r f}{\partial\nu^r}\Big|_{\partial\Omega} \right\},$$

is a linear and bounded mapping from $B^{\sigma}_{p,q}(\Omega)$ onto

$$\prod_{j=0}^{r} B^{\sigma-\frac{1}{p}-j}_{p,q}(\partial\Omega),$$

if $0 < p \leqq \infty$ , $0 < q \leqq \infty$ , and $s > r + \frac{1}{p} + \max(0, (n-1)(\frac{1}{p}-1))$. Now it follows that Definition 2 is meaningful and that the spaces defined there coincide with the corresponding spaces in the sense of the preceding step.

Step 6. (A-priori estimate). If p and s satisfy (1) then there exist two positive constants $c_1$ and $c_2$ such that for all $f \in C^{\infty}(\overline{\Omega})$

$$\begin{aligned} c_1 \|f\|_{B^{s+2m}_{p,p}(\Omega)} &\leqq \|Af\|_{B^{s}_{p,p}(\Omega)} + \|f\|_{B^{s}_{p,p}(\Omega)} + \\ &+ \sum_{j=1}^{m} \|B_j f\|_{B^{s+2m-m_j-\frac{1}{p}}_{p,p}(\partial\Omega)} \leqq c_2 \|f\|_{B^{s+2m}_{p,p}(\Omega)} \end{aligned} \tag{7}$$

Here $\{A, B_1, \ldots, B_m\}$ is regular elliptic. For the proof of (7) it is not necessary that the above Hypothesis is true. The idea is to carry over Arkeryd's proof, cf. [2] , of a corresponding a-priori estimate in the framework of an $L_p$-theory with $1 < p < \infty$ , to the above basic spaces $B^{s}_{p,p}(\Omega)$ instead of $L_p(\Omega)$. We use the version of Arkeryd's proof given in [4] , pp. 364 - 378. An examination of that proof shows that many arguments can be carried over from $L_p$ to $B^{s}_{p,p}$ if one uses the 5 main assertions for general Besov spaces mentioned above: extension properties (Step 1 and Step 4), Fourier multiplier properties (Step 2), diffeomorphic mappings (Step 3), multiplication properties (Step 3), and traces (Step 5). However there remain essentially two points which are trivial for $L_p$-spaces but non-trivial for $B^s_{p,p}$-spaces. (i) If p and s satisfy (1) then S,

$$(Sf)(x) = \begin{cases} f(x) & \text{if } x \in \Omega \\ 0 & \text{if } x \in R_n - \Omega \end{cases}$$

is a linear and bounded operator from $B^s_{p,p}(\Omega)$ into $B^s_{p,p}(R_n)$. This assertion follows from the method of local coordinates and the considerations in [5] , 2.6.4. Cf. also [8] , Proposition 3.5. (ii) Let $a_\alpha(x)$ be the coefficients of A and let K be a ball in $R_n$ with the centre $x^0$ and the radius $\tau$ , where we assume $0 < \tau < 1$. If p and s satisfy (1), then there exists a constant c, which is independent of $x^0$ and $\tau$ such that for all $\psi \in C_0^\infty(K)$ and all $f \in B^s_{p,p}(\Omega)$

$$\sum_{|\alpha|=2m} \| (a_\alpha(x) - a_\alpha(x^0))\, D^\alpha(\psi f) \|_{B^s_{p,p}(\Omega)} \leq c\tau \, \| \psi f \|_{B^{s+2m}_{p,p}(\Omega)} + c \, \| \psi f \|_{B^{s+2m-1}_{p,p}(\Omega)} .$$

Using the method of local coordinates then this estimate follows from

$$\| (a_\alpha(x) - a_\alpha(x^0))\, \psi f \|_{B^s_{p,p}(R_n)} \leq c\tau \, \| \psi f \|_{B^s_{p,p}(R_n)} ,$$

where again c is independent of $\tau$ . This inequality coincides essentially with formula (52) in [6] . If one uses the special properties (i) and (ii) and the above-mentioned general properties for the spaces $B^s_{p,p}$ then one obtains (7) in the same way as in [4] , pp. 364 - 378, where $L_p$ is replaced by $B^s_{p,p}$.

Step 7. (Proof of Theorem 1). If the Hypothesis is satisfied then the term $\| f \|_{B^s_{p,p}(\Omega)}$ in (7) can be omitted (standard arguments). Now, Theorem 1 is a consequence of the classical theory for $\{A, B_1, \ldots, B_m\}$ and (7).

## 3. Outline of the Proof of Theorem 2

Step 1. (Mappings in the nuclear space $C_0^\infty(\bar{\Omega})$). If

$$C_0^\infty(\bar{\Omega}) = \{ f \mid f \in C_0^\infty(R_n), \ \operatorname{supp} f \subset \bar{\Omega} \} ,$$

then $A + \lambda E$ with a sufficiently large real part of $\lambda$ yields an isomorphic mapping from $C_0^\infty(\bar{\Omega})$ onto itself. This is a special case of Theorem 1 in [4] , p. 420.

Step 2. (Decomposition). Next we need some properties of the spaces $B^s_{p,p}(\Omega)$, where s and p satisfy (1). There exists a constant c such that for all $f \in C_0^\infty(\Omega)$

$$(8)\quad \int_\Omega d^{-sp}(x)|f(x)|^p\,dx \leq c(\int_\Omega |f(x)|^p dx + \int_{\Omega\times\Omega} \frac{|f(x)-f(y)|^p}{|x-y|^{n+sp}}\,dx\,dy)^{\frac{1}{p}}.$$

This is a fractional Hardy inequality. A proof of (8) for $1<p<\infty$ may be found in [4], p. 259. Using this result, one can extend (8) with $1<p<\infty$ to all couples (s,p) satisfying (1). In particular,

$$(9)\quad (\int_\Omega d^{-sp}(x)|f(x)|^p\,dx + \int_{\Omega\times\Omega} \frac{|f(x)-f(y)|^p}{|x-y|^{n+sp}}\,dx\,dy)^{\frac{1}{p}}$$

is an equivalent quasi-norm in $B^s_{p,p}(\Omega)$. Now we have a situation which is similar to that one in [4], Subsection 3.2.3 and Subsection 6.3.1. The decomposition methods developed there can be applied ( however some non-trivial additional considerations are necessary, for details we refer to [7] ). Let $K_{j,l} = \{x \mid |x-x_{j,l}| < \tau\, 2^{-j}\}$ be balls such that $x_{j,l} \in \{y \mid y\in\Omega\ ,\ 2^{-j-1} \leq d(y) \leq 2^{-j}\}$ if j = 1,2, 3,..., with a sufficiently small $\tau$. It is assumed that $\Omega = \bigcup_{j=0}^{\infty}\bigcup_{l=1}^{N_j} K_{j,l}$ (modification for j = 0 or for small values of j if necessary). Let $\varphi = \{\varphi_{j,l}\}_{\substack{j=0,1,2,\dots\\ l=1,\dots,N_j}}$ be a smooth resolution of unity with respect to the balls $K_{j,l}$, i. e. if $x\in\Omega$ then

$$0 \leq \varphi_{j,l}(x),\quad \sum_{j=0}^{\infty}\sum_{l=1}^{N_j} \varphi_{j,l}(x) = 1,\quad \operatorname{supp}\varphi_{j,l} \subset K_{j,l}\,.$$

Furthermore, for any multi-index $\gamma$ there exists a number $c_\gamma$ such that

$$|D^\gamma \varphi_{j,l}(x)| \leq c_\gamma\, 2^{j|\gamma|} \quad \text{if } j = 0,1,2,\dots \text{ and } l = 1,\dots,N_j.$$

Now it can be proved that for any system $\varphi$ with the indicated properties

$$(10)\qquad (\sum_{j=0}^{\infty}\sum_{l=1}^{N_j} \|\varphi_{j,l}f\|^p_{B^s_{p,p}(R_n)})^{\frac{1}{p}}$$

is an equivalent quasi-norm in $B^s_{p,p}(\Omega)$ (here s and p satisfy (1)). Similarly one obtains that for the spaces $B^{s+2m}_{p,p}(\Omega,\ d^{-\nu p}(x))$ described in Definition 3 and formula (5)

$$(11)\qquad (\sum_{j=0}^{\infty}\sum_{l=1}^{N_j} \|\varphi_{j,l}f\|^p_{B^{s+2m}_{p,p}(R_n)} + 2^{j\nu p}\|\varphi_{j,l}f\|^p_{B^s_{p,p}(R_n)})^{\frac{1}{p}}$$

is an equivalent quasi-norm.

<u>Step 3.</u> (A-priori estimate). Now we have a situation which is simi-

lar as in [4] , Section 6.3. The proofs given there can be extended to the above case, where (10) and (11) play a decisive role. Again some non-trivial modifications are necessary, cf. [7] , where details are given. One obtains the following a-priori estimate. If p and s satisfy (1) then there exists a real number $\lambda_0$ and two positive numbers $c_1$ and $c_2$ such that for all complex numbers $\lambda$ with Re $\lambda \geqq \lambda_0$ and for all $f \in C_0^\infty(\overline{\Omega})$

$$c_1 \| (A + \lambda E) f \|_{B^s_{p,p}(\Omega)} \leqq \| f \|_{B^{s+2m}_{p,p}(\Omega, d^{-\nu p}(x))} + |\lambda| \, \| f \|_{B^s_{p,p}(\Omega)}$$

$$\leqq c_2 \| (A + \lambda E) f \|_{B^s_{p,p}(\Omega)} \cdot$$

Step 4. (Proof of Theorem 2). Since $C_0^\infty(\overline{\Omega})$ is dense in $B^s_{p,p}(\Omega)$ and dense in $B^{s+2m}_{p,p}(\Omega, d^{-\nu p}(x))$, Theorem 2 is an easy consequence of Step 1 and the a-priori estimate of the preceding step.

## References


[1] S. Agmon. On the eigenfunctions and on the eigenvalues of general elliptic boundary value problems. Comm. Pure Appl. Math. 15 (1962), 119 - 147.

[2] L. Arkeryd. On the $L^p$ estimates for elliptic boundary problems. Math. Scand. 19 (1966), 59 - 76.

[3] J. Peetre. New Thoughts on Besov Spaces. Duke Univ. Math. Series. I. Duke Univ., Durham, 1976.

[4] H. Triebel. Interpolation Theory, Function Spaces, Differential Operators. North - Holland Publ. Comp., Amsterdam, New York, Oxford, 1978.

[5] H. Triebel. Spaces of Besov - Hardy - Sobolev Type. Teubner - Texte Math., Teubner, Leipzig, 1978.

[6] H. Triebel. On the Dirichlet problem for regular elliptic differential operators of second order in the Besov spaces $B^s_{p,q}$. The case $0 < p \leqq \infty$ . Proc. Conf. "Elliptische Differentialgleichungen" held in Rostock, Sept. 1977 (to appear).

[7] H. Triebel. Strongly degenerate elliptic differential operators in Besov spaces. The case $0 < p \leqq \infty$ . Beiträge Anal. 13 (1978).

[8] H. Triebel. On Besov - Hardy - Sobolev spaces in domains and regular elliptic boundary value problems. The case $0 < p \leqq \infty$. (to appear)

Author's address: Sektion Mathematik, Universität Jena,
DDR - 69 Jena, Universitäts - Hochhaus

# A NEW DEFINITION AND SOME MODIFICATIONS OF FILIPPOV CONE

I.Vrkoč, Praha

Let $G$ be a region in an n+1-dimensional Euclidean space $R_{n+1}$ and let $f(t,x)$ be a measurable function $f : G \to R_n$ . Denote by $F(t,x)$ the set

$$F(t,x) = \bigcap_{d>0} \bigcap_{N,m(N)=0} \overline{\text{Conv}}\, f(t,U(x,d)-N) \tag{1}$$

where $U(x,d) = \{y : \|y-x\|<d\} \subset R_n$ , $N \subset U(x,d)$, $m$ is the Lebesgue measure in $R_n$ , $\overline{\text{Conv}}\ A$ is the closed convex hull of the set $A$. The set $\{[q,qy] : q \geq 0 , y \in F(t,x)\}$ is called the Filippov cone at the point $[t,x] \in G$ and mapping $[t,x] \to F(t,x)$ the Filippov mapping.

A vector function $x(t)$ is a generalized solution in the Filippov sense of $\dot{x} = f(t,x)$ if $x(t)$ is defined on a nondegenerate interval $I$, $[t,x(t)] \in G$ for $t \in I$, $x(t)$ is absolutely continuous on $I$ and $\dot{x}(t) \in F(t,x(t))$ for almost all $t \in I$. The notion of Filippov's generalized solutions depends essentially on formula (1). J.Kurzweil established a certain minimum property of the mapping $F$ in his book about ordinary differential equations not yet published. We shall mention the property in more detail later. The purpose of the lecture is to show that the Filippov mapping can be constructed on the basis of this property and to present some modifications offered by this approach.

## The autonomous case

First some definitions and notation. Let $\mathcal{A}_0$ be the class of all subsets of $R_n$ and let $h$ be a mapping $h : G \to \mathcal{A}_0$ where $G$ is a region in $R_n$ . The mapping $h$ is called locally essentially bounded if to every $x_0 \in G$ there exist numbers $d>0$ and $c \geq 0$ such that $m\{x : x \in U(x_0,d),\ h(x) \not\subset U(0,c)\} = 0$ .

The mapping h is called upper semi-continuous if to every $d>0$ and $x\in G$ there exists $r>0$ such that $h(y)\subset U(h(x),d)$ for $y\in \in U(x,r)$ where $U(A,d)$ is the d-neighbourhood of the set A with $U(\emptyset,d)=\emptyset$.

Denote by $\mathcal{C}_0$ the class of all compact subsets of $R_n$ and by $\mathcal{E}_0$ a class fulfilling

a) to every $A\in\mathcal{C}_0$ there exists a set $B\in\mathcal{E}_0$, $A\subset B$;

b) if $B_p\in\mathcal{E}_0$ then $\cap B_p\in\mathcal{E}_0$;

c) if $A\in\mathcal{E}_0$ then $A\in\mathcal{C}_0$.

Let $\mathcal{C}$, $\mathcal{E}$ and $\mathcal{A}$ be classes originating respectively from $\mathcal{C}_0$, $\mathcal{E}_0$ and $\mathcal{A}_0$ by excluding the empty set.

Further, let $\{h_z, z\in Z\}$, $Z\neq\emptyset$ be a family of mappings $h_z: G\to\mathcal{A}_0$. The greatest lower bound $h: G\to\mathcal{A}_0$ of the family is the mapping h defined by $h(x)=\bigcap_{z\in Z} h_z(x)$. We shall write $h=\bigwedge_{z\in Z} h_z$. The mapping $h_1(x)$ is before $h_2(x)$ ($h_1\leqq h_2$) if $h_1(x)\subset h_2(x)$ for all $x\in G$.

<u>Definition.</u> Let f be a mapping $f: G\to\mathcal{A}$ and $\mathcal{E}$ a class such that $\mathcal{E}_0$ fulfils a) to c). Denote by $R(f,\mathcal{E})$ the family of all mappings h fulfilling

i) $h(x)\in\mathcal{E}$ for all $x\in G$;

ii) h is upper semi-continuous on G;

iii) $f(x)\subset h(x)$ for almost all $x\in G$.

The condition under which the set $R(f,\mathcal{E})$ is nonempty is given in

<u>Theorem 1.</u> Let a class $\mathcal{E}_0$ fulfil a) to c). The set $R(f,\mathcal{E})$ is nonempty if and only if the mapping f is locally essentially bounded.

Given a class $\mathcal{E}$ and a mapping f, Theorem 1 enables us to construct the greatest lower bound $S=\bigwedge_{h\in R(f,\mathcal{E})} h$. Basic properties of S are given in

<u>Theorem 2.</u> Let a class $\mathcal{E}_0$ fulfil a) to c). If the mapping $f: G\to\mathcal{A}$

is locally essentially bounded then $S \in R(f, \mathcal{E})$ .

The mapping $S$ depends on the class $\mathcal{E}$ . The most important classes are subclasses of $\mathcal{K}$ where $\mathcal{K}$ is the class of all compact, convex and nonempty subsets of $R_n$. In these cases the existence theorem [2] can be applied due to Theorem 2 to a differential relation $\dot{x} \in S(x)$ . Thus if a locally essentially bounded mapping $f$ and a class $\mathcal{E} \subset \mathcal{K}$ are given, then the mapping $S$ exists and its properties guarantee that the set of all solutions of the differential relation $\dot{x} \in S(x)$ is nonempty. These solutions can be called $\mathcal{E}$-generalized solutions of the differential relation $\dot{x} \in f(x)$. If $\mathcal{E} = \mathcal{K}$ then the $\mathcal{E}$-generalized solutions of $\dot{x} \in f(x)$ can be called the generalized solutions of $\dot{x} \in f(x)$ in the Filippov sense. This definition is justified by the following theorem.

<u>Theorem 3</u>. Let $f$ be a measurable and locally essentially bounded function. If $\mathcal{E} = \mathcal{K}$ then $S = F$ .

This theorem directly implies that the $\mathcal{K}$-generalized solutions of $\dot{x} = f(x)$ are exactly Filippov's generalized solutions of the equation. Theorem 3 together with the definition of $S$ yield $F \leqq h$ for $h \in R(f, \mathcal{K})$. This means that $F$ is the minimum mapping from those fulfilling i) to iii) and this is the minimum property mentioned in the introduction.

Let $\mathcal{Q}$ be the class of all Cartesian products $\prod_{i=1}^{n} J_i$ of compact, nonempty intervals and put $\mathcal{Q}_0 = \mathcal{Q} \cup \{\emptyset\}$ . Certainly $\mathcal{Q}_0$ fulfils conditions a) to c). Another interesting choice of $\mathcal{E}$ is $\mathcal{E} = \mathcal{Q}$.

<u>Theorem 4</u>. Let $f$ be a measurable and locally essentially bounded function. Assume $\mathcal{E} = \mathcal{Q}$ . Then a vector function $x(t)$ is an $\mathcal{E}$-generalized solution of $\dot{x} = f(x)$ if and only if $x(t)$ is a generalized solution of $\dot{x} = f(x)$ in the sense of Viktorovskii.

The generalized solutions in the sense of Viktorovskii are defined in [3] : A vector function $x(t)$ is a generalized solution of $\dot{x} = f(x)$ in the sense of Viktorovskii if $x(t)$ is defined on a nondegenerate interval $I$, $x(t)$ is absolutely continuous on $I$, if to every $d>0$ and to every subset $N$ of $I\times G$ with zero n+1-dimensional Lebesgue measure exist vector functions $z^{(i)}(t)$, $i=1,\dots,n$, defined on $I$ such that $z^{(i)}(t)\in G$ for $t\in I$, $f_i(z^{(i)}(t))$ are integrable on $I$, $\|x(t) - z^{(i)}(t)\| < d$ on $I$, $\left| x_i(t) - x_i(t_o) - \int_{t_o}^{t} f_i(z^{(i)}(s))ds \right| < d$ for $t, t_o \in I$ and $[t, z^{(i)}(t)] \notin N$ for almost all $t\in I$ and $i=1,\dots,n$. Theorem 4 is a consequence of theorems from [4].

Let us sketch the proofs of the previous theorems. If $R(f,\mathcal{E})$ is nonempty, i.e. there exists $h\in R(f,\mathcal{E})$, then items ii) and iii) immediately imply the local essential boundedness of $f$. On the other hand, let $f$ be locally essentially bounded. Choose $x_o\in G$. By definition there exists $d>0$, $c>0$ such that

$$m\left\{x : x\in U(x_o,d),\ h(x)\not\subset U(0,c)\right\} = 0 .$$

Due to a) there exists $B\in\mathcal{E}$, $\overline{U(0,c)}\subset B$. Denote by $h_{x_o}$ the mapping $h_{x_o}(x) = B$ for $x\in U(x_o,d)$ and $h_{x_o}(x) = R_n$ for $x\notin U(x_o,d)$. The mapping $h_{x_o}$ fulfils ii) and iii). We can easily construct a set $X_o$ such that the greatest lower bound $h = \bigwedge_{x_o\in X_o} h_{x_o}$ fulfils all conditions of the definition. Theorem 1 is proved.

We pass to the proof of Theorem 2. First we mention that the set $R(f,\mathcal{E})$ is closed with respect to the greatest lower bounds of countably many mappings, i.e. we have

Lemma 1. If $h_p\in R(f,\mathcal{E})$ then $\bigwedge h_p\in R(f,\mathcal{E})$.

The second step consists in approximating $S$ by a sequence of mappings from $R(f,\mathcal{E})$. The approximation of $S$ at one point is given by

Lemma 2. Let $x \in G$ and $d > 0$ be given. Then there exists $h \in$ $\in R(f, \mathcal{E})$ such that $h(x) \subset U(S(x), d)$ .

This lemma yields that $S$ is upper semi-continuous. Since the mapping $h$ is upper semi-continuous there exists $r > 0$ such that $h(y) \subset U(h(x), d)$ for $y \in U(x,r)$. By the definition of $S$ and by Lemma 2 we have $S(y) \subset h(y) \subset U(h(x), d) \subset U(S(x), 2d)$ for $y \in U(x,r)$. The upper semi-continuity of $S$ is proved.

Let now a point $x \in G$ and a nonnegative integer $p$ be given. By Lemma 2 there exists $h \in R(f, \mathcal{E})$ such that $h(x) \subset U(S(x), 1/p)$. Since $h$ is upper semi-continuous there exists $r(x,p) > 0$ such that $h(y) \subset U(h(x), 1/p) \subset U(S(x), 2/p)$ for $y \in U(x, r(x,p))$. For a given $p$ the balls $U(x, r(x,p))$ cover $G$ and we can choose a countable covering. Denote the corresponding points by $x_p^i$ , $i=1,2,\ldots$ and the corresponding mappings by $h^{(i,p)}$. The upper semi-continuity of $S$ and the properties of $h^{(i,p)}$ imply $\bigwedge_{i,p} h^{(i,p)} \leqq S$. Since $S \leqq$ $\leqq \bigwedge_{i,p} h^{(i,p)}$ by the definition of $S$ we have $S = \bigwedge_{i,p} h^{(i,p)}$. The statement of Theorem 2 now follows from Lemma 1.

Theorem 3 can be now easily proved.

The inclusion $F \in R(f, \mathcal{K})$ follows directly from formula (1). Let now a mapping $h \in R(f, \mathcal{K})$ be given. Choose a point $x_o \in G$ and a number $d > 0$. There exists $r > 0$ such that $h(x) \subset U(h(x_o), d)$ for $x \in U(x_o, r)$ and condition iii) implies $f(x) \in h(x) \subset U(h(x_o), d)$ for almost all $x \in U(x_o, r)$, i.e. the set $N_r = \{x : x \in U(x_o, r), f(x) \notin$ $\notin U(h(x_o), d)\}$ has Lebesgue measure zero. Put $\tilde{N} = \bigcup_{r>0} N_r$ . Consider the formula (1) with the set $\tilde{N}$ instead of $N$. We obtain $F(x_o) \subset$ $\subset \bigcap_{d>0} \overline{\mathrm{Conv}\, f(U(x_o, r) - N_r)} \subset \bigcap_{d>0} \overline{U(h(x_o), d)} = h(x_o)$ . We proved $F \leqq h$ which completes the proof of Theorem 3.

## The nonautonomous case

The construction of $S$ in the nonautonomous case can be reduced to the autonomous case. Denote the points of $R_{n+1}$ by $[t,x]$ where $t \in R_1$ and $x \in R_n$. We shall use the notation $A_t = \{x : [t,x] \in A\}$ for $A \subset R_{n+1}$. Let $f$ be a mapping $f : G \to \mathcal{A}_0$ where $G$ is a region in $R_{n+1}$. We can define mappings $f_t$ on $G_t$ for every $t$ by $f_t(x) = f(t,x)$ for $x \in G_t$.

Definition. A mapping $f : G \to \mathcal{A}$ where $G$ is a region in $R_{n+1}$ is t-locally essentially bounded if to every point $[t_0,x_0] \in G$ there exist $d > 0$ and a function $c(t)$ defined and integrable on the interval $\langle t_0-d, t_0+d \rangle$ such that $m\{x : x \in U(x_0,d), f(t,x) \not\subset \not\subset U(0,c(t))\} = 0$ for almost all $t \in \langle t_0-d, t_0+d \rangle$.

Assume that $f$ is t-locally essentially bounded. Then there exists a set $T$, $T \subset R_1$ with Lebesgue measure zero such that $f_t$ are locally essentially bounded for $t \in R_1 - T$. We can construct tne corresponding $S_t$ for $t \in R_1 - T$. We put $S(t,x) = S_t(x)$. The mapping $S$ is defined on $G - T \times R_n$, i.e. almost everywhere on $G$. As in the autonomous case the solutions of $\dot{x} \in S(t,x)$ will be called the $\mathcal{E}$-generalized solutions of $\dot{x} \in f(t,x)$.

Formula (1) implies also $F_t(x) = F(t,x)$ for almost all $t$ and this allows us to generalize Theorem 3.

Theorem 5. Let $f$ be a measurable and t-locally essentially bounded function, where $G$ is a region in $R_{n+1}$. If $\mathcal{E} = \mathcal{K}$ then $S(t,x) = F(t,x)$ for almost all $t$.

Also Theorem 4 can be generalized.

Theorem 6. Let $f$ fulfil the conditions of Theorem 5. If $\mathcal{E} = \mathcal{Q}$ then an n-dimensional function $x(t)$ is an $\mathcal{E}$-generalized solution of $\dot{x} = f(t,x)$ if and only if $x(t)$ is a generalized solution of $\dot{x} = f(t,x)$ in the sense of Viktorovskii.

Nevertheless, the nonautonomous case is more complicated than the autonomous one since a problem of measurability of $S$ may arise.

<u>Definition.</u> Let $h$ be a mapping $h : G \to \mathcal{a}_0$, $G \subset R_{n+1}$. The mapping $h$ is measurable if the sets $\{[t,x] : h(t,x) \cap A \neq \emptyset\}$ are Lebesgue measurable for all closed sets $A$, $A \subset R_n$. The mapping $h$ is t-measurable if the sets $\{t : h(t,x) \cap A \neq \emptyset\}$ are Lebesgue measurable for all closed sets $A$ and all $x \in R_n$.

Generally, both measurability and t-measurability of $f$ imply neither measurability nor t-measurability of $S$ but there is a wide family of classes $\mathcal{E}$ for which the problem has an affirmative answer.

Let a class $\mathcal{E}$ be given. If $A$ is a set in $R_n$, then $\mathcal{E}(A) = \bigcap_{B \supset A, B \in \mathcal{E}} B$ is called the $\mathcal{E}$-closure of $A$. The $\mathcal{E}$-closure exists if and only if $A$ is bounded, and $\mathcal{E}(A) \in \mathcal{E}$ if and only if $\mathcal{E}(A) \neq \emptyset$.

<u>Definition.</u> The $\mathcal{E}$-closure is called continuous if $\bigcap_n \mathcal{E}(A_n) = \mathcal{E}(\bigcap_n A_n)$ for every sequence of nonempty, compact sets $A_1 \supset A_2 \supset \supset \dots$.

<u>Theorem 7.</u> Let $f$ be a measurable and t-locally essentially bounded mapping $f : G \to \mathcal{a}$, where $G$ is a region in $R_{n+1}$. If a class $\mathcal{E}_0$ fulfils a) to c) and the $\mathcal{E}$-closure is continuous then the corresponding mapping $S$ is both measurable and t-measurable.

This theorem can be applied e.g. for the classes $\mathcal{E} = \mathcal{C}$, $\mathcal{E} = \mathcal{K}$, $\mathcal{E} = Q$ etc.

Sketch of the proof of Theorem 7. First we shall investigate the case $\mathcal{E} = \mathcal{C}$. Let $x \in G_t$ denote

$$B_t(x) = \{z : m_e[f_t^{-1}(U(z,d_2)) \cap U(x,d_1)] > 0 \text{ for all } d_1 > 0,\ d_2 > 0\}$$

where $f_t^{-1}(A) = \{x : f(t,x) \cap A \neq \emptyset\}$ and $m_e$ is the Lebesgue outer measure.

Lemma 3. Let $f$ be a t-locally essentially bounded mapping $f : G \to \mathcal{a}$ and $\mathcal{E} = \mathcal{C}$ . Then $S(t,x) = B(t,x) = B_t(x)$ almost everywhere in $G$ .

If $f$ is a measurable function then the asymptotical continuity of $f$ yields the property iii) (i.e. $f_t(x) \in B_t(x)$ ). Let $x_p \to x_o$ , $y_p \in B_t(x_p)$, $y_p \to y_o$ . If numbers $d_1 > 0$, $d_2 > 0$ are given we can find an index $q$ such that $\|x_p - x_o\| < d_1/2$ , $\|y_p - y_o\| < d_2/2$ for $p \geqq q$ . Since $f_t^{-1}(U(y_o,d_2)) \cap U(x_o,d_1) \supset f_t^{-1}(U(y_p,d_2/2)) \cap$ $\cap U(x_p,d_1/2)$ and the measure of the latter set is positive we obtain $y_o \in B_t(x_o)$ as a consequence of $y_p \in B_t(x_p)$ . The last assertion implies that $B_t$ are upper semi-continuous. Certainly $B_t(x) \in \mathcal{C}$ so that $B_t \in R(f_t, \mathcal{C})$ and $S \leqq B$ . The proof of $B \leqq S$ is similar as in the proof of Theorem 3. In the general case ( $f$ is a mapping) the proof is much more complicated.

In the case $\mathcal{E} = \mathcal{C}$ we can prove Theorem 7 using a lemma from measure theory.

Lemma 4. Let $f$ be a measurable and t-locally essentially bounded mapping $f : G \to \mathcal{a}$ . Then the sets

$\{[t,x] : m_e[f_t^{-1}(\mathcal{O}) \cap U(x,d)] > 0\}$ are Lebesgue measurable in $G$ for every open set $\mathcal{O}$ and $d > 0$ .

The sets $\{t : m_e[f_t^{-1}(\mathcal{O}) \cap U(x,d)] > 0\}$ are Lebesgue measurable on $R_1$ for every open set $\mathcal{O}$, $d > 0$ and $x \in G_t$ .

Now we shall use the notation $S^{(\mathcal{E})}$ to stress the dependence of $S$ on $\mathcal{E}$ . Since $\mathcal{E} \subset \mathcal{C}$ we have $S^{(\mathcal{C})} \subset S^{(\mathcal{E})}$ .

Lemma 5. Let $f$ be a t-essentially bounded mapping and let the $\mathcal{E}$ -closure be continuous. Then $S_t^{(\mathcal{E})}(x) = \mathcal{E}(S_t^{(\mathcal{C})}(x))$ for almost all $t$ .

Since $S_t^{(\mathcal{C})}(x) \subset S_t^{(\mathcal{E})}(x)$ we obtain $\mathcal{E}(S_t^{(\mathcal{C})}(x)) \subset \mathcal{E}(S_t^{(\mathcal{E})}(x)) =$ $= S_t^{(\mathcal{E})}(x)$ . On the other hand, to every $d_1 > 0$, $x_o \in G_t$ there exists

$d_2 > 0$ so that $\overline{\mathcal{E}(U(S_t^{(C)}(x_o), d_2))} \subset U(\mathcal{E}(S_t^{(C)}(x_o), d_1))$ since the $\mathcal{E}$-closure is continuous and the set $S_t^{(C)}(x_o)$ is compact. The upper semi-continuity of $S_t^{(C)}$ implies that there exists $d_3 > 0$ so that $S_t^{(C)}(x) \subset U(S_t^{(C)}(x_o), d_2)$ for $x \in U(x, d_3)$. These inclusions yield that the mappings $\mathcal{E}(S_t^{(C)}(.))$ are upper semi-continuous. Condition iii) follows from $\mathcal{E}(S_t^{(C)}(x)) \supset S_t^{(C)}(x)$ so that $\mathcal{E}(S_t^{(C)}(.)) \in R(f_t, \mathcal{E})$ and finally $S_t^{(\mathcal{E})}(x) = \mathcal{E}(S_t^{(C)}(x))$.

Theorem 7 in the general case now follows from

Lemma 6. Let h be a measurable mapping $h : \langle 0,1 \rangle \longrightarrow \mathcal{C}$. Then $\mathcal{E}(h(.))$ is a measurable mapping $\langle 0,1 \rangle \longrightarrow \mathcal{E}$.

The case of discontinuous closure was also investigated. Generally there exists a class $\mathcal{E}$ such that the $\mathcal{E}$-closure is discontinuous in only one set A but the corresponding problem of measurability has no positive answer. Nevertheless, if some regularity conditions are fulfilled then the problem has an affirmative answer.

References

[1] A.F. Filippov: Differential equations with discontinuous right hand sides (Russian), Mat.Sb. 51 (93), 1960, 99-128

[2] A. Pliś: Measurable orientor fields, Bull.Acad.Polon.Sci. (13), 1965, 565-569

[3] E.E. Viktorovskii: On a generalization of the notion of integral curves for discontinuous field of directions (Russian), Mat.Sb. 34 (76), 1954

[4] J. Pelant: Relation between generalized solutions of ordinary differential equations, Časopis Pěst.Mat. (101), 1976, 159-175

Author's address: Matematický ústav ČSAV, Žitná 25, 115 67 Praha 1, Czechoslovakia

Vol. 551: Algebraic K-Theory, Evanston 1976. Proceedings. Edited by M. R. Stein. XI, 409 pages. 1976.

Vol. 552: C. G. Gibson, K. Wirthmüller, A. A. du Plessis and E. J. N. Looijenga. Topological Stability of Smooth Mappings. V, 155 pages. 1976.

Vol. 553: M. Petrich, Categories of Algebraic Systems. Vector and Projective Spaces, Semigroups, Rings and Lattices. VIII, 217 pages. 1976.

Vol. 554: J. D. H. Smith, Mal'cev Varieties. VIII, 158 pages. 1976.

Vol. 555: M. Ishida, The Genus Fields of Algebraic Number Fields. VII, 116 pages. 1976.

Vol. 556: Approximation Theory. Bonn 1976. Proceedings. Edited by R. Schaback and K. Scherer. VII, 466 pages. 1976.

Vol. 557: W. Iberkleid and T. Petrie, Smooth $S^1$ Manifolds. III, 163 pages. 1976.

Vol. 558: B. Weisfeiler, On Construction and Identification of Graphs. XIV, 237 pages. 1976.

Vol. 559: J.-P. Caubet, Le Mouvement Brownien Relativiste. IX, 212 pages. 1976.

Vol. 560: Combinatorial Mathematics, IV, Proceedings 1975. Edited by L. R. A. Casse and W. D. Wallis. VII, 249 pages. 1976.

Vol. 561: Function Theoretic Methods for Partial Differential Equations. Darmstadt 1976. Proceedings. Edited by V. E. Meister, N. Weck and W. L. Wendland. XVIII, 520 pages. 1976.

Vol. 562: R. W. Goodman, Nilpotent Lie Groups: Structure and Applications to Analysis. X, 210 pages. 1976.

Vol. 563: Séminaire de Théorie du Potentiel. Paris, No. 2. Proceedings 1975–1976. Edited by F. Hirsch and G. Mokobodzki. VI, 292 pages. 1976.

Vol. 564: Ordinary and Partial Differential Equations, Dundee 1976. Proceedings. Edited by W. N. Everitt and B. D. Sleeman. XVIII, 551 pages. 1976.

Vol. 565: Turbulence and Navier Stokes Equations. Proceedings 1975. Edited by R. Temam. IX, 194 pages. 1976.

Vol. 566: Empirical Distributions and Processes. Oberwolfach 1976. Proceedings. Edited by P. Gaenssler and P. Révész. VII, 146 pages. 1976.

Vol. 567: Séminaire Bourbaki vol. 1975/76. Exposés 471–488. IV, 303 pages. 1977.

Vol. 568: R. E. Gaines and J. L. Mawhin, Coincidence Degree, and Nonlinear Differential Equations. V, 262 pages. 1977.

**Vol. 569: Cohomologie Etale SGA 4½. Séminaire de Géométrie Algébrique du Bois-Marie. Edité par P. Deligne. V, 312 pages. 1977.**

Vol. 570: Differential Geometrical Methods in Mathematical Physics, Bonn 1975. Proceedings. Edited by K. Bleuler and A. Reetz. VIII, 576 pages. 1977.

Vol. 571: Constructive Theory of Functions of Several Variables, Oberwolfach 1976. Proceedings. Edited by W. Schempp and K. Zeller. VI. 290 pages. 1977

**Vol. 572: Sparse Matrix Techniques, Copenhagen 1976. Edited by V. A. Barker. V, 184 pages. 1977.**

Vol. 573: Group Theory, Canberra 1975. Proceedings. Edited by R. A. Bryce, J. Cossey and M. F. Newman. VII, 146 pages. 1977.

**Vol. 574: J. Moldestad, Computations in Higher Types. IV, 203 pages. 1977.**

**Vol. 575: K-Theory and Operator Algebras, Athens, Georgia 1975. Edited by B. B. Morrel and I. M. Singer. VI, 191 pages. 1977.**

Vol. 576: V. S. Varadarajan, Harmonic Analysis on Real Reductive Groups. VI, 521 pages. 1977.

Vol. 577: J. P. May, $E_\infty$ Ring Spaces and $E_\infty$ Ring Spectra. IV, 268 pages. 1977.

Vol. 578: Séminaire Pierre Lelong (Analyse) Année 1975/76. Edité par P. Lelong. VI, 327 pages. 1977.

Vol. 579: Combinatoire et Représentation du Groupe Symétrique, Strasbourg 1976. Proceedings 1976. Edité par D. Foata. IV, 339 pages. 1977.

Vol. 580: C. Castaing and M. Valadier, Convex Analysis and Measurable Multifunctions. VIII, 278 pages. 1977.

Vol. 581: Séminaire de Probabilités XI, Université de Strasbourg. Proceedings 1975/1976. Edité par C. Dellacherie, P. A. Meyer et M. Weil. VI, 574 pages. 1977.

Vol. 582: J. M. G. Fell, Induced Representations and Banach *-Algebraic Bundles. IV, 349 pages. 1977.

Vol. 583: W. Hirsch, C. C. Pugh and M. Shub, Invariant Manifolds. IV, 149 pages. 1977.

Vol. 584: C. Brezinski, Accélération de la Convergence en Analyse Numérique. IV, 313 pages. 1977.

Vol. 585: T. A. Springer, Invariant Theory. VI, 112 pages. 1977.

Vol. 586: Séminaire d'Algèbre Paul Dubreil, Paris 1975–1976 (29ème Année). Edited by M. P. Malliavin. VI, 188 pages. 1977.

Vol. 587: Non-Commutative Harmonic Analysis. Proceedings 1976. Edited by J. Carmona and M. Vergne. IV, 240 pages. 1977.

Vol. 588: P. Molino, Théorie des G-Structures: Le Problème d'Equivalence. VI, 163 pages. 1977.

Vol. 589: Cohomologie l-adique et Fonctions L. Séminaire de Géométrie Algébrique du Bois-Marie 1965–66, SGA 5. Edité par L. Illusie. XII, 484 pages. 1977.

Vol. 590: H. Matsumoto, Analyse Harmonique dans les Systèmes de Tits Bornologiques de Type Affine. IV, 219 pages. 1977.

Vol. 591: G. A. Anderson, Surgery with Coefficients. VIII, 157 pages. 1977.

Vol. 592: D. Voigt, Induzierte Darstellungen in der Theorie der endlichen, algebraischen Gruppen. V, 413 Seiten. 1977.

Vol. 593: K. Barbey and H. König, Abstract Analytic Function Theory and Hardy Algebras. VIII, 260 pages. 1977.

Vol. 594: Singular Perturbations and Boundary Layer Theory, Lyon 1976. Edited by C. M. Brauner, B. Gay, and J. Mathieu. VIII, 539 pages. 1977.

Vol. 595: W. Hazod, Stetige Faltungshalbgruppen von Wahrscheinlichkeitsmaßen und erzeugende Distributionen. XIII, 157 Seiten. 1977.

Vol. 596: K. Deimling, Ordinary Differential Equations in Banach Spaces. VI, 137 pages. 1977.

Vol. 597: Geometry and Topology, Rio de Janeiro, July 1976. Proceedings. Edited by J. Palis and M. do Carmo. VI, 866 pages. 1977.

Vol. 598: J. Hoffmann-Jørgensen, T. M. Liggett et J. Neveu, Ecole d'Eté de Probabilités de Saint-Flour VI – 1976. Edité par P.-L. Hennequin. XII, 447 pages. 1977.

Vol. 599: Complex Analysis, Kentucky 1976. Proceedings. Edited by J. D. Buckholtz and T. J. Suffridge. X, 159 pages. 1977.

Vol. 600: W. Stoll, Value Distribution on Parabolic Spaces. VIII, 216 pages. 1977.

Vol. 601: Modular Functions of one Variable V, Bonn 1976. Proceedings. Edited by J.-P. Serre and D. B. Zagier. VI, 294 pages. 1977.

Vol. 602: J. P. Brezin, Harmonic Analysis on Compact Solvmanifolds. VIII, 179 pages. 1977.

Vol. 603: B. Moishezon, Complex Surfaces and Connected Sums of Complex Projective Planes. IV, 234 pages. 1977.

Vol. 604: Banach Spaces of Analytic Functions, Kent, Ohio 1976. Proceedings. Edited by J. Baker, C. Cleaver and Joseph Diestel. VI, 141 pages. 1977.

Vol. 605: Sario et al., Classification Theory of Riemannian Manifolds. XX, 498 pages. 1977.

Vol. 606: Mathematical Aspects of Finite Element Methods. Proceedings 1975. Edited by I. Galligani and E. Magenes. VI, 362 pages. 1977.

Vol. 607: M. Métivier, Reelle und Vektorwertige Quasimartingale und die Theorie der Stochastischen Integration. X, 310 Seiten. 1977.

Vol. 608: Bigard et al., Groupes et Anneaux Réticulés. XIV, 334 pages. 1977.

Vol. 609: General Topology and Its Relations to Modern Analysis and Algebra IV. Proceedings 1976. Edited by J. Novák. XVIII, 225 pages. 1977.

Vol. 610: G. Jensen, Higher Order Contact of Submanifolds of Homogeneous Spaces. XII, 154 pages. 1977.

Vol. 611: M. Makkai and G. E. Reyes, First Order Categorical Logic. VIII, 301 pages. 1977.

Vol. 612: E. M. Kleinberg, Infinitary Combinatorics and the Axiom of Determinateness. VIII, 150 pages. 1977.

Vol. 613: E. Behrends et al., $L^p$-Structure in Real Banach Spaces. X, 108 pages. 1977.

Vol. 614: H. Yanagihara, Theory of Hopf Algebras Attached to Group Schemes. VIII, 308 pages. 1977.

Vol. 615: Turbulence Seminar, Proceedings 1976/77. Edited by P. Bernard and T. Ratiu. VI, 155 pages. 1977.

Vol. 616: Abelian Group Theory, 2nd New Mexico State University Conference, 1976. Proceedings. Edited by D. Arnold, R. Hunter and E. Walker. X, 423 pages. 1977.

Vol. 617: K. J. Devlin, The Axiom of Constructibility: A Guide for the Mathematician. VIII, 96 pages. 1977.

Vol. 618: I. I. Hirschman, Jr. and D. E. Hughes, Extreme Eigen Values of Toeplitz Operators. VI, 145 pages. 1977.

Vol. 619: Set Theory and Hierarchy Theory V, Bierutowice 1976. Edited by A. Lachlan, M. Srebrny, and A. Zarach. VIII, 358 pages. 1977.

Vol. 620: H. Popp, Moduli Theory and Classification Theory of Algebraic Varieties. VIII, 189 pages. 1977.

Vol. 621: Kauffman et al., The Deficiency Index Problem. VI, 112 pages. 1977.

Vol. 622: Combinatorial Mathematics V, Melbourne 1976. Proceedings. Edited by C. Little. VIII, 213 pages. 1977.

Vol. 623: I. Erdelyi and R. Lange, Spectral Decompositions on Banach Spaces. VIII, 122 pages. 1977.

Vol. 624: Y. Guivarc'h et al., Marches Aléatoires sur les Groupes de Lie. VIII, 292 pages. 1977.

Vol. 625: J. P. Alexander et al., Odd Order Group Actions and Witt Classification of Innerproducts. IV, 202 pages. 1977.

Vol. 626: Number Theory Day, New York 1976. Proceedings. Edited by M. B. Nathanson. VI, 241 pages. 1977.

Vol. 627: Modular Functions of One Variable VI, Bonn 1976. Proceedings. Edited by J.-P. Serre and D. B. Zagier. VI, 339 pages. 1977.

Vol. 628: H. J. Baues, Obstruction Theory on the Homotopy Classification of Maps. XII, 387 pages. 1977.

Vol. 629: W. A. Coppel, Dichotomies in Stability Theory. VI, 98 pages. 1978.

Vol. 630: Numerical Analysis, Proceedings, Biennial Conference, Dundee 1977. Edited by G. A. Watson. XII, 199 pages. 1978.

Vol. 631: Numerical Treatment of Differential Equations. Proceedings 1976. Edited by R. Bulirsch, R. D. Grigorieff, and J. Schröder. X, 219 pages. 1978.

Vol. 632: J.-F. Boutot, Schéma de Picard Local. X, 165 pages. 1978.

Vol. 633: N. R. Coleff and M. E. Herrera, Les Courants Résiduels Associés à une Forme Méromorphe. X, 211 pages. 1978.

Vol. 634: H. Kurke et al., Die Approximationseigenschaft lokaler Ringe. IV, 204 Seiten. 1978.

Vol. 635: T. Y. Lam, Serre's Conjecture. XVI, 227 pages. 1978.

Vol. 636: Journées de Statistique des Processus Stochastiques, Grenoble 1977, Proceedings. Edité par Didier Dacunha-Castelle et Bernard Van Cutsem. VII, 202 pages. 1978.

Vol. 637: W. B. Jurkat, Meromorphe Differentialgleichungen. VII, 194 Seiten. 1978.

Vol. 638: P. Shanahan, The Atiyah-Singer Index Theorem, An Introduction. V, 224 pages. 1978.

Vol. 639: N. Adasch et al., Topological Vector Spaces. V, 125 pages. 1978.

Vol. 640: J. L. Dupont, Curvature and Characteristic Classes. X, 175 pages. 1978.

Vol. 641: Séminaire d'Algèbre Paul Dubreil, Proceedings Paris 1976–1977. Edité par M. P. Malliavin. IV, 367 pages. 1978.

Vol. 642: Theory and Applications of Graphs, Proceedings, Michigan 1976. Edited by Y. Alavi and D. R. Lick. XIV, 635 pages. 1978.

Vol. 643: M. Davis, Multiaxial Actions on Manifolds. VI, 141 pages. 1978.

Vol. 644: Vector Space Measures and Applications I, Proceedings 1977. Edited by R. M. Aron and S. Dineen. VIII, 451 pages. 1978.

Vol. 645: Vector Space Measures and Applications II, Proceedings 1977. Edited by R. M. Aron and S. Dineen. VIII, 218 pages. 1978.

Vol. 646: O. Tammi, Extremum Problems for Bounded Univalent Functions. VIII, 313 pages. 1978.

Vol. 647: L. J. Ratliff, Jr., Chain Conjectures in Ring Theory. VIII, 133 pages. 1978.

Vol. 648: Nonlinear Partial Differential Equations and Applications, Proceedings, Indiana 1976–1977. Edited by J. M. Chadam. VI, 206 pages. 1978.

Vol. 649: Séminaire de Probabilités XII, Proceedings, Strasbourg, 1976–1977. Edité par C. Dellacherie, P. A. Meyer et M. Weil. VIII, 805 pages. 1978.

Vol. 650: C*-Algebras and Applications to Physics. Proceedings 1977. Edited by H. Araki and R. V. Kadison. V, 192 pages. 1978.

Vol. 651: P. W. Michor, Functors and Categories of Banach Spaces. VI, 99 pages. 1978.

Vol. 652: Differential Topology, Foliations and Gelfand-Fuks-Cohomology, Proceedings 1976. Edited by P. A. Schweitzer. XIV, 252 pages. 1978.

Vol. 653: Locally Interacting Systems and Their Application in Biology. Proceedings, 1976. Edited by R. L. Dobrushin, V. I. Kryukov and A. L. Toom. XI, 202 pages. 1978.

Vol. 654: J. P. Buhler, Icosahedral Golois Representations. III, 143 pages. 1978.

Vol. 655: R. Baeza, Quadratic Forms Over Semilocal Rings. VI, 199 pages. 1978.

Vol. 656: Probability Theory on Vector Spaces. Proceedings, 1977. Edited by A. Weron. VIII, 274 pages. 1978.

Vol. 657: Geometric Applications of Homotopy Theory I, Proceedings 1977. Edited by M. G. Barratt and M. E. Mahowald. VIII, 459 pages. 1978.

Vol. 658: Geometric Applications of Homotopy Theory II, Proceedings 1977. Edited by M. G. Barratt and M. E. Mahowald. VIII, 487 pages. 1978.

Vol. 659: Bruckner, Differentiation of Real Functions. X, 247 pages. 1978.

Vol. 660: Equations aux Dérivée Partielles. Proceedings, 1977. Edité par Pham The Lai. VI, 216 pages. 1978.

Vol. 661: P. T. Johnstone, R. Paré, R. D. Rosebrugh, D. Schumacher, R. J. Wood, and G. C. Wraith, Indexed Categories and Their Applications. VII, 260 pages. 1978.

Vol. 662: Akin, The Metric Theory of Banach Manifolds. XIX, 306 pages. 1978.

Vol. 663: J. F. Berglund, H. D. Junghenn, P. Milnes, Compact Right Topological Semigroups and Generalizations of Almost Periodicity. X, 243 pages. 1978.

Vol. 664: Algebraic and Geometric Topology, Proceedings, 1977. Edited by K. C. Millett. XI, 240 pages. 1978.

Vol. 665: Journées d'Analyse Non Linéaire. Proceedings, 1977. Edité par P. Bénilan et J. Robert. VIII, 256 pages. 1978.

Vol. 666: B. Beauzamy, Espaces d'Interpolation Réels: Topologie et Géometrie. X, 104 pages. 1978.

Vol. 667: J. Gilewicz, Approximants de Padé. XIV, 511 pages. 1978.

Vol. 668: The Structure of Attractors in Dynamical Systems. Proceedings, 1977. Edited by J. C. Martin, N. G. Markley and W. Perrizo. VI, 264 pages. 1978.

Vol. 669: Higher Set Theory. Proceedings, 1977. Edited by G. H. Müller and D. S. Scott. XII, 476 pages. 1978.